AF358639

Applied Microorganisms and Industrial/Food Enzymes

Applied Microorganisms
and Industrial/Food Enzymes

Applied Microorganisms and Industrial/Food Enzymes

Xian Zhang
Zhiming Rao

Basel • Beijing • Wuhan • Barcelona • Belgrade • Novi Sad • Cluj • Manchester

Editors
Xian Zhang
School of Bioengineering
Jiangnan University
Wuxi
China

Zhiming Rao
School of Bioengineering
Jiangnan University
Wuxi
China

Editorial Office
MDPI AG
Grosspeteranlage 5
4052 Basel, Switzerland

This is a reprint of articles from the Special Issue published online in the open access journal *Fermentation* (ISSN 2311-5637) (available at: www.mdpi.com/journal/fermentation/special_issues/industrial_enzymes1).

For citation purposes, cite each article independently as indicated on the article page online and as indicated below:

Lastname, A.A.; Lastname, B.B. Article Title. *Journal Name* **Year**, *Volume Number*, Page Range.

ISBN 978-3-7258-2258-4 (Hbk)
ISBN 978-3-7258-2257-7 (PDF)
doi.org/10.3390/books978-3-7258-2257-7

Contents

 *fermentation*

Article

Improved Fermentation Yield of Doramectin from *Streptomyces avermitilis* N72 by Strain Selection and Glucose Supplementation Strategies

Xiaojun Pan [1,2,3,4] and Jun Cai [1,2,3,4,*]

[1] Key Laboratory of Fermentation Engineering (Ministry of Education), Hubei University of Technology, Wuhan 430068, China
[2] Hubei Key Laboratory of Industrial Microbiology, Hubei University of Technology, Wuhan 430068, China
[3] Cooperative Innovation Center of Industrial Fermentation (Ministry of Education & Hubei Province), Hubei University of Technology, Wuhan 430068, China
[4] College of Bioengineering and Food, Hubei University of Technology, Wuhan 430068, China
* Correspondence: hgdcaijun@hbut.edu.cn

Abstract: Doramectin is a macrolide antiparasitic that is widely used in the treatment of mammalian parasitic diseases. Doramectin is usually produced by *Streptomyces avermitilis* fermentation using cyclohexanecarboxylic acid (CHC) as a precursor; however, the growth of *S. avermitilis* is usually inhibited by CHC, resulting in a low fermentation yield of doramectin. In this study, a high-yielding strain XY-62 was obtained using the *S. avermitilis* mutant strain *S. avermitilis* N72 as the starting strain, then combined with a CHC tolerance screening strategy using ultraviolet and nitrosoguanidine mutagenesis, and a 96 microtiter plate solid-state fermentation primary sieving and shake flask fermentation rescreening method. Compared with *S. avermitilis* N72, the doramectin fermentation yield increased by more than 1.3 times, and it was more adaptable to temperature, pH, and CHC concentration of the culture; additionally, the viability of the mycelial growth was enhanced. In addition, further studies on the high-yielding strain XY-62 revealed that the accumulation of doramectin could be further increased by glucose supplementation during the fermentation process, and the yield of doramectin reached 1068 µg/mL by scaling up the culture in 50 L fermenters; this has the potential for industrial production. Therefore, mutagenesis combined with CHC tolerance screening is an effective way to enhance the fermentation production of doramectin by *S.* avermitilis. Our strategy and findings can help to improve the production of doramectin in industrial strains of *S. avermitilis*.

Keywords: *Streptomyces avermitilis*; doramectin; mutagenic breeding; CHC; fermentation characteristics

Citation: Pan, X.; Cai, J. Improved Fermentation Yield of Doramectin from *Streptomyces avermitilis* N72 by Strain Selection and Glucose Supplementation Strategies. *Fermentation* **2023**, *9*, 121. https://doi.org/10.3390/fermentation9020121

Academic Editor: Fabrizio Beltrametti

Received: 15 December 2022
Revised: 22 January 2023
Accepted: 23 January 2023
Published: 26 January 2023

1. Introduction

Doramectin is a macrolide disaccharide produced by the fermentation of *S. avermitilis* mutant strains using CHC as a precursor, and is a potent avermectin-like drug [1–3]. Doramectin is an anthelmintic active against internal and external parasites in animals, especially nematodes and mites. Doramectin does not easily cross the blood–brain barrier, and is safe for the treatment of parasitic diseases in mammals as it causes minimal damage to the central nervous system. It is used for the treatment of parasites in horses, cattle, sheep, pigs, and dogs [4–7]. It has a wider antiparasitic range and higher efficacy than avermectin [8,9], ivermectin [10,11] and other avermectin-based drugs [12,13]. It has also been shown to have an inhibitory effect on tumor cells [14].

The functions of the avermectin biosynthetic pathway and its biosynthetic genes have been well documented [15–18]. The biosynthetic pathway of doramectin is similar to that of avermectin, except that the starting unit for the synthesis of the doramectin macrolide backbone is cyclohexanol coenzyme A, rather than methylbutanoyl coenzyme A and isobutyryl coenzyme A [2,19–21].

Doramectin is a targeted biosynthetic metabolite of a mutant strain of *S. avermitilis*, and is not a natural product. Doramectin is produced using CHC-CoA as a starter unit, which can be achieved by adding CHC to the fermentation of *S. avermitilis* or by introducing a CHC-CoA biosynthetic gene cassette (PAC12) [2]. Yields of strains using the introduction of CHC-CoA are extremely low, and production still uses CHC in addition to fermentation. However, the addition of CHC had a significant inhibitory effect on the growth of the strain. In actual production, CHC is added in small amounts and several times during the stable period of mycelial growth or by microflow addition to reduce the inhibition of strain growth; however, flow addition requires additional equipment and increases the risk of bacterial contamination. The addition of CHC at 0.2~0.4 g/L in doramectin fermentation studies is likely to cause strain mortality [2,21]. To reduce the impact of CHC on strain survival, screening for CHC-tolerant strains is of interest. In addition, the production of doramectin still faces many technical problems. One of the difficulties in the production of doramectin is the very high requirements of the strain, which requires not only a high yield of doramectin (CHC-B1), but also a low level of CHC-B2 impurities; this makes the isolation and purification of the product more difficult [20]. In addition, the relatively long production cycle of doramectin, of which the fermentation and incubation time is 12~18 days, leads to increased production costs, thus limiting its application.

Currently, mutant selection and precursor addition strategies are effective methods to improve the yield of avermectin [22–25]. In industrial production, CHC is the starting precursor substance for the biosynthesis of doramectin by *S. avermitilis*, but CHC has an inhibitory effect on the growth of the bacterium, thus affecting the yield of doramectin [21,26]. Therefore, screening for strains with a high tolerance to CHC and reducing their growth inhibition is an effective way to increase the yield of doramectin; however, few applications of this strategy have been reported in the literature. Alternatively, high throughput screening of strains using liquid fermentation in microtiter plates is commonly used as an efficient method [27,28]. The primary screening of high-yielding strains of doramectin by surface culture in microtiter plates is a new endeavor.

In this study, the doramectin-producing strain *S. avermitilis* N72 was screened after ultraviolet and nitrosoguanidine mutagenesis for CHC tolerance, and high-yielding strains were selected by surface culture primary sieving in 96 microtiter plates and rescreening in shake flasks; subsequently, the stability of the passages was investigated. Preliminary studies on the growth in the fermentation characteristics of the high-yielding strains were then carried out, and combined with studies of their metabolic characteristics for glucose supplementation experiments and fermenter culture implementation in a 50 L fermenter.

2. Materials and Methods

2.1. Microbial Strains, Culture Media and Culture Conditions

Strain: *S. avermitilis* N72 is a mutant strain of *S. avermitilis* ATCC 31267 with doramectin production capacity (300 μg/mL), selected and conserved by our laboratory and used as the starting strain in this study (labeled N72 in the diagram).

Medium: Solid medium G consisted of the following (g/L): soybean meal, 4; mannitol, 4; agar, 20; pH 7.0~7.2. Seed medium S consisted of (g/L): glucose, 5; maltodextrin, 20; soybean cake flour, 10; cottonseed cake flour, 10; pH 7.0~7.2. Fermentation medium F consisted of (g/L): soluble starch, 90; bean cake flour, 15; cottonseed flour, 15; yeast extract, 5; sodium chloride, 1; dipotassium hydrogen phosphate, 2.5; calcium carbonate, 7; magnesium sulfate, 5; CHC, 0.8; pH 7.0~7.2; for surface culture agar 20 g/L added to F. All media were autoclaved at 121 °C for 25 min.

Culture conditions: *S. avermitilis* was cultured on solid medium G for 7 days, and a spore suspension of 10^6~10^7 spores/mL was prepared in saline. The spore suspension was inoculated (1 mL) into seed medium S (30 mL in a 250 mL flask) and incubated for 48 h at 200 rpm in a shaking flask. Then, the inoculum (8%, *v/v*) was inoculated into fermentation medium F (40 mL in a 250 mL flask) in shaking flasks at 220 rpm for 12 days. The incubation temperature was 30 °C. Surface culture was carried out in 96 microtiter plates, and fermenter culture implementation was carried out in 10 L–50 L fermenter systems (EastBio, Zhenjiang, China).

2.2. Assessment of Strain CHC Tolerance and Mutagenic Lethality

2.2.1. CHC Tolerance Assessment

S. avermitilis N72 spore suspensions were inoculated on solid medium G plates containing different concentrations of CHC (0, 0.4, 0.8, 1.0, 1.2, 1.4, 1.6, 1.8 g/L) for 7 days (30 °C). Plates without CHC were used as a blank control, and the number of growing colonies was recorded to calculate the survival rate. The calculation formula is given as Equation 1 below.

$$Survival\ rate = 100\% \times \frac{a}{b} \tag{1}$$

where a is the CFU of the CHC supplemented plate, and b is the CFU of the control plate.

2.2.2. Mutagenic Lethality

S. avermitilis N72 spore suspensions were mutagenized at a distance of 20 cm from a 15 W UV lamp at various times (0, 30, 60, 90, 120, 150, 180 s). For nitrosoguanidine (NTG) mutagenesis, *S. avermitilis* N72 spore suspensions were mutagenized by adding NTG (Macklin, Shanghai, China) to a concentration of 600 µg/mL, and NTG was mutagenized for various times (0,10, 20, 30, 40, 50, 60 min). For combined UV and NTG mutagenesis, *S. avermitilis* N72 spore suspensions were first treated with UV for different times (30 s, 60 s) and then mutagenized with NTG for different times (0, 10, 20, 30, 40, 50, 60 min). After the spore mutagenesis treatment, the spores were inoculated into solid medium G plates (100 µL per dish) and incubated in the dark for 7 days (30 °C). Spore suspension plates without mutagenesis were used as blank controls, and the number of growing colonies (CFU) was recorded to calculate the lethality [29]. The calculation formula is given in Equation 2 below.

$$Lethality\ \ rate = 100\% \times \left(1 - \frac{a}{b}\right) \tag{2}$$

where a is the CFU of the plate seeded with the mutagenized spores, and b is the CFU of the control plate.

2.3. Screening and Genetic Stability of High-Yielding Strains

2.3.1. Testing of the Analytical Method for Doramectin Content

Doramectin standards (Sigma-Aldrich, USA) were diluted to a range of concentrations (25, 50, 100, 200, 400 µg/mL) and analyzed (detection wavelength was 245 nm) by HPLC (LC-20AD, Shimadzu, Japan), and the standard curve plotted. The doramectin standards were diluted to a range of concentrations (25, 50,100,150, 200, and 250 µg/mL) and analyzed (detection wavelength was 245 nm) by ELISA (Synergy2, Gene Company Limited, USA) and the standard curve plotted. Twenty single colonies of mutant strains were randomly selected and inoculated in 96 microplates for surface culture (30 °C, 12 d), and the samples were analyzed by HPLC and ELISA, respectively. Sixteen single colonies of mutant strains were randomly selected and incubated in seed flasks (30 °C, 48 h), then inoculated in 96 microplates and fermentation flasks for surface culture (30 °C, 12 d) and fermentation flask culture (30 °C, 12 d); the samples were analyzed by HPLC.

2.3.2. 96-Well Plate Surface Culture Primary Sieve

Three different mutagenesis treatments of *S. avermitilis* N72 spore suspensions ($10^6 \sim 10^7$ spores/mL) were inoculated (100 µL per dish) onto CHC-tolerant plates (30 °C, 7 days), and the growing single colonies were inoculated simultaneously into 96 microplate A (G medium, 30 °C, 7 days) and 96 microplate B incubated at (F solid medium, 30 °C, 12 days). Microplate A (for growth and screening) and microplate B (for fermentation and detection) were inoculated with the same mutant strain at the same serial number of wells. Samples from microplate B were processed and analyzed for doramectin content by

ELISA for preliminary screening to identify positive mutant strains (over 20% increase in doramectin production compared to control), and the corresponding strains in microplate A were further studied.

2.3.3. Shake Flask Fermentation Rescreening

Positive mutant strains, after initial screening, were passed through slant expansion, inoculated onto shake flask medium S, and cultured (30 °C, 48 h), and transferred to shake flask medium F (30 °C, 12 days). The fermentation broth was tested by HPLC for doramectin yield, and the strain with a high yield was selected.

2.3.4. Assessing the Genetic Stability of High-Yielding Strains

The high-yielding strains obtained in this study were passaged six times on solid medium G. The strains from each passaging were expanded in S seed flasks under the same conditions and then inoculated into F fermentation shake flasks for 12 days to detect doramectin production.

2.4. Comparison of the Differences in Growth and Fermentation Characteristics between S. avermitilis N72 and the High-Yielding Strain XY-62

2.4.1. Growth Characteristics of the Strains in Seed Shake Flasks

For the examination of fermentation temperature, XY-16 and *S. avermitilis* N72 seed solutions were inoculated into shake flasks of medium S and incubated at different temperatures (20, 25, 30, 35 and 40 °C) for 48 h. For the initial pH examination, XY-16 and *S. avermitilis* N72 seed solutions were inoculated into medium S with different initial pH values (6.0, 6.5, 7.0, 7.5and 8.0) and incubated at 30 °C for 48 h. For the CHC tolerance study, XY-16 and *S. avermitilis* N72 seed solutions were inoculated into medium S with different concentrations of CHC (0.6, 0.8, 1.0, 1.2, 1.4 g/L) and incubated at 30 °C for 48 h. At the end of the incubation, the seed solution PMV was tested and growth was analyzed.

2.4.2. Metabolic Characteristics of the Strains in Fermentation Shake Flasks

XY-16 and *S. avermitilis* N72 seed solutions were inoculated with fermentation medium F in shake flasks (30 °C) and assayed for total sugars, reducing sugars, PMV, and doramectin production at different time points of the culture (0, 48, 96, 144, 192, 240, 288 h). The fermentation metabolic characteristics of the strains were compared at different growth periods.

2.5. Supplementary Fermentation of High-Yielding Strain XY-62 with a 50 L Fermenter Scale-Up Culture

2.5.1. Shake Flask Fermentation with Glucose Supplementation

XY-62 inoculated fermentation shake flasks were supplemented with glucose (0, 0.5%, 1%, 1.5%, 2%, 2.5%, w/v) at different fermentation time points (0, 96, 144, 192, 240 h). A total of 12 days of fermentation incubation at 30 °C and assayed for doramectin production. The effect of glucose on doramectin yield was analyzed.

2.5.2. Scaled-Up Culture in 50 L Fermenters

XY-62 was expanded in seed bottles and inoculated into 10 L fermenters (seed medium S, filling volume 6 L, inoculum 5%) for 48 h. The seed solution was transferred to 50 L fermenters (fermentation medium F, filling volume 30 L, inoculum 8%) for incubation. The tank temperature was 30 °C, tank pressure 0.05 MPa, stirring speed 60~180 r/min, air flow rate 0.8~1.2 vvm, and the fermentation-dissolved oxygen value remained above 35% by controlling the aeration flow rate and stirring speed. The fermentation was supplemented with 1.5% glucose at 192 h. The culture cycle was 12 days, and samples were taken every 24 h. The fermentation broth was tested for pH, total sugars, reducing sugars, amino nitrogen, bacterial concentration, and doramectin production. The metabolic characteristics of the strains were analyzed at different growth periods.

2.6. Analysis Methods

Microscopy was used for mycelial observation and spore counting. Biomass was estimated as the mycelial volume (PMV) of 10 mL of culture medium by centrifugation at 5000 rpm for 5 min [30]. The pH was determined by an acidity meter. Reducing sugars and total sugars in the fermentation broth were determined by titration with Felling's reagent [31]. The amino nitrogen content of the fermentation broth was determined by titration with formalin solution [32]. Surface culture samples were analyzed in 96-well microplates using ELISA. Then, 800 μL of methanol was added to each sample for 2 h. The supernatant was collected by centrifugation (8000 rpm, 10 min), and an aliquot (200 μL) of the supernatant was transferred to a 96 microplate and measured at 245 nm. Shake flask and fermenter samples were measured by HPLC; 1 mL of fermentation broth was mixed with 4 mL of methanol, and the cells were extracted by sonication for 2 h. The supernatant was then collected by centrifugation, and filtered through 0.22 μm; 10 μL was separated on a Waters C18 column. The elution was carried out at 35 °C with methanol/water (90:10, *v/v*) as the mobile phase at a flow rate of 1 mL/min for 25 min. The UV detection wavelength was 245 nm. The doramectin content in the samples was calculated from the doramectin standard curve.

2.7. Statistical Analysis

Experiments were performed in triplicate. The error lines in the graphs indicate the standard deviation of the three replicates of the corresponding experiments. All analyses were performed using SPSS software (version 26, IBM Inc., Armonk, NY, USA), and plots were processed using Origin software (version 2021b, OriginLab Corporation, Northampton, MA, USA).

3. Results

3.1. Assessment of Strain CHC Tolerance and Mutagenic Conditions

3.1.1. Effect of CHC Concentration on the Viability of *S. avermitilis* N72 Cells

The growth inhibition of CHC was significant, with cell mortality increasing with increasing CHC concentrations. Less than 50% and less than 4% survivals were detected when the concentration of CHC was increased to 1.2 and 1.8 g/L, respectively (Figure 1A). In the CHC resistance screening assay, a CHC concentration of 1.4~1.6 g/L was chosen as reasonable.

3.1.2. Effect of UV Light and NTG on the Viability of *S. avermitilis* N72 Cells

Cell mortality increased with increasing UV irradiation time. Less than 40% survival was detected when the spores were irradiated for 90 s, while cell mortality reached 96% after exposure to UV for 180 s (Figure 1B). A 600 μg/mL concentration of NTG affected the growth of the strain, and the mortality rate increased with increasing NTG mutagenesis time. The survival rate detected after 20 min of NTG mutagenesis was less than 50%, while the cell mortality rate detected after 50 min of NTG mutagenesis was 98% (Figure 1C). The combination of UV irradiation and NTG treatment increased the lethality of *S. avermitilis*. Cell mortality could reach 80%~90% after UV irradiation for 30~60 s and 600 μg/mL NTG mutagenesis for 20 min, while cell mortality could reach 100% after UV irradiation for 30 s and 600 μg/mL NTG mutagenesis for 30 min (Figure 1D). Some investigators have suggested that mutagenesis with a lethality of 80% to 90% is more effective (Wang et al., 2011 [21]; Cao et al., 2018 [27]). Considering the operability of the experiment, the UV mutagenesis time was therefore chosen to be 120~150 s, and the NTG mutagenesis time was chosen to be 30~40 min. The combination of two mutagens was chosen, to be followed by UV mutagenesis for 30~60 s and NTG mutagenesis for 20~30 min.

3.2. Selection and Breeding of High-Yielding Strains and Their Genetic Stability

3.2.1. Analysis of Doramectin Content and Feasibility of Fermentation Methods

The HPLC method for doramectin was reliable, with good linearity between concentration and peak area for doramectin standards over a fairly wide range ($R^2 = 0.9987$)

(Figure 2A). The analytical efficiency can be improved by using the microtiter-plate based ELISA method over a wide range (R^2 = 0.9979); a good linear relationship exists between the concentration of doramectin standards and OD245 (Figure 2B). Extracts from 96 microtiter plates were selected for analysis by HPLC and showed a good correlation between OD245 and concentrations measured by HPLC (R^2 = 0.9856) (Figure 2C). In addition, there was a good correlation between single-colony solid fermentation and shake flask fermentation (R^2 = 0.9597) (Figure 2D). Therefore, the ELISA assay methods and HPLC-estimated concentration of doramectin are linearly correlated.

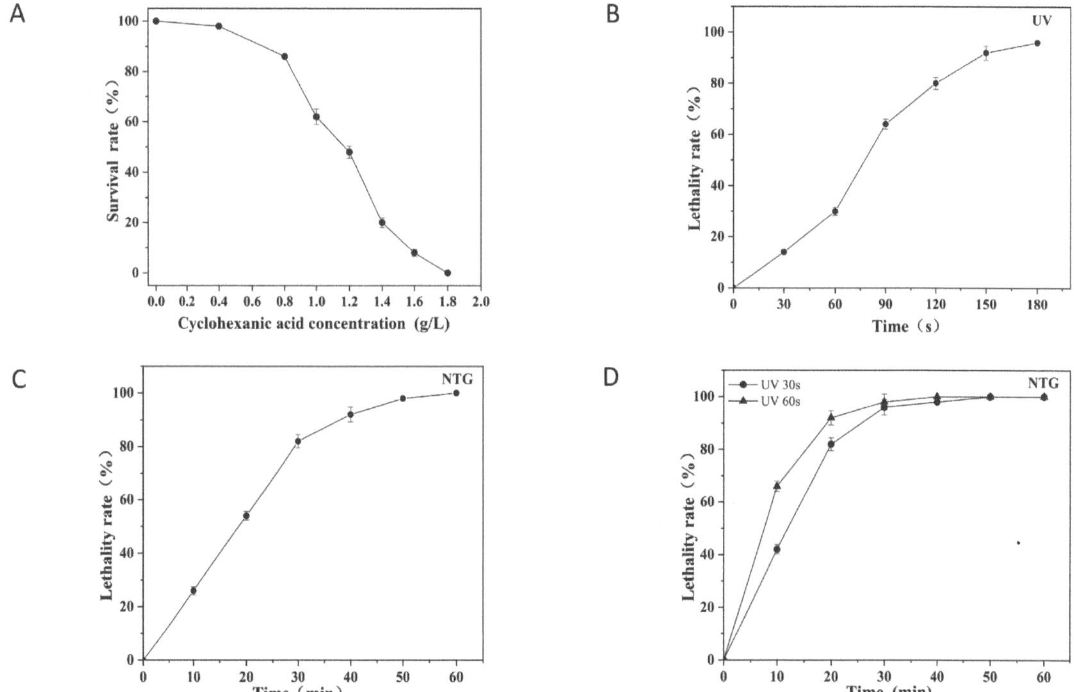

Figure 1. Assessment of CHC tolerance and the effect of mutagenic conditions on the viability of *S. avermitilis* N72. (**A**) The survival rate of *S. avermitilis* N72 at different concentrations of CHC, (**B**) lethality of *S. avermitilis* N72 by UV, (**C**) lethality of *S. avermitilis* N72 by NTG, (**D**) lethality of *S. avermitilis* N72 by the combined effect of UV and NTG.

3.2.2. 96-Well Plate Primary Sieve for Surface Culture

The fermentation yield of doramectin was measured in 96-well plates by selecting 450 UV-mutagenized and CHC-tolerant single colonies. Fifteen strains displayed improved doramectin yields ranging from 120 to 180 µg/mL in surface culture. This corresponded to a 20~80% increase in doramectin yield compared to *S. avermitilis* N72 (Figure 3A). The fermentation yield of doramectin was measured also in 443 NTG-mutagenized 30 min CHC-tolerant single colonies, and they were fermented in 96-well plates. Ten strains with higher yields ranging from 120 to 190 µg/mL in surface culture were identified, and displayed a 20~90% increase in doramectin yield compared to *S. avermitilis* N72 (Figure 3B). The fermentation yield of doramectin was also measured by selecting 446 UV-mutagenized 30 s and NTG-mutagenized 20 min tolerant single colonies in 96-well plates. Twelve strains with higher yields had solid fermentation yields of 120~230 µg/mL compared to *S. avermitilis* N72. This corresponded to a 20~130% higher doramectin yield (Figure 3C).

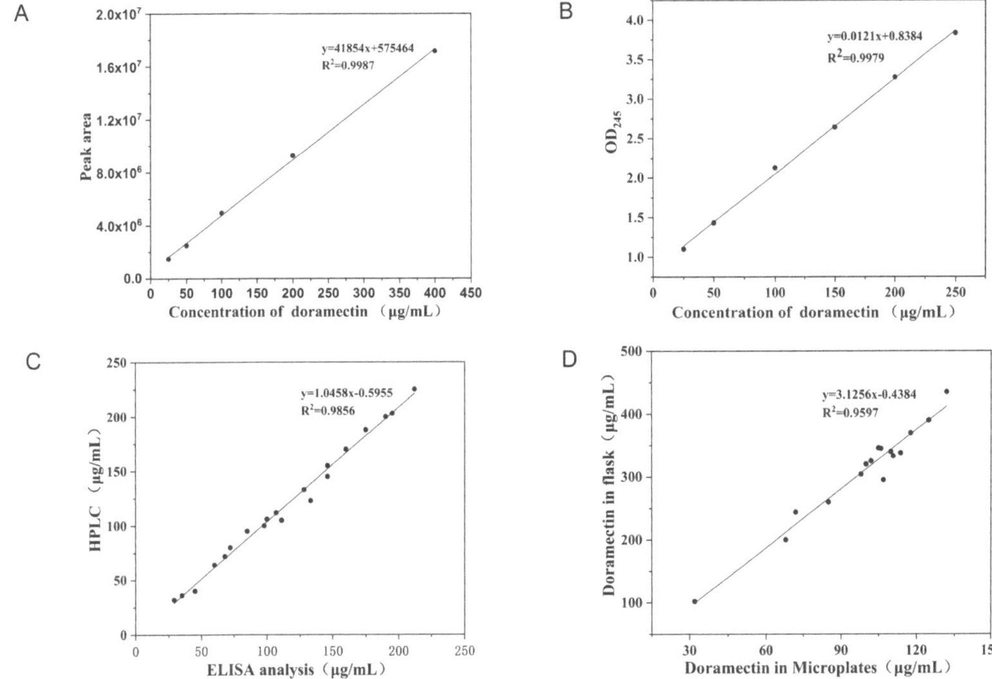

Figure 2. Analysis of the correlation between doramectin ELISA assay and HPLC quantification. (**A**) doramectin HPLC standard curve, (**B**) doramectin ELISA standard curve, (**C**) correlation of HPLC assay with ELISA method, (**D**) correlation of single-colony microplate surface culture with shake flask fermentation.

3.2.3. Shake Flask Fermentation Rescreening and Genetic Stability of High-Yielding Strains

Thirty-seven positive mutant strains from the initial screening were seeded on slants and fermented in a shake flask (15 strains were UV mutagenized, 10 strains were NTG mutagenized and 12 strains were compound mutagenized) for 12 days, with *S. avermitilis* N72 fermentation as a control. The yield of strain XY-62 doramectin reached 700 µg/mL, representing a 1.3-fold increase compared to *S. avermitilis* N72 (Figure 3D). The morphologies of XY-62 and *S. avermitilis* N72 differed slightly in solid medium G, with XY-62 colonies being more distinctly raised, and there was no significant difference in growth patterns between the two in liquid medium S (Figure 3E). The ability of XY-62 to produce doramectin is relatively genetically stable, as indicated by the fact that there was no significant change in doramectin production when the high-yielding strain XY-62 was passed through six consecutive generations (Figure 3F).

3.3. Comparison of the Differences in Growth and Fermentation Characteristics between the High-Yield Strains XY-62 and S. avermitilis N72

3.3.1. Growth Characteristics of the Strain

S. avermitilis N72 growth on seed medium S (Figure 4A) was characterized by a rapid increase in PMV from 0 to 48 h. After a peak in PMV between 48 and 72 h, it gradually decreased, and the growth of the bacterium was seen to slow down; the bacterium gradually declined as the incubation time increased. In comparison, the PMV of XY-62 was significantly higher than that of *S. avermitilis* N72 when cultured on seed medium S. The effect of different temperatures on the growth of mutant strains XY-62 and *S. avermitilis* N72 was similar (Figure 4B). This was characterized by incubation at 20~40 °C, with a gradual increase in PMV with temperature reaching a maximum at 30 °C, and a gradual decrease with increasing temperature. Comparing XY-62 and *S. avermitilis* N72, the growth activity of the mutant strain XY-62 was significantly higher than that of *S. avermitilis* N72 at 30~40 °C. XY-62 has better adaptability to culture temperature. The effect of pH on the growth of mutant strains XY-62 and *S. avermitilis* N72 was similar (Figure 4C). pH values of 6.5~7.5 showed relatively stable growth. Beyond this range, the growth of the strains was reduced, but the growth of XY-62 was less inhibited. Therefore, the mutant strain XY-62 was more stable in its adaptation to environmental pH. The growth of mutant strains XY-62 and *S. avermitilis* N72 was affected by different concentrations of CHC (Figure 4D). At CHC concentrations above 0.8 g/L, *S. avermitilis* N72 growth was reduced, while XY-62 was able to grow at 1.2 g/L CHC and above, and *S. avermitilis* N72 was barely able to grow. This showed that XY-62 was more tolerant of CHC.

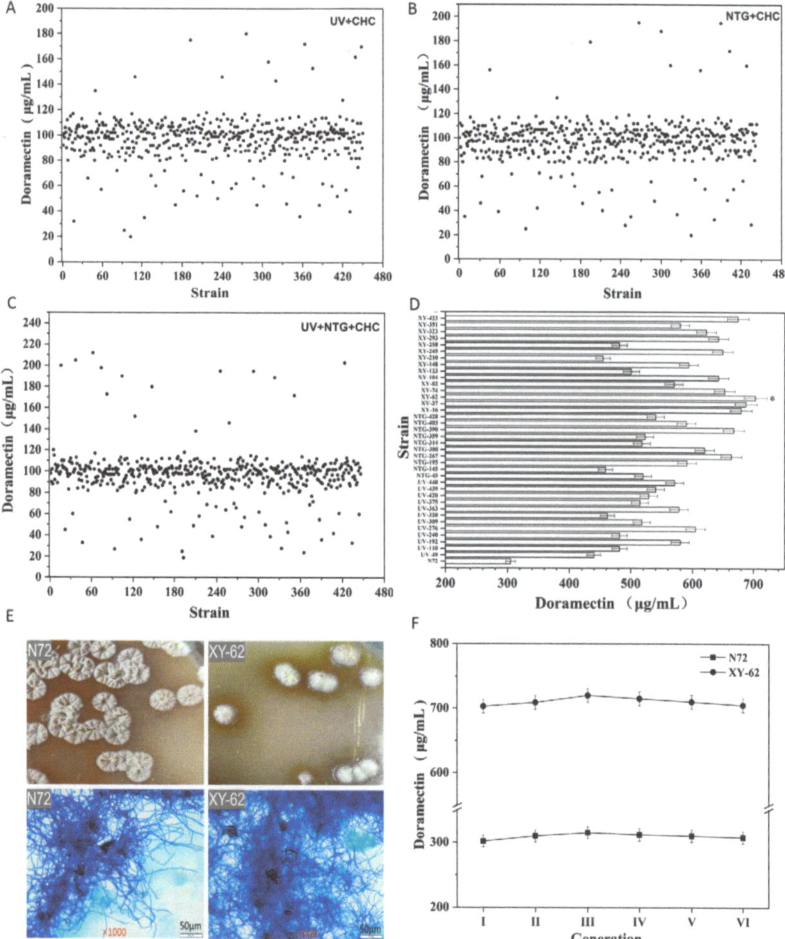

Figure 3. Stability of mutagenic selection and high-yielding strains of *S. avermitilis* N72. (**A**) single colonies of *S. avermitilis* N72 after UV mutagenesis and CHC-tolerance screening in a 96-microplate surface culture primary sieve, (**B**) single colonies of *S. avermitilis* N72 after NTG mutagenesis and CHC-tolerance screening in a 96-microplate surface culture primary sieve, (**C**) single colonies of *S. avermitilis* N72 after UV and NTG mutagenesis and CHC-tolerance screening in 96-microplate solid-state fermentation primary sieve, (**D**) shake flask rescreening of positive mutant strains (* represents $p < 0.05$), (**E**) growth pattern of *S. avermitilis* N72 and high-yielding strain XY-62 in solid medium G and seed medium S, (**F**) enetic stability of *S. avermitilis* N72 and the high-yielding strain XY-62.

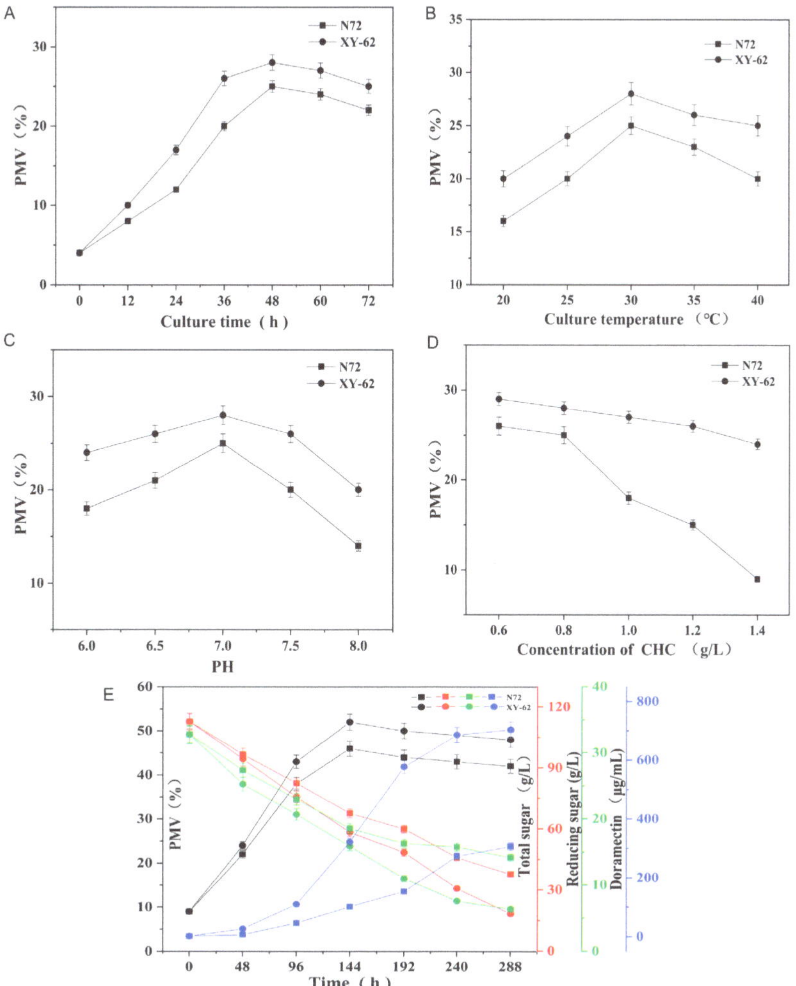

Figure 4. Differences in growth and metabolism between *S. avermitilis* N72 and XY-62 shake flasks. (**A**) effect of different incubation times on biomass, (**B**) effect of different incubation temperatures on biomass, (**C**) effect of different pH values on biomass, (**D**) effect of different CHC concentrations on biomass, (**E**) fermentation shake flask growth and metabolism curves.

3.3.2. Metabolic Characteristics of the Strains

The growth metabolism curves of XY-62 and *S. avermitilis* N72 in fermentation medium F were similar (Figure 4E). From 0 to 144 h, PMV increased rapidly, and the total and reducing sugar contents decreased, implying rapid growth of the organism and accelerated sugar utilization. Production of doramectin started at 96 h, followed by a rapid increase in yield, which slowed down at 240~288 h. Compared to *S. avermitilis* N72, the biomass and sugar consumption of XY-62 were higher, the viability of the organism was stronger and the production capacity for doramectin was greater.

3.4. Supplementary Fermentation of High-Yielding Strain XY-62 and Scale-Up Culture in 50 L Fermenters

Based on the metabolic characteristics of XY-62 fermentation, sugar consumption was enhanced, and the amount of reducing sugars was low in the late stage, so glucose was supplemented. The supplementation of glucose promoted the accumulation of doramectin (Figure 5A), especially at the 192nd hour of fermentation when supplemented with 1.5% glucose. The doramectin yield reached 810 μg/mL, which was 14.8% and 10.2% higher than that of the control group without glucose supplementation and with 1.5% glucose added to the base medium, respectively.

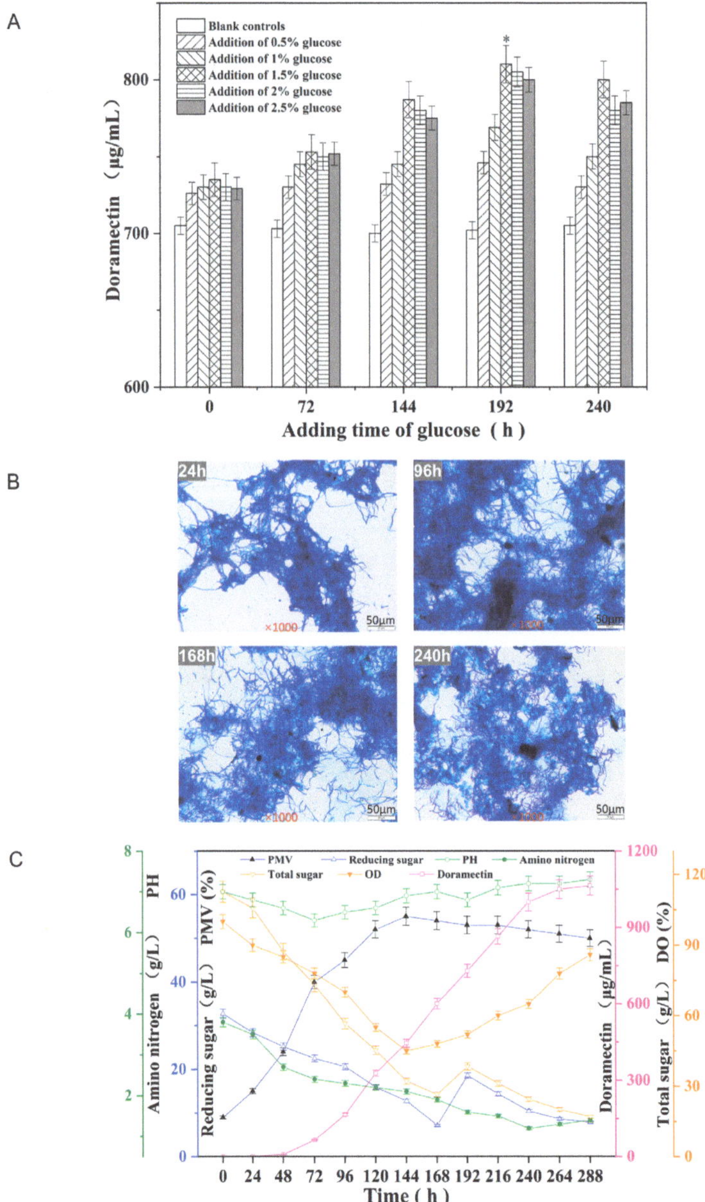

Figure 5. Fermentation of high-yielding strain XY-62 with increased glucose amount and culture in a 50 L fermenter. (**A**) effect of glucose supplementation on the yield of doramectin synthesis by XY-62 (* represents $p < 0.05$), (**B**) mycelial morphology of XY-62 at different growth periods in the 50 L fermenter, (**C**) growth and metabolic curves of XY-62 in 50 L fermenter.

The metabolic characteristics of the growth of the bacterium in the 50 L fermenter magnified culture (Figure 5B,C) were roughly divided into three stages: the early stage (0~48 h), the middle stage (48~192 h) and the late stage (192~288 h). In the early stage, PMV increased, the growth of the bacterium was faster, total sugar and reducing sugar consumption was rapid, amino nitrogen decreased, pH decreased, DO decreased, and doramectin was not produced. In the middle stage, PMV reached a maximum, total sugar and reduced sugar consumption accelerated, the pH increased to 7.0, and the bacterium started to produce a large amount of doramectin. At the later stage, the PMV decreased, the growth of the bacteria gradually slowed down, the consumption of total sugars and reducing sugars decreased, the pH increased, the DO increased, and the increase in doramectin slowed down. The yield of doramectin reached 1068 μg/mL, representing an increase of 30% compared with shake flask fermentation.

4. Discussion

We report a method for screening high-yielding strains of doramectin. Mutagenesis is a cost-effective method for the creation of high-yielding strains, Song et al. studied a genetic mutant *Streptomyces avermitilis* S-233 with high-avermectins B1a by comobtained treatment with carbon heavy ion irradiation and sodium nitrite, and Xu et al. obtained a strain *Streptomyces viridochromogenes* F-23 with high-avilamycin production by combined mutagenesis with UV and ARTP [23,33]. The growth of the strains was inhibited by CHC, which is consistent with literature reports [21]. In addition, mutagenesis is a viable method for selecting strains with increased adaptability to the environment [34]. The significant increase in doramectin levels in the CHC-tolerant mutant strain in this study may be due to the enhanced stress of the strain to the precursor substance CHC. Mutagenic selection can further enhance the adaptability of the strain in CHC-containing media, resulting in a strain more suitable for industrial production.

Microorganisms have an enhanced capacity to consume nutrients and supplemental fermentation is an effective method. Supplementation experiments have shown that glucose has a positive effect on doramectin yield enhancement. From the biosynthetic pathway of doramectin [2,17], it is known that glucose is not only used as a carbon source to promote microbial growth but also as a raw material for the direct synthesis of the doramectin disaccharide side chain. In addition, the degradation of glucose to form malonyl-CoA and methyl malonyl-CoA can provide more precursors for doramectin biosynthesis and facilitate the formation of macrolides. Glucose supplementation promoted the accumulation of doramectin, especially in the middle and late stages of supplementation.

The growth and metabolism of *S. avermitilis* are aerobic, and compared to shake flask culture, fermenter culture can better regulate the fermentation process, especially the regulation of dissolved oxygen and replenishment operation. The dissolved oxygen level has a great influence on the production of secondary metabolites during the fermentation of *S. avermitilis* [33,35,36]. The 30% increase in fermentation yield in doramectin fermenters compared to shake flask fermentation may be caused by the better-dissolved oxygen conditions provided by the fermenters. In addition, the doramectin yield of strain XY-62 was relatively high compared to those reported in the reviewed literature for the same fermentation time [2,19–21].

In conclusion, our approach has been successful in enhancing the ability of *S. avermitilis* N72 to produce doramectin. This is mainly based on the selection and breeding of the CHC-resistant strain XY-62 to improve its survival in CHC-containing media and on the enhanced viability of the strain and enhanced sugar metabolism, especially glucose metabolism, which promotes the accumulation of doramectin. Based on this study, the fermentation process can be further optimized, and the mechanism of high doramectin production by mutant strain XY-62 can be investigated to better enhance the fermentation potential of high-yielding strain XY-62.

5. Conclusions

In this study, *S. avermitilis* N72 was treated for mutagenesis and CHC tolerance, and a high-yielding strain XY-62 was obtained using an efficient screening strategy of solid fermentation in 96-well microplates. Comparing the growth and metabolic characteristics of the strains before and after screening indicated that XY-62 was more suitable for industrial production. The strategy of adding 1.5% glucose in the middle and late stages of fermentation could further promote the accumulation of doramectin. It was also validated in a 50 L fermenter. In conclusion, the screening of CHC-tolerant strains and the glucose supplementation strategy are effective methods to significantly enhance the production of doramectin by *S. avermitilis* fermentation.

Author Contributions: X.P.: conceived and designed the study, performed the research, and formal analysis, and wrote the original draft. J.C.: conceived and designed the study, resources, supervision, writing—review, and editing. All authors have read and agreed to the published version of the manuscript.

Funding: This research received no external funding.

Institutional Review Board Statement: Not applicable.

Informed Consent Statement: Not applicable.

Data Availability Statement: Not applicable.

Conflicts of Interest: The authors declared no conflict of interest.

References

1. Dutton, C.; Gibson, S.; Kinns, M.; Swanson, A.; Bordner, J. Structure of doramectin. *J. Chem. Soc. Peak T2* **1995**, *2*, 403. [CrossRef]
2. Cropp, T.; Wilson, A.; Reynolds, K. Identification of a cyclohexylcarbonyl CoA biosynthetic gene cluster and application in the production of doramectin. *Nat. Biotechnol.* **2000**, *18*, 980–983. [CrossRef] [PubMed]
3. Thuan, N.; Pandey, R.; Sohng, J. Recent advances in biochemistry and biotechnological synthesis of avermectins and their derivatives. *Appl. Microbiol. Biot.* **2014**, *98*, 7747–7759. [CrossRef] [PubMed]
4. Pérez, R.; Cristina, P.; Lgnacio, C.; Luis, R.; Margarita, A. The influence of gastrointestinal parasitism on fecal elimination of doramectin, in lambs. *Ecotoxicol. Environ. Saf.* **2010**, *73*, 2017–2021. [CrossRef] [PubMed]
5. Cordero, A.; Quek, S.; Mueller, R. Doramectin in the treatment of generalized demodicosis. *Vet. Dermatol.* **2018**, *29*, 104-e41. [CrossRef]
6. Lye, G.; Jacob, A.; Pomroy, W.; Stafford, K.; Singh, P. Pharmacokinetics of subcutaneously administered doramectin in alpacas. *J. Vet. Pharmacol. Ther.* **2020**, *43*, 123–128. [CrossRef]
7. Myers, M.; Howard, K.; Kawalek, J. Pharmacokinetic comparison of six anthelmintics in sheep, goats, and cattle. *J. Vet. Pharmacol. Ther.* **2021**, *44*, 58–67. [CrossRef]
8. Siddique, S.; Syed, Q.; Adnan, A.; Qureshi, F. Isolation, Characterization and Selection of Avermectin-Producing *Streptomyces avermitilis* Strains From Soil Samples. *Jundishapur J. Microbiol.* **2014**, *7*, e10366. [CrossRef]
9. Salman, M.; Abbas, R.; Mehmood, K.; Hussain, R.; Shah, S.; Faheem, M.; Zaheer, T.; Abbas, A.; Morales, B.; Aneva, I.; et al. Assessment of Avermectins-Induced Toxicity in Animals. *Pharmaceuticals* **2022**, *15*, 332. [CrossRef]
10. Crump, A.; Omura, S. Ivermectin, 'Wonder drug' from Japan: The human use perspective. *Proc. Jpn. Acad. Ser. B* **2011**, *87*, 13–28. [CrossRef]
11. Singh, L.; Singh, K. Ivermectin: A Promising Therapeutic for Fighting Malaria. Current Status and Perspective. *J. Med. Chem.* **2021**, *64*, 9711–9731. [CrossRef]
12. Molinari, G.; Soloneski, S.; Larramendy, M. New Ventures in the Genotoxic and Cytotoxic Effects of Macrocyclic Lactones, Abamectin and Ivermectin. *Cytogenet. Genome Res.* **2010**, *128*, 37–45. [CrossRef]
13. El-Saber Batiha, G.; Alqahtani, A.; Ilesanmi, O.; Saati, A.; El-Mleeh, A.; Hetta, H.; Magdy Beshbishy, A. Avermectin Derivatives, Pharmacokinetics, Therapeutic and Toxic Dosages, Mechanism of Action, and Their Biological Effects. *Pharmaceuticals* **2020**, *13*, 196. [CrossRef]
14. Chen, C.; Liang, H.; Qin, R.; Li, X.; Wang, L.; Du, S.; Chen, Z.; Meng, X.; Lv, Z.; Wang, Q.; et al. Doramectin inhibits glioblastoma cell survival via regulation of autophagy *in vitro* and *in vivo*. *Int. J. Oncol.* **2022**, *60*, 1–16. [CrossRef]
15. Ikeda, H.; Omura, S. Avermectin Biosynthesis. *Chem. Rev.* **1997**, *97*, 2591–2610. [CrossRef]
16. Ikeda, H.; Ishikawa, J.; Hanamoto, A.; Shinose, M.; Kikuchi, H.; Shiba, T.; Sakaki, Y.; Hattori, M.; Ōmura, S. Complete genome sequence and comparative analysis of the industrial microorganism *Streptomyces avermitilis*. *Nat. Biotechnol.* **2003**, *21*, 526–531. [CrossRef]
17. Tang, Y.; Wang, M.; Qin, H.; An, X.; Guo, Z.; Zhu, G.; Zhang, L.; Chen, Y. Deciphering the Biosynthesis of TDP-beta-L-oleandrose in Avermectin. *J. Nat. Prod.* **2020**, *83*, 3199–3206. [CrossRef]

18. Lee, Y.; Lee, N.; Hwang, S.; Kim, W.; Cho, S.; Palsson, B.; Cho, B. Genome-scale analysis of genetic regulatory elements in *Streptomyces avermitilis* MA-4680 using transcript boundary information. *BMC Genom.* **2022**, *23*, 1–16. [CrossRef]
19. Dutton, C.; Gibson, S.; Goudie, A.; Holdom, K.; Pacey, M.; Ruddock, J.; Bulock, J.; Richards, M. Novel avermectins produced by mutational biosynthesis. *J. Antibiot.* **1991**, *44*, 357–365. [CrossRef]
20. Stutzman-Engwall, K.; Conlon, S.; Fedechko, R.; McArthur, H.; Pekrun, K.; Chen, Y.; Jenne, S.; La, C.; Trinh, N.; Kim, S.; et al. Semi-synthetic DNA shuffling of aveC leads to improved industrial scale production of doramectin by *Streptomyces avermitilis*. *Metab. Eng.* **2005**, *7*, 27–37. [CrossRef]
21. Wang, J.; Pan, H.; Tang, G. Production of doramectin by rational engineering of the avermectin biosynthetic pathway. *Bioorgan. Med. Chem. Lett.* **2011**, *21*, 3320–3323. [CrossRef] [PubMed]
22. Gao, Q.; Tan, G.; Xia, X.; Zhang, L. Learn from microbial intelligence for avermectins overproduction. *Curr. Opin. Biotechnol.* **2017**, *48*, 251–257. [CrossRef] [PubMed]
23. Song, X.; Zhang, Y.; Zhu, X.; Wang, Y.; Chu, J.; Zhuang, Y. Mutation breeding of high avermectin B-1a-producing strain by the combination of high energy carbon heavy ion irradiation and sodium nitrite mutagenesis based on high throughput screening. *Biotechnol. Bioprocess Eng.* **2017**, *22*, 539–548. [CrossRef]
24. Wang, W.; Li, S.; Li, Z.; Zhang, J.; Fan, K.; Tan, G.; Ai, G.; Lam, S.; Shui, G.; Yang, Z.; et al. Harnessing the intracellular triacylglycerols for titer improvement of polyketides in Streptomyces. *Nat. Biotechnol.* **2020**, *38*, 76–83. [CrossRef]
25. Hao, Y.; You, Y.; Chen, Z.; Li, J.; Liu, G.; Wen, Y. Avermectin B1a production in *Streptomyces avermitilis* is enhanced by engineering aveC and precursor supply genes. *Appl. Microbiol. Biot.* **2022**, *106*, 2191–2205. [CrossRef]
26. Zhang, J.; Wang, X.; Diao, J.; He, H.; Zhang, Y.; Xiang, W. Streptomycin resistance-aided genome shuffling to improve doramectin productivity of *Streptomyces avermitilis* NEAU1069. *J. Microbiol. Biot.* **2013**, *40*, 877–889. [CrossRef]
27. Cao, X.; Luo, Z.; Zeng, W.; Xu, S.; Zhao, L.; Zhou, J. Enhanced avermectin production by *Streptomyces avermitilis* ATCC 31267 using high-throughput screening aided by fluorescence-activated cell sorting. *Appl. Microbiol. Biot.* **2018**, *102*, 703–712. [CrossRef]
28. Yu, F.; Zhang, M.; Sun, J.; Wang, F.; Li, X.; Liu, Y.; Wang, Z.; Zhao, X.; Li, J.; Chen, J.; et al. Improved Neomycin Sulfate Potency in *Streptomyces* fradiae Using Atmospheric and Room Temperature Plasma (ARTP) Mutagenesis and Fermentation Medium Optimization. *Microorganisms* **2022**, *10*, 94. [CrossRef]
29. Wang, X.; Wang, X.; Xiang, W. Improvement of milbemycin-producing *Streptomyces* bingchenggensis by rational screening of ultraviolet- and chemically induced mutants. *World J. Microbiol. Biotechnol.* **2009**, *25*, 1051–1056. [CrossRef]
30. Yu, Z.; Shen, X.; Wu, Y.; Yang, S.; Ju, D.; Chen, S. Enhancement of ascomycin production via a combination of atmospheric and room temperature plasma mutagenesis in *Streptomyces hygroscopicus* and medium optimization. *AMB Express* **2019**, *9*, 1–15. [CrossRef]
31. Du, Z.; Zhang, Y.; Qian, Z.; Xiao, H.; Zhong, J. Combination of traditional mutation and metabolic engineering to enhance ansamitocin P-3 production in *Actinosynnema pretiosum*. *Biotechnol. Bioeng.* **2017**, *114*, 2794–2806. [CrossRef]
32. Choi, Y.; Lee, H.; Yang, J.; Hong, S.; Park, S.; Lee, M. Changes in quality properties of kimchi based on the nitrogen content of fermented anchovy sauce, Myeolchi Aekjeot, during fermentation. *Food Sci. Biotechnol.* **2018**, *27*, 1145–1155. [CrossRef]
33. Yu, G.; Peng, H.; Cao, J.; Liao, A.; Long, P.; Huang, J.; Hui, M. Avilamycin production enhancement by mutagenesis and fermentation optimization in *Streptomyces viridochromogenes*. *World J. Microbiol. Biotechnol.* **2022**, *38*, 50. [CrossRef]
34. Che, J.; Liu, B.; Liu, G.; Chen, Q.; Huang, D. Induced mutation breeding of *Brevibacillus brevis* FJAT-0809-GLX for improving ethylparaben production and its application in the biocontrol of Lasiodiplodia theobromae. *Postharvest Biol. Technol.* **2018**, *146*, 60–67. [CrossRef]
35. Cheng, L.; Zhao, C.; Yang, X.; Song, Z.; Lin, C.; Zhao, X.; Wang, J.; Wang, J.; Wang, L.; Xia, X.; et al. Application of a dissolved oxygen control strategy to increase the expression of *Streptococcus suis* glutamate dehydrogenase in *Escherichia coli*. *World J. Microbiol. Biotechnol.* **2021**, *37*, 1–9. [CrossRef]
36. Liang, J.; Chu, X.; Xiong, Z.; Chu, J.; Wang, Y.; Zhuang, Y.; Zhang, S. Oxygen uptake rate regulation during cell growth phase for improving avermectin B1a batch fermentation on a pilot scale (2 m^3). *World J. Microbiol. Biotechnol.* **2011**, *27*, 2639–2644. [CrossRef]

 *fermentation*

Article

Insight into the Substrate Specificity of *Lactobacillus paracasei* Aspartate Ammonia-Lyase

Yi-Hao Huang [1,†], Weir-Chiang You [2,†], Yung-Ju Chen [1], Jhih-Ying Ciou [1] and Lu-Sheng Hsieh [1,*]

[1] Department of Food Science, College of Agriculture and Health, Tunghai University, Taichung 40704, Taiwan
[2] Department of Radiation Oncology, Taichung Veterans General Hospital, Taichung 40705, Taiwan
* Correspondence: lshsieh@thu.edu.tw; Tel.: +886-4-23590121 (ext. 37331)
† These authors contributed equally to this work.

Abstract: Aspartate ammonia-lyase (AAL) catalyzes the reversible conversion reactions of aspartate to fumaric acid and ammonia. In this work, *Lactobacillus paracasei LpAAL* gene was heterologously expressed in *Escherichia coli*. As well as a recombinant His-tagged LpAAL protein, a maltose-binding protein (MBP) fused LpAAL protein was used to enhance its protein solubility and expression level. Both recombinant proteins showed broad substrate specificity, catalyzing aspartic acid, fumaric acid, phenylalanine, and tyrosine to produce fumaric acid, aspartic acid, *trans*-cinnamic acid, and *p*-coumaric acid, respectively. The optimum reaction pH and temperature of LpAAL protein for four substrates were measured at 8.0 and 40 °C, respectively. The K_m values of LpAAL protein for aspartic acid, fumaric acid, phenylalanine, and tyrosine as substrates were 5.7, 8.5, 4.4, and 1.2 mM, respectively. The k_{cat} values of LpAAL protein for aspartic acid, fumaric acid, phenylalanine, and tyrosine as substrates were 6.7, 0.45, 4.96, and 0.02 s^{-1}, respectively. Therefore, aspartic acid, fumaric acid, phenylalanine, and tyrosine are *bona fide* substrates for LpAAL enzyme.

Keywords: aspartate ammonia-lyase; *Lactobacillus paracasei*; maltose-binding protein; phenylalanine ammonia-lyase; substrate specificity; tyrosine ammonia lyase

Citation: Huang, Y.-H.; You, W.-C.; Chen, Y.-J.; Ciou, J.-Y.; Hsieh, L.-S. Insight into the Substrate Specificity of *Lactobacillus paracasei* Aspartate Ammonia-Lyase. *Fermentation* **2023**, *9*, 49. https://doi.org/10.3390/fermentation9010049

Academic Editors: Xian Zhang and Zhiming Rao

Received: 26 November 2022
Revised: 1 January 2023
Accepted: 5 January 2023
Published: 6 January 2023

1. Introduction

Aspartate ammonia-lyase (AAL, EC 4.3.1.1), which catalyzes the reversible non-oxidative deamination of L-aspartic acid to fumaric acid (Figure 1), is an important part of the link between amino acid metabolism and organic acid metabolism [1]. AAL enzyme activity is widely present in various microorganisms and has been purified from thermophilic bacteria, *Escherichia coli*, *Pseudomonas fluorescent*, *Bacillus subtilis*, and *Bacillus thermophilus* [2–5]. *Escherichia coli* AAL (eAAL) protein has been studied extensively; the eAAL protein is composed of 477 amino acids with a molecular weight of approximately 52 kDa, which is a tetrameric enzyme consisting of four identical subunits [2,3]. Its crystal structure has been elucidated, and it has been shown that the catalysis reaction must be carried out under Mg^{2+} and alkaline conditions [2,6,7]. The bAAL protein purified from *Bacillus thermophilus* is similar in molecular weight to other AAL proteins from mesophilic microorganisms and is a homotetrameric protein, but shows different biochemical properties with eAAL protein [8]. One of the most obvious features is that the bAAL protein is structurally more stable under high temperature conditions, and the catalytic reaction also requires the participation of Mg^{2+} under alkaline conditions [8]. Previous studies have shown that AAL proteins only accept L-aspartic acid as a substrate to catalyze the formation of fumaric acid and ammonia under deamination reaction, and lack the activity of catalyzing other L-amino acids, other unsaturated acids, or D-aspartic acid [9,10]. In other words, L-aspartic acid is a very specific substrate for AAL enzymes [9,10]. However, it is later shown that the AAL proteins are capable of catalyzing other substrates. The AAL protein purified from *Pseudomonas aeruginosa* not only catalyzes the reverse reaction of

L-aspartic acid to generate fumaric acid and ammonia, but also catalyzes L- phenylalanine to form *trans*-cinnamic acid [11].

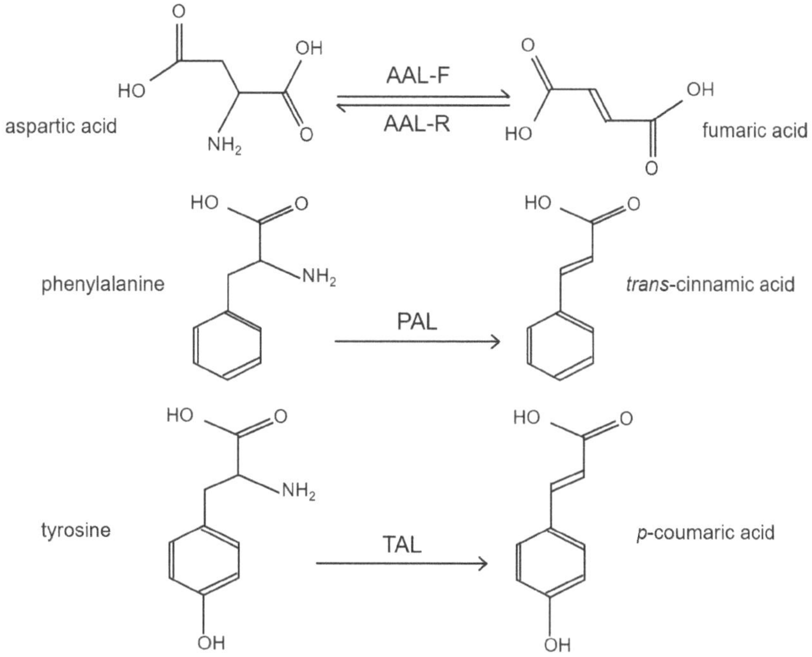

Figure 1. The enzyme reactions catalyzed by the LpAAL protein. AAL-F: aspartate ammonia-lyase forward reaction; AAL-R: aspartate ammonia-lyase reverse reaction; PAL: phenylalanine ammonia-lyase; TAL: tyrosine ammonia-lyase.

Phenylalanine ammonia-lyase (PAL, EC 4.3.1.24) can utilize L-phenylalanine as a substrate to generate *trans*-cinnamic acid and ammonia under non-oxidative deamination reaction [12,13]. Its product, *trans*-cinnamic acid, is then functioned as an important intermediate in the plant phenylpropanoid pathway, being utilized as precursors for thousands of compounds, such as tannins, flavonoids, lignin, etc. [12,13]. PAL proteins from dicot plants are reported to exhibit high specificity to its substrate phenylalanine, whereas, in our previous studies, that monocot plant *Bambusa oldhamii* phenylalanine ammonia-lyase 4 protein (BoPAL4, EC 4.3.1.25) is a dual functional phenylalanine-tyrosine ammonia-lyase (PTAL) enzyme, catalyzing *trans*-cinnamic acid and *p*-coumaric acid production from L-phenylalanine and L-tyrosine via its PAL and tyrosine ammonia-lyase (TAL) activities [13].

Lactobacillus paracasei is a Gram-positive, heterotrophic *Lactobacillus*, distributed in a variety of media, including many fermented dairy products, vegetables, and the human gastrointestinal tract [14,15]. Because of its industrial value and health-promoting potential, *Lactobacillus paracasei* has been extensively studied and used as probiotic [16] and is believed to have various health-promoting properties for humans, including antitumor, anti-inflammatory, and biologics such as microbiological modulation intestinal bacteria [14]. Although AAL protein function was studied in *Lactobacilli* [17,18], our understanding of AAL enzyme in this Genus is still very limited. In addition, *Pseudomonas aeruginosa* AAL protein is reported to have PAL enzymatic activity in 2017 [11], which aroused our interest in the hypothesis that AAL may have similar substrate specificity as bamboo PALs [12,13]. In this study, we further examined and proved that *Lactobacillus paracasei* AAL (LpAAL) has both PAL and TAL activities by using the BoPAL4 protein as a control group to comprehensively study the substrate specificity in the LpAAL protein.

2. Materials and Methods

2.1. Chemicals

L-aspartic acid, fumaric acid, L-phenylalanine, L-tyrosine, *trans*-cinnamic acid, and *p*-coumaric acid were obtained from MilliporeSigma (Burlington, MA, USA). Reagents for protein electrophoresis and molecular biology manipulations were described in previous publications [12,13,19,20].

2.2. Construction of the Expression Vectors

The full-length *LpAAL* gene was chemically synthetic by MDBio, Inc. (New Taipei City, Tai wan) and then subcloned into pET28a plasmid for making pET28a-LpAAL plasmid (Table 1). The *LpAAL* gene fragment was amplified from pET28a-LpAAL plasmid by a PCR reaction and then subcloned into pMAL-c2x plasmid for making a maltose-binding protein fused LpAAL protein (Table 1).

Table 1. Expression plasmids used for recombinant proteins expressions in *Escherichia coli*.

Plasmids	Relevant Characteristics	Source/Reference
pET28a	*E. coli* expression vector with N-terminal His$_6$-tag fusion	Invitrogen
pET28a-LpAAL	*LpAAL* coding sequence inserted into pET28b	This study
pMAL-c2x	*E. coli* expression vector with N-terminal MBP fusion	New England Biolab
pMAL-LpAAL	*LpAAL* coding sequence inserted into pMAL-c2x	This study
pTrcHis-BoPAL4	*BoPAL4* coding sequence inserted into pTrcHisA	[13]

2.3. E. coli Strains and Protein Expression Conditions

E. coli Top10 strain was used for recombinant BoPAL4 protein expression [13]. *E. coli* DH5α and BL21(DE3) strains were used for plasmid storage and for LpAAL and MBP-LpAAL expression, respectively. *E. coli* BL21(DE3) cells carrying the expression plasmids (Table 1) were grown at 30 °C in Luria-Bertani medium (1% tryptone, 0.5% yeast extract, 1% NaCl, and pH 7.0) supplemented with 100 mg/mL ampicillin or 50 mg/mL *kanamycin*. The expression of LpAAL and MBP-LpAAL proteins was induced at 30 °C with 0.1 mM isopropyl-β-D-thiogalactoside (IPTG) for 4 h. Cells were centrifuged at 6000× *g* for 10 min, and pellets were stored at −20 °C freezer before use.

2.4. Preparations of LpAAL and MBP-LpAAL Enzymes

E. coli cells were re-dissolved with 20 mL of 1× lysis buffer (50 mM Tris-HCl, pH 7.5, 10 mM imidazole, 100 mM NaCl, 1 mM PMSF) and disrupted by sonication using a Branson cell disruptor [19,20]. Cell lysates with LpAAL protein or MBP-LpAAL protein were placed in a Ni-NTA column for affinity column chromatography, and finally eluted with buffer containing 250 imidazole.

2.5. SDS-Polyacrylamide Gel Electrophoresis

Purity of target proteins were analyzed by polyacrylamide gel electrophoresis (PAGE) using a Mini-PROTEAN Tetra Cell system (Bio-Rad, Hercules, CA, USA). Protein samples on gels were stained and decolorized with Coomassie Brilliant Blue R-250 and 20% methanol. The image of the gel was captured by a Gel Doc XR+ Imaging System (Bio-Rad, Hercules, CA, USA).

2.6. AAL-F Activity Assay

The aspartate ammonia-lyase forward reaction activity (AAL-F) was assayed by measuring the production of fumaric acid as the absorbance increased at 240 nm [21] using fumaric acid as the standard. The reaction mixture for the AAL-F assay contained 50 mM Tris−HCl, pH 8.0, 11 mM fumaric acid, and 50 μL of enzyme solution in a total volume of 1.0 mL. The reaction mixture was incubated at 40 °C for 3 min. AAL-F activity unit was defined as nanomole of fumaric acid formatted per minute.

2.7. AAL-R Activity Assay

The aspartate ammonia-lyase reverse reaction activity (AAL-R) was assayed by measuring the production of L-aspartic acid as the absorbance increased at 293 nm [22] using aspartic acid as the standard. The reaction mixture for the AAL-R assay contained 50 mM Tris−HCl, pH 8.0, 40 mM L-aspartic acid, and 50 μL of enzyme solution in a total volume of 1.0 mL. The reaction mixture was incubated at 40 °C for 6 min. AAL-R activity unit was defined as nanomole of aspartic acid formed per minute.

2.8. PAL Activity Assay

The PAL activity was assayed by measuring the production of *trans*-cinnamic acid as the absorbance increased at 290 nm [12] using *trans*-cinnamic acid as the standard. The reaction mixture for the PAL assay contained 50 mM Tris−HCl, pH 8.0, 40 mM L-phenylalanine, and 50 μL of enzyme solution in a total volume of 1.0 mL. The reaction mixture was incubated at 40 °C for 6 min. PAL activity was defined as nanomole of *trans*-cinnamic acid formed per minute.

2.9. TAL Activity Assay

The TAL activity was assayed by measuring the production of *p*-coumaric acid as the absorbance increased at 310 nm [12] using *p*-coumaric acid as the standard. The reaction mixture for the AAL-R assay contained 50 mM Tris−HCl, pH 8.0, 40 mM L-tyrosine, and 50 μL of enzyme solution in a total volume of 1.0 mL. The reaction mixture was incubated at 40 °C for 6 min. TAL activity was defined as nanomole of *p*-coumaric acid formed per minute.

2.10. Enzyme Biochemical Properties and Kinetics

To measure the optimum reaction temperature, activities' assays were conducted at standard reaction condition in a range of temperatures from 30 to 70 °C. To measure the optimum reaction pH, activities' assays were conducted at standard reaction condition in a range of pH from 5.0 to 9.0. To determine the kinetic parameter of AAL-F, AAL-R, PAL, and TAL, a range of concentration of L-aspartic acid, fumaric acid, L-phenylalanine, and L-tyrosine were varied from 0~20 mM, from 0~11 mM, from 0~12 mM, and from 0~2 mM, respectively. K_m and k_{cat} values were calculated from substrate saturation curves [12,13,19,20].

3. Results

3.1. Analysis of the LpAAL Gene

The *LpAAL* gene (GenBank Accession No. EEQ60018) has an open-reading frame (ORF) of 1395 bp and can be deduced into a 464 amino acid or a 50 kDa polypeptide. The three-dimensional structure of the AAL protein monomer subunit is dominated by the α-helix, and, as a homotetramer, it contains a total of four fusion active sites, and each active site consists of three monomer sequences identified to be conserved, in which about 10 rings (SS-loop) are formed by amino acid residues with the characteristic sequence GSSxxPxKxN [2,21,23]. The sequence of the LpAAL SS-loop is G_{308}SSIMPGKVN$_{317}$.

3.2. Expression and Purification of Recombinant LpAAL and MBP-LpAAL Proteins in Escherichia coli

Initially, full length *LpAAL* gene was inserted in pET28a plasmid for expressing an N-terminal His-tagged LpAAL recombinant protein (Figure 2A). The solubility of recombinant LpAAL protein was low, and most of the LpAAL proteins became inclusion bodies.

This result was similar to the expression of *Entamoeba histolytica* aspartate ammonia-lyase protein [24]. Therefore, we constructed another expression vector for expressing a maltose-binding protein (MBP)-tagged LpAAL with C-terminal His-tag for facilitating purification (Figure 2A) and stepped forward the induction conditions, including reduced induction temperature, prolonged induction time, and decreased IPTG concentration, to overcome the scenario of insoluble LpAAL proteins. The results showed that the recombinant proteins of LpAAL and MBP-LpAAL purified with Ni-resin resin were separated SDS-PAGE followed by Coomassie Brilliant Blue R-250 staining (Figure 2B). Both recombinant proteins, LpAAL and MBP-LpAAL, were migrated to their anticipated positions, and molecular weights were approximately 50 and 92.5 kDa, respectively. Therefore, *E. coli* expression system is capable of synthesizing LpAAL and MBP-LpAAL recombinant proteins [25].

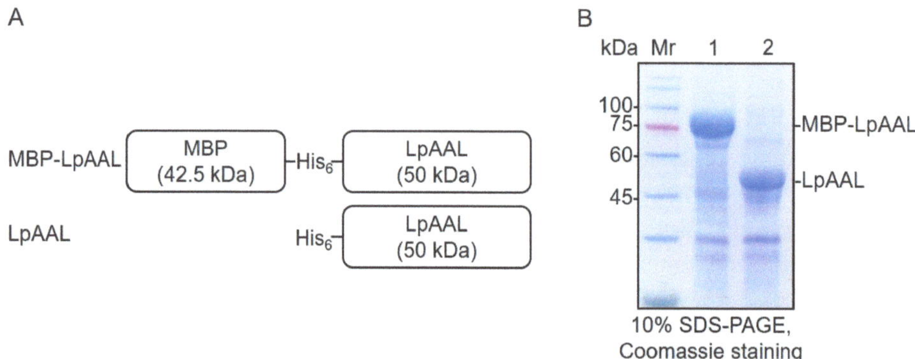

Figure 2. Preparation of recombinant MBP-LpAAL and LpAAL proteins. (**A**) LpAAL protein and MBP-LpAAL fusion protein were expressed in *Escherichia coli*. (**B**) Recombinant proteins were purified using Ni-NTA affinity column and separated using 10% SDS–PAGE, and then stained with Coomassie Blue dye. Mr, molecular weight SDS–PAGE marker; lane 1, MBP-LpAAL; lane 2, LpAAL. MBP: maltose-binding protein.

3.3. Optimum Temperature of AAL and AAL-R Activities of LpAAL and MBP-LpAAL Proteins

A previous study had shown that two different AAL proteins purified from *Aeromonas medium* and *Pseudomonas aeruginosa* had optimal reaction temperatures of 35 °C and 40 °C [7,11]. Singh and Yadav further showed that AAL protein was more stable between 25 and 40 °C, after which it may lose AAL activity, presumably due to denaturation at higher temperatures [7,11]. Recombinant LpAAL and MBP-LpAAL proteins purified from *E. coli* were both active enzymes. Our results showed that the optimal reaction temperatures of AAL-F and AAL-R activities of LpAAL enzyme were both 40 °C (Figure 3A), while the optimal reaction temperatures of AAL-F and AAL-R activities of MBP-LpAAL enzyme are 45 °C and 35 °C, respectively (Figure 3B). Taken together, the optimal reaction temperature of LpAAL and MBP-LpAAL recombinant proteins were comparable with other microbial AAL proteins.

3.4. Optimum pH for AAL-F and AAL-R Activities of LpAAL and MBP-LpAAL Proteins

Normally, AAL enzymes are used to react in alkaline conditions, where the optimum pH is between 7.0 and 9.0 [23]. The optimal pH of the AAL enzyme purified from *Escherichia coli* and *Pseudomonas aeruginosa* were 8.0 and 8.5, respectively [11,21]. Our result showed that the optimal reaction pH of both AAL-F and AAL-R activities of LpAAL enzyme was 8.0 (Figure 4A), while the optimal pHs of AAL-F and AAL-R activities of MBP-LpAAL enzyme were 8.5 and 8.0, respectively (Figure 4B). As a result, the optimal reaction pH of LpAAL and MBP-LpAAL proteins was comparable with other microbial AAL proteins.

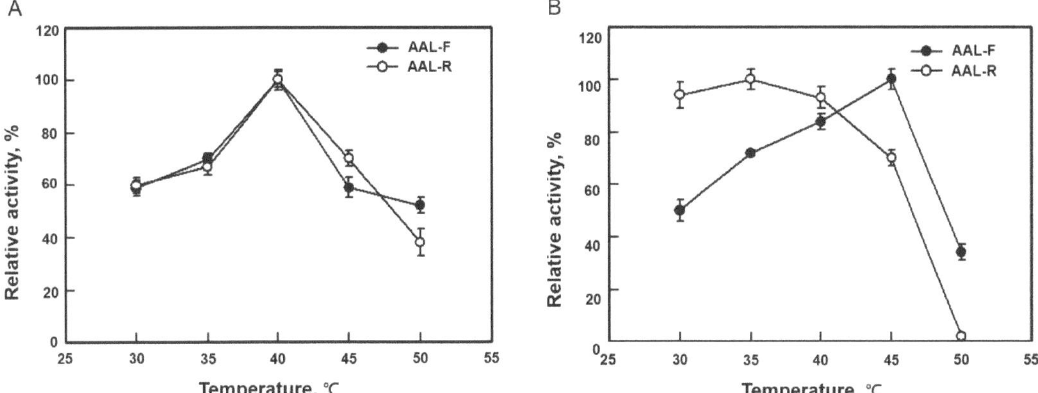

Figure 3. Optimum reaction temperature of AAL-F and AAL-R activities of LpAAL (**A**) and MBP-LpAAL (**B**) enzymes. Activities were measured under standard assay conditions in a range of temperatures from 30 to 50 °C. All experiments were performed in triplicate and expressed as average ± standard deviation (S.D., error bars). The average of maximum activity was normalized as 100%.

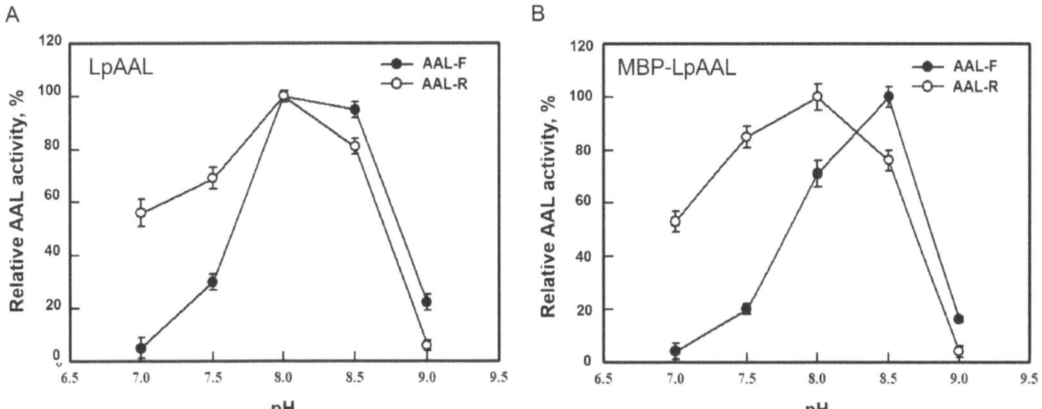

Figure 4. Optimum reaction pH of AAL-F and AAL-R activities of LpAAL (**A**) and MBP-LpAAL (**B**) enzymes. Activities were measured under standard assay conditions in a range of pH from 7.0 to 9.0. All experiments were performed in triplicate and expressed as average ± standard deviation (S.D., error bars). The average of maximum activity was normalized as 100%.

3.5. Kinetic Parameters for AAL-F and AAL-R Activities of LpAAL and MBP-LpAAL Proteins

The kinetics parameters of the LpAAL and MBP-LpAAL enzymes were studied and the K_m values of the enzymes were calculated using saturation curves (Figure 5) and the Michaelis–Menten equation. After calculation, the result showed that the K_m values of LpAAL and MBP-LpAAL proteins using aspartic acid as a substrate were 5.7 (Figure 5A) and 5.2 mM (Figure 5B) at pH 8.0, respectively, and were similar to that of *Pseudomonas fluorescens* AAL (K_m = 5.1 mM) [21]. The K_m values of LpAAL and MBP-LpAAL proteins using fumaric acid as a substrate were both 8.50 mM at pH 8.0 (Figure 5C,D), which is significantly different from the study showing the K_m value of the *Pseudomonas aeruginosa* AAL enzyme using fumaric acid as a substrate to be 21,234 mM [11].

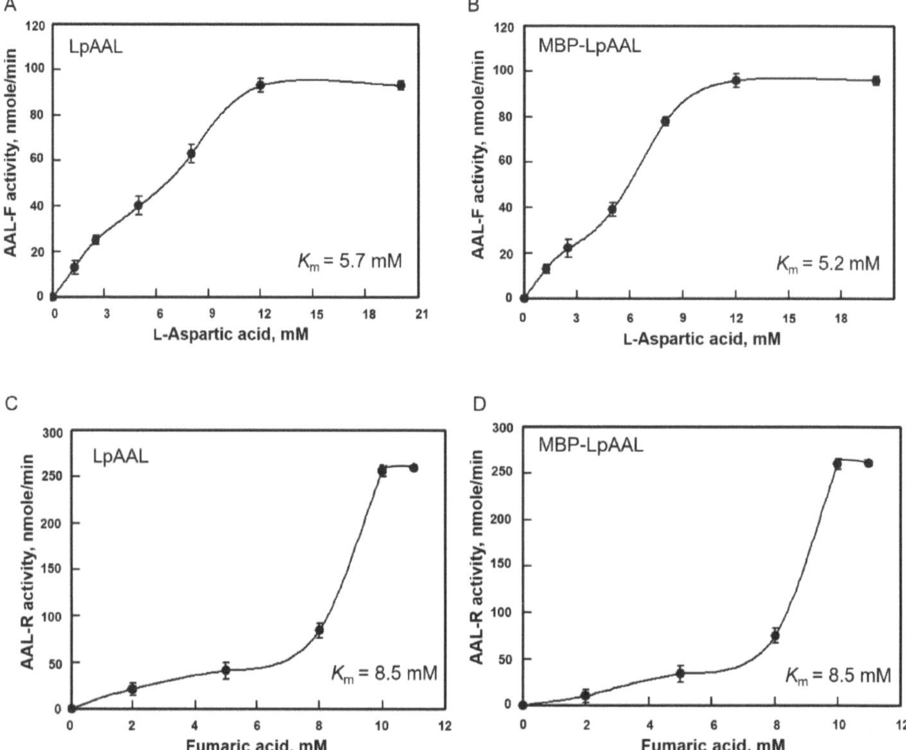

Figure 5. Kinetic parameters of LpAAL and MBP-LpAAL with L-aspartic acid and fumaric acid as substrate. (**A**), to determine LpAAL kinetic parameters using L-aspartic acid as substrate, substrate saturation curve. (**B**), to determine MBP-LpAAL kinetic parameters using L-aspartic acid as substrate, substrate saturation curve. (**C**), to determine LpAAL kinetic parameters using fumaric acid as substrate, substrate saturation curve. (**D**), to determine MBP-LpAAL kinetic parameters using fumaric acid as substrate, substrate saturation curve. All experiments were performed in triplicate and expressed as average ± standard deviation (S.D., error bars). The average of maximum activity was normalized as 100%.

The V_{max} values were obtained from double reciprocal plots derived from substrate saturation curve (Figure 5, Table 1). The V_{max} values were divided by protein amounts (nmole) to obtain the k_{cat} values. The k_{cat} values of LpAAL for L-aspartic acid and fumaric acid were estimated as 6.70 and 0.45 s^{-1}, respectively. Moreover, the k_{cat} values of MBP-LpAAL for L-aspartic acid and fumaric acid were estimated as 13.1 and 37.6 s^{-1}, respectively (Table 2). In conclusion, both recombinant LpAAL and MBP-LpAAL proteins were undoubtedly functional AAL enzymes, and LpAAL and MBP-LpAAL enzyme activities were comparable.

Table 2. Comparison of biochemical properties and kinetic parameters of AAL activities in various AAL proteins.

Enzyme	Substrate	Specific Activity (U/mg)	Optimum pH	Optimum Temp (°C)	V_{max} (nmole/min)	k_{cat} (s^{-1})	K_m (mM)	k_{cat} / K_m (s^{-1} mM)	Ref.
LpAAL	L-Asp	7.7×10^7	8.0	40	93	6.7	5.70	1.18	This study
	fumaric acid	6.09	8.0	40	257	0.45	8.50	0.05	

Table 2. *Cont.*

Enzyme	Substrate	Specific Activity (U/mg)	Optimum pH	Optimum Temp (°C)	V_{max} (nmole/min)	k_{cat} (s^{-1})	K_m (mM)	k_{cat} / K_m (s^{-1} mM)	Ref.
MBP-LpAAL	L-Asp	8.3×10^7	8.5	45	97	13.1	5.20	5.20	This study
	fumaric acid	5.73	8.0	35	267	37.64	8.50	8.5	
pAAL [1]	L-Asp	—	—	—	—	4.91	3×10^{-4}	—	[11]
BAAL [2]	L-Asp	—	—	—	287 (mM/h)	—	213	—	[26]
pfAAL [3]	L-Asp	—	8.8	—	—	130	5.1	25	[21]

[1] pAAL: Pseudomonas aeruginosa PAO1; [2] BAAL: Bacillus sp. YM55-1; [3] pfAAL: Pseudomonas fluorescens R124.

3.6. Optimum Temperature and pH between PAL and TAL for LpAAL and MBP-LpAAL

Recently, Patel et al. reported that *Pseudomonas aeruginosa* AAL protein can utilize phenylalanine as a substrate to yield cinnamic acid [11]. To investigate whether LpAAL/MBP-LpAAL proteins can also use L-phenylalanine or can further use L-tyrosine as a substrate, the optimum temperature of PAL and TAL activities were determined to between 30~50 °C. The optimal PAL and TAL activities of LpAAL protein were both at 40 °C (Figure 6A), similar to the optimal temperature of MBP-LpAAL (Figure 6B). However, the optimal temperatures of BoPAL4 protein for PAL (50 °C) and TAL activities (60 °C) were significantly higher than that of LpAAL and MBP-LpAAL proteins (Figure 6A,B).

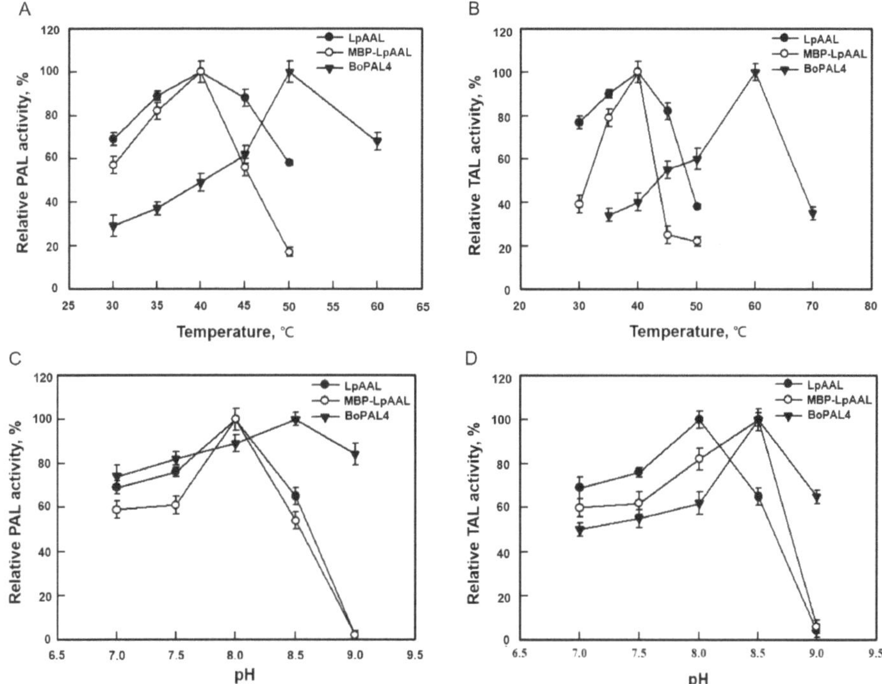

Figure 6. Optimum reaction temperature and pH of PAL and TAL activities of LpAAL, MBP-LpAAL and BoPAL4 enzymes. (**A**) PAL activities were measured under standard assay conditions in a range

of temperatures from 30 to 60 °C. (**B**) TAL activities were measured under standard assay conditions in a range of temperatures from 30 to 60 °C. (**C**) PAL activities were measured under standard assay conditions in a range of pH from 7.0 to 9.0. (**D**) TAL activities were measured under standard assay conditions in a range of pH from 7.0 to 9.0. All experiments were performed in triplicate and expressed as average ± standard deviation (S.D., error bars). The average of maximum activity was normalized as 100%.

Accordingly, the optimum pH for PAL and TAL activities were determined between 7.0 and 9.0, and the optimum pH for PAL and TAL activities of LpAAL protein were both 8.0 (Figure 6C), which is similar to the optimum pH for MBP-LpAAL protein (Figure 6D). In addition, the optimum pH of BoPAL4 protein for both PAL and TAL activities was 8.5, which was comparable with that of LpAAL and MBP-LpAAL proteins (Figure 6C,D). LpAAL and MBP-LpAAL proteins exhibited better PAL/TAL relative activities at lower temperature and pH than that of BoPAL4 protein. Taken together, LpAAL and MBP-LpAAL proteins were able to produce cinnamic acid and *p*-coumaric acid through their PAL and TAL activities.

3.7. Kinetics Parameters for LpAAL and MBP-LpAAL with L-Phenylalanine and L-Tyrosine as Substrates

The K_m of LpAAL protein with L-phenylalanine and L-tyrosine as substrates were 4.43 and 1.17 mM, respectively, and the K_m of MBP-LpAAL protein of L-phenylalanine and L-tyrosine as substrates were 4.07 and 1.10 mM, respectively (Table 3). This was not similar to the result previously reported by Patel et al. with a K_m value with L-phenylalanine of 1.7 mM for AAL [11]. In addition, the k_{cat} of LpAAL enzyme with L-phenylalanine and L-tyrosine were 0.45 and 0.02 s^{-1} respectively, and the k_{cat} of MBP-LpAAL enzyme based on L-phenylalanine and L-tyrosine were 0.84, 0.04, and s^{-1}, respectively. By comparing with the kinetic parameters of BoPAL4 protein, LpAAL and MBP-LpAAL proteins were truly TAL and PAL enzymes in spite of showing lower catalytic capability than that of BoPAL4 protein (Table 3).

Table 3. Comparison of biochemical properties and kinetic parameters of PAL and TAL activities in LpAAL, MBP-LpAAL, and BoPAL4 proteins.

Proteins	Substrate	Optimum pH	Optimum Temp (°C)	k_{cat} (s^{-1})	K_m (mM)	Ref.
LpAAL	L-Phe	8.0	40	0.45	4.43	This study
	L-Tyr	8.0	40	0.02	1.17	
MBP-LpAAL	L-Phe	8.0	40	0.84	4.07	This study
	L-Tyr	8.5	40	0.04	1.10	
BoPAL4	L-Phe	9.0	50	1.14	2.10	[13]
	L-Tyr	8.5	60	0.18	0.10	

3.8. Comparison of Specific Activities in LpAAL, MBP-LpAAL, and BoPAL4 Proteins

Besides enzyme kinetic parameters, the specific activities of AAL-F, AAL-R, PAL, and TAL were compared (Figure 7). The specific AAL-F activity of LpAAL protein was similar to that of MBP-LpAAL protein (Figure 7A). The specific AAL-F activities of both recombinant proteins were slightly higher (1.3-fold) than that of the specific AAL-R activities (Figure 7A). Accordingly, BoPAL4 protein was used as control to evaluate the PAL and TAL activities of LpAAL and MBP-LpAAL proteins. The results showed that the PAL (Figure 7B) and TAL (Figure 7C) activities of LpAAL and MBP-LpAAL proteins were about one-half and one-sixth of BoPAL4 protein. Although the PAL and TAL activities of LpAAL and MBP-LpAAL proteins were lower than that of BoPAL4, phenylalanine and tyrosine were *bona fide* substrates for both LpAAL and MBP-LpAAL proteins.

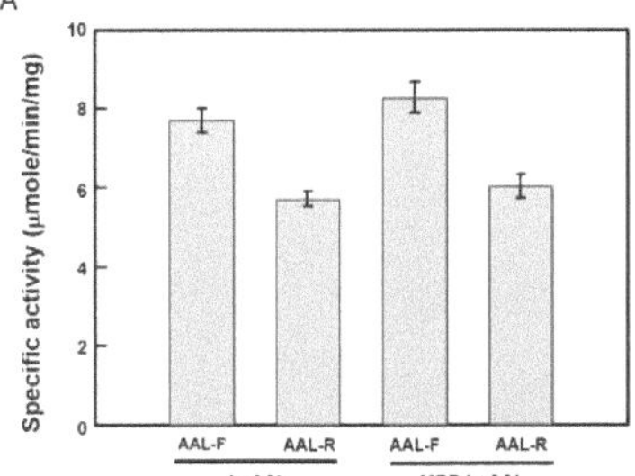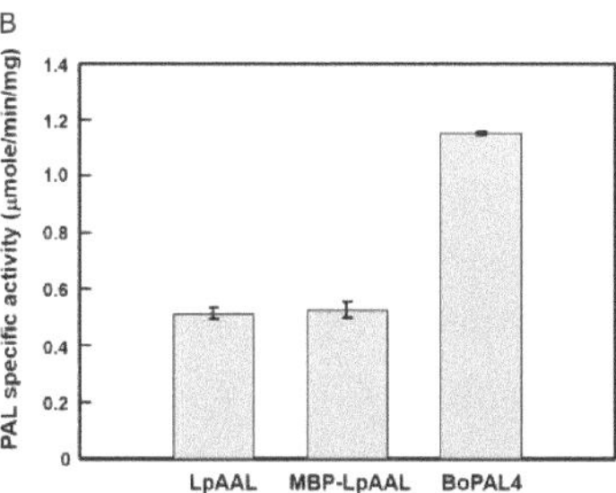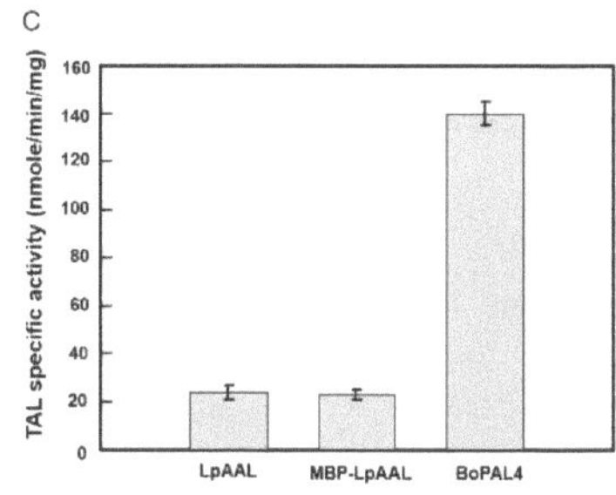

Figure 7. Comparison of specific activities in LpAAL, MBP-LpAAL, and BoPAL4. (**A**) Specific AAL-F and AAL-R activities in LpAAL and MBP-LpAAL were measured under standard assay condition. (**B**) Specific PAL activities in LpAAL, MBP-LpAAL and BoPAL4 were measured under standard assay condition. (**C**) Specific TAL activities in LpAAL, MBP-LpAAL and BoPAL4 were measured under standard assay condition. All experiments were performed in triplicate and expressed as average ± standard deviation (S.D., error bars).

4. Discussion

In this study, we proved that *Lactobacillus paracasei* aspartate ammonia-lyase, LpAAL, has broad substrate specificity and can be used to synthesize at least four products from four substrates. LpAAL protein is a multifunctional enzyme that is active under alkaline and mesophilic conditions, catalyzing the nonoxidative deamination of L-aspartic acid, fumaric acid, L-phenylalanine, and L-tyrosine to yield fumaric acid, L-aspartic acid, *trans*-cinnamic acid, and *p*-coumaric acid. Recombinant LpAAL expressed in *E. coli* was prone to form inclusion bodies, and MBP can significantly increase its solubility without affecting enzymatic activities. In Figure 7A, when the same protein amount of LpAAL and MBP-LpAAL enzymes were subjected to the same reaction, there was no significant difference in the specific activity of both proteins. The lower mole ratio of MBP-LpAAL protein compared to LpAAL could catalyze substrates more efficiently, presumably due to MBP can stabilize LpAAL protein structure. Most of the AAL enzymes were highly specific for using aspartic acid as the substrate, but a few other substrate-catalyzing cases were also observed, including an AAL protein from *Pseudomonas aeruginosa* with PAL activity [11,21]. In this study, LpAAL and MBP-LpAAL proteins were comprehensively elevated to have four enzyme activities, namely AAL-F, AAL-R, PAL, and TAL activities. To our knowledge, our result is the first finding that AAL enzyme can also function as TAL enzyme using a dual function PAL and TAL enzyme, BoPAL4, as the control protein. In the future, other putative aromatic substrates can be tested to figure out if they could be utilized as substrate by LpAAL protein.

Fumaric acid can be used as a medicine for skin diseases [27,28]; L-aspartic acid is used as a raw material for the production of aspartame [29]; *trans*-cinnamic acid has the protective effects in the nervous system [30] and cardiovascular system [31]; and *p*-coumaric acid exhibits antimicrobial and anti-inflammatory effects [32]. Immobilization of microbial AAL proteins has been applied for L-aspartate synthesis [21,26]. All four metabolites of LpAAL enzyme are applied in food, cosmetic, and medical industries. Therefore, LpAAL protein can also be immobilized on electrospun nanofibers to further characterize its recyclability, storage, and kinetic properties.

Author Contributions: Methodology, Y.-H.H., W.-C.Y., Y.-J.C., J.-Y.C. and L.-S.H.; Software, Y.-H.H. and J.-Y.C.; Validation, W.-C.Y. and L.-S.H.; Formal analysis, W.-C.Y. and L.-S.H.; Resources, W.-C.Y. and L.-S.H.; Data curation, L.-S.H.; Writing—original draft preparation, Y.-H.H. and L.-S.H.; Supervision, project administration, and funding acquisition, W.-C.Y. and L.-S.H. All authors have read and agreed to the published version of the manuscript.

Funding: This study was supported by research grants from the Taichung Veterans General Hospital and Tunghai University Joint Research Program (TCVGH-T1097806 to W.-C.Y. and L.-S.H.) and from the National Science and Technology Council, Taiwan (NSTC 110-2313-B-029-001-MY3 to LSH).

Institutional Review Board Statement: Not applicable.

Informed Consent Statement: Not applicable.

Data Availability Statement: Data are contained in the main article.

Conflicts of Interest: The authors declare no conflict of interest.

Abbreviations

AAL	aspartate ammonia-lyase
AAL-F	aspartate ammonia-lyase forward reaction
AAL-R	aspartate ammonia-lyase reverse reaction
bAAL	B: *Bacillus* sp. YM55-1 aspartate ammonia-lyase
BoPAL4	*Bambusa oldhamii* phenylalanine ammonia-lyase 4
eAAL	e: *E. coli* aspartate ammonia-lyase
IPTG	isopropyl β-D-1-thiogalactopyranoside
LB	Luria-Bertani
LpAAL	Lp: *Lactobacillus paracasei* aspartate ammonia-lyase
MBP	maltose-binding protein
pAAL	p: *Pseudomonas aeruginosa* PAO1 aspartate ammonia-lyase
PAL	phenylalanine ammonia-lyase
pfAAL	pf: *Pseudomonas fluorescens* R124 aspartate ammonia-lyase
PTAL	phenylalanine-tyrosine ammonia-lyase
TAL	tyrosine ammonia-lyase

References

1. Kazuoka, T.; Masuda, Y.; Oikawa, T.; Soda, K. Thermostable aspartase from a marine psychrophile, *Cytophaga* sp. KUC-1: Molecular characterization and primary structure. *J. Biochem.* **2003**, *133*, 51–58. [CrossRef] [PubMed]
2. Fibriansah, G.; Veetil, V.P.; Poelarends, G.J.; Thunnissen, A.M.W. Structural basis for the catalytic mechanism of aspartate ammonia lyase. *Biochemistry* **2011**, *50*, 6053–6062. [CrossRef] [PubMed]
3. Karsten, W.E.; Hunsley, J.R.; Viola, R.E. Purification of aspartase and aspartokinase-homoserine dehydrogenase I from *Escherichia coli* by dye-ligand chromatography. *Anal. Biochem.* **1985**, *147*, 336–341. [CrossRef]
4. Kawata, Y.; Tamura, K.; Yano, S.; Mizobata, T.; Nagai, J.; Esaki, N.; Soda, K.; Tokushige, M.; Yumoto, N. Purification and characterization of thermostable aspartase from *Bacillus* sp. YM55-1. *Arch. Biochem. Biophy.* **1999**, *366*, 40–46. [CrossRef]
5. Nuiry, I.I.; Hermes, J.D.; Weiss, P.M.; Chen, C.Y.; Cook, P.F. Kinetic mechanism and location of rate-determining steps for aspartase from *Hafnia alvei*. *Biochemistry* **1984**, *23*, 5168–5175. [CrossRef]
6. Lu, J.; Zhang, J.; Zhang, H.; Wang, X. Studies on the properties of mutants of aspartase from *Escherichia coli* Wa. *Ann. N. Y. Acad. Sci.* **1998**, *864*, 631–635. [CrossRef] [PubMed]
7. Singh, R.S.; Yadav, M. Single-step purification and characterization of recombinant aspartase of *Aeromonas media* NFB-5. *Appl. Biochem. Biotechnol.* **2012**, *167*, 991–1001. [CrossRef]
8. de Villiers, M.; Veetil, V.P.; Raj, H.; de Villiers, J.; Poelarends, G.J. Catalytic mechanisms and biocatalytic applications of aspartate and methylaspartate ammonia lyases. *ACS Chem. Biol.* **2012**, *7*, 1618–1628. [CrossRef]
9. Veetil, V.P.; Raj, H.; Quax, W.J.; Janssen, D.B.; Poelarends, G.J. Site-directed mutagenesis, kinetic and inhibition studies of aspartate ammonia lyase from *Bacillus* sp. YM55-1. *FEBS J.* **2009**, *276*, 2994–3007. [CrossRef]
10. Parmeggiani, F.; Weise, N.J.; Ahmed, S.T.; Turner, N.J. Synthetic and therapeutic applications of ammonia-lyases and aminomutases. *Chem. Rev.* **2018**, *118*, 73–118. [CrossRef]
11. Patel, A.T.; Akhani, R.C.; Patel, M.J.; Dedania, S.R.; Patel, D.H. Bioproduction of L-aspartic acid and cinnamic acid by L-aspartate ammonia lyase from *Pseudomonas aeruginosa* PAO1. *Appl. Biochem. Biotechnol.* **2017**, *182*, 792–803. [CrossRef] [PubMed]
12. Hong, P.Y.; Huang, Y.H.; Lim, G.C.W.; Chen, Y.P.; Hsiao, C.J.; Chen, L.H.; Ciou, J.Y.; Hsieh, L.S. Production of *trans*-cinnamic acid by immobilization of the *Bambusa oldhamii* BoPAL1 and BoPAL2 phenylalanine ammonia-lyases on electrospun nanofibers. *Int. J. Mol. Sci.* **2021**, *22*, 11184. [CrossRef]
13. Hsieh, C.Y.; Huang, Y.H.; Yeh, H.H.; Hong, P.Y.; Hsiao, C.J.; Hsieh, L.S. Phenylalanine, tyrosine, and DOPA are *bona fide* substrates for *Bambusa oldhamii* BoPAL4. *Catalysts* **2021**, *11*, 1263. [CrossRef]

14. Cui, Y.; Qu, X. Genetic mechanisms of prebiotic carbohydrate metabolism in lactic acid bacteria: Emphasis on *Lacticaseibacillus casei* and *Lacticaseibacillus paracasei* as flexible, diverse and outstanding prebiotic carbohydrate starters. *Trends Food Sci. Technol.* **2021**, *115*, 486–499. [CrossRef]
15. Azam, M.; Mohsin, M.; Ijaz, H.; Tulain, U.R.; Ashraf, M.A.; Fayyaz, A.; Kamran, Q. Lactic acid bacteria in traditional fermented Asian foods. *Pak. J. Pharm. Sci.* **2017**, *30*, 1803–1804. [PubMed]
16. Hill, D.; Sugrue, I.; Tobin, C.; Hill, C.; Stanton, C.; Ross, R.P. The *Lactobacillus casei* group: History and health related applications. *Front. Microbiol.* **2018**, *9*, 2107. [CrossRef] [PubMed]
17. Rollan, G.C.; Manca de Nadra, M.C.; Pesce de Ruiz Holgado, A.A.; Oliver, G. Aspartate metabolism in *Lactibacillus murinus* CNRS 313. I. Aspartase. *J. Gen. Appl. Microbiol.* **1985**, *31*, 403–409. [CrossRef]
18. Fernández, M.; Zúñiga, M. Amino acid catabolic pathways of lactic acid bacteria. *Crit. Rev. Microbiol.* **2006**, *32*, 155–183. [CrossRef]
19. Hsiao, C.J.; Hsieh, C.Y.; Hsieh, L.S. Cloning and characterization of the *Bambusa oldhamii BoMDH*-encoded malate dehydrogenase. *Protein Expr. Purif.* **2020**, *174*, 105665. [CrossRef]
20. Hsu, W.H.; Huang, Y.H.; Chen, P.R.; Hsieh, L.S. NLIP and HAD-like domains of Pah1 and Lipin 1 phosphatidate phosphatases are essential for their catalytic activities. *Molecules* **2021**, *26*, 5470. [CrossRef]
21. Csuka, P.; Molnár, Z.; Tóth, V.; Imarah, A.O.; Balogh-Weiser, D.; Vértessy, B.G.; Poppe, L. Immobilization of the aspartate ammonia-lyase from *Pseudomonas fluorescens* R124 on magnetic nanoparticles: Characterization and kinetics. *ChemBioChem* **2022**, *23*, e202100708. [CrossRef] [PubMed]
22. Ida, N.; Tokushige, M. L-Aspartate-induced activation of aspartase. *J. Biochem.* **1985**, *98*, 35–39. [CrossRef] [PubMed]
23. Zhang, J.; Liu, Y. A QM/MM study of the catalytic mechanism of aspartate ammonia lyase. *J. Mol. Graph. Model.* **2014**, *51*, 113–119. [CrossRef]
24. Eze, S.O.O.; Ghulam, J.; Husain, A.; Nozaki, T. Molecular cloning and characterization of aspartate ammonia-lyase from *Entamoeba histolytica*. *Nigerian J. Biochem. Mol. Biol.* **2009**, *24*, 1–7.
25. Wang, L.J.; Kong, X.D.; Zhang, H.Y.; Wang, X.P.; Zhang, J. Enhancement of the activity of L-aspartase from *Escherichia coli* W by directed evolution. *Biochem. Biophy. Res. Commun.* **2000**, *276*, 346–349. [CrossRef] [PubMed]
26. Cárdenas-Fernández, M.; López, C.; Alvaro, G.; López-Santín, J. Immobilized L-aspartate ammonia-lyase from *Bacillus* sp. YM55-1 as biocatalyst for highly concentrated L-aspartate synthesis. *Bioprocess Biosyst. Eng.* **2012**, *35*, 1437–1444. [CrossRef]
27. Mrowietz, U.; Christophers, E.; Altmeyer, P. Treatment of psoriasis with fumaric acid esters: Results of a prospective multicentre study. *Br. J. Dermatol.* **1998**, *138*, 456–460. [CrossRef]
28. Roa Engel, C.A.; Straathof, A.J.; Zijlmans, T.W.; van Gulik, W.M.; van der Wielen, L.A. Fumaric acid production by fermentation. *Appl. Microbiol. Biotechnol.* **2008**, *78*, 379–389. [CrossRef]
29. Werpy, T.; Petersen, G. *Top Value Added Chemicals from Biomass: Volume I-Results of Screening for Potential Candidates from Sugars and Synthesis Gas (No. DOE/GO-102004-1992)*; National Renewable Energy Lab.: Golden, CO, USA, 2004. [CrossRef]
30. Prorok, T.; Jana, M.; Patel, D.; Pahan, K. Cinnamic acid protects the nigrostriatum in a mouse model of Parkinson's disease via peroxisome proliferator-activated receptorα. *Neurochem. Res.* **2019**, *44*, 751–762. [CrossRef]
31. Song, F.; Li, H.; Sun, J.; Wang, S. Protective effects of cinnamic acid and cinnamic aldehyde on isoproterenol-induced acute myocardial ischemia in rats. *J. Eethnopharmacol.* **2013**, *150*, 125–130. [CrossRef]
32. Liu, Y.; Xu, W.; Xu, W. Production of *trans*-cinnamic and *p*-coumaric acids in engineered *E. coli*. *Catalysts* **2022**, *12*, 1144. [CrossRef]

Review

Research Progress on the Effect of Autolysis to *Bacillus subtilis* Fermentation Bioprocess

Kexin Ren [†], Qiang Wang [†], Mengkai Hu, Yan Chen, Rufan Xing, Jiajia You, Meijuan Xu, Xian Zhang * and Zhiming Rao *

Key Laboratory of Industrial Biotechnology, Ministry of Education, School of Biotechnology, Jiangnan University, Wuxi 214000, China
* Correspondence: zx@jiangnan.edu.cn (X.Z.); raozhm@jiangnan.edu.cn (Z.R.)
† These authors contributed equally to this work.

Abstract: *Bacillus subtilis* is a gram-positive bacterium, a promising microorganism due to its strong extracellular protein secretion ability, non-toxic, and relatively mature industrial fermentation technology. However, cell autolysis during fermentation restricts the industrial application of *B. subtilis*. With the fast advancement of molecular biology and genetic engineering technology, various advanced procedures and gene editing tools have been used to successfully construct autolysis-resistant *B. subtilis* chassis cells to manufacture various biological products. This paper first analyses the causes of autolysis in *B. subtilis* from a mechanistic perspective and outlines various strategies to address autolysis in *B. subtilis*. Finally, potential strategies for solving the autolysis problem of *B. subtilis* are foreseen.

Keywords: *Bacillus subtilis*; cell autolysis; gene editing; fermentation technology; chassis cell

Citation: Ren, K.; Wang, Q.; Hu, M.; Chen, Y.; Xing, R.; You, J.; Xu, M.; Zhang, X.; Rao, Z. Research Progress on the Effect of Autolysis to *Bacillus subtilis* Fermentation Bioprocess. *Fermentation* **2022**, *8*, 685. https://doi.org/10.3390/fermentation8120685

Academic Editor: Mohammad Taherzadeh

Received: 2 November 2022
Accepted: 25 November 2022
Published: 28 November 2022

Publisher's Note: MDPI stays neutral with regard to jurisdictional claims in published maps and institutional affiliations.

1. Introduction

B. subtilis is a generally recognized as safe (GRAS) microorganism [1], which is famous for its strong ability for heterologous expression, clear physiological and biochemical characteristics, and legacy relatively simple to operate. In addition, they have the advantages of being easy to cultivate and robust in industrial fermentations [2]. Using *B. subtilis* as the chassis cell, the exogenous introduction of the desired synthetic pathway and the systematic optimization of the strain's own global metabolism and the coordination of the balance between the strain's metabolic network and the exogenous pathway allow for the efficient production of a large number of products. As a crucial industrial biotechnology powerful chassis cell, *B. subtilis* produces valuable enzymes and biopolymers [3]. For example, xylanase can be fermented in a recombinant strain of *Bacillus subtilis* to obtain an enzymatic activity of 38 U/mL [4], and L-asparaginase (ASN) can reach a yield of 407.6 U/mL (2.5 g/L) by *Bacillus subtilis* fermenters [5]. Biopolymers are primarily represented by riboflavin, surface activator, ethylene coupling, phytase, xylanase, L-asparaginase, chondroitin, N-ethyleneglycolate phytase, xylanase, L-asparaginase, chondroitin, N-acetyl acyl glucosamine, etc. [6] (shown in Figure 1). The yield of riboflavin in *Bacillus subtilis* was ahead of other strains of the chassis. As early as 1999, Perkins et al. applied genetic engineering technology to construct a high-yielding strain of riboflavin using *Bacillus subtilis* as the substrate cells, and the yield of riboflavin could reach 15 g/L at 56 h of fermentation [7]. In the agriculture field, *B. subtilis* is not only used to treat plant and animal illnesses and to replace hazardous chemical fungicides to minimize the danger to other species and the environment [8]. Regarding probiotic foods, *B. subtilis* can enter animals in the form of spores [9]. After germination, it can produce many antibacterial substances to inhibit the growth of harmful intestinal microorganisms such as *Escherichia coli* to achieve the effect of gastrointestinal protection [10].

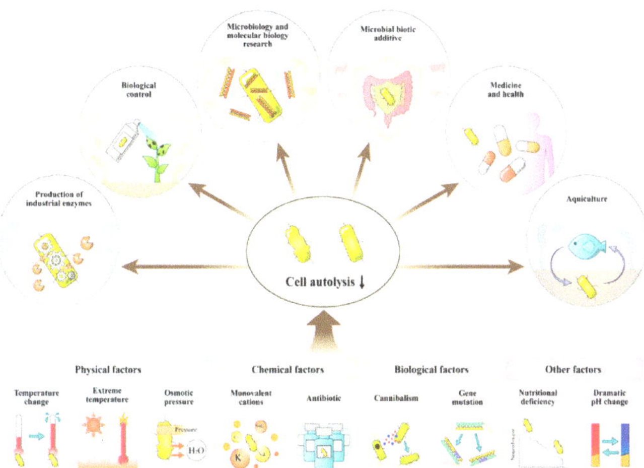

Figure 1. Application of *B. subtilis* as industrial production chassis and factors related to autolysis of *B. subtilis*.

However, there is a problem with cell autolysis in fermentation cultures of *B. subtilis*. Cell autolysis is widespread in microorganisms, such as bacteria, actinomycetes, and fungi. The researchers defined the self-structural degradation process that bacteria undergo at the end of their life cycle as "autolysis" and found that "autolysins" such as cytosolic hydrolases and peptidoglycan hydrolases, which catalyze the hydrolysis of the cell wall peptidoglycan layer, are closely related to the autolysis process. Reduced cell autolysis in *B. subtilis* leads to a significant increase in cell secretion of recombinant proteins [11] and other productions. In a later stage of fermentation or under fermentation conditions unsuitable for cell growth, the cells will begin to autolyze. This effect leads to a substantial reduction in cell biomass, which seriously affects the expression of products and fermentation efficiency [12]. In particular, high-yield strains modified by metabolic engineering are more likely to occur during cell autolysis. For example, WB800 [13] (*B. subtilis* with eight extracellular proteases knocked out), most often applied to express extracellular proteases, has more severe cell autolysis [14]. Therefore, it is essential to study the autolysis of *B. subtilis*. Studies have shown that the root cause of *B. subtilis* autolysis is that the activity of autolysins is controlled by various conditions to perform their respective functions and thus hydrolyze the peptidoglycan of the cells. These autolysis enzymes include those closely related to the growth and development of *B. subtilis* and those from prophages.

This review analyzes the relevant factors of *B. subtilis* autolysis in terms of mechanism, summarizes the functions of critical autolysins in its autolysis process, and outlines the various effective strategies that have been used in recent years to address autolysis in *B. subtilis* and the results they have achieved. Finally, future research directions for solving the autolysis problem of *B. subtilis* have been prospected.

2. Mechanism and Causative Factors of Autolysis Phenomenon

Microbial autolysis is mainly manifested in the action of enzymes on the cell wall. As shown in Figure 2, the cell wall of *B. subtilis* consists primarily of peptidoglycan and anionic polymers. The anionic polymers provide chemical bonds that can be hydrolyzed, thus generating a more significant number of potential active sites for bacterial autolysins action [15]. These actions eventually lead to disrupting the cross-linked structure of peptidoglycan and the peptidoglycan cleavage [16]. The following figure shows the structural units and autolysis action sites (four in total) of the *B. subtilis* metrocyte cell wall and spore cortex.

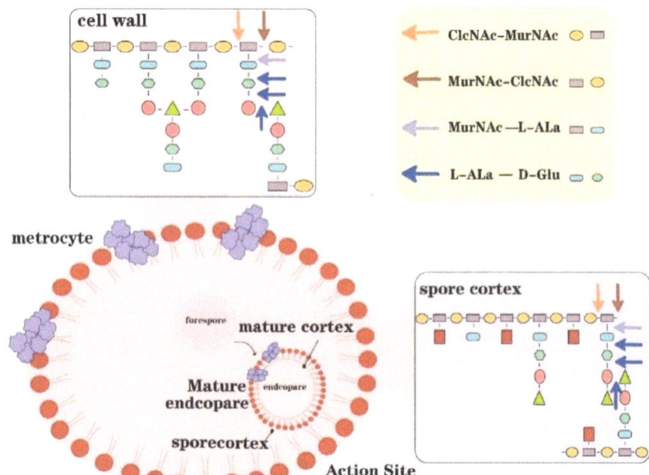

Figure 2. Mechanisms of autolysis in *Bacillus subtilis* site of action of autolysin. The peptidoglycan PG in the cell wall structure is the backbone component of the cell wall, and the side chains of the peptidoglycan are linked to anionic polymers, which provide the sites of action for autolysis enzymes.

In addition to the normal senescence and death of cells, any factors that are unsuitable for the growth and physiological metabolism of microorganisms and cause extreme disturbance in their physiological state can effectively induce autolysis expression and cause microbial autolysis. Comprehensive of the existing reports at home and abroad, we summarize the various factors that can cause cell autolysis in *B. subtilis*, as shown in Figure 1.

2.1. Physical Factors That Induce Autolysis

The main physical factors are temperature, osmotic pressure, and pH. Among them, temperature variation is an essential environmental parameter affecting microbial cell growth and reproduction. High-temperature fermentation has some advantages in production because it creates a harsh environment (preventing the growth of hybrid bacterium) that promotes the growth of the target strain. Nonetheless, raising the high temperature of the culture might have a detrimental impact on microorganism development and lead to autolysis of the cultured cells in certain situations.

B. subtilis can grow at 25~40 °C, and the optimum temperature for its growth is 35 °C. As mentioned in the article by Regamey and Karamat et al. [17], after *B. subtilis* was subjected to thermal excitation at 50 °C, prophage *spβ* was induced, and many cells underwent autolysis. Similarly, Nandy et al. [18] found that moving *B. subtilis* from 37 °C to 48 °C water bath conditions caused massive autolysis of *B. subtilis* cells. According to Antelmann et al. [19], high-temperature incubation caused the production of autolysis enzymes in some bacteria, leading to massive cell autolysis of these microorganisms. This phenomenon poses a severe threat to microbial culture. Therefore, using a higher temperature to culture microorganisms and avoid cell autolysis has become one of the main obstacles to high-temperature culture microorganisms. One of the most successful strategies to overcome this issue is to improve microbial strains to adapt to high-temperature fermentation settings [20]. In addition to this, autolysis of *B. subtilis* induced by supercooling was also found by Yamanaka et al. [21]. Following a rapid decrease in temperature, the physiology of *B. subtilis* cells changes profoundly; this phenomenon is called cold shock [22]. *B.subtilis* vegetative cells undergo autolysis when exposed to cold shock treatment.

In the aspect of osmotic pressure, the suitable osmotic pressure for *Bacillus subtilis* is converted to a culture speed of approximately 200–250 rpm. Svarachorn et al. [23]

discovered that when *B. subtilis* cells lose their capacity to regulate osmotic pressure, the intracellular potassium ions of the bacterium will leak, and the cells will not take up glucose. If the bacterium is continuously incubated at 37 °C in high concentrations of near-monovalent cations, autolysis of cells will be observed. Sahoo et al. [24] and Yamanaka et al. [21] exposed *B. subtilis* to drastic pressure change conditions (rate of 1.482/s). After 10 h incubation, they found that the biomass of the bacterium decreased drastically. The effect of high pressure on the autolysis of *B. subtilis* can also be understood during the decontamination of HHP (high hydrostatic pressure) [25].

pH is a very important parameter for microbial growth, a comprehensive indicator of metabolic activity, and closely related to microbial autolysis. Different species of microorganisms have different pH requirements. The optimum pH for most bacteria is 6.5–7.5. The optimal culture pH for *B. subtilis* is about 7. Jolliffe et al. [26] found that *Bacillus subtilis* showed the highest autolysis of its cells when it was in alkaline medium (pH = 9).

2.2. Chemical Factors That Induce Autolysis

Chemical factors are mainly the effects of monovalent cations and antibiotics on cell autolysis. These substances affect the normal expression of autolytic enzymes and proteases in cells. For example, Svarachorn et al. [27] found that monovalent cations such as K^+, NH_4^+, Na^+, Cs^+, and Li^+ could cause intracell autolysis of *B. subtilis* 168 when the concentration of K^+, NH_4^+, Na^+, Cs^+, and Li^+ was 100 mM. The autolysis rate of cells in the above buffer was in the range of 2.8–4.0 k $\times$ 10^2/min. This shows that high concentrations of monovalent cations in buffers can promote autolysis in *B. subtilis* 168, mainly due to activating important, internal, and significant N-acetylmuramoyl-L-alanine amidase inside this bacterium, e.g., NH^+, Li^+, Rb^+, Cs^+, and Na^+ at 100 mM. Their enzyme activities were increased by 6.4, 3.9, 3.9, 3.6, and 3.4 times, respectively. Autolysis is more likely to occur when protein synthesis is active, but peptidoglycan synthesis is inadequate.

The role of proteases in antibiotic-induced cell lysis is also a more complex process [28]. Jolliffe et al. [29] investigated the impact of nafcillin on *B. subtilis* intracell autolysis and discovered that mutants that release many proteases are incredibly resistant to nafcillin-induced fatal and lytic effects. Under nafcillin induction, however, protease-deficient bacteria were more sensitive to mortality and lysis.

2.3. Biological Factors That Induce Autolysis

Biological factors mainly include external and internal factors, which are the microbial interactions and alterations in the autolysin gene expression level of *B. subtilis*. Among the external factors, *B. subtilis* in the same culture system compete with each other for nutrients. The interaction between microorganisms causes autolysis. As for internal factors, mutations in some genes that can affect the expression of autolysis enzymes can also have a consequence on cell autolysis.

2.3.1. Cell Autolysis Caused by Cannibalism

The phenomenon of "kill one's kind" in *B. subtilis* was reported by Pastor et al. [30]. They found that *B. subtilis* cells forming spores are prone to cannibalism, whereby the bacteria delay the formation of spores in other cells by cannibalism. The nature of cannibalism is that *B. subtilis* that have entered sporulation can produce and export a killing factor and a signaling protein. Synergistic action prevents other sister cells from forming spores and causes sister cell autolysis. Immediately after Allenby et al. [31] discovered the significance of the phenomenon of cannibalism, when *B. subtilis* was subjected to phosphate-deficient conditions, the harsh conditions significantly induced the lysis of non-spores, thus providing a good source of nutrients for other cells. To test this idea, the researchers transferred *B. subtilis* to a carbon-deficient medium and the bacteria underwent rapid cell autolysis [31,32]. Nandy et al. [33] revealed that this crucial regulatory protein of *B. subtilis*, Spo0A, which controls spore development, uses a certain antimicrobial factor to lyse *E. coli* cells under its regulation.

Gonzalez-Pastor et al. [34] analyzed the nature of *B. subtilis* "conspecific" and concluded that those cells entering the spore pathway produce and export a killing factor (SkfA) and a signal peptide protein (SdpC). The genes encoding these two proteins are located on the *skfA* operon and the *sdpC* operon, respectively, where the *skf* operon element is responsible for the output of SkfA and the product of immune function against SkfA [30]. Together, *skfA* and *sdpC* cause sister cell autolysis and prevent the formation of spores. The processing and secretion of the *sdpC* gene also produces a sporulation-delaying protein (SDP). SDP is a peptide toxin that kills cells outside the biofilm to support continued growth. SDP rapidly collapses the proton motive force (PMF) of *B. subtilis*, which induces dramatic autolysin-mediated lysis [35,36].

The *sdpC* operon is also expressed in Spo0A-active cells, producing a signaling protein (called delaying the formation of spore protein, SdpC) that causes transcription of the *yvbA* gene into the protein YvbA. The YvbA protein turns on the operon containing the ATP synthase gene (*atp*) and the lipid catabolism inhibitor gene (*yusLKJ*), thereby delaying the entry of Spo0A-activated cells to delay entry into the bud formation process. However, in Spo0A-inactivated cells, SdpC signaling protein also induces the production of YvbA protein, which promotes the death of Spo0A-inactivated cells by inhibiting a σ^W factor. The opened element of the metabolic inhibitor gene (*yusLKJ*) decomposes, thereby delaying the entry of Spo0A-inactivated cells into the formation process of spores. However, in Spo0A-inactivated cells, the SdpC signaling protein also induces the production of YvbA protein, which promotes the death of Spo0A-inactivated cells by inhibiting a σ^W factor. (This σ^W factor is associated with antibiotic resistance and detoxification and protects cells from lysis by killing aspects.) Thus, in Spo0A-inactivated cells, YvbA assists the killing factor in causing cell autolysis. The final result of the above process is that the nutrients released by the dead cells are used as food for the Spo0A-activated cells, keeping the Spo0A-activated cells growing instead of entering the formation process of spores [34].

2.3.2. Gene Mutations Cause Cell Autolysis

In *B. subtilis*, the expression of autolysis enzymes is regulated by a variety of genes. However, with the use of *B. subtilis* in industrial development, many of the mutants produced by genetic modification to improve fermentation yield are more prone to autolysis because they disrupt this regulatory balance. According to Kodama, the *B. subtilis* gene mutant is an excellent host for exogenous protease synthesis [37]. Stephenson [14] reported in the early years that mutants deficient in the extracellular proteases NprE and AprE were more likely to increase the intracellular hydrolysis of *B. subtilis* between the transition and stabilization phases of batch cultures of the bacterium. Afterward, the Wong group [38] constructed deletion mutants of eight major exogenous protease genes in *B. subtilis*. Although the production of cellulase was improved in this mutant, the cells were highly susceptible to autolysis during culture. In recent studies, a number of other genes have also been found to be associated with autolysis. Palomino et al. [39] found that high salt and *pbpE* mutations lead to cell wall defects. The resulting mutant peptidoglycan shows increased solubility and sensitivity to mutant autolysins. Perez [40] found that *spoVG* mutations decreased and *spoVS* mutations increased *sigma*(D)-directed gene expression, causing cell separation and autolysis.

2.4. Other Factors That Induce Autolysis

Autolysis of bacterial cells is readily caused by the severe circumstances to which bacteria are exposed, including nutritional deficits. Such harsh conditions act as an inducing factor, resulting in the cannibalism mentioned above. For example, Allenby et al. [30] reported that when *B. subtilis* was subjected to phosphate-deficient conditions, the harsh conditions significantly induced transcription of the *skfA* operon element in *B. subtilis*, which made it enter a state of cannibalism, leading to cell autolysis. It is worth adding that Jolliffe et al. [41] discovered that adding chemicals that distribute charge or pH gradients to *B. subtilis* cells might lyse the bacterium.

3. *B. subtilis* Key Autolysins Enzymes

From the above studies, it is clear that autolysis in *B. subtilis* is ultimately caused by autolysins. The final results of various evoked autolytic factors are stimulated autolysins expression. Therefore, the research conducted on autolysins is the key to solving the autolysis problem. *B. subtilis* produces several hydrolases during the trophic growth phase [42–44]. Smith et al. [45] nicely summarized the autolysins associated with *B. subtilis* growth and development. Based on the specificity of autolysins' hydrolysis of *B. subtilis* cell wall peptidoglycan and the specificity of the hydrolytic chemical bonds of the autolysins, autolysins and peptidoglycan hydrolases can be classified as muramidase, acetyl-glucosaminidase, acetylmuramoyl alanine amidase, and endopeptidase. The functions of several of the most critical enzymes and their mechanisms of action in autolysis are detailed in Figure 3.

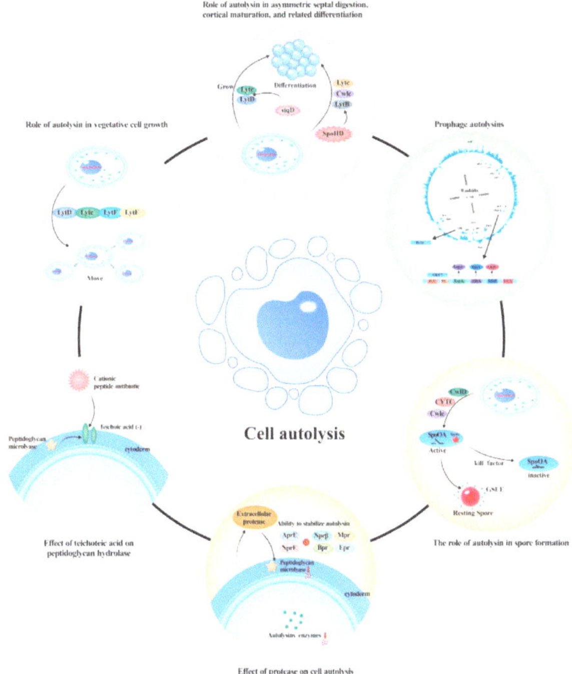

Figure 3. The role of key autolysis enzymes of *B. subtilis* in life processes and the factors affecting them.

3.1. Analysis of the Mechanism of Action of Related Enzymes

The respective functions of the life cycle of *B. subtilis* autolysis enzymes are demonstrated in Figure 3. Among these hydrolases, N-acetylmuramoyl-L-alanine amidase (amidase, LytC or CwlB) and N-acetyl-β-glucosaminidase (glucosaminidase, LytD or CwlG) are two of the significant autolysins [46–48]. During the nutritional development stage of bacteria, they carry the majority of the autolysis activity. Inactivation of LytC and LytD show that these two autolysins bear 95% of the autolytic activity of the cells [49,50]. In addition to the main autolysins described above, other less significant enzymes were discovered individually in nutritional cell surface extracts, including LytF (CwlE) and LytE (CwlF) [51–53]. Table 1 summarizes the peptidoglycan hydrolase-related genes. People are still researching the roles of various additional, less significant, but potentially vital autolysins. Different groups studying the functions of the above autolysins have come to different conclusions due to subtle differences in strain genetic background and experimental conditions. Still, they all have the same impression about these autolysins. There is a high degree of functional redundancy in these various multifunctional autolysins.

3.1.1. The Role of Autolysins in the Formation of Spores

Autolysins aid cell differentiation. Regarding *B. subtilis* differentiation activities, including the formation of spores and germination, which need substantial peptidoglycan rearrangements and alterations, almost all of them are affected by autolysins [54,55].

Even though spores have considerable resistance and dormancy, they possess a stress-sensitive system that reacts to germination agents. Maternal cell autolysis aids in the dispersion of mature spores to a more favorable environment. Foster et al. [56] found the presence of a 30 kDa sized protein associated with maternal cell autolysis in the formation of spores that are well expressed before maternal cell autolysis occurs. It is presumed to be primarily responsible for maternal cell autolysis. Then, Kuroda et al. [57] cloned and identified the gene responsible for encoding this protein and named it *"cwlC."* They tested the function of the *cwlC* deletion mutant. They showed that the mutant did not undergo significant phenotypic changes in shoot cell formation, germination, and shoot resistance to heat and autolysins. Moreover, in the *cwlC* mutant, the mother cell still underwent cell autolysis. They also found that CwlC proteins hydrolyze both the nutrient cell wall peptidoglycan of *B. subtilis* and the peptidoglycan of spores. Mother cell autolysis required for the release of mature endosperms during the formation of spores is mainly dependent on two amidases of the LytC family, CwlC and LytC; CwlC and LytC play complementary roles in hydrolyzing the cell wall peptidoglycan of mother cells [44,56].

In addition, endospore formation in *B. subtilis* produces a germination-specific lyase (GSLE) that can only be isolated from germinating spores in its active form and can only hydrolyze the stress peptidoglycan of permeabilized, undamaged spores in vitro. GSLE is present in dormant spores as an inactive preprotein [58].

3.1.2. Role of Autolysins in the Digestion of the Asymmetric Septum, Cortical Maturation, and Related Differentiation Processes

The *B. subtilis* autolysins play an essential role in asymmetric septum digestion, cortical maturation, and related differentiation processes [15]. *spoIID* is involved in the digestion of the asymmetric septum, as strains carrying mutations in this gene are found when the septum is only partially digested and remains in an encircled state around the cell. *spoIID* encodes a modified protein similar to the LytB sequence. Studies claim that *spoIID* may control the movement of an autolysin to digest peptidoglycan from asymmetric septa [59].

3.1.3. The Role of Autolysins during the Vegetative Phase

The main autolysins during the vegetative phase of *B. subtilis* are LytC and LytD; the *lytC* gene encodes LytC, and *lytC* is located within the operon element *lytLABC* [50]. *lytC* is active at the end of the logarithmic growth phase of *B. subtilis*, and this autolytic activity is maintained until late in the stabilization phase. It remains active even during the formation of spores (referred to in Section 3.1.1 mentioned above). LytC and LytD are the main autolysins produced by *B. subtilis* during growth. Both of them are transcriptionally regulated by *sigD*. sigma D is expressed during vegetative growth and is associated with exponential growth and transcription of genes expressed during early stationary periods [60,61]. Furthermore, both proteins are required for bacterial motility. Knocking out the *lytC* gene in *B. subtilis* ATCC 6051 increased bacteriophage density and amylase activity by 1.6-fold. *lytC* and *lytD* deletion also inhibited cell lysis, but the inhibition was less than in the *lytC*-only mutant.

lytE and *lytF* encode peptide chain endopeptidases and are involved in bacterial segregation. Deleting *lytE* and *lytF* alone resulted in mutants with defective growth segregation (division or separation). Still, the inhibitory effect on cell autolysis was not significant. In addition, during cell growth, cells must expand covalently closed macromolecules of the cell membrane [62]. *B. subtilis* contains two functionally redundant D, L-endopeptidases

(CwlO and LytE) that cleave peptide crosslinks to allow expansion of PG (the cell wall peptidoglycan) [63]. Double deletion of *lytE* with the *cwlO* mutant, which affects cell length increase, leads to a bacteriophage lethal effect [64].

3.2. Prophage Autolysins

In addition to the numerous autolysins involved in the growth and development, there are also autolysins from *B. subtilis* prophage. When the expression of autolysin encoded by prophages (*PBSX*) is induced, pores with auxiliary or unknown functions are produced in the peptidoglycan cell wall. The cytoplasmic membrane material protrudes through these openings and is released as a membrane sac (MV) [65]. The induced cells eventually die due to loss of membrane integrity. The vesicle-producing cell leads to cell autolysis through the enzymatic action of releasing autolysin, which induces the formation of MV in neighboring cells [66]. The study of prophage sequences has been studied since as early as 1997; Presecan [67] identified many potentially non-essential chromosomal motifs in the whole genome nucleotide sequence of *B. subtilis*, particularly 10 relatively large AT-rich regions representing known prophages and prophage-like regions scattered across the genome. Analog sequences (*pro*1-7) are obtained by horizontal gene transfer [68–70]. Prophage *PBSX*, *spβ*, and *skin* remain lysogenic [17,71,72]. Lysogenic phages usually encode peptidoglycan hydrolases, called prophage autolysins. Under certain conditions, it can lyse host cell walls and release phage particles after replication.

3.2.1. Prophage Autolysins PBSX

PBSX, a phage-like phibacin of *B. subtilis* 168, carried by bacteria can be induced to lyse DNA-damaging agents and produce *PBSX* particles; however, these particles cannot transmit the *PBSX* genome [73]. This suicide reaction produces particles that kill non-lysogenic strains of *PBSX*. The prophage *PBSX* includes *xlyA*, which encodes the peptidoglycan hydrolase XlyA mentioned above, *xepA* encodes an export protein XepA, *xhlA* encodes a membrane-associated protein XhlA, and *xhlB* encodes a perforin protein XhlB. The proteins encoded by the latter three genes play an assisting role in cell autolysis. In subsequent studies conducted by Krogh et al. [74], it was found that a late operon located within *B. subtilis PBSX* was used to express autolysins associated with the prophage *PBSX*, as shown in Figure 3. Four different genes on late operon encode these autolysins proteins [73].

Table 1. Summary of peptidoglycan hydrolase-related genes.

	Gene	Comments	Function	Ref
1	*cwlC*	Amidase, mother cell-specific	Cell separation, spore	[43,55]
2	*lytC*	Amidase, vegetative/sporulation expressed only during vegetative growth	Cell separation, motility, cell lysis, mother cell lysis	[46–48]
3	*lytD*	Glucosaminidase, vegetative	Motility	[48,49]
4	*lytE lytF*	Encoding peptide chain endopeptidase	Participation in bacteriophage isolationCell separation, motility, cell lysis, mother cell lysis	[46–48,50–52]
5	*gslEgslE*	GSLE-related proteinsGSLE-related proteins	Cortex hydrolysisCortex hydrolysis	[57]
6	*spoIIDlytE lytF*	Control of the activity of an autolysin to digest peptidoglycan from asymmetric septaEncoding peptide chain endopeptidase	digestion of the asymmetric septumParticipation in bacteriophage isolation	[50–52,59]
7	*xlyAspoIID*	Amidase, mitomycin C-inducibleControl of the activity of an autolysin to digest peptidoglycan from asymmetric septa	PBSX lysisdigestion of the asymmetric septum	[59,75]
8	*cwlAcwlC*	Amidase, silent gene in the skin elementAmidase, mother cell-specific	Cryptic prophageCell separation, spore	[43,55,76,77]

3.2.2. XlyA Amidase Family

As further research into it, Longchamp et al. [72] divided and identified two peptidoglycan hydrolases of the XlyA amidase family, XlyA and YomC, and found YomC is within the *spβ*, respectively, and belongs to the *B. subtilis* prophage autolysins. Another enzyme from the XlyA family, N-acetylmuramoyl-L-alanine amidase, is an active peptidoglycan hydrolase [75]. The protein N-acetylmuramoyl-L-alanine amidase (CwlA) is related to cell autolysis. Later studies showed that CwlA is an autolysin gene located within the *skin* component of the prophage [76]. Kunst et al. [77] also found that the two open reading frames *yqxH* and *yqxG* immediately adjacent to *cwlA* encode two other proteins, respectively, which they hypothesized to be related to the autolytic activity of CwlA and named them "Yqxh" (similar to holin) and "Yqxg" (similar to phage-related lytic exoenzyme), respectively.

3.2.3. Prophage Autolysins BlyA

Regamey and Karamata [17] identified and characterized a DNA fragment in *B. subtilis* and named it *"blyA"* gene, which encodes a protein with a molecular weight of 39.6 kDa and contains 367 amino acids (BlyA). BlyA exhibited N-acetylcytidyl-L-alanine amidase activity associated with *spβ* phage-mediated cell autolysis. They conclusively demonstrated that BlyA belongs to the prophage *spβ* autolysins and that heat stimulation of *B. subtilis* CU1147 (CU1147 is an *Spβ*c2 lysogenic strain, and *Spβ*c2 is a temperature-sensitive phage) would induce this *Spβ*c2 to express BlyA and lead to cell autolysis.

3.3. The Effect of Phosphopiridic Acid on Peptidoglycan Hydrolase

It has been shown that the peptidoglycan hydrolase associated with *B. subtilis* autolysis is regulated by lipoteichoic acid in the cell wall and the action of extracellular proteases. Lipoteichoic acid (LTA) can absorb Mg2+ to support the activity of several synthetic enzymes in the cell membrane. It also functions as a storage element and regulates the activity of intracell autolysis enzymes [78]. In *B. subtilis*, LTA modification is controlled by the *dltA-E* operon. Previous studies have shown that deletion of the *dltA-D* gene disrupts LTA modification, which further alters the cell surface microenvironment and enhances cell autolysis [79–81].

3.4. Effect of Proteases on Cell Autolysis

B. subtilis produces at least eight extracellular proteases [82]. Some examples suggest that these extracellular proteases are associated with cell autolysis. Jolliffe et al. reported that *B. subtilis* protease mutants have autolysin levels in the cell wall equal to or higher than wild-type strains [35]. Extracellular proteases have been identified as involved in degrading peptidoglycan hydrolases in the cell walls. Extracellular protease deficient strains showed increased lysis, where the extracellular proteases AprE and NprE had a greater ability to stabilize autolysins than NprB, Bpr, Mpr, and Epr [83]. Coxon et al. [84] found that the *B. subtilis* protease deletion mutant had a higher peptidoglycan folding rate while increasing susceptibility to intracell autolysis. In his research, Stephenson found that *B. subtilis* strains with inactivated protease genes exhibited cell autolysis, and mutants lacking multiple extracellular proteases became more susceptible to autolysis [14].

4. Strategies to Inhibit Cell Autolysis in *B. subtilis*

4.1. Inactivation-Associated Autolysins Genes

Inactivating single or few autolysin genes in *B. subtilis* cells becomes the first choice on the road to solving the autolysis problem. The strategies and results for modifying the autolysis aspects are presented in Table 2. Starting from the main control of autolysis gene *lytC*, Kuroda and Sekiguchi et al. [49] reported that *lytC* mutants are resistant to cell autolysis even after 6 days of culture at 37 °C. Similarly, Yamamoto et al. [64] found that inactivation of the *lytC* gene effectively inhibited the cell autolysis in wild-type *B. subtilis* cells and found that it can reduce the cell wall hydrolysis activity in stable cells by approximately 90%. On this basis, Smith and Foster et al. [44] found that dual mutants of the *lytC* and *cwlC* genes were effective against autolysis of mother cells during late *B. subtilis* bacteriophage formation. Nugroho et al. [85] reported that a double mutant of the *cwlC* and *cwlH* genes inhibited autolysis in *B. subtilis* cells. In addition, other knockout strategies for spore-associated autolysin genes are gradually emerging. Gonzalez-Pastor et al. [34] found that a mutant of a killing factor (SkfA) in *B. subtilis* protects non-spores from lysis by cells forming spores. Kodama et al. [37] found that introducing a significant autolysin gene mutation (*lytC* mutation) into *B. subtilis* mutant strain with a deletion of *spo0A* effectively suppressed cell autolysis caused by the *spo0A* mutation. Chen R. et al. [60] demonstrated that appropriate deletion of autolysin genes could improve recombinant protein yield in extracellular protease-deficient strains, and knockdown of autolysin-related genes *lytC* and *sigD* in tandem with eight extracellular proteases could effectively increase exogenous protein yield and delay autolysis.

4.2. Inactivation Operons and Regulatory Genes

Inhibition of autolysis can also be achieved by regulating the operon of autolysin, phibacin, or other regulatory genes. The *dlt* operon mediates lysophosphatidic acid-D-alanylation, the primary net adverse charge modification process on the cell surface. *dlt* operon deficiency can improve the net negative charge of the cell wall. *dlt*ABCD deletion studied by Chen et al. [78] increased nattokinase, α-amylase, and β-mannanase of *B. subtilis* by 37.13%, 44.53%, and 53.06%.

Genevieve S Dobihal et al. [86] enabled artificial control of *B. subtilis* cell wall hydrolysis by regulating the WalRK two-component signaling pathway, homeostasis of cell membrane cleavage, and elongation. They also found that the cleavage products of PG hydrolases (LytE and CwlO) in the WalR regulon could effectively inhibit WalK signaling, thus offering the potential for therapeutic development.

Zhou et al. [87] knocked down *spo0A*, a regulatory gene regulating *skf* and *sdp*. They found that the knockdown *spo0A* strains showed significantly faster autolysis rate and lower production of exogenous alkaline protease. The results also confirmed the complexity of the regulatory gene *spo0A* in the whole metabolic network of the bacterium, and further exploration of the critical regulators and signaling of the genes related to similar feeding is needed.

In the operon of prophage autolysins, it was reported that inactivation of *Pcf* factor would inhibit the expression of the prophage *PBSX* autolysin genes *xlyA*, *xhlA*, *xhlB*, and *xepA*, thereby inhibiting the cell autolysis of the bacterium [88].

4.3. Deletion of the Prophage Sequence

As previously stated, prophage sequences include many hydrolytic enzymes related to cell autolysis. Therefore, the elimination of these sequences improves autolysis significantly. Analysis of the genome of *B. subtilis* revealed that 4.2 Mb *B. subtilis* genome contains 10 horizontally obtained prophages (*PBSX* and *spβ*) and prophage-like sequences (*pro1-7* and *skin*) [89]. In addition, 2.8% of the genome contains two large operons that produce secondary metabolites (*pks* and *pps*). Westes reported a 7.7% (0.53 Mb) reduction in the genome of *B. subtilis* Δ6 mutant strains produced by deleting two prophages (*spβ* and *PBSX*), three primary phage-like sequences (*pro1*, *pro6*, and *skin*), and the *pks* operon [12]. Takuya Morimoto et al. [90] generated a strain MGB469 in which all prophage and prophage-like sequences except *pro7* and *pks* and *pps* operons were deleted. In this case, cell growth was normal. Li et al. [91] deleted all known prophages (prophage1-7, *spβ*, *skin*, and *PBSX*) on the *B. subtilis* 168 genomes, and the autolysis rate of the deleted strain *B. subtilis* BSK756 was significantly reduced compared to that of the parents, which further confirms that the deletion of the prophage sequence containing autolysin has an ameliorating effect on cell autolysis.

4.4. Tool-Mediated Genome-Wide Editing by CRISPR and Others

As the research of the cell hydrolytic enzymes of *B. subtilis* continues, key genes are being explored and genetic editing is being carried out in large quantities to improve autolysis and produce a variety of products efficiently. In this process, CRISPR, a tool more suitable for large-scale gene editing, has played an important role. Gerald E. M et al. [92] found that sequential deletion of *skfA*, *sdpC*, and *xpf* genes reduced cell lysis and increased biomass. Zhao et al. [93] inactivated peptidoglycan hydrolase-related genes alone or in different combinations, including *sigD, lytE, lytF, lytC, lytD*, and lytG. Compared to *B. subtilis* 168, mutants with multiple gene inactivations (e.g., Δ*sigD* Δ*lytE* Δ*lytD*) exhibited easier sedimentation, significantly increased growth rate, improved sensitivity to antibiotics, and improved α-amylase production. Mutants Δ*sigD* Δ*lytE* Δ*lytD* and Δ*sigD* Δ*lytE* Δ*lytC* Δ*lytD* also showed increased tolerance to the high osmotic pressure of sodium chloride. They allowed unsterilized fermentation, all of which contributed to lower processing costs.

Table 2. Strategies and results of the modification of autolysis aspects.

	Gene	Remodeling Method	Remodeling Results	Ref
1	*skfA*	Deactivation	Protects cells that do not form budding spores from lysis	[29]
2	*lytC, cwlC*	Deactivation	The cells were still resistant to autolysis after six days of incubation at 37 °C.	[56]
3	*lytC + sigD*	Knockout	Slowed autolysis, improved exogenous protein production	[59]
4	*dlt* operon	Knockout	Nattokinase, α-amylase, and β-mannanase increased by 37.13%, 44.53%, and 53.06%, respectively.	[77]
5	*bylA, cwlH*	Deactivation	Reduction of heat stress-induced autolysis	[84]
6	*Pcf*	Deactivation	Inhibition of autolysis genes and thus cell autolysis.	[87]
7	prophage1-7 + *spβ* + *skin* + PBSX	Delet	Significantly lower autolysis rate	[88]
8	*sigD + lytE + lytD + lytC*	Knockout	Easier sedimentation, significantly increased growth rate, improved sensitivity to antibiotics, and increased alpha-amylase production. Tolerance to the high osmotic pressure of sodium chloride was improved.	[93]

Liu et al. [94] used *B. subtilis* ATCC6051 as the expression host and deleted eight extracellular proteases (*aprE, nprE, nprB, epr, mpr, bpr, vpr*, and *wprA*), *spoIIAC*, and *srfAC* to produce mutant *B. subtilis* ATCC6051Δ10. The final maximum extracellular PUL activity (625.5 U/mL) showed the highest expression level (stem cell weight 18.7 g/L).

Zhang et al. [95] used *B. subtilis* strain WS5 (already deficient in the protease-encoding genes *nprE* and *aprE*) as a starting strain and disrupted the six protease-encoding genes (*nprB, bpr, mpr, epr, vpr*, and *wprA*) in the genome in sequence using the CRISPR/Cas9 system to finally obtain a recombinant strain WS9PUL that achieved 5951.8 U/mL of branched-chain amylase activity.

Wang et al. [11] reduced cell lysis and enhanced biomass by deleting *skfA, sdpC, xpf*, and *lytC* alone. A multiple deletion mutant LM2531 (*skfA sdpC lytC xpf*) was constructed. After 4 h of incubation, its biomass production was significantly increased compared to the prototype *B. subtilis* 168 (wild type) strain, lysing 15% and 92% of cells in LM2531 and wild-type cultures, respectively. In addition, two expression vectors were constructed under the control of the P43 promoter to produce recombinant proteins (β-galactosidase and nattokinase). Cultures of LM2531 and wild-type transformants had 13,741 U/mL and 7991 U/mL of intracellular β-galactosidase, respectively (1.72-fold increase). In addition, strain LM2531 produced a 2.6-fold increase in secreted nattokinase compared to the wild type (5226 IU/mL and 2028 IU/mL, respectively).

4.5. Minimal Genome of B. subtilis

As more and more autolysis-related genes are deleted, it has been found that "minimal genome" may also be an effective approach for addressing autolysis. Since completing the whole genome sequencing of *B. subtilis*, Kunst et al. [77] found that the construction of the minimal genome of *B. subtilis* has also started to develop rapidly. The *B. subtilis* genome has been deleted from 7.7% in the beginning to 36.5% in the present. The corresponding non-essential genes (e.g., spore, motility, antibiotic synthesis, secondary carbon sources metabolism, and unknown function genes) have been heavily depleted [96]. The strains obtained by deleting different fragment lengths have various performances. Specialized *B. subtilis* genomic databases have been established, such as DBTBS, SubiWiki, MetaCyc, SubtiList, SubtiPathways, and SubtInteract databases. Researchers can extract the required information from these databases and analyze data, including DNA sequences, metabolic pathways, protein interactions, and gene transcriptional regulation, to enable genomic modification of *B. subtilis* as a chassis host.

5. Conclusions

This review is devoted to the summary of autolysis in *B. subtilis*. We proceed from the mechanism of autolysis in *B. subtilis* (effect of autolysin on the cell wall) to (1) analyze the external and internal causes of autolysis, (2) introduce the currently known autolysis function, and (3) propose the solution measures to resist autolysis in recent decades. Researchers have been working on the autolysis problem of *B. subtilis* because the ability to produce recombinant proteins is a major advantage of using it, and the yield of these heterologous proteins is often tied to cellular biomass. In the case of the β-galactosidase described in the article by Wang et al. [11], wild bacteria, for example, and enzyme production began to decrease along with biomass after 7 h of incubation, while the biomass of the modified autolysis-resistant strain was still increasing and accompanied by enzyme production. Therefore, it can also be further understood that reducing the autolysis problem of *B. subtilis* can also extend the fermentation cycle and meet the needs of energy saving and efficient production. Although the external factors that induce autolysis in *B. subtilis* are largely understood, controlling these factors also provides researchers with a challenge. Moreover, multiple genes control autolysis in *B. subtilis*; therefore, it is currently difficult to accurately assess the specific contribution of each autolysis enzyme in the cell autolysis process. Researchers have been working to inactivate multiple autolysis-related genes simultaneously. With the in-depth study of autolysis enzyme function in *B. subtilis*, more and more regulatory factors controlling autolysis have been discovered and applied, and the study of prophages has led to the gradual implementation of minimal genome construction. In the future, more precise metabolic regulation techniques and holographic analysis will help to overcome the bottleneck of cell autolysis, and by combining techniques and strategies from the histological investigation, laboratory evolution, metabolic engineering, and synthetic biology, *B. subtilis* will be able to maintain stable biomass under high temperature and other environments, especially in industrial large-scale high-density fermentation production. New metabolic engineering tools such as the CRISPR-Cas system will be the mainstay to achieve this [97]. As research advances and bioinformatics, structural biology, and other fields advance, adopting innovative methods will be even more effective in improving the *B. subtilis* autolysis issue. In the future, with the deepening of the autolysis problems investigation and research, the precise autolyzed-related gene research and control, to further improve the fermentation production of *B. subtilis* such as chassis cell production capacity and efficiency, make the training cost reduced in the process of its application which is expected to achieve even closely integrated biomass and yield of high-efficiency production.

Author Contributions: K.R., Q.W., M.H., Y.C., R.X., J.Y., M.X., X.Z. and Z.R. conducted the literature and drafted the manuscript. All authors have read and agreed to the published version of the manuscript.

Funding: National Key Research and Development Program of China (No. 2021YFC2100900), National Natural Science Foundation of China (No. 32171471, No. 32071470), Project funded by China Postdoctoral Science Foundation (No. 2022M711365), Key Research and Development Program of Ningxia Hui Autonomous Region (No. 2020BFH01001), the Project Funded by the Priority Academic Program Development of Jiangsu Higher Education Institutions, and Top-notch Academic Programs Project of Jiangsu Higher Education Institutions.

Institutional Review Board Statement: Not applicable.

Informed Consent Statement: Not applicable.

Conflicts of Interest: The authors declare no conflict of interest.

References

1. Harwood, C.R. *Bacillus subtilis* and its relatives: Molecular biological and industrial workhorses. *Trends Biotechnol.* **1992**, *10*, 247. [CrossRef] [PubMed]
2. Young, E.J. Engineering the Bacterial Microcompartment Domain for Molecular Scaffolding Applications. *Front. Microbiol.* **2017**, *8*, 1441. [CrossRef]
3. Liu, Y. Pathway engineering of *Bacillus subtilis* for microbial production of N-acetylglucosamine. *Metab. Eng.* **2013**, *19*, 107–115. [CrossRef] [PubMed]
4. Panahi, R. Auto-inducible expression system based on the SigB-dependent ohrB promoter in *Bacillus subtilis*. *Mol. Biol.* **2014**, *48*, 852–857. [CrossRef]
5. Feng, Y. Enhanced extracellular production of L-asparaginase from *Bacillus subtilis* 168 by B. subtilis WB600 through a combined strategy. *Appl. Microbiol. Biotechnol.* **2017**, *101*, 1509–1520. [CrossRef]
6. Schallmey, M.; Singh, A.; Ward, O.P. Developments in the use of Bacillus species for industrial production. *Can. J. Microbiol.* **2004**, *50*, 1–17. [CrossRef]
7. Perkins, J. Genetic engineering of *Bacillus subtilis* for the commercial production of riboflavin. *J. Ind. Microbiol. Biotechnol.* **1999**, *22*, 8–18. [CrossRef]
8. Montesinos, E. Development, registration and commercialization of microbial pesticides for plant protection. *Int. Microbiol.* **2003**, *6*, 245–252. [CrossRef] [PubMed]
9. Hoa, T.T. Fate and dissemination of *Bacillus subtilis* spores in a murine model. *Appl. Environ. Microbiol.* **2001**, *67*, 3819–3823. [CrossRef]
10. D'Arienzo, R. *Bacillus subtilis* spores reduce susceptibility to Citrobacter rodentium-mediated enteropathy in a mouse model. *Res. Microbiol.* **2006**, *157*, 891–897. [CrossRef] [PubMed]
11. Wang, Y. Deleting multiple lytic genes enhances biomass yield and production of recombinant proteins by *Bacillus subtilis*. *Microb. Cell Factories* **2014**, *13*, 129. [CrossRef]
12. Westers, H. Genome engineering reveals large dispensable regions in *Bacillus subtilis*. *Mol. Biol. Evol.* **2003**, *20*, 2076–2090. [CrossRef]
13. Nguyen, T.T.; Quyen, T.D.; Le, H.T. Cloning and enhancing production of a detergent- and organic-solvent-resistant nattokinase from *Bacillus subtilis* VTCC-DVN-12-01 by using an eight-protease-gene-deficient *Bacillus subtilis* WB800. *Microb. Cell Fact.* **2013**, *12*, 79. [CrossRef]
14. Stephenson, K.; Bron, S.; Harwood, C.R. Cellular lysis in *Bacillus subtilis*; the affect of multiple extracellular protease deficiencies. *Lett. Appl. Microbiol.* **1999**, *29*, 141–145. [CrossRef]
15. Smith, T.J.; Blackman, S.A.; Foster, S.J. Autolysins of *Bacillus subtilis*: Multiple enzymes with multiple functions. *Microbiology* **2000**, *146*, 249–262. [CrossRef] [PubMed]
16. Ghuysen, J.-M. Penicillin and beyond: Evolution, protein fold, multimodular polypeptides, and multiprotein complexes. *Microb. Drug Resist.* **1996**, *2*, 163–175. [CrossRef] [PubMed]
17. Regamey, A.; Karamata, D. The N-acetylmuramoyl-L-alanine amidase encoded by the *Bacillus subtilis* 168 prophage SP beta. *Microbiology* **1998**, *144*, 885–893. [CrossRef] [PubMed]
18. Nandy, S.K.; Prasad, V.; Venkatesh, K.V. Effect of Temperature on the Cannibalistic Behavior of *Bacillus subtilis*. *Appl. Environ. Microbiol.* **2008**, *74*, 7427–7430. [CrossRef] [PubMed]
19. Antelmann, H. The extracellular proteome of *Bacillus subtilis* under secretion stress conditions. *Mol. Microbiol.* **2003**, *49*, 143–156. [CrossRef] [PubMed]
20. Abdel-Monem, M.O.; Al-Zubeiry, A.H.S.; Al-Gheethi, A.A.S. Biosorption of nickel by Pseudomonas cepacia 120S and *Bacillus subtilis* 117S. *Water Sci. Technol.* **2010**, *61*, 2994–3007. [CrossRef]
21. Yamanaka, K. Characterization of *Bacillus subtilis* mutants resistant to cold shock-induced autolysis. *FEMS Microbiol. Lett.* **1997**, *150*, 269–275. [CrossRef]
22. Graumann, P.L.; Marahiel, M.A. Cold shock response in *Bacillus subtilis*. *J. Mol. Microbiol. Biotechnol.* **1999**, *1*, 203–209. [PubMed]
23. Svarachorn, A. Autolysis of *Bacillus-subtilis* induced by low-temperature. *J. Ferment. Bioeng.* **1991**, *71*, 281–283. [CrossRef]
24. Sahoo, S.; Rao, K.K.; Suraishkumar, G.K. Reactive oxygen species induced by shear stress mediate cell death in *Bacillus subtilis*. *Biotechnol. Bioeng.* **2006**, *94*, 118–127. [CrossRef]
25. Inaoka, T. Characterization of high hydrostatic pressure-injured *Bacillus subtilis* cells. *Biosci. Biotechnol. Biochem.* **2017**, *81*, 1235–1240. [CrossRef]
26. Svarachorn, A. Autolysis of *Bacillus subtilis* 168 induced by monovalent cations and effects of mono- and divalent cations on autolysin activity in vitro. *Appl. Microbiol. Biotechnol.* **1989**, *30*, 299–304. [CrossRef]
27. Rogers, H.J.; Thurman, P.F.; Burdett, I.D. The bactericidal action of beta-lactam antibiotics on an autolysin-deficient strain of *Bacillus subtilis*. *J. Gen. Microbiol.* **1983**, *129*, 465–478.
28. Jolliffe, L.K.; Doyle, R.J.; Streips, U.N. Extracellular proteases increase tolerance of *Bacillus subtilis* to nafcillin. *Antimicrob. Agents Chemother.* **1982**, *22*, 83–89. [CrossRef] [PubMed]

29. Gallardo, O.; Diaz, P.; Pastor, F.I.J. Cloning and production of Xylanase B from Paenibacillus barcinonensis in *Bacillus subtilis* hosts. *Biocatal. Biotransformation* **2007**, *25*, 157–162. [CrossRef]
30. Allenby, N.E.E. Phosphate starvation induces the sporulation killing factor of *Bacillus subtilis*. *J. Bacteriol.* **2006**, *188*, 5299–5303. [CrossRef] [PubMed]
31. Schmeisser, F. A new mutation in *spo0A* with intragenic suppressors in the effector domain. *FEMS Microbiol. Lett.* **2000**, *185*, 123–128. [CrossRef] [PubMed]
32. Rice, K.C.; Bayles, K.W. Death's toolbox: Examining the molecular components of bacterial programmed cell death. *Mol. Microbiol.* **2003**, *50*, 729–738. [CrossRef] [PubMed]
33. Nandy, S.K.; Bapat, P.M.; Venkatesh, K.V. Sporulating bacteria prefers predation to cannibalism in mixed cultures. *FEBS Lett.* **2007**, *581*, 151–156. [CrossRef] [PubMed]
34. Gonzalez-Pastor, J.E.; Hobbs, E.C.; Losick, R. Cannibalism by sporulating bacteria. *Science* **2003**, *301*, 510–513. [CrossRef] [PubMed]
35. Jolliffe, L.K.; Doyle, R.J.; Streips, U.N. Extracellular proteases modify cell wall turnover in *Bacillus subtilis*. *J. Bacteriol.* **1980**, *141*, 1199–1208. [CrossRef]
36. Tipper, D.J. Mechanism of autolysis of isolated cell walls of Staphylococcus aureus. *J. Bacteriol.* **1969**, *97*, 837–847. [CrossRef] [PubMed]
37. Kodama, T. Effect of *Bacillus subtilis spo0A* mutation on cell wall lytic enzymes and extracellular proteases, and prevention of cell lysis. *J. Biosci. Bioeng.* **2007**, *103*, 13–21. [CrossRef] [PubMed]
38. Cho, H.-Y. Production of minicellulosomes from Clostridium cellulovorans in *Bacillus subtilis* WB800. *Appl. Environ. Microbiol.* **2004**, *70*, 5704–5707. [CrossRef]
39. Palomino, M.M.; Sanchez-Rivas, C.; Ruzal, S.M. High salt stress in *Bacillus subtilis*: Involvement of PBP4*as a peptidoglycan hydrolase. *Res. Microbiol.* **2009**, *160*, 117–124. [CrossRef]
40. Perez, A.R.; Abanes-De Mello, A.; Pogliano, K. Suppression of engulfment defects in *Bacillus subtilis* by elevated expression of the motility regulon. *J. Bacteriol.* **2006**, *188*, 1159–1164. [CrossRef] [PubMed]
41. Jolliffe, L.K.; Doyle, R.J.; Streips, U.N. The energized membrane and cellular autolysis in *Bacillus subtilis*. *Cell* **1981**, *25*, 753–763. [CrossRef] [PubMed]
42. Fukushima, T. A polysaccharide deacetylase gene (pdaA) is required for germination and for production of muramic delta-lactam residues in the spore cortex of *Bacillus subtilis*. *J. Bacteriol.* **2002**, *184*, 6007–6015. [CrossRef] [PubMed]
43. Rashid, M.H.; Sato, N.; Sekiguchi, J. Analysis of the minor autolysins of *Bacillus-subtilis* during vegetative growth by zymographY. *FEMS Microbiol. Lett.* **1995**, *132*, 131–137. [CrossRef]
44. Smith, T.J.; Foster, S.J. Characterization of the involvement of 2 compensatory autolysins in mother cell-lysis during sporulation of *Bacillus-subtilis-168*. *J. Bacteriol.* **1995**, *177*, 3855–3862. [CrossRef]
45. Smith, T.J.; Blackman, S.A.; Foster, S.J. Peptidoglycan hydrolases of *Bacillus subtilis* 168. *Microb. Drug Resist. Mech. Epidemiol. Dis.* **1996**, *2*, 113–118. [CrossRef] [PubMed]
46. Herbold, D.R.; Glaser, L. *Bacillus subtilis* N-acetylmuramic acid L-alanine amidase. *J. Biol. Chem.* **1975**, *250*, 1676–1682. [CrossRef] [PubMed]
47. Rogers, H.J. Purification and properties of autolytic endo-beta-N-acetylglucosaminidase and the N-acetylmuramyl-L-alanine amidase from *Bacillus subtilis* strain 168. *J. Gen. Microbiol.* **1984**, *130*, 2395–2402.
48. Margot, P.; Mauel, C.; Karamata, D. The gene of the n-acetylglucosaminidase, a *Bacillus-subtilis-168* cell-wall hydrolase not involved in vegetative cell autolysis. *Mol. Microbiol.* **1994**, *12*, 535–545. [CrossRef]
49. Kuroda, A.; Sekiguchi, J. Molecular-cloning and sequencing of a major *Bacillus-subtilis* autolysin gene. *J. Bacteriol.* **1991**, *173*, 7304–7312. [CrossRef]
50. Lazarevic, V. Sequencing and analysis of the *Bacillus-subtilis* lytrabc divergon—A regulatory unit encompassing the structural genes of the n-acetylmuramoyl-l-alanine amidase and its modifier. *J. Gen. Microbiol.* **1992**, *138*, 1949–1961. [CrossRef] [PubMed]
51. Ishikawa, S.; Yamane, K.; Sekiguchi, J. Regulation and characterization of a newly deduced cell wall hydrolase gene (cwlJ) which affects germination of *Bacillus subtilis* spores. *J. Bacteriol.* **1998**, *180*, 1375–1380. [CrossRef] [PubMed]
52. Margot, P. The *lytE* gene of *Bacillus subtilis* 168 encodes a cell wall hydrolase. *J. Bacteriol.* **1998**, *180*, 749–752. [CrossRef]
53. Ohnishi, R.; Ishikawa, S.; Sekiguchi, J. Peptidoglycan hydrolase LytF plays a role in cell separation with Cw1F during vegetative growth of *Bacillus subtilis*. *J. Bacteriol.* **1999**, *181*, 3178–3184. [CrossRef] [PubMed]
54. Park, S.S. *Bacillus subtilis* subtilisin gene (aprE) is expressed from a sigma A (sigma 43) promoter in vitro and in vivo. *J. Bacteriol.* **1989**, *171*, 2657–2665. [CrossRef] [PubMed]
55. Tipper, D.J.; Linnett, P.E. Distribution of peptidoglycan synthetase activities between sporangia and forespores in sporulating cells of Bacillus sphaericus. *J. Bacteriol.* **1976**, *126*, 213–221. [CrossRef]
56. Foster, S.J. Analysis of the autolysins of *Bacillus-subtilis-168* during vegetative growth and differentiation by using renaturing polyacrylamide-gel electrophoresis. *J. Bacteriol.* **1992**, *174*, 464–470. [CrossRef] [PubMed]
57. Kuroda, A.; Asami, Y.; Sekiguchi, J. Molecular-cloning of a sporulation-specific cell-wall hydrolase gene of *Bacillus-subtilis*. *J. Bacteriol.* **1993**, *175*, 6260–6268. [CrossRef] [PubMed]
58. Foster, S.J.; Johnstone, K. Purification and properties of a germination-specific cortex-lytic enzyme from spores of Bacillus megaterium KM. *Biochem. J.* **1987**, *242*, 573–579. [CrossRef] [PubMed]
59. Illing, N.; Errington, J. Genetic-regulation of morphogenesis in *Bacillus-subtilis*—roles of sigma-e and sigma-f in prespore engulfment. *J. Bacteriol.* **1991**, *173*, 3159–3169. [CrossRef] [PubMed]
60. Chen, R. Role of the sigma(D)-Dependent Autolysins in *Bacillus subtilis* Population Heterogeneity. *J. Bacteriol.* **2009**, *191*, 5775–5784. [CrossRef]

61. Márquez, L.M. Studies of sigma D-dependent functions in *Bacillus subtilis*. *J. Bacteriol.* **1990**, *172*, 3435–3443. [CrossRef]
62. Rohs, P.D.A.; Bernhardt, T.G. Growth and Division of the Peptidoglycan Matrix. *Annu. Rev. Microbiol.* **2021**, *75*, 315–336. [CrossRef] [PubMed]
63. Buist, G. LysM, a widely distributed protein motif for binding to (peptido)glycans. *Mol. Microbiol.* **2008**, *68*, 838–847. [CrossRef]
64. Yamamoto, H.; Kurosawa, S.I.; Sekiguchi, J. Localization of the vegetative cell wall hydrolases LytC, LytE, and LytF on the *Bacillus subtilis* cell surface and stability of these enzymes to cell wall-bound or extracellular proteases. *J. Bacteriol.* **2003**, *185*, 6666–6677. [CrossRef] [PubMed]
65. Abe, K. Autolysis-mediated membrane vesicle formation in *Bacillus subtilis*. *Environ. Microbiol.* **2021**, *23*, 2632–2647. [CrossRef] [PubMed]
66. Toyofuku, M. Prophage-triggered membrane vesicle formation through peptidoglycan damage in *Bacillus subtilis*. *Nat. Commun.* **2017**, *8*, 1–10. [CrossRef]
67. Presecan, E. The *Bacillus subtilis* genome from gerBC (311 degrees) to licR (334 degrees). *Microbiology* **1997**, *143*, 3313–3328. [CrossRef] [PubMed]
68. Zahler, R.S.; Sussmann, H.J. Claims and accomplishments of applied catastrophe theory. *Nature* **1977**, *269*, 759–763. [CrossRef]
69. Wood, H.E.; Devine, K.M.; McConnell, D.J. Characterisation of a repressor gene (xre) and a temperature-sensitive allele from the *Bacillus subtilis* prophage, PBSX. *Gene* **1990**, *96*, 83–88. [CrossRef] [PubMed]
70. Takemaru, K.-i. Complete nucleotide sequence of a *skin* element excised by DNA rearrangement during sporulation in *Bacillus subtilis*. *Microbiology* **1995**, *141*, 323–327. [CrossRef]
71. Arigoni, F. SpoIIE governs the phosphorylation state of a protein regulating transcription factor sigma F during sporulation in *Bacillus subtilis*. *Proc. Natl. Acad. Sci. USA* **1996**, *93*, 3238–3242. [CrossRef] [PubMed]
72. Longchamp, P.F.; Mauel, C.; Karamata, D. Lytic enzymes associated with defective prophages of *Bacillus subtilis*: Sequencing and characterization of the region comprising the N-acetylmuramoyl-L-alanine amidase gene of prophage PBSX. *Microbiology* **1994**, *140*, 1855–1867. [CrossRef]
73. Buxton, R.S. Selection of *Bacillus subtilis* 168 Mutants with Deletions of the PBSX Prophage. *J. Gen. Virol.* **1980**, *46*, 427–437. [CrossRef] [PubMed]
74. Krogh, S.; Jørgensen, S.T.; Devine, K.M. Lysis genes of the *Bacillus subtilis* defective prophage PBSX. *J. Bacteriol.* **1998**, *180*, 2110–2117. [CrossRef]
75. Sekiguchi, J. Nucleotide sequences of the *Bacillus subtilis* flaD locus and a B. licheniformis homologue affecting the autolysin level and flagellation. *J. Gen. Microbiol.* **1990**, *136*, 1223–1230. [CrossRef]
76. Foster, S.J. Cloning, expression, sequence-analysis and biochemical-characterization of an autolytic amidase of *Bacisllus-subtilis* 168 trpc2. *J. Gen. Microbiol.* **1991**, *137*, 1987–1998. [CrossRef] [PubMed]
77. Kunst, F. The complete genome sequence of the Gram-positive bacterium *Bacillus subtilis*. *Nature* **1997**, *390*, 249–256. [CrossRef] [PubMed]
78. Chen, Y.Z. Enhanced production of heterologous proteins by Bacillus licheniformis with defective d-alanylation of lipoteichoic acid. *World J. Microbiol. Biotechnol.* **2018**, *34*, 135. [CrossRef] [PubMed]
79. Hyyrylainen, H.L. D-Alanine substitution of teichoic acids as a modulator of protein folding and stability at the cytoplasmic membrane/cell wall interface of *Bacillus subtilis*. *J. Biol. Chem.* **2000**, *275*, 26696–26703. [CrossRef]
80. Kiriukhin, M.Y.; Neuhaus, F.C. D-alanylation of lipoteichoic acid: Role of the D-alanyl carrier protein in acylation. *J. Bacteriol.* **2001**, *183*, 2051–2058. [CrossRef]
81. Kovács, M. A functional *dlt* operon, encoding proteins required for incorporation of d-alanine in teichoic acids in gram-positive bacteria, confers resistance to cationic antimicrobial peptides in Streptococcus pneumoniae. *J. Bacteriol.* **2006**, *188*, 5797–5805. [CrossRef] [PubMed]
82. Antelmann, H. Stabilization of cell wall proteins in *Bacillus subtilis*: A proteomic approach. *Proteomics* **2002**, *2*, 591–602. [CrossRef] [PubMed]
83. Ferrari, E.; Howard, S.M.; Hoch, J.A. Effect of stage 0 sporulation mutations on subtilisin expression. *J. Bacteriol.* **1986**, *166*, 173–179. [CrossRef] [PubMed]
84. Coxon, R.D.; Harwood, C.R.; Archibald, A.R. Protein export during growth of *Bacillus-subtilis*—The effect of extracellular protease deficiency. *Lett. Appl. Microbiol.* **1991**, *12*, 91–94. [CrossRef]
85. Nugroho, F.A. Characterization of a new sigma-K-dependent peptidoglycan hydrolase gene that plays a role in *Bacillus subtilis* mother cell lysis. *J. Bacteriol.* **1999**, *181*, 6230–6237. [CrossRef] [PubMed]
86. Dobihal, G.S. Homeostatic control of cell wall hydrolysis by the WalRK two-component signaling pathway in *Bacillus subtilis*. *Elife* **2019**, *8*, e52088. [CrossRef] [PubMed]
87. Zhou, C.X. Optimization of alkaline protease production by rational deletion of sporulation related genes in Bacillus licheniformis. *Microb. Cell Factories* **2019**, *18*, 127. [CrossRef] [PubMed]
88. McDonnell, G.E. Genetic-control of bacterial suicide—regulation of the induction of *pbsx* in *Bacillus-subtilis*. *J. Bacteriol.* **1994**, *176*, 5820–5830. [CrossRef]
89. Westers, L.; Westers, H.; Quax, W.J. *Bacillus subtilis* as cell factory for pharmaceutical proteins: A biotechnological approach to optimize the host organism. *Biochim. Et Biophys. Acta Mol. Cell Res.* **2004**, *1694*, 299–310. [CrossRef] [PubMed]
90. Morimoto, T. Enhanced Recombinant Protein Productivity by Genome Reduction in *Bacillus subtilis*. *DNA Res.* **2008**, *15*, 73–81. [CrossRef]
91. Li, Y. Characterization of genome-reduced *Bacillus subtilis* strains and their application for the production of guanosine and thymidine. *Microb. Cell Factories* **2016**, *15*, 94. [CrossRef] [PubMed]

92. McDonnell, G.E.; McConnell, D.J. Overproduction, isolation, and DNA-binding characteristics of Xre, the repressor protein from the *Bacillus subtilis* defective prophage *PBSX*. *J. Bacteriol.* **1994**, *176*, 5831–5834. [CrossRef] [PubMed]
93. Zhao, L. Engineering peptidoglycan degradation related genes of *Bacillus subtilis* for better fermentation processes. *Bioresour. Technol.* **2018**, *248 Pt A*, 238–247. [CrossRef]
94. Liu, X. Efficient production of extracellular pullulanase in *Bacillus subtilis* ATCC6051 using the host strain construction and promoter optimization expression system. *Microb. Cell Factories* **2018**, *17*, 163. [CrossRef] [PubMed]
95. Zhang, K.; Su, L.; Wu, J. Enhanced extracellular pullulanase production in *Bacillus subtilis* using protease-deficient strains and optimal feeding. *Appl. Microbiol. Biotechnol.* **2018**, *102*, 5089–5103. [CrossRef] [PubMed]
96. Reuß, D.R. Large-scale reduction of the *Bacillus subtilis* genome: Consequences for the transcriptional network, resource allocation, and metabolism. *Genome Res.* **2017**, *27*, 289–299. [CrossRef]
97. Liu, D.Y. Development and characterization of a CRISPR/Cas9n-based multiplex genome editing system for Bacillus subtilis. *Biotechnol. Biofuels* **2019**, *12*, 197. [CrossRef] [PubMed]

Article

Galactitol Transport Factor GatA Relieves ATP Supply Restriction to Enhance Acid Tolerance of *Escherichia coli* in the Two-Stage Fermentation Production of D-Lactate

Jinhua Yang [1,2], Zheng Peng [1,2], Xiaomei Ji [1,2], Juan Zhang [1,2,*] and Guocheng Du [1,2,*]

[1] Key Laboratory of Industrial Biotechnology, Ministry of Education, School of Biotechnology, Jiangnan University, Wuxi 214122, China
[2] Science Center for Future Foods, Jiangnan University, Wuxi 214122, China
* Correspondence: zhangj@jiangnan.edu.cn (J.Z.); gcdu@jiangnan.edu.cn (G.D.)

Abstract: *Escherichia coli* is a major contributor to the industrial production of organic acids, but its production capacity and cost are limited by its acid sensitivity. Enhancing acid resistance in *E. coli* is essential for improving cell performance and production value. Here, we propose a feasible strategy for improving cellular acid tolerance by reducing ATP supply restriction. Transcriptome assays of acid-tolerant evolved strains revealed that the galactitol phosphotransferase system transporter protein GatA is an acid-tolerance factor that assists *E. coli* in improving its resistance to a variety of organic acids. Enhanced GatA expression increased cell survival under conditions of lethal stress due to D-lactic acid, itaconic acid and succinic acid by 101.8-fold, 29.4-fold and 41.6-fold, respectively. In addition, fermentation patterns for aerobic growth and oxygen-limited production of D-lactic acid were identified, and suitable transition and induction stages were evaluated. GatA effectively compensated for the lack of cellular energy during oxygen limitation and enabled the D-lactic acid producing strain to exhibit more sustainable productivity in acidic fermentation environments with a 55.7% increase in D-lactic acid titer from $9.5\,g\cdot L^{-1}$ to $14.8\,g\cdot L^{-1}$ and reduced generation of by-product. Thus, this study developed a method to improve the acid resistance of *E. coli* cells by compensating for the energy gap without affecting normal cell metabolism while reducing the cost of organic acid production.

Keywords: *Escherichia coli*; acid sensitivity; GatA; D-lactic acid; sustainable productivity

Citation: Yang, J.; Peng, Z.; Ji, X.; Zhang, J.; Du, G. Galactitol Transport Factor GatA Relieves ATP Supply Restriction to Enhance Acid Tolerance of *Escherichia coli* in the Two-Stage Fermentation Production of D-Lactate. *Fermentation* **2022**, *8*, 665. https://doi.org/10.3390/fermentation8120665

Academic Editors: Frank Vriesekoop and Nancy N. Nichols

Received: 1 October 2022
Accepted: 16 November 2022
Published: 23 November 2022

Publisher's Note: MDPI stays neutral with regard to jurisdictional claims in published maps and institutional affiliations.

1. Introduction

Well-defined genetic backgrounds and the ease with which genetic manipulation can be accomplished, coupled with the rapidity of transitions between production runs, have made *Escherichia coli* a desirable host with broad applicability to fermentation [1,2]. For example, engineered *E. coli* has become an important platform for organic acid production [3]. The global market for lactic acid, one of the three major organic acids, is expected to reach USD 5.02 billion by 2028 [3]. D-lactic acid (D-LA) is an important monomer feedstock for industrial syntheses, including that of the thermoplastic polyester polylactic acid Poly(D-, L-lactic acid) synthesized with L-lactic acid in a specific ratio is optimal in many aspects [4,5]. D-LA fermented by *E. coli* can maximize the titer and purity required for the synthesis of polymers to meet industrial production requirements [6–9]. However, the accumulation of acidic metabolites leads to a rapid shift in environmental pH away from neutral, limiting the productivity and production of *E. coli* which is acid-sensitive.

Decreased pH or high proton concentrations can interfere with ATP production [10]. Cells initiate response mechanisms to defend against an adverse acidic environment and devote more energy to regulating the expression of factors that act to counter the effects of the acid [11–13]. An insufficient ATP supply greatly affects the fermentation capacity of the cells. To maintain fermentation stability and improve productivity, exogenous neutralizers (e.g., KOH, NH4OH and Ca(OH)2) are commonly used to maintain the pH in a dynamic and stable neutral environment [14,15]. However, fermentation is prone to the accumulation of microsoluble by-products such as salts that increase osmotic stress, potentially influencing cell growth and metabolism [16]. In addition, the recovery cost is increased by the dissociation of the acid product at higher pH [17,18], and strict maintenance of a neutral fermentation environment increases the production costs [3,11]. Therefore, improving the tolerance of *E. coli* to organic acids to increase production is urgently required.

Various strategies have been used to improve the acid tolerance of cells to acidic metabolites. Adaptive evolution, or even genomic shuffling, have been the predominant strategies [19–21]. The combined strategy of mutagenesis using atmospheric and room temperature plasma (ARTP) and adaptive laboratory evolution increased the growth of *E. coli* cells by 3.12-fold and the yield of succinate under anaerobic conditions up to 0.69 g·g^{-1} glucose [22]. Multiple *E. coli* strains show improved 3-hydroxypropionic acid tolerance and production performance following mutation or deletion of the transcription factor *yieP* [23]. Glutamic acid decarboxylase *gadBC* as an anti-acid factor in the *E. coli* acid-resistant system (AR2), assists in the production of higher succinic acid titers in *E. coli* at low pH conditions [24]. Although some success has been reported, it is impossible to generalize and fully predict strain responses. In addition, the applicability of anti-acid factors remains a challenge, as overexpression may disrupt or compete with metabolic pathways that are already enhanced in the producing strains.

The aim of this study was to develop an effective strategy that promotes tolerance to a wide range of organic acids in *E. coli*, and increases the ability to produce organic acids without affecting normal cellular metabolism. In a previous study, we used an adaptive evolutionary strategy to obtain an evolved strain of *E. coli* DLA3, with an increased tolerance to D-LA from 3.4 g·L^{-1} to 4.2 g·L^{-1} [25]. We compared the differences in transcript levels between DLA3 and the parental strain *E. coli* MG1655 before and after D-LA lethal stress, and found that the transporter factor GatA, which acts in the transmembrane transport metabolic pathway of galactitol has potential as a universal anti-acid component. Overexpression of GatA compensated for the stress of limited cellular ATP production at low pH and greatly improved the ability of *E. coli* to survive the lethal threat of multiple organic acids. By comparing the effects of different fermentation patterns on yield, a two stage fermentation strategy was determined to improve the productivity of D-LA by *E. coli*, and the fermentation conditions during the transition period and period of induced cell growth were investigated. Overexpression of GatA relieved restrictions on ATP supply at low pH in the strain engineered for D-LA production and noticeably improved productivity and glucose utilization. This study provides a reference for improving the acid tolerance of strains for organic acid production and reducing the cost of production.

2. Material and Methods

2.1. Strains, Plasmids, and Culture Conditions

E. coli MG1655 was used as the parent strain. The MG1655-derived strain was obtained by knocking out the genes *pflB*, *adhE* and *frdA* using the λRed recombinant system tool [6], and named LBBE317. LBBE317 produced more D-LA without changing growth rate and biomass accumulation [26]. All the strains and plasmids used in this study are listed in Table S1. Using *E. coli* MG1655 genomic DNA was used as a template to amplify the gene linearized fragment *gatA* with homology arms of 18 and 22 bp by PCR. The linearized fragment was ligated with the linearized vector pTrc99a (One-step Cloning Kit, Vazyme Biotech, Nanjing, China) to form a recombinant plasmid, which was then transformed into *E. coli*. DNA purification, gel extraction and plasmid preparation were performed using kits from Sangon Biotech Co. Ltd. (Shanghai, China). *E. coli* cells were cultured in Luria–Bertani (LB) medium at 37 °C with shaking at 220 rpm. The inoculation rate is usually 2%. When appropriate, LB medium was supplemented with 100 µg·mL^{-1} ampicillin. Overexpression of the recombinant plasmid was induced at an OD$_{600}$ of between 0.5–0.6 by the addition of 200 µM isopropyl β-D-1-thiogalactopyranoside (IPTG).

2.2. Growth Curve Assays

Overnight seed cultures were transferred to 500 mL shake flasks containing 50 mL of fresh LB medium with a 2% inoculum, and 2 mL samples were taken every 2 h. Cell suspensions diluted to the appropriate concentration (OD_{600} less than 0.8) were measured using a spectrophotometer (723N, Shanghai Precision Scientific Instruments Co., Ltd., Shanghai, China), and the cell density was marked by an value OD_{600} of [26]. The growth curves of the cells were plotted with the sampling time as the horizontal coordinate and absorbance value as the vertical coordinate. The experiment was repeated three times, and the results were averaged.

2.3. Stress-Tolerance Assays

Cells were cultivated in 10 mL LB medium in a 100 mL shake flask and collected during the exponential phase. The collected cells were washed twice with 100 mM PBS solution (pH 7.2). The mixture was centrifuged ($6000\times g$) for 3 min, and the supernatant was discarded. The cells were resuspended in an equal volume of lethal stress LB medium (with adjustment of organic acids), and the stress time was started. The appropriate pH of lethal stress was considered as one that would permit the culture to reach a cell survival rate of $10^{-4}\%$ after 4 h of incubation [27]. In this experiment, the pH of the lethal stress LB medium for D-LA, succinic acid and itaconic acid was 4.0, 4.3 and 4.2, respectively. Samples were withdrawn at 0, 1, 2, 3, and 4 h, washed twice by the same procedure, and resuspended in an equal volume of 100 mM PBS. Cell suspensions (10 µL of a serial gradient dilution) were dropped onto agar LB plates and incubated for 24 h at 37 °C. Plates containing 30–300 colony-forming units (CFU) were counted. Each sample was analyzed in triplicate. Stress tolerance is shown as the survival rate, and the transmission of errors was calculated according to Equations (1) and (2). The experiment was repeated three times, and the results were averaged.

$$C = \left(\frac{A_n}{A_0}\right) \div \left(\frac{B_n}{B_0}\right) \times 100\% \tag{1}$$

$$\Delta C = C \times \sqrt{\left(\frac{\Delta A}{A}\right)^2 + \left(\frac{\Delta B}{B}\right)^2} \tag{2}$$

A: Number of viable bacteria after stress of the engineered strain (A_0, number of viable bacteria at 0 h; A_n, number of viable bacteria at 1, 2, 3, or 4 h); ΔA, error in the number of viable bacteria after stress of the engineered strain; B: number of viable bacteria after stress of the control strain (B_0, number of viable bacteria withdrawn at 0 h; B_n, number of viable bacteria withdrawn at 1, 2, 3, or 4 h); ΔB: error of viable bacteria after stress of the control strain; C: survival rate (%); ΔC: error of survival rate (%).

2.4. Intracellular ATP Detection

Cell cultures were sampled at 2 mL per hour under incubation conditions of D-LA lethal stress (D-LA concentration, pH and sampling time settings were the same as for stress-tolerance assays), and rapidly transferred to liquid nitrogen to prevent metabolism (5 min). Frozen cells were placed on ice to thaw and collected by centrifugation ($8000\times g$, 4 °C) for 10 min. Intracellular ATP content was determined using an ATP assay kit (Beyotime, Biotechnology, Shanghai, China). Whilst keeping the ice surface manipulated, cells were gently ejected and added to the specific lysis cytosol reagent in the kit without additional disruption/lysis of cells. After lysis, cells were centrifuged ($12,000\times g$, 4 °C) for 5 min and 1 mL of supernatant was used for subsequent assays, of which 0.5 mL was used to determine intracellular ATP concentration and the remaining 0.5 mL was used to detect intracellular protein concentration using the Protein Concentration Assay Kit (Beyotime Biotechnology, Shanghai, China). ATP content was expressed as nmol·mg^{-1} of intracellular protein.

2.5. Intracellular pH Detection

Intracellular pH was measured fluorescently using 2′,7′-bis(2-carboxyethyl)-5(and 6)-carboxyfluorescein acetoxymethyl ester (BCECF AM, Beyotime Biotechnology, Shanghai, China) as a fluorescent probe. Three mL of cells cultured under different conditions were harvested by centrifugation at $7000\times g$ for 3 min and washed three times with 50 mM HEPES-K (Sigma-Aldrich, Shanghai, China) buffer (pH 8.0) [25]. The cell precipitate was resuspended in 3 mL of the same buffer and 1 µL of BCECF AM was added and mixed in a water bath at 30 °C for 20 min. Cells were washed three times with 50 mM potassium phosphate buffer (pH 7.0) and resuspended. The fluorescence intensity of bacterial suspension (S) and filtrate (F) was measured by fluorescence spectrophotometer with excitation spectra of 490 nm (pH sensitive) and 440 nm (pH insensitive). The luminescence was measured at 525 nm. The width of the excitation and emission slits was set to 5 nm. The ratio of the emission intensity was calculated according to Guan et al. [28]. The extracellular pH (pH_{ex}) was used as the data of the medium measured by the pH meter.

2.6. Fermentation Analysis

Cells were precultured as described above and transferred to 150 mL of fresh LB medium (1000 mL flasks). Seed cultures were incubated for 9 h at 37 °C with shaking at 220 rpm. Seed cultures were inoculated into a 5 L bioreactor (T&J—Atype; Parallel -Bioreactor. Co, Shanghai, China) with an initial OD_{600} of 0.08 containing 3 L of AM2 medium [6], the inoculation rate was approximately 5%. Subsequently, single-phase or two-phasefermentation was performed, as shown in Table S2. Fermentation was stopped when glucose utilization was below 0.2 $g \cdot L^{-1} \cdot h^{-1}$. The pH was controlled with 15% ($w \cdot v^{-1}$) ammonia and the temperature was maintained at 37 °C. IPTG (200 µM) was used to induce fermentation of the recombinant strains.

2.7. Metabolite Detection

Samples from the cultures were centrifuged ($12,000\times g$) for 3 min. The concentrations of D-LA, acetic acid, and succinic acid in the sample supernatant were analyzed by HPLC (1260 Infinity LC, Agilent, Santa Clara, CA, USA) using an Aminex HPX-87H ion-exclusion chromatography column (300 × 7.8 mm). 5.0 mM H_2SO_4 solution was used as the mobile phase (0.5 mL min^{-1}) at column temperature 40 °C, and using a 210 nm photodiode array detector.

2.8. Statistical Analysis

Statistical analysis was performed using GraphPad Prism 6 (GraphPad Software, San Diego, California, U.S.A). Statistical significance was set at p-values < 0.05 and is indicated with an asterisk: *, $p < 0.05$; **, $p < 0.01$; ***, $p < 0.005$ and ****, $p < 0.001$.

3. Results and Discussion

3.1. Identification and Functional Verification of GatA

Candidate proteins with the potential to enhance lactic acid tolerance in *E. coli* were identified by comparing transcriptomic data before and after evolution under D-LA lethal stress [25]. From this analysis, *gatA* was regarded as a candidate acid-tolerant functional factor that enhances *E. coli* lactate tolerance. Its expression was significantly upregulated by 2.62- and 2.09-fold, respectively, under conditions of before and after evolution under D-LA lethal stress. The RNA-seq raw reads were submitted to NCBI under BioProject number PRJNA675154.

An overexpression recombinant strain of *E. coli* (GatA) was constructed based on the gene *gatA* sequence (Figure 1A). Expression of GatA was induced to turn on at a pre-log growth OD_{600} of 0.5–0.6 (Incubation for 2.5 h). This recombinant strain grew normally in LB medium without acid stress, and the cell density at the stabilization phase was not obviously different from that of the control strain (Figure 1B), indicating that the overexpression of GatA did not affect the physiological metabolism of the cells; therefore, the difference in survival performance of the cells under lethal stress culture conditions could be assessed at the same growth level. After 4 h of incubation, the pH value that decreases the cell viability by 6 orders of magnitude (from 10^6 to 10^0) could obviously reflect the variability of cells [27,29]. In this study, we follow the previous pH setup value 4.0 of D-LA lethal stress medium [25,26]. When cells at mid-log growth were reacquired and transferred to D-LA lethal stress medium (pH 4.0), the number of viable cells decreased rapidly. The difference in survival between the recombinant strain *E. coli* (GatA) and the control strain *E. coli* (Vector) gradually increased with culture time, and when cultured for 4 h, the survival of *E. coli* (GatA) increased 101.8-fold (Figure 1C).

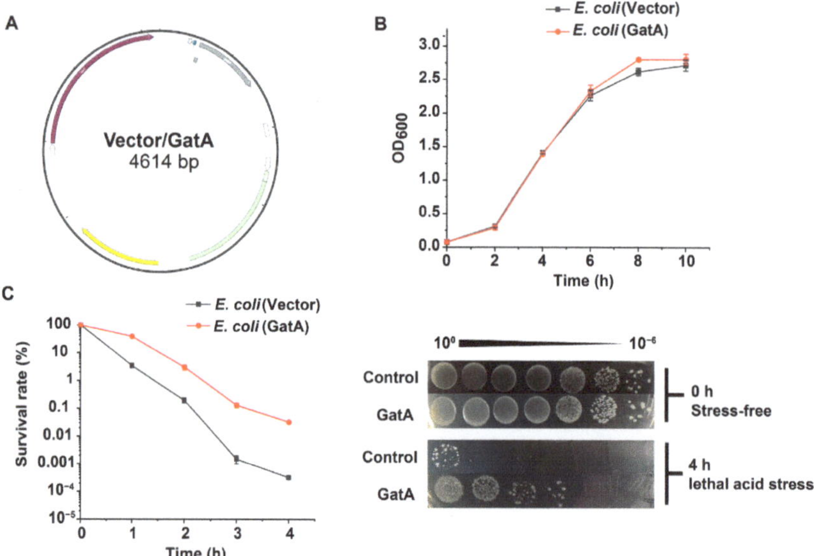

Figure 1. Construction and performance of recombinant strain *E. coli* (Vector) and *E. coli* (GatA). (**A**) The recombinant plasmid pTrc99a/GatA was constructed by inserting the galactitol transport factor *gatA* into plasmid pTrc99a. (**B**) Growth curve of recombinant strains in normal growth condition without acid stress. (**C**) Cell viability of recombinant strains under conditions of D-LA lethal stress.

gatA encodes the small hydrophilic peptide GatA, which plays an important role in the transport and phosphorylation of the phosphotransferase system (PTS) [30]. PTS is dependent on phosphoenolpyruvate (PEP), which mainly phosphorylates various sugars and their derivatives and transports them intracellularly through a cascade of phosphorylation reactions [31]. Of the three cellular transport mechanisms, PTS belongs to the group translocation category, which means that substrates are metabolized simultaneously during transport [32]. Polypeptide chain GatA (EII A) and GatB (EII B) equipped with transmembrane protein GatC (EII C), are a specific transporter protein complexes of galactitol [32]. Previously, GatC was shown to eliminate ATP limitation and replace the original ATP-dependent transporter protein to transport xylose [33]. As is well-known, ATP is used as an energy source by cells, and cells consume energy to maintain their growth and metabolism, while withstanding adverse factors in the extracellular environment. Therefore, we hypothesized that the anti-acid factor GatA enhances cell survival under conditions of organic acid stress, with the beneficial effects due to a globalized ATP repletion mechanism (Figure 2A).

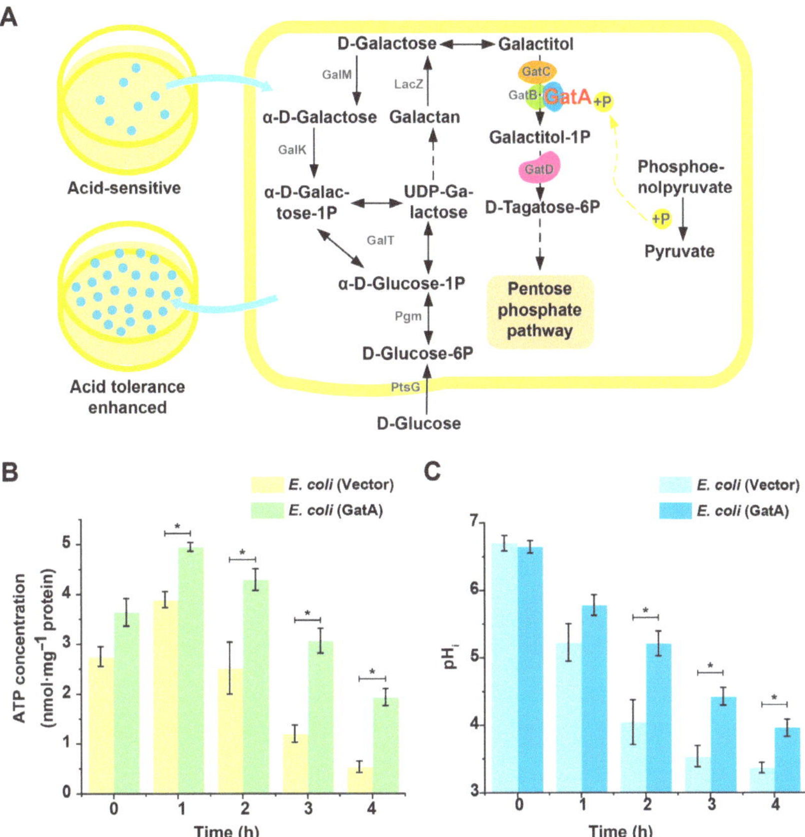

Figure 2. Overexpression of galactitol transport factor GatA enhances cell survival under D-LA lethal stress (pH 4.0) through ATP supply. (**A**) GatA is involved in the galactitol PTS pathway, the overexpression of GatA enhances cell survival number under D-LA lethal stress. (**B**) Overexpression of GatA enhances intracellular ATP of the strain against D-LA lethal stress. (**C**) Overexpression of GatA maintains intracellular pH homeostasis in strains that were incubated under lethal stress conditions of D-LA. *, $p < 0.05$.

We then monitored the intracellular ATP concentration under the same LB medium condition of D-LA stress (pH 4.0), and the same time of induction of GatA expression with Figure 1C condition. Intracellular ATP content accumulated with cell growth, so that a higher ATP content of approximately 1.06 nmol·mg⁻¹ protein accumulated in the cells of the recombinant strain *E. coli* (GatA) when it was first transferred to a lethal stress culture environment (0 h stress). Cells appear to initiate ATP repletion mechanism during the initial phase of acid-stress stimulation. The intracellular ATP concentrations of *E. coli* (GatA) and *E. coli* (Vector) peaked at 4.95 and 3.89 nmol·mg⁻¹ protein at 1 h of lethal stress incubation, and then decreased to 1.93 and 0.56 nmol·mg⁻¹ protein after 4 h of lethal stress incubation (Figure 2B). ATP concentration peaked at 1 h of stress and then decreased. This may be due to the rapid accumulation of ATP to resist stress when cells in the exponential phase are forced to leave a neutral environment and are placed in a lethal acidic culture environment that severely exceeds cellular tolerance. However, the physiological activity of cells is inevitably weakened and may even undergo apoptosis. The recombinant strain *E. coli* (GatA) maintained a high ATP content throughout the stressful environment culture, indicating that the intervention with *gatA* helped the cells accumulate ATP to counteract the organic acid stress. The H^+-ATPase involved in proton transfer is essential for maintaining intracellular pH (pH_i) [34], so it can be speculated that the higher energy status in the overexpressing GatA recombinant strain provides more energy for the catalytic action of H^+-ATPase to pump the accumulated protons out of the cell to maintain intracellular pH homeostasis and relieve acid stress [35,36]. When the cells were first exposed to D-LA lethal stress, there was no significant difference in pH_i between strains. The pH_i decreased sharply with increasing stress time, but the pH_i of *E. coli* (GatA) decreased less rapidly than that of *E. coli* (Vector). After 4 h of stress, the pH_i of *E. coli* (GatA) was higher than that of *E. coli* (Vector) by 0.6. This indicated that the recombinant strain had a higher ability to maintain pH_i under acidic stress environment to alleviate acid stress (Figure 2C). Acid-tolerant production strains can metabolize more products [23,26]. The candidate acid-tolerant factor GatA showed the potential to increase D-LA yield by promoting acid tolerance in D-LA-producing strains.

3.2. Aerobic Growth and Oxygen-Limited Fermentation Pattern Facilitate D-LA Production

Lactic acid is one of the main metabolites associated with biological growth processes [37]. Our initial findings suggest that GatA can act as a functional factor conferring acid resistance. However, before confirming whether this gene, which helps improve survival of strains under conditions of exogenous organic acid stress, can be applied in actual production experiments, it is necessary to select the optimal fermentation pattern in D-LA-producing strains to characterize the usefulness of this effect. The strong link between cell growth and acid production promotes the production of D-LA fermentation. Three main types of fermentation patterns have been reported for D-LA: single-phase fermentation (SPF, low-speed stirring 100 rpm for full process oxygen limitation incubation) [6,7], two-phase fermentation I (TPF I, growth phase with oxygen supply and high-speed agitation 400–600 rpm, switching to restricted oxygen supply and low-speed stirring 100 rpm at transition to acid production phase but no aeration) [8] and two-phase fermentation II (TPF II, same growth phase as TPF I, transition to microaerobic state with $1\ L\cdot min^{-1}$ aeration and 200 rpm agitation) [9]. In this study, we used each of these three patterns for the D-LA producing strain LBBE317 to determine the optimal fermentation pattern in a 5 L bioreactor.

In this study, the D-LA titer was preferred as the screen for optimal fermentational standard under normal pH condition value 7.0. Pattern TPF accumulated biomass rapidly during the 12 h of aerobic growth, and metabolized higher D-LA 51.7–53.8 $g\cdot L^{-1}$ in the 40 h acid production phase with low-speed stirring (Figure 3A), but a 0.19 $g\cdot g^{-1}$ lower yield of TPF2 was accompanied by the accumulation of cell density (Figure 3B). Pattern SPF accumulated D-LA 46.4 $g\cdot L^{-1}$ and was not selected, although full-scale SPF demonstrated a higher productivity of 1.03 $g\cdot L^{-1}\ h^{-1}$ and yield of 0.89 $g\cdot g^{-1}$ glucose in the whole 48 h process (Figure 3C), which was not beneficial for the accumulation of cell mass, resulting in lower productivity later in the fermentation. Therefore, two-phase fermentation TPFI with aerobic growth and oxygen-limited fermentation is recommended as the fermentation process.

3.3. Effect of Biomass at the Transition Point between Growth and Acid Production Stages

During the transition to the acid-producing stage, cellular activity is dominated by acid accumulation. The biomass content during the aerobic phase of growth influences the flow of glucose metabolism to lactate [9]. Therefore, biomass at the transition point between the aerobic growth and acid-producing stages may be correlated with the rate of D-LA production. The two-stage fermentation pattern was found to be optimal, and recombinant plasmid-induced expression of the acid-resistance factor GatA was turned on at the aerobic growth stage, consistent with the settings for inducing expression during the growth stage when acid tolerance was analyzed. Strain LBBE317, which contained an empty plasmid, was constructed for further examination of fermentation points.

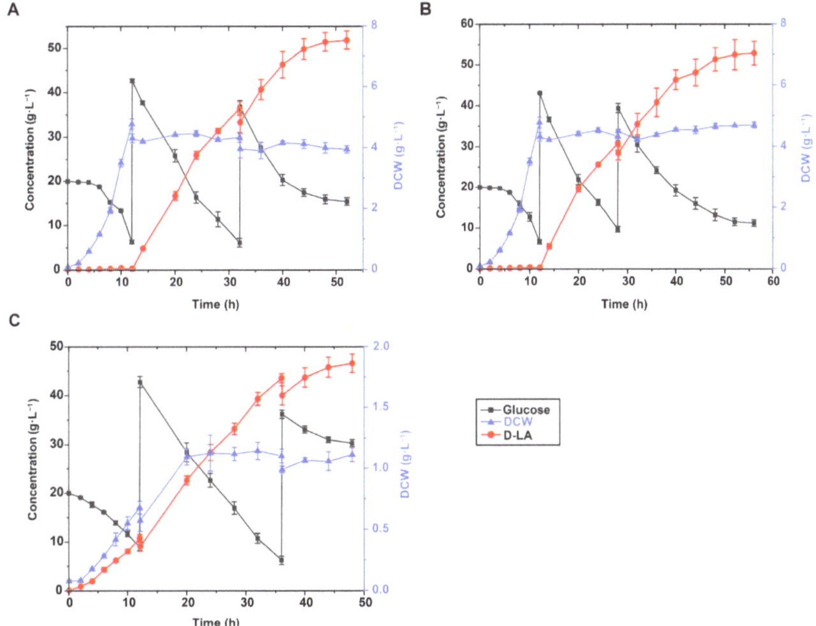

Figure 3. Effect of three different fermentation patterns on the synthesis of D-LA utilizing glucose by strain LBBE317. (**A**) Two-stage fermentation I (TPF1, growth stage with oxygen supply and high-speed agitation, switching to restricted oxygen supply and low low-speed stirring at transition to acid production stage). (**B**) Two-stage fermentation II (TPF2, same growth stage as TPFI, transition to microaerobic state with little aeration and low speed agitation). (**C**) Single-stage fermentation (SPF, low-speed stirring for full process oxygen limitation incubation).

Different transition points have a noticeable effect on the accumulation of D-LA under normal pH condition value 7.0. When the biomass accumulated to about 50% (OD_{600} ~8.6) of the maximum biomass (OD_{600}~17) at the transition to the acid-producing stage (Figure S1), both D-LA production 33.4 g·L^{-1} and productivity 0.56 g·L^{-1} h^{-1} were affected and the fermentation cycle was prolonged. This inefficiency may have been caused by the low activity of key enzymes in the cellular D-LA production pathway. D-LA titer could reach 45.6–48.7 g·L^{-1} when the biomass accumulated to approximately 70% (OD_{600}~12) or 90% (OD_{600}~15.5) of the maximum biomass for the transition. However, 90% biomass accumulation results in a 12.5% longer fermentation cycle, which is a negative factor for rapid equipment turnaround. In conclusion, a biomass accumulation of 70% of the maximum biomass was the optimal condition used for subsequent studies.

3.4. Inductional Effect on D-LA Synthesis at Different Growth Stages

Conditions under which a culture is at its highest growth rate is considered the optimal time to induce expression in a strain [38]. IPTG enables control of the *lac* promoter to regulate recombinant protein manufacturing in *E. coli*, but inappropriate induction times affect cell proliferation [39] and thus product accumulation. Therefore, after determining the optimal transition point for D-LA production, the effect of adding inducers during three different growth periods on D-LA synthesis was evaluated: late lag phase (LLP, OD_{600}~3.0), pre exponential phase (PEP, OD_{600}~4.5) and mid-pre exponential phase (MPEP, OD_{600}~6.0).

When IPTG is added at low cell densities, the metabolism of the target protein is assumed to harm normal bacterial growth under normal pH condition value 7.0, leading to a decrease in OD_{600} [40,41]. Similarly, in this study, when 200 mM IPTG was added during

LLP, the recombinant plasmid was activated prematurely and cell growth was retarded thereby negatively affecting D-LA biosynthesis, and the concentration of D-LA was reduced to 36.4 g·L^{-1} (Figure 4A). In contrast, the growth rate and biomass accumulation of cells in the exponential phase were not significantly affected by addition of the inducer. When the cells were induced with PEP, the D-LA titer reached a maximum of 48.9 g·L^{-1} (Figure 4B), which was higher than that of 46.6 g·L^{-1} when the inducer was added at MPEP (Figure 4C). Shorter fermentation cycle and outstanding productivity were more favorable for D-LA production. Theses results suggest that the addition of IPTG at the pre-exponential phase is suitable for the synthesis of D-LA, allowing the acid-resistant effect of the functional element to be observed more clearly.

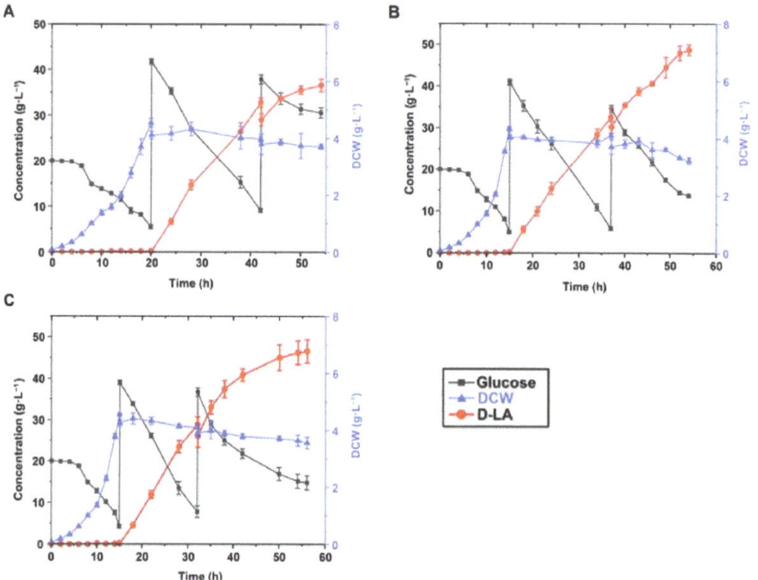

Figure 4. Effect of IPTG addition on D-LA synthesis during growth to different growth stages. (**A**) Addition of IPTG at LLP (late lag phase) had a shorter fermentation cycle but lower D-LA synthesis. (**B**) Addition of IPTG at PEP (pre-exponential) had the maximal amount of D-LA synthesis. (**C**) Addition of IPTG at MPEP (mid pre-exponential period) had D-LA synthesis close to that at PEP period, but with slightly longer fermentation cycle.

3.5. LBBE317PGA Enhanced D-LA Production under Low pH Incubation

When oxygen is abundant, *E. coli* can use one molecule of glucose to produce 36–38 ATP molecules. However, D-LA synthesis occurs during anaerobic metabolism, and under these conditions, only two ATP molecules accumulate after one molecule of glucose produces two molecules of D-LA. Although a moderately low rate of oxygen supply to the medium can promote consumption of the carbon source [42], suprathreshold oxygen tension may affect D-LA production and productivity [16], and the synthesis of by-products is also not conducive to efficient isolation of the downstream product. Previously, it has been suggested that deletion of the ATP-dependent transporter protein within the D-LA-producing engineered strain CL3 leads to an improvement in cell growth rate, indicating an increase in theoretical ATP production [6]. Subsequently, Utrilla et al. confirmed that a mutation in GatC (GatCS184L) was responsible for increased xylose consumption by whole-genome sequencing, but the mechanism was not clear [33]. In this study, we clarified that the overexpression of GatA, another necessary component of the galactitol transport complex, effectively replenishes the ATP content of cellular metabolic processes and relieves

ATP supply restriction. This suggests that GatA may bring about cellular tolerance to D-LA while maintaining metabolic production.

Stress caused by the accumulation of acidic products threatens the productivity of cells. To maximize the effectiveness of the application of anti-acid components, it is necessary to determine the pH at which the strain is most affected by acid stress. Based on the pKa of lactate (3.86), the pH values of the acid-producing phase medium settings (6.69, 5.47, 4.53 and 3.53) [11] were considered in a previous study, we found that with increasing concentrations of D-LA, the stress to cells by acid stress peaked when the pH was maintained at pH 5.5 [26]. Overexpression of GatA showed greatly enhanced survival under D-LA stress and was therefore transferred into LBBE317 as LBBE317PGA. During the aerobic growth stage, LBBE317PGA exhibiteda higher growth rate of 0.26 h^{-1}. When transitioning to the lactic acid production stage, the acid production rate of LBBE317PGA was not visibly inhibited compared with that of the control strain LBBE317P, indicating that the enhanced expression of GatA did not affect the cellular activity of metabolizing D-LA. When the environmental pH dropped to 5.5, the ammonia pump was turned on to maintain the pH. The D-LA production of the control strain LBBE317P was approximately 9.5 g·L^{-1} (Figure 5A), while LBBE317PGA produced approximately 14.8 g·L^{-1} D-LA, which represents an increase of 55.7%, and maintained a higher cell density (Figure 5B). The titers of succinic and acetic acids were reduced by 67.2% and 58.5%, respectively. These results confirmed that the overexpression of the galactitol transport factor GatA markedly enhanced the metabolic activity of *E. coli* under acid stress. With the assistance of the transporter protein GatA, the limitation of ATP supplementation was effectively compensated for while simultaneously ensuring the metabolism of D-LA and normal physiological activities.

To further validate the tolerance brought about by overexpression of GatA, we examined the effects of GatA on exogenous organic acids with different numbers of carbons and not limited to three-carbon D-LA. We tested the survival rate under stress with four-carbon succinic acid and five-carbon itaconic acid. Cell survival was calculated in the same way as for the validation of D-LA stress. After 4 h of itaconic acid lethal stress (pH 4.2, itaconic acid adjusted), the survival rate of the recombinant strain was 29.4 times higher than that of the control (Figure 6A). It is possible that *E. coli* has a weaker baseline tolerance to succinic acid, and the strain survived for a shorter time. After 3 h of succinic acid lethal stress (pH 4.3, succinic acid adjusted), the survival rate of the recombinant strain was 41.6-fold higher than that of the control strain (Figure 6B).

The trends of ATP concentration under itaconic acid and succinic acid lethal stress conditions were similar to the trends under D-LA lethal stress conditions, both exhibiting an increasing trend followed by a decreasing trend. Under itaconic acid lethal stress conditions, the intracellular ATP concentrations of *E. coli* (GatA) and *E. coli* (Vector) peaked at 4.18 and 3.72 nmol·mg^{-1} protein at 1 h incubation, and then decreased to 1.16 and 0.45 nmol·mg^{-1} protein after 4 h stress incubation. The pH_i also decreased sharply with increasing stress time. After 4 h stress incubation, the pH_i of *E. coli* (GatA) was higher than that of *E. coli* (Vector) by 0.44 (Figure S2A). Under succinic acid lethal stress conditions, the intracellular ATP concentrations of *E. coli* (GatA) and *E. coli* (Vector) peaked at 3.56 and 3.34 nmol·mg^{-1} protein at 1 h incubation, and then decreased to 0.76 and 0.56 nmol·mg^{-1} protein after 4 h incubation. After 4 h stress incubation, the pH_i of *E. coli* (GatA) was higher than that of *E. coli* (Vector) by 0.12 (Figure S2B).

These results suggest that GatA is an effective anti-acid factor with wide applications in enhancing organic acid tolerance in *E. coli*.

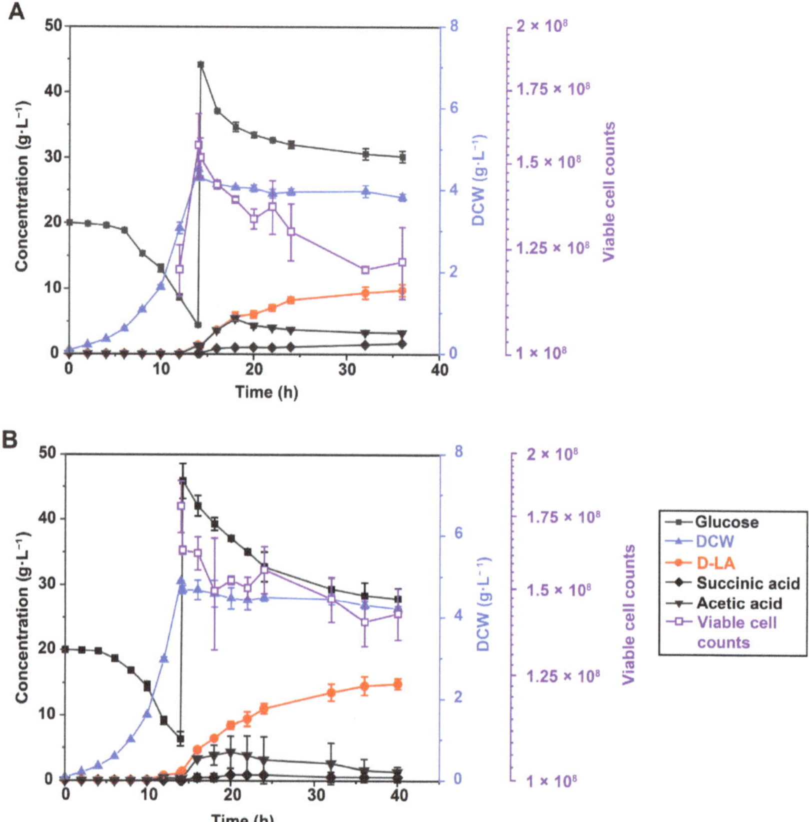

Figure 5. DCW, viable cell counts, glucose, and product concentrations during fermentation process under pH 5.5 in 5 L bioreactor. (**A**) *E. coli* LBBE31P. (**B**) *E. coli* LBBE317PGA.

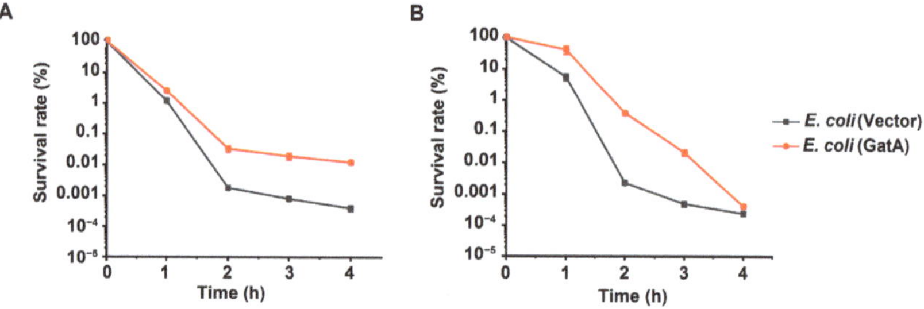

Figure 6. Cell viability of recombinant strains under different organic acid lethal stress conditions with incubation time. (**A**) Recombinant strains incubated under itaconic acid lethal stress. (**B**) Recombinant strain incubated in succinic acid lethal stress.

4. Conclusions

In this study, overexpression of the galactitol-specific PTS enzyme IIA component GatA improved the tolerance of *E. coli* at low pH to D-LA, itaconic acid and succinic acid at low pH. Furthermore, regulated overexpression of GatA regulation allowed *E. coli* to ferment D-LA at lower pH environments, increasing the titer by 55.7%. To the best of our knowledge, this is the first report showing that the galactitol transporter GatA can enhance the stress tolerance of *E. coli* to a variety of organic acids. Based on the commonality of PTS, we believe that this anti-acid element may also have applications in improving the production efficiency of other organic acids.

Supplementary Materials: The following supporting information can be downloaded at: https://www.mdpi.com/article/10.3390/fermentation8120665/s1, Figure S1: Effect of different biomass during the transition from the growth phase to the acid production phase; Figure S2 Intracellular ATP concentration and pH under itaconic and succinic acid lethal stress; Table S1 Strains and plasmids used in this study; Table S2 Bioreactor experimental fermentation conditions.

Author Contributions: Conceptualization, J.Z. and G.D.; formal analysis, J.Y.; investigation, J.Y.; methodology, G.D.; resources, J.Z.; supervision, J.Z. and G.D.; validation, X.J.; writing—original draft, J.Y.; writing—review and editing, Z.P. The final manuscript read and approved by all authors. All authors have read and agreed to the published version of the manuscript.

Funding: This research was funded by the National Key Research and Development Program of China (2019YFC1605800), the Independent Research Project of State Key Laboratory of Food Science and Technology (SKLF-ZZB-202015).

Institutional Review Board Statement: Not applicable.

Informed Consent Statement: Not applicable.

Data Availability Statement: The original RNA-seq reads have been submitted to NCBI under the number PRJNA675154.

Conflicts of Interest: The authors declare no competing financial interest.

References

1. Shi, Y.H.; Sun, H.Y.; Lu, D.M.; Le, Q.; Chen, D.; Zhou, Y. Separation of glycolic acid from glycolonitrile hydrolysate by reactive extraction with tri-n-octylamine. *Sep. Purif. Technol.* **2006**, *49*, 20–26.
2. Ahn, J.H.; Jang, Y.S.; Lee, S.Y. Production of succinic acid by metabolically engineered microorganisms. *Curr. Opin. Biotechnol.* **2016**, *42*, 54–66. [CrossRef] [PubMed]
3. Yang, J.; Zhang, J.; Zhu, Z.; Du, G. The challenges and prospects of *Escherichia coli* as an organic acid production host under acid stress. *Appl. Microbiol. Biot.* **2021**, *105*, 8091–8107. [CrossRef]
4. Datta, R.; Henry, M. Lactic acid: Recent advances in products, processes and technologies—A review. *J. Chem. Technol. Biot.* **2006**, *81*, 1119–1129. [CrossRef]
5. Tsuji, F. Autocatalytic hydrolysis of amorphous-made polylactides: Effects of L-lactide content, tacticity, and enantiomeric polymer blending. *Polymer* **2002**, *43*, 1789–1796. [CrossRef]
6. Utrilla, J.; Gosset, G.; Martinez, A. ATP limitation in a pyruvate formate lyase mutant of *Escherichia coli* MG1655 increases glycolytic flux to d-lactate. *J. Ind. Microbiol. Biot.* **2009**, *36*, 1057–1062. [CrossRef]
7. Liu, Y.; Gao, W.; Zhao, X.; Wang, J.H.; Garza, E.; Manow, R.; Zhou, S.D. Pilot scale demonstration of D-lactic acid fermentation facilitated by Ca(OH)(2) using a metabolically engineered *Escherichia coli*. *Bioresour. Technol.* **2014**, *169*, 559–565. [CrossRef]
8. Zhou, L.; Zuo, Z.R.; Chen, X.Z.; Niu, D.D.; Tian, K.M.; Prior, B.A.; Shen, W.; Shi, G.Y.; Singh, S.; Wang, Z.X. Evaluation of Genetic Manipulation Strategies on d-Lactate Production by *Escherichia coli*. *Curr. Microbiol.* **2011**, *62*, 981–989. [CrossRef]
9. Zhou, L.; Tian, K.M.; Niu, D.D.; Shen, W.; Shi, G.Y.; Singh, S.; Wang, Z.X. Improvement of d-lactate productivity in recombinant *Escherichia coli* by coupling production with growth. *Biotechnol. Lett.* **2012**, *34*, 1123–1130. [CrossRef]
10. Olson, E.R. Influence of Ph on Bacterial Gene-Expression. *Mol. Microbiol.* **1993**, *8*, 5–14. [CrossRef]
11. Zhong, W.; Yang, M.H.; Hao, X.M.; Sharshar, M.M.; Wang, Q.H.; Xing, J.M. Improvement of D-lactic acid production at low pH through expressing acid-resistant gene IoGAS1 in engineered Saccharomyces cerevisiae. *J. Chem. Technol. Biot.* **2021**, *96*, 732–742. [CrossRef]
12. Valli, M.; Sauer, M.; Branduardi, P.; Borth, N.; Porro, D.; Mattanovich, D. Improvement of lactic acid production in Saccharomyces cerevisiae by cell sorting for high intracellular pH. *Appl. Environ. Microb.* **2006**, *72*, 5492–5499. [CrossRef] [PubMed]
13. Piper, P.; Calderon, C.O.; Hatzixanthis, K.; Mollapour, M. Weak acid adaptation: The stress response that confers yeasts with resistance to organic acid food preservatives. *Microbiology* **2001**, *147*, 2635–2642. [CrossRef] [PubMed]
14. Lu, P.L.; Ma, D.; Chen, Y.L.; Guo, Y.Y.; Chen, G.Q.; Deng, H.T.; Shi, Y.G. L-glutamine provides acid resistance for *Escherichia coli* through enzymatic release of ammonia. *Cell Res.* **2013**, *23*, 635–644. [CrossRef]
15. Roe, A.J.; O'Byrne, C.; McLaggan, D.; Booth, I.R. Inhibition of *Escherichia coli* growth by acetic acid: A problem with methionine biosynthesis and homocysteine toxicity. *Microbiology* **2002**, *148*, 2215–2222. [CrossRef]
16. Singhvi, M.; Zendo, T.; Sonomoto, K. Free lactic acid production under acidic conditions by lactic acid bacteria strains: Challenges and future prospects. *Appl. Microbiol. Biot.* **2018**, *102*, 5911–5924. [CrossRef]

17. Fletcher, E.; Feizi, A.; Bisschops, M.M.M.; Hallstrom, B.M.; Khoomrung, S.; Siewers, V.; Nielsen, J. Evolutionary engineering reveals divergent paths when yeast is adapted to different acidic environments. *Metab. Eng.* **2017**, *39*, 19–28. [CrossRef]

18. Fu, X.M.; Wang, Y.X.; Wang, J.H.; Garza, E.; Manow, R.; Zhou, S.D. Semi-industrial scale (30 m(3)) fed-batch fermentation for the production of D-lactate by *Escherichia coli* strain HBUT-D15. *J. Ind. Microbiol. Biot.* **2017**, *44*, 221–228. [CrossRef] [PubMed]

19. Lee, J.K.; Kim, S.; Kim, W.; Kim, S.; Cha, S.; Moon, H.; Hur, D.H.; Kim, S.Y.; Na, J.G.; Lee, J.W.; et al. Efficient production of d-lactate from methane in a lactate-tolerant strain of *Methylomonas* sp. DH-1 generated by adaptive laboratory evolution. *Biotechnol. Biofuels* **2019**, *12*, 234. [CrossRef]

20. Park, H.J.; Bae, J.H.; Ko, H.J.; Lee, S.H.; Sung, B.H.; Han, J.I.; Sohn, J.H. Low-pH production of d-lactic acid using newly isolated acid tolerant yeast Pichia kudriavzevii NG7. *Biotechnol. Bioeng.* **2018**, *115*, 2232–2242. [CrossRef]

21. Zheng, H.; Gong, J.; Chen, T.; Chen, X.; Zhao, X. Strain improvement of Sporolactobacillus inulinus ATCC 15538 for acid tolerance and production of D-lactic acid by genome shuffling. *Appl. Microbiol. Biotechnol.* **2010**, *85*, 1541–1549. [CrossRef]

22. Ma, J.F.; Wu, M.K.; Zhang, C.Q.; He, A.Y.; Kong, X.P.; Li, G.L.; Wei, C.; Jiang, M. Coupled ARTP and ALE strategy to improve anaerobic cell growth and succinic acid production by *Escherichia coli. J. Chem. Technol. Biot.* **2016**, *91*, 711–717. [CrossRef]

23. Nguyen-Vo, T.P.; Liang, Y.; Sankaranarayanan, M.; Seol, E.; Chun, A.Y.; Ashok, S.; Chauhan, A.S.; Kim, J.R.; Park, S. Development of 3-hydroxypropionic-acid-tolerant strain of *Escherichia coli* W and role of minor global regulator yieP. *Metab. Eng.* **2019**, *53*, 48–58. [CrossRef] [PubMed]

24. Wu, M.; Li, X.; Guo, S.; Lemma, W.D.; Zhang, W.; Ma, J.; Jia, H.; Wu, H.; Jiang, M.; Ouyang, P. Enhanced succinic acid production under acidic conditions by introduction of glutamate decarboxylase system in *E. coli* AFP111. *Bioprocess Biosyst. Eng.* **2017**, *40*, 549–557. [CrossRef] [PubMed]

25. Yang, J.; Zhang, J.; Zhu, Z.; Jiang, X.; Zheng, T.; Du, G. Revealing novel synergistic defense and acid tolerant performance of *Escherichia coli* in response to organic acid stimulation. *Appl. Microbiol. Biotechnol.* **2022**, *106*, 1–18. [CrossRef] [PubMed]

26. Yang, J.; Peng, Z.; Zhu, Q.; Zhang, J.; Du, G. [NiFe] Hydrogenase Accessory Proteins HypB-HypC Accelerate Proton Conversion to Enhance the Acid Resistance and d-Lactic Acid Production of *Escherichia coli. ACS Synth. Biol.* **2022**, *11*, 1521–1530. [CrossRef] [PubMed]

27. Zhu, Z.M.; Yang, J.H.; Yang, P.S.; Wu, Z.M.; Zhang, J.; Du, G.C. Enhanced acid-stress tolerance in Lactococcus lactis NZ9000 by overexpression of ABC transporters. *Microb. Cell Factories* **2019**, *18*, 1–14. [CrossRef]

28. Guan, N.Z.; Liu, L.; Shin, H.D.; Chen, R.R.; Zhang, J.; Li, J.H.; Du, G.C.; Shi, Z.P.; Chen, J. Systems-level understanding of how Propionibacterium acidipropionici respond to propionic acid stress at the microenvironment levels: Mechanism and application. *J. Biotechnol.* **2013**, *167*, 56–63. [CrossRef]

29. Zhu, Z.; Ji, X.; Wu, Z.; Zhang, J.; Du, G. Improved acid-stress tolerance of Lactococcus lactis NZ9000 and *Escherichia coli* BL21 by overexpression of the anti-acid component recT. *J. Ind. Microbiol. Biotechnol.* **2018**, *45*, 1091–1101. [CrossRef]

30. Nobelmann, B.; Lengeler, J.W. Molecular analysis of the gat genes from *Escherichia coli* and of their roles in galactitol transport and metabolism. *J. Bacteriol.* **1996**, *178*, 6790–6795. [CrossRef]

31. Saier, M.H.; Hvorup, R.N.; Barabote, R.D. Evolution of the bacterial phosphotransferase system: From carriers and enzymes to group translocators. *Biochem. Soc. Trans.* **2005**, *33*, 220–224. [CrossRef] [PubMed]

32. Volpon, L.; Young, C.R.; Matte, A.; Gehring, K. NMR structure of the enzyme GatB of the galactitol-specific phosphoenolpyruvate-dependent phosphotransferase system and its interaction with GatA. *Protein Sci.* **2006**, *15*, 2435–2441. [CrossRef] [PubMed]

33. Utrilla, J.; Licona-Cassani, C.; Marcellin, E.; Gosset, G.; Nielsen, L.K.; Martinez, A. Engineering and adaptive evolution of *Escherichia coli* for D-lactate fermentation reveals GatC as a xylose transporter. *Metab. Eng.* **2012**, *14*, 469–476. [CrossRef] [PubMed]

34. Cotter, P.D.; Hill, C. Surviving the acid test: Responses of gram-positive bacteria to low pH. *Microbiol. Mol. Biol. R.* **2003**, *67*, 429–453. [CrossRef]

35. Baek, S.H.; Kwon, E.Y.; Kim, S.Y.; Hahn, J.S. GSF2 deletion increases lactic acid production by alleviating glucose repression in Saccharomyces cerevisiae. *Sci. Rep.* **2016**, *6*, 34812. [CrossRef]

36. Guan, N.Z.; Shin, H.D.; Chen, R.R.; Li, J.H.; Liu, L.; Du, G.C.; Chen, J. Understanding of how Propionibacterium acidipropionici respond to propionic acid stress at the level of proteomics. *Sci. Rep.* **2014**, *4*, 6951. [CrossRef]

37. Luedeking, R.; Piret, E.L. A kinetic study of the lactic acid fermentation. Batch process at controlled pH (Reprinted from Journal of Biochemical and Microbiological Technology and Engineering, vol 1, pg 393, 1959). *Biotechnol. Bioeng.* **2000**, *67*, 636–644. [CrossRef]

38. Yao, P.; You, S.P.; Qi, W.; Su, R.X.; He, Z.M. Investigation of fermentation conditions of biodiesel by-products for high production of beta-farnesene by an engineered *Escherichia coli. Environ. Sci. Pollut. R.* **2020**, *27*, 22758–22769. [CrossRef]

39. Lalwani, M.A.; Ip, S.S.; Carrasco-Lopez, C.; Day, C.; Zhao, E.M.; Kawabe, H.; Avalos, J.L. Optogenetic control of the lac operon for bacterial chemical and protein production. *Nat. Chem. Biol.* **2021**, *17*, 71–79. [CrossRef]

40. Lopes, C.; dos Santos, N.V.; Dupont, J.; Pedrolli, D.B.; Valentini, S.R.; Santos-Ebinuma, V.D.; Pereira, J.F.B. Improving the cost effectiveness of enhanced green fluorescent protein production using recombinant *Escherichia coli* BL21 (DE3): Decreasing the expression inducer concentration. *Biotechnol. Appl. Bioc.* **2019**, *66*, 527–536. [CrossRef]

41. Omoya, K.; Kato, Z.; Matsukuma, E.; Li, A.L.; Hashimoto, K.; Yamamoto, Y.; Ohnishi, H.; Kondo, N. Systematic optimization of active protein expression using GFP as a folding reporter. *Protein Expr. Purif.* **2004**, *36*, 327–332. [CrossRef] [PubMed]
42. Fernandez-Sandoval, M.T.; Galindez-Mayer, J.; Moss-Acosta, C.L.; Gosset, G.; Martinez, A. Volumetric oxygen transfer coefficient as a means of improving volumetric ethanol productivity and a criterion for scaling up ethanol production with *Escherichia coli*. *J. Chem. Technol. Biot.* **2017**, *92*, 981–989. [CrossRef]

 *fermentation*

Article

Changes of Physicochemical Properties in Black Garlic during Fermentation

Xinyu Yuan [1,†], Zhuochen Wang [2,†], Lanhua Liu [1], Dongdong Mu [1], Junfeng Wu [3], Xingjiang Li [1] and Xuefeng Wu [1,4,*]

[1]   Anhui Fermented Food Engineering Research Center, School of Food and Biological Engineering, Hefei University of Technology, Hefei 230601, China

[2]   Institution of Agricultural Products Processing, Anhui Academy of Agricultural Sciences, Hefei 230001, China

[3]   Linquan County Hengda Food Co., Ltd., Fuyang 236400, China

[4]   Anhui Oriental Orchard Biotechnology Co., Ltd., Suzhou 234000, China

*   Correspondence: wuxuefeng@hfut.edu.cn; Tel.: +86-551-62901507

†   These authors contributed equally to this work.

Abstract: To investigate the changes of the main ingredients in black garlic (BG) during fermentation, the contents of moisture, total acids and reducing sugars were determined. Allicin, 5-Hydroxy methylfurfural (5-HMF), and total phenols were also determined as bioactive substances. DPPH scavenging capacity was determined to indicate the antioxidant activity of BG. The changes in hardness and color were detected as well. The results showed that the moisture content decreased from 66.13% to 25.8% during the fermentation. The content of total acids, total phenols, and reducing sugars increased from 0.03 g/g to 0.29 g/g, from 0.045 μg/g to 0.117 μg/g, and from 0.016 g/g to 0.406 g/g, respectively. The content of 5-HMF increased from 0 to 4.12 μg/mL continuously, while the content of allicin increased from 0.09 mmol/100 g to 0.30 mmol/100 g and then decayed to 0.00 mmol/100 g. The L*, a*, and b* values of BG were 23.65 ± 0.44, 0.64 ± 0.06, and 0.85 ± 0.05, respectively. There was a higher intensity of dark color in BG than that in fresh garlic. The hardness values decreased first and then increased in later fermentation from 465.47 g to 27,292.38 g. Principal component analysis (PCA) showed that the samples were divided into three clusters, including cluster1 (fresh garlic, S0), cluster2 (S1), and cluster3 (S3−S9). This research effectively clarified the various stage of the BG fermentation process, and it is expected to supply references for reducing production time in industrial BG fermentation.

Keywords: black garlic; physicochemical properties; antioxidant activity; principal components analysis

Citation: Yuan, X.; Wang, Z.; Liu, L.; Mu, D.; Wu, J.; Li, X.; Wu, X. Changes of Physicochemical Properties in Black Garlic during Fermentation. *Fermentation* **2022**, *8*, 653. https://doi.org/10.3390/fermentation8110653

Academic Editors: Xian Zhang and Zhiming Rao

Received: 11 October 2022
Accepted: 17 November 2022
Published: 20 November 2022

Publisher's Note: MDPI stays neutral with regard to jurisdictional claims in published maps and institutional affiliations.

1. Introduction

Garlic is widely used as a condiment around the world. According to authoritative data, China is the main garlic-producing area all over the world, and the yield of garlic reaches 2 million tons annually. Previous research has shown that garlic has a series of biologically active functions, e.g., improving cardiovascular function [1], preventing cancer [2], regulating blood sugar [3], and possessing antibacterial effects [4] and antihypertension [5]. Generally considered, organosulfur compounds are the main flavor ingredients in garlic, and they are the most important contributor to the biological activity as well [6,7].

Nowadays, black garlic (BG) is becoming more and more popular around the world as a derivative of fresh garlic (FG). BG is made by heat-treating FG at 60–80 °C for 1−3 months under humidity control [8]. With high-temperature treatment, biochemical reactions include the Maillard reaction, caramelization reaction, and enzyme reaction, with the color of garlic turned from white to black. Meanwhile, the composition and taste have also changed tremendously [9].

With the growing popularity of BG, more and more research has been performed on it. A lot of studies have shown that the antioxidant capacity of BG is stronger than that of

FG [10–15]. Most of the current research is devoted to the influence of the total amount of polyphenols and flavonoids on the scavenging ability of free radicals [16–19].

Several studies are devoted to exploring the changes of certain specific compounds during heat treatment in BG. For example, Yuan et al. [20] studied the changes in various sugar contents in FG and BG. Sembiring et al. [21] reviewed the contents of components in BG. Some studies used omics methods to comprehensively determine the ingredients in BG. William [9] used UHPLC-HRMS to quantitatively analyze organic acids, organosulfur compounds, glycerophospholipids, and Maillard reaction products. Ding et al. [11] used NMR to quantitatively analyze the ingredients in BG with different packaging materials.

As has been proved, LCMS is a mighty tool for characterizing the compositions of different plant substrates, and it also can be used to analyze the compositions of FG and BG. For example, UPLC-Q-TOF/MS [22] was used to determine the content in the compositions of glycerophospholipids, flavonoids, organic oxygen compounds, and fatty acids in Pu'er tea. Wang et al. [23] used UPLC-Q-TOF/MS to analyze the changes of bitter metabolites in beer fermentation; 1239 characteristic peaks were obtained, and 32 biochemical markers were further determined.

At present, most studies focus on the differences between BG and FG, and the improvement of biological activity caused by these differences is also interesting. There are relatively few studies on the changes in the content of various substances. In this work, the contents of moisture, total acids, total phenols, reducing sugars, allicin, and 5-Hydroxymethylfurfural (5-HMF) were determined to explore the changes in various physicochemical properties of BG during fermentation. In addition, the DPPH scavenging capacity, hardness, and color of BG were detected.

2. Materials and Methods

2.1. Overall Strategy

The workflow of this study is shown in Figure 1. Samples were taken every other day during the fermentation of BG. The samples were treated with liquid nitrogen first. Then, the samples were ground. Finally, extraction was conducted according to the method mentioned below. We determined the contents of various ingredients in samples. The data we obtained were analyzed statistically. We classified the differences between black garlic and fresh garlic into three aspects: sensory, physicochemical properties, and bioactive functions.

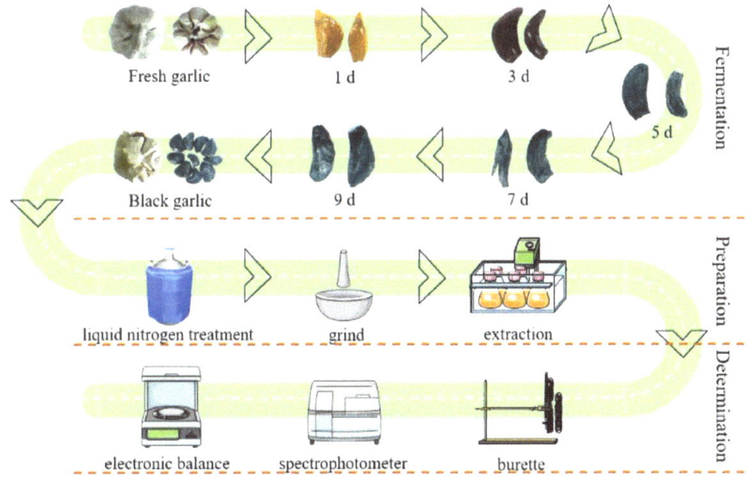

Figure 1. Overall workflow of this study.

2.2. Material

Fresh garlic collected in 2021 at the optimal ripening stage was provided by RT-Mart (Hefei, Anhui, China). The bulbs were stored at −20 °C until used. Cysteine, barbituric acid, and 5-HMF were from Shyuanye (Shanghai, China). Other reagents were supplied by HUSHI (Shanghai, China). Deionized water (18 MΩ cm^{-1}) was obtained through a Milli-Q water purification system from Millipore (Bedford, MA).

2.3. Sample Preparation

In the BG fermentation, samples were taken on 0, 1, 3, 5, 7, and 9 days as S0, S1, S3, S5, S7, and S9, respectively. Sensory evaluation and colorimetric and hardness measurements were performed immediately. Then, every 5 g sample were ground in a mortar. They were transferred into a 100 mL beaker containing 50 mL of deionized water and were ultrasonically extracted at 30 °C for 20 min. Then, the solution was filtered, and the filter residue was discarded. The filtrate was moved into a volumetric flask, followed by dilution with deionized water to 100 mL. It was diluted with deionized water to the appropriate concentration, when it was used.

2.4. Sensory Evaluation

The sensory properties of samples were analyzed by sensory evaluation. Sensory evaluation was conducted according to the method proposed by different researchers [24–27]. Healthy panelists (8 females and 8 males, 20–25 years of age), who had good cognition on the quality attributes on black garlic, were recruited from School of Food and Biological engineering, Hefei University of Technology (Hefei, China). Before sensory evaluation, commercial BG was used to help participants get familiar with sensory characteristics.

For sensory analysis, a whole BG sample was put into a glass container which was marked with a random number. The containers were placed on a plate for each panelist to evaluate them at the same time. The evaluations of taste, color, texture, smell, and dryness were recorded (0 = unsatisfactory and 10 = satisfactory extremely for taste, smell, color, dryness, and texture) by the same panel. The average scores for all descriptors of samples obtained from each panelist were taken. Water was drunk as a taste neutralizer before the evaluation of the next sample.

2.5. Basic Analysis of BG

The sample was dissolved using deionized water to 50 mg/mL. The contents of moisture, total phenolic, total acids, reducing sugars, allicin, and 5-HMF were analyzed. The content of moisture and reducing sugars were determined by the method conducted by Zambrano [28] and Huang [29]. The content of total acids was determined in keeping with the Chinese National Standard (GB). The content of total phenols was determined with the Folin−Ciocalteu method and gallic acid was used as a standard.

In addition, the contents of allicin and 5-HMF were determined by the spectrophotometric method, according to the method conducted by Joan [30] and Feng [31].

2.6. Hardness and Color

The colors of FG and BG surfaces were determined using a CHROMA METER, CR-400 (Minolta, Osaka, Japan). The results were presented in the CIE system, where L*, a*, and b* represent lightness, redness/greenness, and yellowness/blueness, respectively.

The hardness values of FG and BG were determined using a Texture Analyzer (Stable Micro Systems, Godalming, Surrey, UK). The maximum force required to break samples individually was presented as gram (g).

2.7. Measurement of Oxidation Resistance

The DPPH scavenging activity assay was conducted in the light of the method reported by Ren et al. [32]. DPPH was dissolved in ethanol to 0.25 mg/mL in an Erlenmeyer flask. To ensure the stability of the DPPH solution, the reagents were prepared 2 h in advance.

To protect the DPPH solution from the light, the flask was covered with aluminum foil and stored in a fridge. During measurement, the DPPH solution was diluted to one-fifth of its original concentration. All experiments were repeated three times. One milliliter of appropriately diluted samples was added to 1 mL DPPH solutions. As a control, 1 mL of ethanol was added instead of 1 mL of appropriately diluted samples. As for the blank, 1 mL samples were mixed with 1 mL ethanol, and the absorbance was measured at 571 nm.

2.8. Statistical Analysis

Statistical analyses of the data were conducted by SPSS Statistics 19.0 (IBM Corporation, Chicago, IL, USA) and Origin 2021b (OriginLab, Northampton, Massachusetts, USA). All experiments were performed in triplicate. All data were expressed as mean $\pm$ standard deviation (SD). ANOVA tests, followed by the LSD multiple range test, were conducted for chroma parameters. Differences were considered statistically significant at $p < 0.05$.

3. Results

3.1. Sensory Evaluation

The results of the sensory evaluation are shown in Table 1. In general, the acceptability of the garlic was progressively higher, as fermentation progressed. The quality was at its best on the seventh day, after which there was no significant change in product quality. In terms of color, the garlic turned completely black in the first three days of fermentation. Regarding dryness and texture, there were no significant changes throughout the fermentation process, and the scores were high. In regard to odor, the score gradually increased as the fermentation progressed, but there was no significant difference. In terms of taste, the scores on the first three days of fermentation showed significantly lower scores than those on the fifth day and beyond, reaching a score of 14 on the fifth day and continuing to increase with no significant change thereafter. The product scores improved significantly on the first five days, reaching the maximum score on the seventh day and maintaining it. The final product score stayed around 86 points.

Table 1. Sensory scores of samples during the fermentation.

Samples	Color	Dry Degree	Texture	Smell	Taste	Total Score
S1	6.67 [b]	16.67 [a]	17.67 [a]	13.33 [a]	3.67 [b]	58.00 [d]
S3	17.67 [a]	17.00 [a]	15.00 [a]	13.67 [a]	7.00 [b]	70.33 [c]
S5	17.67 [a]	17.00 [a]	15.67 [a]	15.00 [a]	14.67 [a]	80.00 [b]
S7	18.00 [a]	18.33 [a]	16.00 [a]	17.00 [a]	17.33 [a]	86.67 [a]
S9	18.00 [a]	18.33 [a]	16.00 [a]	16.67 [a]	17.33 [a]	86.33 [ab]

[a–d] Different superscripts in the same column indicate significant differences in the comparison of sensory parameters at different fermentation times.

3.2. Physicochemical Properties

3.2.1. Moisture

As can be seen from Figure 2A, the moisture content showed a decreasing trend during the production process. The moisture content decreased faster at the beginning of fermentation. The moisture content in fresh garlic (S0) was 66.13%, and after one day of fermentation (S1), the moisture content was 53.52%. The rate of decline was 19.07%. In the subsequent production process, the rates of the moisture content decline slowed down during the nine-day fermentation, which decreased from 53.52% in S1 to 25.8% in S9, with an average decline rate of 8.7%. The average rate of the moisture decline throughout the fermentation process was approximately 9.93%.

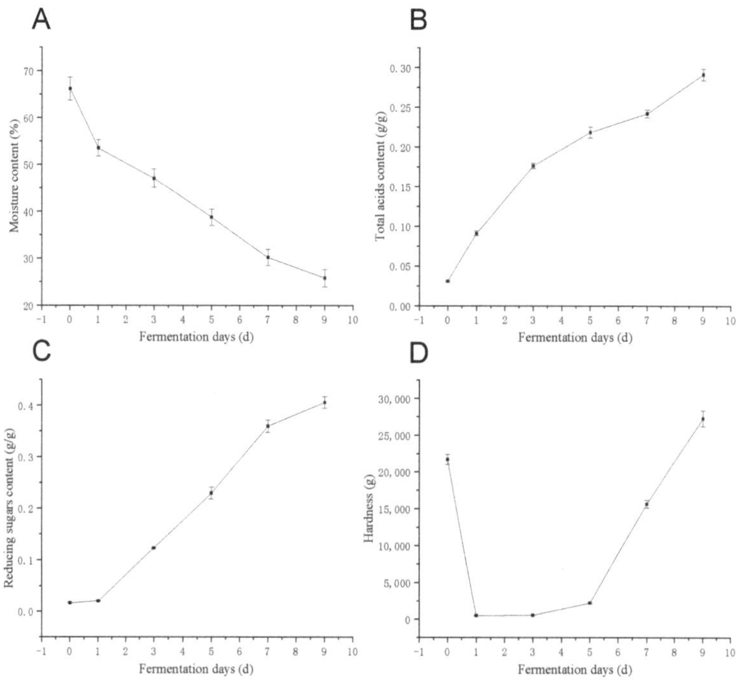

Figure 2. The change of moisture content (**A**), total acids content (**B**), reducing sugars content (**C**), and hardness (**D**) during BG fermentation. All data were expressed as mean ± standard deviation (SD).

3.2.2. Total Acids

During the fermentation process, the total acid content showed a significant upward trend, as shown in Figure 2B. The rate of increase was faster in the first three days. The total acid content in fresh garlic (S0) was low at 0.0309 g/g, and then, it increased to 0.0911 g/g after the first day of fermentation (S1), with a growth rate of 195%. The average growth rate for the first three days of fermentation was about 78.76%. In the following four days, the total acid content increased from 0.1765 g/g (S3) to 0.2420 g/g (S7), with an average growth rate of about 8.21%. The total acid content in the product at the end of fermentation was 0.2914 g/g (S9), which was nine times higher than in fresh garlic (S0), with an average growth rate of 28.32%.

3.2.3. Reducing Sugars

As can be seen from Figure 2C, the reducing sugar content showed a significant increasing trend during the entire fermentation process. The increase in reducing sugar content was not significant at the beginning of fermentation. There was no significant difference in the reducing sugar content for the first two days (1.58% in S0 and 1.99% in S1). From then on, the reducing sugar content increased dramatically, from 1.99% in S1 to 40.57% in S9, with an average growth rate of 45.77%.

3.2.4. Hardness

As can be seen from Figure 2D, the hardness of garlic decreased significantly immediately after the beginning of fermentation. The hardness of the fresh garlic (S0) was 21,711.77 g, and the hardness of the sample at the end of the first day of fermentation (S1) was 465.47 g, a 98% decrease compared to that of S0. During the next four days, the hardness remained at a lower level and increased slightly (465.47 g for S1, 530.41 for S3, and 2182.23 for S5). From the fifth day until the end of fermentation, the hardness showed a sharp increase, with 2186.23 g for S5, 15,653.56 g for S7, and 27,292.38 g for S9, an 11-fold increase compared to that of S5, with an average growth rate of 87.97%.

3.2.5. Color

As can be seen from Figure S1, the white color of the garlic disappeared after one day of fermentation. By the third day, the garlic changed to dark brown, and after five days the color of the samples became completely black. Table 2 shows the chromaticity values of the products. The L* values of the samples kept decreasing with the extension of the fermentation time. The L* values of S0, S1, and S3 were 63.73, 37.00, and 22.86, respectively, indicating that the colors of the samples became closer to black. The L* of S5 was the lowest at 19.34, after which the L* values of the samples increased slightly but remained stable, indicating that the blackening of the samples was completed and the color stabilized. a* and b* values at the fermentation a* changed from −2.17 in S0 to 12.77 in S1 and then to 1.86 in S3. b* changed from 21.94 in S0 to 19.84 in S1 and then to 2.95 in S3. Correspondingly, the ΔE values of the samples showed an increasing trend in the first three days, from 43.00 in S0 to 72.61 in S3, and remained stable in the subsequent production process, indicating that the total color difference of the samples was increased.

Table 2. Chromaticity of samples with different fermentation days.

Chroma	S0	S1	S3	S5	S7	S9
L*	63.73 ± 0.67 [a]	37.00 ± 0.40 [b]	22.86 ± 0.32 [c]	19.34 ± 0.24 [e]	21.33 ± 0.32 [d]	23.65 ± 0.44 [c]
a*	−2.17 ± 0.18 [e]	12.77 ± 0.17 [a]	1.86 ± 0.09 [b]	1.11 ± 0.06 [c]	1.01 ± 0.05 [c]	0.64 ± 0.06 [d]
b*	21.94 ± 0.39 [a]	19.84 ± 0.41 [b]	2.95 ± 0.17 [c]	1.79 ± 0.03 [d]	1.46 ± 0.06 [d]	0.85 ± 0.05 [e]
ΔL	37.28 ± 0.62 [a]	−55.70 ± 1.00 [b]	−72.56 ± 1.09 [d]	−78.74 ± 1.00 [e]	−72.45 ± 0.77 [d]	−70.39 ± 0.88 [c]
Δa	−2.28 ± 0.20 [e]	12.44 ± 0.44 [a]	2.15 ± 0.20 [b]	1.17 ± 0.06 [cd]	1.26 ± 0.08 [c]	0.77 ± 0.06 [d]
Δb	21.31 ± 0.28 [a]	15.23 ± 0.41 [b]	−1.69 ± 0.15 [c]	−2.66 ± 0.15 [d]	−3.10 ± 0.18 [d]	−3.62 ± 0.10 [e]
ΔE	43.00 ± 0.63 [d]	59.08 ± 0.93 [c]	72.61 ± 1.09 [b]	77.13 ± 1.89 [a]	72.52 ± 0.77 [b]	70.48 ± 0.89 [b]

* All data were expressed as mean ± standard deviation (SD). L*, lightness; a*, redness/greenness; b*, blueness/yellowness. Different superscripts (a–e) within the row indicate significant differences of each color parameter compared for the same fermentation time.

3.3. Bioactive Functions

3.3.1. Allicin

As shown in Figure 3A, the fresh garlic (S0) contained a very small amount of allicin, with a specific content of 0.09 mmol/100 g. On the first day of fermentation, the allicin content increased significantly, with 0.30 mmol/100 g in S1. After the first day, the allicin content decreased sharply, with the allicin contents from S3 to S9 decreasing to nearly 0.

3.3.2. 5-HMF

The change of 5-HMF content during fermentation is shown in Figure 3B. In the early stage of fermentation, no 5-HMF was detected in S0, S1, and S3. From S5, 5-HMF started to be produced gradually and accumulated rapidly. 5-HMF contents were found to be 0.14 μg/mL in S5 and 4.12 μg/mL in S9, with an average growth rate of 133%.

3.3.3. Polyphenols

Overall, the polyphenol content showed a clear increasing trend during the fermentation process, as shown in Figure 3C. The total phenolic content in fresh garlic (S0) was 0.045 μg/g and the content increased extremely rapidly in the first three days, while the phenolic content in S3 was 0.10 μg/g with an average growth rate of 30.50%. At the later stage of fermentation, the growth rate of the polyphenol content slowed down but still showed an obvious increasing trend. For S9, the polyphenol content was 0.12 μg/g, with an average growth rate of 11.51% throughout the fermentation stage, which exceeded the polyphenol content in fresh garlic by 25 times.

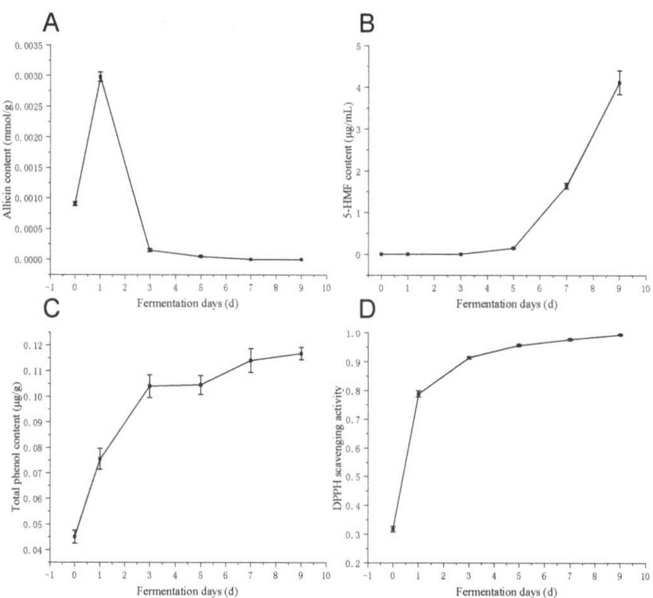

Figure 3. The change of allicin content (**A**), 5-HMF (5-Hydroxymethylfurfural) content (**B**), total phenols content (**C**), and antioxidant properties of black garlic (DPPH scavenging activity) (**D**) during the fermentation. All data were expressed as mean ± standard deviation (SD).

3.3.4. Antioxidation Property

The antioxidant capacity can be characterized by DPPH scavenging activity. Figure 3D shows the changes in the DPPH scavenging activity of BG during the fermentation. The DPPH scavenging rate of fresh garlic was relatively low at 31.72%. The DPPH scavenging rate increased to a relatively high level after the first day of fermentation, and the DPPH scavenging rate of S1 was 78.79%, which was 2.5 times higher than that of fresh garlic. From S3 to S9, the DPPH scavenging rate increased slowly. At the end of fermentation, the DPPH clearance rate of S9 was 99.47%, which was about three times that of fresh garlic (S0).

3.4. Principal Components Analysis

The obtained PCA biplot is shown in Figure 4. It can be observed from the figure that the two principal components explained 88.4% of the total variability, of which PC1 was the most significant, explaining 77.7% of the total variability. A significant separation caused by the fermentation time was observed along PC1, especially the separation of the samples at the beginning of the fermentation from the later stages of the fermentation. What influenced PC1 positively were DPPH scavenging activity, total phenol, total acids, reducing sugars, and 5-HMF. Their concentrations discriminated BG in the later stages of production from the rest of the fermentation times. PC2 also affected the separation of FG and samples at the beginning of fermentation. In this case, allicin and moisture scored the highest, indicating that the separation of S0 and S1 samples was mainly caused by changes in allicin and moisture content. In contrast, these two compounds were the only compounds with negative values in PC1. Obviously, the samples can be grouped into three clusters, including cluster1 (fresh garlic, S0), cluster2 (S1), and cluster3 (S3−S9), respectively. This showed that the BG during fermentation was mainly divided into three stages by the changes of physicochemical properties, which included FG (unfermented), initial fermentation, and late fermentation. It is conducive to the formation of a better quality of BG by proper adjustment of production conditions.

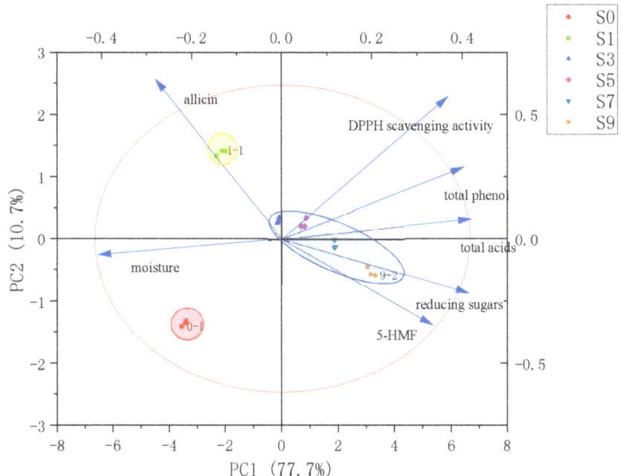

Figure 4. Principal component analysis based on the contents of main components in black garlic at different processing times. The biplot describes the evolution of garlic during fermentation and the contribution of compounds to the variability among samples.

The data obtained in the fermentation experiment were statistically analyzed by PCA to discriminate among fermentation times and establish the physicochemical factors with the largest contribution to separation.

4. Discussion

Black garlic is produced from fresh garlic in a long-time fermentation with a high-temperature and high-humidity environment. Compared with fresh garlic, black garlic not only has no harsh odor, but also has improved texture and taste, especially some special physicochemical properties, and it is well received by consumers [33].

The black garlic obtained from the treatment differs significantly from the fresh garlic in all dimensions, and all the differences are ultimately reflected in the increase in consumer acceptance. According to the results of sensory evaluation, the most significant changes in the sensory evaluation of each dimension during the fermentation process were taste and color. The production process of black garlic is macroscopically a dynamic combination of various biochemical reactions, and the final effect reflected in the product is the browning reaction, including enzymatic browning and the Maillard reaction. The Maillard reaction is the most important reaction in black garlic production. The Maillard reaction, also known as a non-enzymatic browning reaction, is a widespread non-enzymatic browning in food processing. The Maillard reaction is an important reaction in the formation of flavor and color in foods [34]. The end product of the Maillard reaction is melanoidin, a high-molecular-weight brown compound, which is the reason why black garlic turns black [15]. During the heat treatment of black garlic, the cells are broken under high temperature. Enzymes and substrates come into full contact, and various enzymatic reactions occur at an accelerated rate. Alliin is degraded to allicin, which is the origin of the pungent taste of garlic. Allicin continues to degrade under the action of enzymes and high temperature to various organic sulfides, and the pungency disappears [9]. Likewise, polysaccharides are degraded by enzymes and high temperatures to the corresponding monosaccharides and oligosaccharides, producing a sweet taste. In this way, the spiciness of the garlic disappear, and a sweetness emerge. Therefore, the taste of black garlic is better and more acceptable than that of fresh garlic.

During the production of black garlic, the physicochemical properties change considerably. The production process of black garlic is a high-temperature process in which the moisture of the BG continued to evaporate, which means that the moisture content continues to decrease. The change in moisture content can have both physical and chemical effects on black garlic. Physically, the decrease in moisture causes a change in texture. BG becomes sticky and waxy rather than crunchy, as the moisture content decreases, providing the consumer with a completely different taste than fresh garlic. Chemically, the moisture content is one of the most important factors affecting the rate of the Maillard reaction. The strength of the Maillard reaction depends heavily on the hydration of the medium. Firstly, studies have shown that too low moisture content inhibits the Maillard reaction, and it is difficult for the Maillard reaction to occur in completely dry foods. Within a certain range of moisture content, the higher the moisture content, the faster the Maillard reaction. Secondly, too high moisture content also inhibits the Maillard reaction [35]. In our research, 5-HMF, a marker of Maillard reaction, started to be produced in large quantities on the fifth day. We speculated that the moisture content above 40% is not conducive to the Maillard reaction. As the moisture content decreased, the texture of the sample inevitably changed. Hardness is relevant to the strength of the sample, which is expressed as the maximum force of the first compression [36]. The maximum compressive force during crushing was used to explain the texture of samples from the point of hardness. The hardness of fresh garlic suddenly decreased at the beginning of production and gradually increased during the fermentation process. The results are in agreement with the study reported by Karnjanapratum [12], which found a significant increase in hardness during the fermentation process. Several studies have reported that steam pretreatment softens the garlic matrix and controls color production in BG at a higher rate during fermentation [12,37,38]. As the Maillard reaction proceeds, the alkaline amino groups in the BG decrease gradually [35]. Acids are also produced because of the degradation of free amino acids, peptides, and sugars [39]. This results in a significant increase in the total acid content of black garlic compared to that of fresh garlic, giving it a sour taste as well. Fructans are the most abundant polysaccharides in garlic [24]. During the heat treatment of garlic, both the total polysaccharide and fructan contents decrease, and the fructan content increases. Fructan hydrolases are easily inactivated at high temperatures [40], so it is speculated that heating is the main cause of polysaccharide degradation. It has been shown that fructose is the main reducing sugar in black garlic [24], and glucose is not detected in black garlic. Therefore, it is the Heyns rearrangement rather than the Amadori rearrangement that occurs in the Murad reaction in black garlic. During the fermentation of BG, the content of reducing sugars continuously increases. According to the changing trend of reducing sugar content, Li et al. proposed the concept of the reducing sugar balance point (RSBP) to characterize the degree of the Maillard reaction [24]. The generation rate of reducing sugar was greater than the consumption rate before RSBP. Oppositely, the generation rateis less than the consumption rate after RSBP. The key point of RSBP indicates that the reducing sugar required for the Maillard reaction is gradually reduced. The Maillard reaction may be restricted after RSBP, and the reaction rate decreases accordingly. The end product of the Maillard reaction is melanoidin, which causes the garlic to turn black [41]. Moreover, blackening is an important indicator of black garlic ripening. According to our observations, the garlic turned dark brown on the third day and was completely black by the fifth day, indicating that the product entered the ripening stage. The observation of the product was consistent with the colorimetric measurements. The blackening of BG was significantly and positively correlated with 5-HMF content [42]. BG started to darken, when 5-HMF started to accumulate in large amounts. In addition to the Maillard reaction, enzymatic browning is also a cause of black formation.

Allicin is an important characteristic flavor substance in fresh garlic, and it is the source of the pungent flavor of garlic. In fresh garlic, allicin is produced by the degradation of alliin in the presence of alliinase. Under the function of relevant enzymes and high temperature, allicin further generated a series of organosulfur compounds such as S-allyl-L-cysteine (SAC), S-propenyl-L-cysteine (SPC), γ-glutamyl-S-allyl-L-cysteine (GSAC), and γ-glutamyl-S-1-propenyl-L-cysteine (GSPC), among which SAC is considered as the main biologically active substance in BG [9]. This is the reason why the allicin content rose sharply at the beginning of black garlic production and then fell sharply. 5-HMF in black garlic is formed by fructose through the Maillard reaction, which is one of the important intermediate products of the Maillard reaction and an important indicator to mark the stage of the Maillard reaction [43]. It is also an important indicator to characterize the maturation of BG and is closely related to the blackening of BG. Our results revealed that 5-HMF was not detected in the early stage of black garlic fermentation but accelerated accumulation from the fifth day, indicating that the Maillard reaction was in the primary stage in the first four days, entered the intermediate stage and started to accelerate from the fifth day. Polyphenols are an important class of bioactive substances in black garlic. It is reported that polyphenols have several biologically active functions such as improving cardiovascular and cerebrovascular function, kidney function, and eye health [44]. Some researchers have performed in vitro gastrointestinal digestion simulations on black garlic [45]. They found that polyphenols decreased at the beginning of the digestion process and were mainly affected in gastrointestinal digestion. The Folin−Ciocalteu method is currently the classical method for the detection of polyphenols, although its selectivity and specificity are questionable [46]. In the above experiments of in vitro digestion, researchers used UHPLC-HRMS to analyze phenolics in black garlic [45]. Caeic acid and gallic acid are the main compounds in the fresh garlic, while these compounds plus coumaric acid are the main ones found in the black garlic. In our study, the polyphenol content in black garlic at the end of fermentation was 2.45 times higher than that in fresh garlic. In some other studies, the polyphenol content was increased by 1.5 to 10 times [16,47]. We assumed that it is the difference in raw materials that causes the difference in polyphenol content. The phenolic content of black garlic is greatly influenced by the raw material variety, cultivation, and environment [48]. There are several explanations for why the polyphenol content increases. First, bound forms are broken under high temperature, leading to the increase of free forms. Second, enzymatic oxidation involving the antioxidant compounds may be inhibited in the raw plant material. Third, at the later phase of the browning reaction, the levels of the complex polyphenols increase [17]. It is generally believed that the antioxidant properties of BG come from polyphenols, in which sulfhydryl and electrophilic groups have the effects of scavenging oxygen and free radicals. Our research found that black garlic had three times the DPPH scavenging capacity of fresh garlic, approaching 100%. Relevant studies have shown that polysaccharides are also partly responsible for excellent antioxidant properties. Research by Cheng et al. [49] showed that the chemical modification of polysaccharides is an effective method to improve antioxidant capacity.

Overall, compared to fresh garlic, black garlic has a significant improvement in organoleptic, physicochemical properties, and bioactive functions, which leads to a higher acceptance of black garlic in certain groups of consumers. Although there is a general understanding of the content changes of various substances in the BG production process, the mutual transformation mechanism is still unclear and further research is needed. Currently, metabolomics and proteomics are widely used to analyze the changes of metabolites in BG. Further research on BG metabolites will help clarify the formation mechanism of BG, shorten the production cycle and improve the quality.

5. Conclusions

This work studied the changes in the physicochemical properties and the contents of various characteristic substances of homemade BG during the fermentation with statistical methods. The results showed that during the whole heating production process, the moisture content slightly reduced, and the total acid and the reducing sugar content continued to increase, endowing the product with a sweet and sour taste. Polyphenols and 5-HMF were used as biologically active substances in BG, whose contents presented a trend of continuous increase. The content of allicin increased drastically and then decreased sharply, so the pungent smell disappeared. In terms of color, because of the Maillard reaction, 5-HMF concentration gradually accumulated, and the samples started to turn black on the 3rd day and completely black on the 5th day. The beginning of maturity of BG was marked by the formation of black color, which was consistent with experimental results. The results of PCA showed that the samples at each fermentation stage were quite different. The contributions of the two main components were 71.5% and 14.1%, respectively. The samples were obviously separated along time. The products on the 3rd to 9th days were clustered together, indicating that the product began to mature after three days of fermentation.

Although there is general understanding of the content changes of various substances in the BG production process, the mutual transformation mechanism is still unclear and further research is needed. Currently, metabolomics and proteomics are widely used to analyze the changes of metabolites in BG. Further research on BG metabolites will help clarify the formation mechanism of BG, shorten the production cycle and improve the quality.

Supplementary Materials: The following supporting information can be downloaded at: https://www.mdpi.com/article/10.3390/fermentation8110653/s1, Figure S1: Changes of the colors of the samples (S0–S9) during the fermentation.

Author Contributions: Conceptualization, X.W. and X.L.; methodology, X.Y.; software, Z.W.; validation, X.Y., Z.W. and L.L.; formal analysis, X.Y.; investigation, X.Y.; resources, J.W.; data curation, D.M.; writing—original draft preparation, X.Y.; writing—review and editing, X.W.; visualization, X.Y.; supervision, X.W.; project administration, X.W.; funding acquisition, X.W. All authors have read and agreed to the published version of the manuscript.

Funding: This work was supported by the Project of Anhui Province (grant numbers: 202107d06020011, 2108085MC123, and 202103b06020019), the Fundamental Research Funds for the Central Universities of China (grant number: PA2021KCPY0048), and the Project of Suzhou City in Anhui Province (grant number: 2021054).

Institutional Review Board Statement: Not applicable.

Informed Consent Statement: Not applicable.

Data Availability Statement: The data presented in this study are available on request from the corresponding author.

Conflicts of Interest: The authors declare that they have no known competing financial interest or personal relationship that could have appeared to influence the work reported in this paper.

References

1. Zhang, X.; Shi, Y.; Wang, L.; Li, X.; Zhang, S.; Wang, X.; Jin, M.; Hsiao, C.-D.; Lin, H.; Han, L. Metabolomics for biomarker discovery in fermented black garlic and potential bioprotective responses against cardiovascular diseases. *J. Agric. Food Chem.* **2019**, *67*, 12191–12198. [CrossRef]
2. Atun, S.; Aznam, N.; Arianingrum, R.; Devi, Y.; Melasari, R. Characterization and biological activity test of garlic and its fermentation as antioxidant, analgesic, and anticancer. In Proceedings of the 7th International Conference on Research, Implementation, and Education of Mathematics and Sciences (ICRIEMS 2020), Yogyakarta, Indonesia, 25–26 September 2020; Atlantis Press: Paris, France; pp. 159–165. [CrossRef]
3. Isnaini, F.; Yudistia, R.; Faradilla, A.; Rahman, M. Effect of black garlic extract on blood glucose, lipid profile, and sgpt-sgot of wistar rats diabetes mellitus model. *Maj. Kedokt. Bdg.* **2019**, *51*, 82–87.
4. Halimah, S.L.; Hasan, K. Differences of bio (chemical) characterization of garlic and black garlic on antibacterial and antioxidant activities. *J. Phys. Conf. Ser.* **2021**, *1764*, 012005. [CrossRef]
5. Andres, R.; Karin, R.; Sobenin, I.A.; Bucher, H.C.; Nordmann, A.J. A systematic review and metaanalysis on the effects of garlic preparations on blood pressure in individuals with hypertension. *Am. J. Hypertens.* **2015**, *28*, 414–423. [CrossRef]
6. Locatelli, D.A.; Nazareno, M.A.; Fusari, C.M.; Camargo, A.B. Cooked garlic and antioxidant activity: Correlation with organosulfur compound composition. *Food Chem.* **2017**, *220*, 219–224. [CrossRef]
7. Ramirez, D.A.; Locatelli, D.A.; González, R.E.; Cavagnaro, P.F.; Camargo, B.A. Analytical methods for bioactive sulfur compounds in allium: An integrated review and future directions. *J. Food Compos. Anal.* **2017**, *61*, 4–19. [CrossRef]
8. Qiu, Z.; Zheng, Z.; Zhang, B.; Sun-Waterhouse, D.; Qiao, X. Formation, nutritional value, and enhancement of characteristic components in black garlic: A review for maximizing the goodness to humans. *Compr. Rev. Food Sci. Food Saf.* **2020**, *19*, 801–834. [CrossRef]
9. Chang, W.C.; Chen, Y.T.; Chen, H.J.; Hsieh, C.W.; Liao, P.C. Comparative uhplc-q-orbitrap hrms-based metabolomics unveils biochemical changes of black garlic during aging process. *J. Agric. Food Chem.* **2020**, *68*, 14049–14058. [CrossRef]
10. Dewi, A.N.N.; Mustika, I.W. Nutrition content and antioxidant activity of black garlic. *Int. J. Health Sci.* **2018**, *2*, 11–20. [CrossRef]

11. Ding, Y.; Zhou, X.; Zhong, Y.; Wang, D.; Dai, B.; Deng, Y. Metabolite, volatile and antioxidant profiles of black garlic stored in different packaging materials. *Food Control* **2021**, *127*, 108131. [CrossRef]
12. Karnjanapratum, S.; Supapvanich, S.; Kaewthong, P.; Takeungwongtrakul, S. Impact of steaming pretreatment process on characteristics and antioxidant activities of black garlic (*Allium sativum* L.). *J. Food Sci. Technol.* **2021**, *58*, 1869–1876. [CrossRef]
13. Rios-Rios, K.L.; Gaytan-Martinez, M.; Rivera-Pastrana, D.M.; Morales-Sanchez, E.; Villamiel, M.; Montilla, A.; Mercado-Silva, E.M.; Vazquez-Barrios, M.E. Ohmic heating pretreatment accelerates black garlic processing. *LWT–Food Sci. Technol.* **2021**, *151*, 112218. [CrossRef]
14. Sun, -E.Y.; Wang, W. Changes in nutritional and bio-functional compounds and antioxidant capacity during black garlic processing. *J. Food Sci. Technol.* **2018**, *55*, 479–488. [CrossRef]
15. Wu, J.F.; Jin, Y.; Zhang, M. Evaluation on the physicochemical and digestive properties of melanoidin from black garlic and their antioxidant activities in vitro. *Food Chem.* **2021**, *340*, 127934. [CrossRef]
16. Choi, I.S.; Cha, H.S.; Lee, Y.S. Physicochemical and antioxidant properties of black garlic. *Molecules* **2014**, *19*, 16811–16823. [CrossRef] [PubMed]
17. Kim, J.S.; Kang, O.J.; Gweon, O.C. Comparison of phenolic acids and flavonoids in black garlic at different thermal processing steps. *J. Funct. Foods* **2013**, *5*, 80–86. [CrossRef]
18. Toledano-Medina, M.A.; Perez-Aparicio, J.; Moreno-Rojas, R.; Merinas-Amo, T. Evolution of some physicochemical and antioxidant properties of black garlic whole bulbs and peeled cloves. *Food Chem.* **2016**, *199*, 135–139. [CrossRef]
19. Wang, X.M.; Liu, R.; Yang, Y.K.; Zhang, M. Isolation, purification and identification of antioxidants in an aqueous aged garlic extract. *Food Chem.* **2015**, *187*, 37–43. [CrossRef]
20. Yuan, H.; Sun, L.; Chen, M.; Wang, J. An analysis of the changes on intermediate products during the thermal processing of black garlic. *Food Chem.* **2018**, *239*, 56–61. [CrossRef]
21. Sembiring, N.B.; Iskandar, Y. A review of component and pharmacology activities of black garlic. *Maj. Obat Tradis.* **2019**, *24*, 178–183. [CrossRef]
22. Ma, Y.; Ling, T.J.; Su, X.Q.; Jiang, B.; Nian, B.; Chen, L.J.; Liu, M.L.; Zhang, Z.Y.; Wang, D.P.; Mu, Y.Y.; et al. Integrated proteomics and metabolomics analysis of tea leaves fermented by aspergillus niger, aspergillus tamarii and aspergillus fumigatus. *Food Chem.* **2021**, *334*, 127560. [CrossRef] [PubMed]
23. Wang, L.; Hong, K.; Agbaka, J.I.; Zhu, G.; Lv, C.; Ma, C. Application of uhplc-q/tof-ms-based metabolomics analysis for the evaluation of bitter-tasting krausen metabolites during beer fermentation. *J. Food Compos. Anal.* **2021**, *99*, 103850. [CrossRef]
24. Li, F.; Cao, J.; Liu, Q.; Hu, X.; Liao, X.; Zhang, Y. Acceleration of the maillard reaction and achievement of product quality by high pressure pretreatment during black garlic processing. *Food Chem.* **2020**, *318*, 126517. [CrossRef]
25. González-Ramírez, P.; Pascual-Mathey, L.; García-Rodríguez, R.; Jiménez, M.; Beristain, C.; Sanchez-Medina, A.; Pascual-Pineda, L. Effect of relative humidity on the metabolite profiles, antioxidant activity and sensory acceptance of black garlic processing. *Food Biosci.* **2022**, *48*, 101827. [CrossRef]
26. Zhang, X.; Li, N.; Lu, X.; Liu, P.; Qiao, X. Effects of temperature on the quality of black garlic. *J. Sci. Food Agric.* **2016**, *96*, 2366–2372. [CrossRef]
27. Zhou, X.; Cui, H.; Zhang, Q.; Hayat, K.; Yu, J.; Hussain, S.; Tahir, M.U.; Zhang, X.; Ho, C.-T. Taste improvement of maillard reaction intermediates derived from enzymatic hydrolysates of pea protein. *Food Res. Int.* **2021**, *140*, 109985. [CrossRef]
28. Zambrano, M.V.; Dutta, B.; Mercer, D.G.; MacLean, H.L.; Touchie, M.F. Assessment of moisture content measurement methods of dried food products in small-scale operations in developing countries: A review. *Trends Food Sci. Technol.* **2019**, *88*, 484–496. [CrossRef]
29. Huang, H.; Hu, X.; Tian, J.; Jiang, X.; Luo, H.; Huang, D. Rapid detection of the reducing sugar and amino acid nitrogen contents of daqu based on hyperspectral imaging. *J. Food Compos. Anal.* **2021**, *101*, 103970. [CrossRef]
30. Han, J.; Lawson, L.; Han, G.; Han, P. A spectrophotometric method for quantitative determination of allicin and total garlic thiosulfinates. *Anal. Biochem.* **1995**, *225*, 157–160. [CrossRef]
31. Feng, H.W.; Xiong, F.U. Determination of 5-hydroxymethyl-2-furfural in molasses by ultraviolet spectrophotometry. *Sci. Technol. Food Ind.* **2010**, *31*, 365–367.
32. Ren, F.; Perussello, C.A.; Zhang, Z.; Gaffney, M.T.; Kerry, J.P.; Tiwari, B.K. Effect of agronomic practices and drying techniques on nutritional and quality parameters of onions (*Allium cepa* L.). *Dry. Technol.* **2017**, *36*, 435–447. [CrossRef]
33. Kimura, S.; Tung, Y.C.; Pan, M.H.; Su, N.W.; Lai, Y.J.; Cheng, K.C. Black garlic: A critical review of its production, bioactivity, and application. *J. Food Drug Anal.* **2017**, *25*, 62–70. [CrossRef] [PubMed]
34. Cui, H.; Yu, J.; Zhai, Y.; Feng, L.; Chen, P.; Hayat, K.; Xu, Y.; Zhang, X.; Ho, C.-T. Formation and fate of amadori rearrangement products in maillard reaction. *Trends Food Sci. Technol.* **2021**, *115*, 391–408. [CrossRef]
35. Ellis, G.P. The maillard reaction. *Adv. Carbohydr. Chem.* **1959**, *14*, 63–134. [CrossRef]
36. Chang, H.-J.; Xu, X.-L.; Zhou, G.-H.; Li, C.-B.; Huang, M. Effects of characteristics changes of collagen on meat physicochemical properties of beef semitendinosus muscle during ultrasonic processing. *Food Bioprocess Technol.* **2009**, *5*, 285–297. [CrossRef]
37. He, Y.; Fan, G.-J.; Wu, C.-E.; Kou, X.; Li, T.-T.; Tian, F.; Gong, H. Influence of packaging materials on postharvest physiology and texture of garlic cloves during refrigeration storage. *Food Chem.* **2019**, *298*, 125019. [CrossRef]
38. Nugrahedi, P.; Dekker, M.; Widianarko, B.; Verkerk, R. Quality of cabbage during long term steaming; phytochemical, texture and colour evaluation. *LWT–Food Sci. Technol.* **2016**, *65*, 421–427. [CrossRef]

39. Eric, K.; Raymond, L.V.; Huang, M.; Cheserek, M.J.; Hayat, K.; Savio, N.D.; Amédée, M.; Zhang, X. Sensory attributes and antioxidant capacity of maillard reaction products derived from xylose, cysteine and sunflower protein hydrolysate model system. *Food Res. Int.* **2013**, *54*, 1437–1447. [CrossRef]

40. Lu, X.; Li, N.; Qiao, X.; Qiu, Z.; Liu, P. Effects of thermal treatment on polysaccharide degradation during black garlic processing. *LWT–Food Sci. Technol.* **2018**, *95*, 223–229. [CrossRef]

41. Kang, O.J. Evaluation of melanoidins formed from black garlic after different thermal processing steps. *Prev. Nutr. Food Sci.* **2016**, *21*, 398–405. [CrossRef]

42. Dursun Capar, T.; Inanir, C.; Cimen, F.; Ekici, L.; Yalcin, H. Black garlic fermentation with green tea extract reduced hmf and improved bioactive properties: Optimization study with response surface methodology. *J. Food Meas. Charact.* **2022**, *16*, 1340–1353. [CrossRef]

43. Rizelio, V.M.; Gonzaga, L.V.; Borges, G.D.S.C.; Micke, G.A.; Fett, R.; Costa, A.C.O. Development of a fast meck method for determination of 5-hmf in honey samples. *Food Chem.* **2012**, *133*, 1640–1645. [CrossRef]

44. Del Bo', C.; Bernardi, S.; Marino, M.; Porrini, M.; Tucci, M.; Guglielmetti, S.; Cherubini, A.; Carrieri, B.; Kirkup, B.; Kroon, P. Systematic review on polyphenol intake and health outcomes: Is there sufficient evidence to define a health-promoting polyphenol-rich dietary pattern? *Nutrients* **2019**, *11*, 1355. [CrossRef]

45. Moreno-Ortega, A.; Pereira-Caro, G.; Ordóñez, J.L.; Moreno-Rojas, R.; Ortíz-Somovilla, V.; Moreno-Rojas, J.M. Bioaccessibility of bioactive compounds of 'fresh garlic' and 'black garlic' through in vitro gastrointestinal digestion. *Foods* **2020**, *9*, 1582. [CrossRef] [PubMed]

46. Bedrníček, J.; Laknerová, I.; Lorenc, F.; Moraes, P.P.d.; Jarošová, M.; Samková, E.; Tříska, J.; Vrchotová, N.; Kadlec, J.; Smetana, P. The use of a thermal process to produce black garlic: Differences in the physicochemical and sensory characteristics using seven varieties of fresh garlic. *Foods* **2021**, *10*, 2703. [CrossRef]

47. Setiyoningrum, F.; Priadi, G.; Afiati, F.; Herlina, N.; Solikhin, A. Composition of spontaneous black garlic fermentation in a water bath. *Food Sci. Technol.* **2021**, *41*, 557–562. [CrossRef]

48. Ríos-Ríos, K.L.; Montilla, A.; Olano, A.; Villamiel, M. Physicochemical changes and sensorial properties during black garlic elaboration: A review. *Trends Food Sci. Technol.* **2019**, *88*, 459–467. [CrossRef]

49. Cheng, H.; Huang, G.; Huang, H. The antioxidant activities of garlic polysaccharide and its derivatives. *Int. J. Biol. Macromol.* **2020**, *145*, 819–826. [CrossRef]

fermentation

MDPI

Article

Biocontrol of Geosmin Production by Inoculation of Native Microbiota during the *Daqu*-Making Process

Hai Du *,†, Junlin Wei †, Xitong Zhang and Yan Xu

Laboratory of Brewing Microbiology and Applied Enzymology, Key Laboratory of Industrial Biotechnology of Ministry of Education, School of Biotechnology, Jiangnan University, 1800 Lihu Avenue, Wuxi 214122, China
* Correspondence: duhai88@126.com; Tel.: +86-510-8591-8201
† These authors contributed equally to this work.

Abstract: Geosmin produced by *Streptomyces* can cause an earthy off-flavor at trace levels, seriously deteriorating the quality of Chinese liquor. Geosmin was detected during the *Daqu* (Chinese liquor fermentation starter)-making process, which is a multi-species fermentation process in an open system. Here, biocontrol, using the native microbiota present in *Daqu* making, was used to control the geosmin contamination. Six native strains were obtained according to their inhibitory effects on *Streptomyces* and then were inoculated into the *Daqu* fermentation. After inoculation, the content of geosmin decreased by 34.40% (from 7.18 ± 0.13 μg/kg to 4.71 ± 0.30 μg/kg) in the early stage and by 55.20% (from 8.86 ± 1.54 μg/kg to 3.97 ± 0.78 μg/kg) in the late stage. High-throughput sequencing combined with an interaction network revealed that the fungal community played an important role in the early stage and the correlation between *Pichia* and *Streptomyces* changed from the original indirect promotion to direct inhibition after inoculation. This study provides an effective strategy for controlling geosmin contamination in *Daqu* via precisely regulating microbial communities, as well as highlights the potential of biocontrol for controlling off-flavor chemicals at trace levels in complex fermentation systems.

Keywords: biocontrol; geosmin; *Streptomyces*; *Daqu*; native microbiota

Citation: Du, H.; Wei, J.; Zhang, X.; Xu, Y. Biocontrol of Geosmin Production by Inoculation of Native Microbiota during the *Daqu*-Making Process. *Fermentation* 2022, 8, 588. https://doi.org/10.3390/fermentation8110588

Academic Editor: Angela Capece

Received: 4 October 2022
Accepted: 20 October 2022
Published: 30 October 2022

Publisher's Note: MDPI stays neutral with regard to jurisdictional claims in published maps and institutional affiliations.

1. Introduction

Geosmin is a volatile metabolite produced by *Streptomyces*, cyanobacteria, and fungi [1,2], which can cause an earthy off-flavor at trace levels (the threshold is at the ng/L level) [3], seriously deteriorating the flavor and quality of multiple products, such as aquatic products [4–6], drinking water [7,8], juice [9,10], and especially Chinese liquor [1,4]. Geosmin contamination has always been a serious problem for the Chinese liquor industry and leaves a dirty and dusty impression. The highest concentrations of geosmin have been determined in the light-aroma-type liquor [2,4]. Geosmin has an extremely low threshold (from 6 to 10 ng/L), and people usually start to feel uncomfortable at 7 ng/L at 45 °C [11,12]. Therefore, although the concentration of geosmin in food and beverages remains at a very low level, it still causes unpleasant feelings. In our previous studies, *Streptomyces* spp. acting as the geosmin producer have been regarded as the most frequent and serious microbial contamination. *Streptomyces* spp. produce a vast array of antibiotics that inhibit the growth and metabolism of brewing functional yeasts and molds, thereby reducing the flavor compounds (alcohols and esters) in Chinese liquor [1,2,4,13].

Chinese liquor is a traditional distilled alcoholic beverage with thousands of years' history [4,14]. Different from distilled liquors in the West, such as whisky and brandy, Chinese liquor is a typical solid-state fermentation product from grains [13], with saccharification and fermentation simultaneously proceeding [14]. *Daqu* acts as the saccharifying and fermenting agent and contributes a large number of functional communities to the fermentation process of liquor [2,15,16]. *Daqu* is manufactured through a spontaneous solid-state fermentation process in an open environment [15]. Functional mold, bacteria, and yeasts form a special microbiota and produce various enzymes and flavor compounds through growth and metabolism during *Daqu* fermentation. The dominant mold existing in the *Daqu* includes the genus *Aspergillus* (*A. terreus*, *A. oryzae*), *Rhizopus* (*R. oryzae*, *R. peka*), *Thermomyces* (*T. crustaceus*), and *Thermoascus* (*T. crustaceus*). Yeasts mainly include *Pichia* (*P. anomala*, *P. fermentans*, *P. kudriavzevii*), *Saccharomyces* (*S. cerevisiae*), *Wickerhamomyces* (*W. anomalus*), *Torulaspora*, and *Candida*. Bacteria includes *Bacillus* (*B. subtilis*, *B. licheniformis*, *B. amyloliquefaciens*, *B. sonorensis*), *Lactobacillus* (*L. helveticus*, *L. fermentum*, and *L.panis*), *Lactococcus* (*L. lactis*), *Weissella* (*W. cibaria*), and acetic acid bacteria species [5,17]. Therefore, microbiota in *Daqu* largely determine the characteristics of *Daqu* and Chinese liquor [13]. However, in an open system, high concentrations of geosmin were found during the *Daqu*-making process, causing damage to the quality of the liquor [13].

Recently, several effective methods have been suggested to control geosmin in drinking water and wine. For example, activated carbon, chlorination, ozonation process, and K_2FeO_4 oxidation were applied in drinking water [8,18–21]. Chitosan, zeolite, and filtration have been used to remove geosmin from white wine [22]. However, these methods are not effective to control geosmin and *Streptomyces* contamination in the practical production of *Daqu* making. Biocontrol is an efficient and environmentally friendly approach to rationally improve the fermentation quality, where some bacteria and yeasts have been used as biocontrol agent in fermented foods production [23–25]. For instance, non- and low-fermenting yeast strains were used in grape juice to control the ochratoxin A producer *Aspergillus* spp. [23]. *Bacillus megaterium* was used in peanuts to control the aflatoxin produced by *Aspergillus flavus* and reduce the post-harvest rot problem of peanut kernels [24]. In our previous study, *Bacillus* showed a good ability to inhibit geosmin production in the simulated fermentation experiments on a laboratory scale [2]. Therefore, biocontrol could be a more promising solution to solve the urgent requirements in geosmin contamination. However, different from the laboratory-scale fermentation, the practical production process of *Daqu* is in an open environment with multi-species and is influenced by multiple factors. Therefore, whether biocontrol effectively controls geosmin production in the complex fermentation process of *Daqu* still needs further verification.

In this study, *Streptomyces albus* F5A-1 was used as the representative strain of *Streptomyces* spp. A medium-temperature *Daqu* was used to select the functional microorganisms that inhibited *Streptomyces* growth. Then, the functional microorganisms were inoculated into the *Daqu*-making process to evaluate their effects on the geosmin production, microbial structure, and flavor metabolites of the *Daqu*. Headspace solid-phase microextraction–gas chromatography–mass spectrometry (HS-SPME-GC-MS) was used to determine the content of geosmin and other volatile metabolites. High-throughput sequencing was used to determine the microbial community structure during *Daqu* fermentation.

2. Materials and Methods

2.1. Strains

The indicator strain of geosmin production was *Streptomyces albus* F5A-1, which was previously isolated from medium-temperature *Daqu* and stored in our lab at $-80\ ^\circ$C in a Luria-Bertani (LB) broth glycerol stock [1,2].

Geosmin-inhibiting microorganisms were isolated by the conventional agar dilution method from medium-temperature *Daqu*. As a type of *Daqu*, the maximum temperature in the production of medium-temperature *Daqu* is 40–50 $^\circ$C. For the isolation of geosmin-inhibiting microorganisms, 10 g of the *Daqu* sample was mixed with 90 mL of sterile saline solution (0.90% (wt/vol)) and then shaken for 1 h at 200 rpm at 30 $^\circ$C. The homogenates were diluted serially 10-fold with sterile saline solution. Finally, 100 µL of the dilution was spread on the agar plates and incubated. The fungi, which included mold and yeast, were incubated in potato dextrose agar (PDA; potato extract 6 g/L, glucose 20 g/L, penicillin–streptomycin solution 100 mg/L, and agar 20 g/L) plates at 28 $^\circ$C for 5 days. Bacteria was incubated in Luria-Bertani (LB) agar (peptone 10 g/L, NaCl 10 g/L, yeast extract 5 g/L, nystain 10 mg/L, and agar 20 g/L, pH: 7.4) plates at 37 °C for 2 days. The colonies of different forms were purified by repetitive streaking until a single-colony morphology. DNA of isolated strains was extracted based on previously described methods [26]. For fungi, primers ITS1 and ITS4 were used to amplify the ITS region [27]. For bacteria, primer sets 27F and 1492R were used to amplify the 16S rRNA genes [28]. PCR products were obtained by GENEWIZ (Suzhou, China). The sequences were used for BLAST searches (http://www.ncbi.nlm.nih.gov/BLAST/, accessed on 1 April 2020) for the identification of isolates. Finally, all the isolates were stored in their corresponding broth glycerol stock at −80 °C.

The growth inhibition assays were carried out according to Zhi et al. [2]. Every inhibition zone on the plate was measured six times with a vernier caliper. The final value is the average of the six values. We quantified the growth inhibition based on the diameter ratio of inhibition zone to colony ($R_{\text{inhibition zone}}/R_{\text{colony}}$). The strains with $R_{\text{inhibition zone}}/R_{\text{colony}} > 1.15$ were chosen as the geosmin-inhibiting microorganisms.

2.2. Preparation of Geosmin-Inhibiting Microorganisms

The yeast strains were separately inoculated into yeast extract–peptone–dextrose (YPD) broth (yeast extract 10 g/L, peptone 20 g/L, and glucose 20 g/L) at 30 °C, spun at 200 rpm for 2 d, and *Bacillus* was inoculated into Luria-Bertani (LB) broth (beef extract 5 g/L, peptone 10 g/L, sodium chloride 10 g/L) at 37 °C, spun at 200 rpm for 2 d. After inoculation, each seed broth was centrifuged at 12,000 rpm/min for 10 min; then the supernatant was discarded, and the precipitated cells were washed three times with sterile normal saline. The precipitated cells were then resuspended in sterile water and counted using a hemocytometer. Finally, each suspension was separately diluted to 10^6 cells/g *Daqu* and then mixed them together to inoculate into the *Daqu* fermentation.

2.3. Experimental Design and Daqu Sample Collection

The experiments were performed in a distillery in Shaanxi Province, China (34°55′ N; 107°32′ E). Four independent batches of *Daqu* fermentation from two workshops (Workshops A and B, which were closely located) were chosen. Two batches from Workshop A were the test group (inoculation with the geosmin-inhibiting microorganisms) and another two batches from Workshop B were the control group. Four batches used the same raw materials (wheat) and technological parameters (such as the water content and machine parameters (pressure and the times of pressing bricks)). The seed broth was evenly mixed with crushed materials. Then, the uninoculated and inoculated *Daqu* were fermented in the same way. According to a previous study, geosmin is mainly produced in the early and the middle stage of fermentation [2,29]. Hence, the *Daqu* samples were collected on Days 0, 3, 5, 10, 15, and 20. Samples were collected from the upper, middle, and lower layers, and three *Daqu* bricks were obtained from each layer. Finally, the rind and the core of the *Daqu* bricks from different points in the same layer were separately crushed and then mixed to form one sample (Figure S1). Six parallel samples were obtained in each fermentation time. *Daqu* samples were then transferred to the lab for the analysis of physicochemical parameters, volatile compound analysis, and microbial community.

2.4. Total DNA Extraction, Amplification, and Sequencing

Each sample (7 g) was treated with 0.1 mol/L sterile phosphate-buffered saline (PBS) and centrifuged at $300 \times g$ for 10 min at 4 °C. The supernatant was then centrifuged again at 4 °C, $13,000 \times g$, for 10 min. An E.Z.N.A. (easy nucleic acid isolation) soil DNA kit (Omega bio-tek, Norcross, GA, USA) was used for isolating DNA, according to the manufacturer's instructions. The DNA isolated from the six parallel samples was mixed to form one sample to reduce the heterogeneity before amplification and sequencing. The bacterial V3–V4 region of the 16S rRNA gene was amplified using sets F338 and barcode-R806 primers [30]. The fungal ITS2 region was amplified with primers ITS3 and ITS4 [31]. A PCR purification kit was used to purify the PCR products. The barcoded PCR products were sequenced on a

MiSeq benchtop sequencer for 250-bp paired-end sequencing (2 × 250 bp; Illumina, San Diego, CA, USA) at Beijing Auwigene Tech. Ltd. (Beijing, China).

The generated raw sequences were processed using QIIME v.1.9.1 and R v.3.3.1, (http://www.r-project.org, accessed on 1 December 2020) [32]. Briefly, the raw sequences were quality-filtered, and sequences with ambiguous bases ('N') were removed using Trimmomatic. Chimeric sequences were removed using the Uchime algorithm. The high-quality sequences were clustered (with a 97% sequence similarity threshold) into OTUs using the Qiime Uparse pipeline [33–35]. The bacterial OTU sequences were annotated using the Silva database (Release 138 http://www.arb-silva.de, accessed on 1 January 2021) [36]. The fungal OTU sequences were compared using a BLAST search against the UNITE fungal ITS database (Release 8.2 http://unite.ut.ee/index.php, accessed on 1 January 2021). Then, the Chao1 richness and Shannon diversity indices were calculated using Qiime [37].

2.5. Real-Time Quantitative PCR (qPCR) to Estimate Microbial Biomass

The biomass of *Streptomyces albus* was determined by real-time quantitative PCR (qPCR). We extracted template DNA for a standard curve from the *Daqu* substate in which 10× serial dilutions of conidial suspensions of *Streptomyces albus* had been inoculated, as described by Hoffman and Winston [38]. ddH$_2$O was used to dissolve the DNA, and then DNA concentrations were quantified using a NanoDrop ND-1000 spectrophotometer (NanoDrop Technologies, Wilmington, DE, USA). A Real-Time PCR System (Applied Biosystems, Foster City, CA, USA) was used for qPCR analysis. The genomic DNA was used as the template to amplify *Streptomyces albus* by primers 245F (5′-TCT TCT TCG ACG ACC ACT TCC T-3′) and 551R (5′-CGG CGC ATC TCG ATG TAC TC-3′) [5]. Each reaction was conducted in 20.0 μL which contained 10.0 μL SYBR Green Supermix (SYBR Premix ExTaq II, Takara, Dalian, China), 0.4 μL of each primer (20 μM), 1.0 μL of DNA template, and 8.2 μL ddH$_2$O. The amplification conditions and calibration curves were as follows: initial denaturation at 98 °C for 5 min, followed by 30 cycles of denaturation at 95 °C for 60 s, annealing and extension at 56 °C for 5 s, and increases of 0.5 °C every 5 s from 65 °C to 95 °C for melting curve analysis. The standard curve was established according to the C$_t$ values obtained from the genomic DNA of the serially diluted spores and the logarithm of the corresponding initial spore concentrations.

2.6. Analysis of Volatile Compounds and Geosmin

We added 5 g of the *Daqu* samples to 20 mL ultrapure water. The samples were treated at 4 °C for 30 min ultrasonically and centrifuged 4000× *g* for 10 min. After centrifugation, 8 mL of supernatant and 20 μL menthol (internal standard, 100 μg/mL) were added into a 20 mL headspace vial with 3 g NaCl. [4]. Volatile compounds and geosmin were determined by headspace solid-phase microextraction–gas chromatography (HS-SPME-GC-MS) on a DB-Wax column (30 m × 0.25 mm i.d., 0.25 μm film thickness; J&W Scientific, Folsom, CA, USA), as described by Gao et al. [39].

2.7. Physicochemical Parameters Detection and Analysis

We detected the total titratable acidity by titration, as described by Li et al. [40]. Liquefying power and saccharifying power were determined by the method in He et al. [17]. The amount of starch liquefied was regarded as one unit of liquefying power. The amount of glucose released per hour by 1 g of *Daqu* was defined as the saccharifying power [17]. The esterifying power and fermenting power were determined based on the national professional standard methods. The content of ethyl caproate produced by 50 g of *Daqu* per 7 days in the mixture of caproic acid and ethanol at 35 °C was regarded as one unit of esterifying power. The amount of CO$_2$ released by 1 g of *Daqu* per 72 h at 30 °C was regarded as one unit of fermenting power [17]. Each detection was conducted in triplicate and indicated by dry weight.

2.8. Statistical Analysis

The statistical analysis and data plots were performed with OriginPro 2018 (OriginLab, Northampton, MA, USA), Microsoft Office Excel 2016 (Microsoft, Redmond, WA, USA), and Adobe Illustrator CC 22.0 (Adobe Systems Incorporated, San Jose, CA, USA). Principal component analysis (PCA) was analyzed via CANOCO (vision 5). The relationships among microbial communities were calculated based on all possible Spearman's rank correlations between the abundant genera. To reduce network complexity, only the genera with relatively high abundance (with average abundance of >0.5%) were considered [41]. The correlations were considered valid if they were significant at $p < 0.05$ (with false discovery rate correction) [42]. The creation and visualization of the network were conducted by Gephi (Web Atlas, Paris, France) [43]. Each genus was represented by a node. A strong and significant correlation between nodes was represented by the edge [44,45].

2.9. Data Availability

All sequences generated were submitted to the NCBI database under the accession number PRJNA691687 and PRJNA886836.

3. Results and Discussion

3.1. Isolation of Geosmin-Inhibiting Microorganisms from Daqu

A total of 50 strains (including 23 bacteria, 22 yeasts, and 5 mold) were isolated from the medium-temperature *Daqu*. These strains were used for antagonism against *Streptomyces albus* by modified agar well-diffusion assay [46]. Finally, six strains, including five yeasts and a *Bacillus* strain, with R $_{inhibition\ zone}$/R $_{colony}$ > 1.50 were chosen as the geosmin-inhibiting microorganisms [2]. These six strains were *Pichia fermentans* Y1B-2, *Pichia fermentans* Y1A-1, *Pichia kudriavzevii* MY5-2, *Saccharomyces cerevisiae* Y8-2, *Issatchenkia orientalis* Y2A-1, and *Bacillus subtilis* J7-4. The sketch map of the antibacterial experiment is shown in Figure 1. The R $_{inhibition\ zone}$/R $_{colony}$ values of these strains are shown in Table 1.

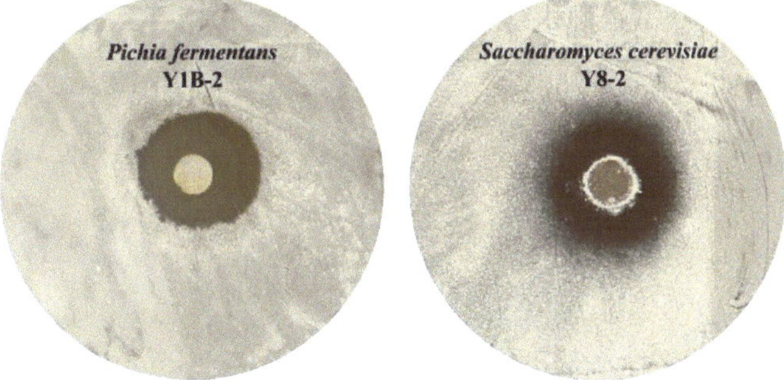

Figure 1. The sketch map of the antibacterial experiment results.

Table 1. The R $_{inhibition}$ zone/R $_{colony}$ values of the six strains.

Strains	R $_{inhibition\ zone}$/R $_{colony}$
Pichia fermentans Y1B-2	2.82 ± 0.08
Pichia fermentans Y1A-1	2.69 ± 0.07
Pichia kudriavzevii MY5-2	2.14 ± 0.03
Saccharomyces cerevisiae Y8-2	2.90 ± 0.06
Issatchenkia orientalis Y2A-1	2.18 ± 0.03
Bacillus subtilis J7-4	2.28 ± 0.04

Saccharomyces cerevisiae Y8-2 and *Pichia fermentans* Y1B-2 had the largest inhibition on the growth of *Streptomyces albus*, and their R $_{inhibition\ zone}$/R $_{colony}$ values reached 2.90 ± 0.06

and 2.82 ± 0.08, respectively. *S. cerevisiae* is one of the most important species in fermented foods and contributes to the production of alcohols and esters [47]. In winemaking, *S. cerevisiae* is widely used as a starter culture [48]. Moreover, inoculation with *S. cerevisiae* has been recommended to enhance flavor complexity and microbial community stability. For example, pomegranate wine fermentation with *S. cerevisiae* inoculation transformed the negative correlations into positive correlations among the fungal communities [48]. Inoculated *S. cerevisiae* can suppress wild microflora, determining the starter's ability to dominate the process [49]. *Pichia* is a very critical non-*Saccharomyces* and plays important roles in fermented food fermentation. *P. fermentans* is one of the most important species of the *Pichia* genus. A *P. fermentans* strain isolated from Chinese liquor increased the esters and fatty acids levels in mixed fermentation of wines [50]. Mixed fermentation with *P. fermentans* and *S. cerevisiae* enhanced the fruity and floral traits in wine [51]. *P. kudriavzevii* has an ecological function and can maintain the diversity of the yeast community and antagonize fungal blooms [52,53]. In addition, *P. kudriavzevii* can also decrease the level of some unsafe compounds, such as ethyl carbamate in *Baijiu* fermentation [54]. However, there have been few studies reporting use of *S. cerevisiae* and *Pichia* for control of *Streptomyces*-caused food or beverage contamination. Therefore, *Daqu* microbes have developed an intrinsic community structure during their natural evolution. Native functional microorganisms can efficiently inhibit the growth of *Streptomyces albus*, indicating that they are effective biocontrol agents to prevent *Streptomyces* contamination and geosmin production during *Daqu* fermentation.

3.2. Geosmin and Physicochemical Parameters Analysis during Daqu Fermentation

Figure 2A shows the geosmin contents in the early stage and late stage during *Daqu* fermentation. Compared to the control group, the inoculation of geosmin-inhibiting microorganisms (test group) significantly ($p < 0.05$) reduced the geosmin content in *Daqu*. In the early stage, the geosmin content in the test group was 4.71 ± 0.30 μg/kg, which was significantly ($p < 0.05$) lower than that in the control group (7.18 ± 0.13 μg/kg), thus being reduced by 34.40%. In the late stage, the geosmin content in the test group was 3.97 ± 0.78 μg/kg, and it was 8.86 ± 1.54 μg/kg in the control group. The geosmin content was thus reduced by 55.20% in the late stage of *Daqu* fermentation by the inoculation of geosmin-inhibiting microorganisms. These results revealed that these inoculation microorganisms quickly reside in the *Daqu* and inhibit geosmin production.

The physicochemical parameters of the rind and the core of *Daqu* are shown in Figure 2B,C and Table S2. In the rind of *Daqu*, the test group exhibited a higher liquefying power, saccharifying power, and esterifying power than the control group. By inoculation of geosmin-inhibiting microorganisms, the liquefying power increased from 0.71 g starch/g.h to 0.82 g starch/g.h; the saccharifying power increased from 0.55 g glucose/g.h to 0.74 g glucose/g.h; and the esterifying power increased from 0.1 g/50 g·7 d to 0.33 g/50 g·7 d. The liquefying power and saccharifying power in the test group was 13.50% and 25.7% higher than the control group, respectively. However, the fermenting power and acidity were decreased after inoculation of geosmin-inhibiting microorganisms. The fermenting power decreased from 0.46 g/0.5 g·72 h to 0.24 g/0.5 g·72 h, and the acidity decreased from 0.78 mmol/10 g to 0.68 mmol/10 g. In the core of the *Daqu*, the physicochemical parameters showed no significant differences between the test group and control group. Only the liquefying power decreased from 0.18 g starch/g.h to 0.13 g starch/g.h after the inoculation of geosmin-inhibiting microorganisms.

The enzymes in *Daqu* have pivotal effects on initiating the ethanol fermentation of Chinese liquor through hydrolyzing macromolecules. A higher liquefying power and saccharifying power in the test group indicated its greater starch hydrolysis capacity. During *Daqu* fermentation, the abundant microorganisms, such as *Bacillus*, *Lactobacillus*, and filamentous fungi, have the ability to secrete starch-degrading enzymes [55]. These results revealed that inoculation with geosmin-inhibiting microorganisms changed the microbial composition in the *Daqu*.

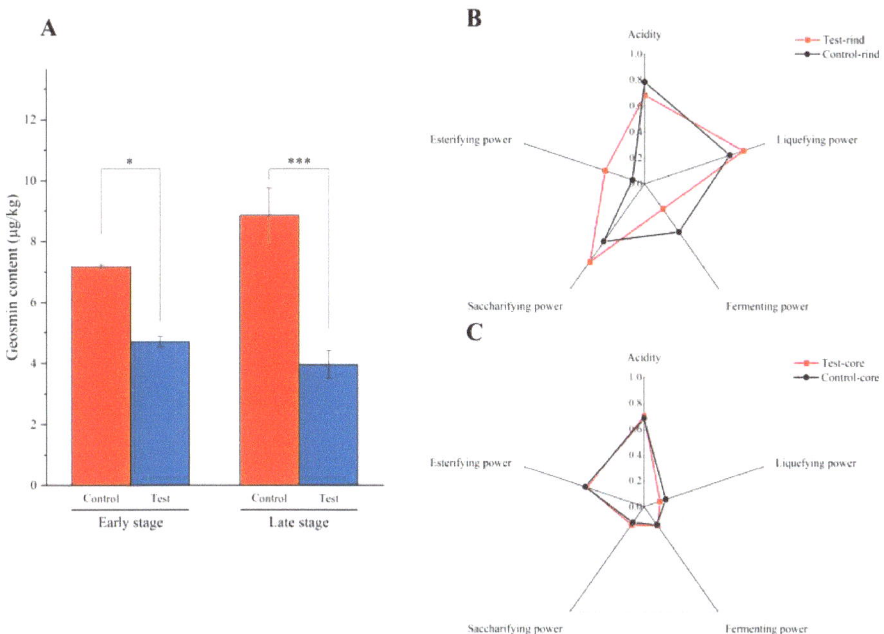

Figure 2. (**A**) Geosmin content in the early stage and late stage during *Daqu* incubation. *, $p < 0.05$; ***, $p < 0.001$. (**B**) The physicochemical characteristics in the rind of the *Daqu*. (**C**) The physicochemical characteristics in the core of the *Daqu*. The red line represents the test group; the black line represents the control group.

3.3. Microbial Community Structure and Interaction during Daqu Fermentation

qPCR was applied to detect the dynamics in the *Streptomyces albus* biomass during *Daqu* fermentation. Figure 3A shows the total biomass of *Streptomyces albus* in the early stage and late stage during *Daqu* fermentation. In the early stage, the biomass of *Streptomyces albus* was 3.02 ± 0.08 log10 spore count/g in the test group, which was significantly ($p < 0.001$) lower than that in the control group (5.51 ± 0.48 log10 spore count/g). *Streptomyces albus* was reduced by 45.20% in the early stage after inoculation with geosmin-inhibiting microorganisms. In the late stage, the biomass of *Streptomyces albus* in the test group was 3.06 ± 0.39 log10 spore count/g, whereas it was 5.62 ± 0.39 log10 spore count/g in the control group; thus, significantly ($p < 0.001$) decreased by 45.60%.

High-throughput sequencing was applied to investigate the microbial community structures during *Daqu* fermentation. A total of 1,065,436 high-quality reads from the V3–V4 region of 16S rRNA gene, and 917,050 high–quality reads from the ITS region were obtained from all 21 samples. For bacteria, we found an average of 50,735 reads per sample, ranging from 17,688 to 93,991 reads; and for fungi, an average of 43,669 reads per sample, ranging from 26,306 to 76,944 reads. The rarefaction curves of the microbial communities reached a saturation plateau, indicating that microbial communities were well represented at the sequencing depth.

Microbial alpha diversity was determined based on the Shannon index and Chao1 index (Tables S3 and S4). Bacterial diversity was significantly higher ($p < 0.05$) in the control group than that in the test group (Figure 3B,C). Fungal diversity showed no significant difference between two groups ($p > 0.05$). At the genus level, a total of 135 bacterial genera and 40 fungal genera were identified from all samples (Tables S5 and S6). Only 12 bacterial genera and 7 fungal genera were abundant (with over 1% average abundance).

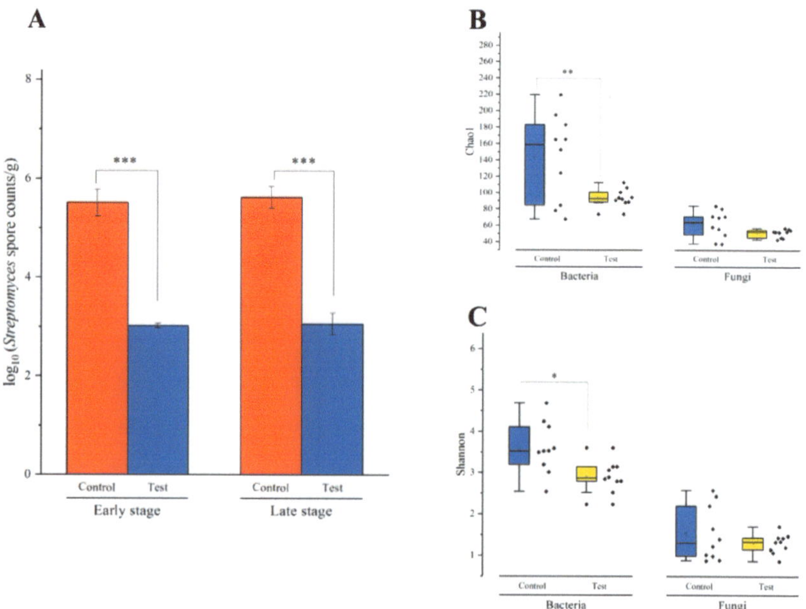

Figure 3. (**A**) The biomass of *Streptomyces albus* during the *Daqu* incubation process. (**B**) Chao1 index of microbial communities in the *Daqu*. Each point represented the Chao1 value. (**C**) Shannon index of microbial communities in the *Daqu*. Each point represented the Shannon value. *, $p < 0.05$; **; $p < 0.01$; ***, $p < 0.001$.

Principal component analysis (PCA) of the bacterial community was carried out based on weighted UniFrac distances (Figure 4B). The two axes explained 34.36% and 53.87% of the variance in bacterial community differentiation, respectively. Results showed that the bacterial community structure in the test group and the control group of *Daqu* in the first five days of fermentation were close and differentiated from the other samples. From the 10th day, the bacterial communities of the control group and the test group began to show greater differences. It can be seen that the bacterial community mainly play an important role in the 5 days to the 20 days during *Daqu* fermentation. *Lactobacillus* and *Weissella* were the most abundant (with over 20% average abundance) genera in *Daqu*. In addition, *Staphylococcus, Acetobacter, Pediococcus, Sebaldella, Acinetobacter, Streptomyces,* and *Pantoea* were also abundant (with over 1% but under 20% average abundance) in the *Daqu*. During fermentation, *Lactobacillus* and *Weissella* in the rind and the core of the *Daqu* were the most abundant at the beginning of fermentation (in the first 3 days of fermentation). After fermenting for 3 days, *Lactobacillus* decreased in the test group, and was lower in the test group than that in the control group at the end of fermentation. In the rind of *Daqu*, the relative abundance of *Lactobacillus, Staphylococcus, Pediococcus, Acetobacter, Streptomyces,* and *Bacillus* were lower in the test group than those in the control group; whereas the relative abundance of *Weissella* and *Acinetobacter* were higher in the test group than those in the control group at the end of fermentation. In the core of the *Daqu*, the relative abundance of *Lactobacillus, Weissella, Streptomyces, Acetobacter, Pediococcus, Sebaldella, Pantoea, Apibacter, Bacillus,* and *Lactococcus* were lower in the test group than those in the control group; whereas the relative abundance of *Staphylococcus, Acinetobacter, Brevibacterium* and *Enterococcus* were higher in the test group than those in the control group at the end of fermentation (Figure 4A).

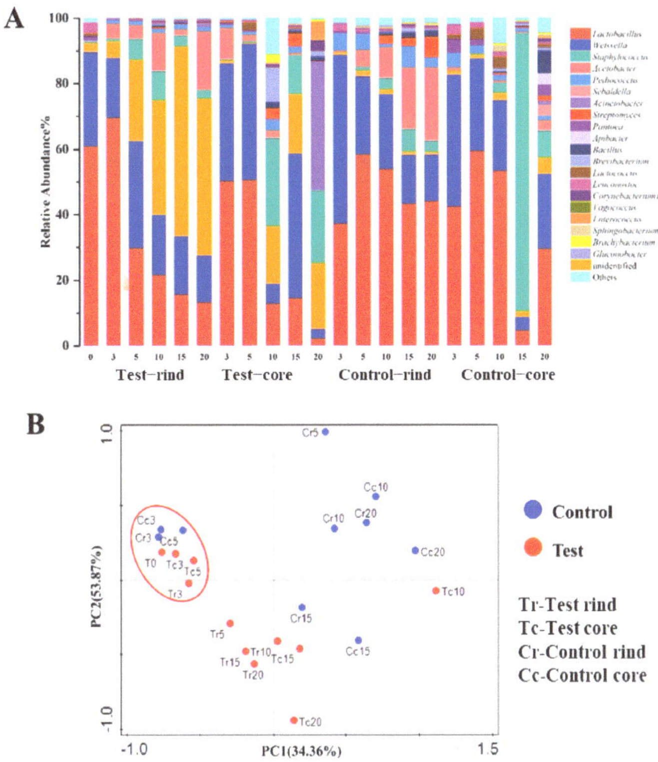

Figure 4. (**A**) The relative abundance of the dominant bacterial communities in *Daqu* at the genus level. (**B**) Principal component analysis of the bacterial communities based on weighted UniFrac distances. The red circle represents clustered results. Tr represents the rind of the test group; Tc represents the core of test group; Cr represents the rind of the control group; Cc represents the core of the control group.

For fungi, the two axes explained 36.51% and 50.49% of the variance in the fungal community, according to the PCA analysis (Figure 5B). Results showed that the fungal community structure of the core of the *Daqu* in both the test group and control group in the first five days differentiated from the other samples, and the fungal community showed obvious clustering in the next 15 days. Combined with the relative abundance analysis of *Pichia* and *Saccharomyces*, it can be seen that the fungal community plays an important role in the first five days of *Daqu* incubation. What is more, seven fungal genera were abundant (with over 1% average abundance) in the *Daqu* samples. *Pichia* and *Thermoascus* were the most abundant (with over 25% average abundance) genera in *Daqu*. Meanwhile, *Lichtheimia*, *Aspergillus*, *Rhizopus*, and *Saccharomyces* were also abundant (with over 1% but under 20% average abundance) in *Daqu*. During fermentation, *Pichia* dominated in the rind of the *Daqu* and *Thermoascus* dominated in the core of the *Daqu*. In the rind of the *Daqu*, the relative abundance of *Pichia* and *Candida* in the test group was lower than in the control group, and the relative abundance of *Thermoascus*, *Lichtheimia*, *Rhizopus*, *Saccharomyces*, and *Aspergillus* was higher in the test group than those in the control group at the end of fermentation. In the core of the *Daqu*, it can be clearly seen that the domination of *Pichia* was gradually replaced by *Thermoascus* as the *Daqu* fermentation proceeded. *Pichia* decreased and *Thermoascus* increased both in the test and control group during fermentation. On the 5th day of *Daqu* fermentation, the relative abundance of *Pichia* was 81.5% in the test rind of the *Daqu*, 73.9% in the control rind of the *Daqu*, 26.7% in the test core of the *Daqu*, and

19.9% in the control core of the *Daqu*. The relative abundance of *Saccharomyces* was 1.4% in the test rind of the *Daqu*, 0.7% in the control rind of the *Daqu*, 0.9% in the test core of the *Daqu*, and 0.4% in the control core of the *Daqu* (Figure 5A).

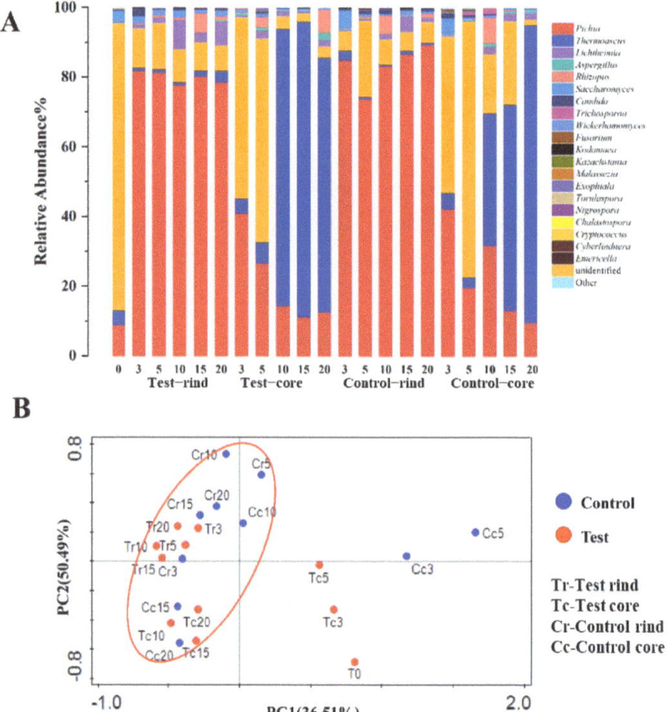

Figure 5. (**A**) The relative abundance of the dominant fungal communities in *Daqu* at the genus level. (**B**) Principal component analysis of the fungal communities based on weighted UniFrac distances. The red circle represents clustered results. Tr represents the rind of the test group; Tc represents the core of test group; Cr represents the rind of the control group; Cc represents the core of the control group.

To illuminate the effect of inoculation of geosmin-inhibiting microorganisms on microbial interaction during *Daqu* fermentation, the correlations among the microbes were explored based on Spearman's rank correlations ($|\rho| > 0.5$ and $p < 0.05$). Figure 6A shows the correlation among microorganisms in the control group and Figure 6B shows the correlation among microorganisms in the test group. We put the *Streptomyces* on the left; the first column to the right of *Streptomyces* represents microorganisms directly related to *Streptomyces*, and the second column to the right of *Streptomyces* represents microorganisms indirectly related to *Streptomyces*. In the control group, 38 pairs of significant and robust correlations, including 20 pairs of positive correlations and 18 pairs of negative correlations, were identified from 19 genera, with an average degree (AD) of 4.75 (Figure 6A). For the network, the average path length (APL) between the nodes was 2.72 edges, with a network diameter (ND; the longest distance between nodes) of four edges, and an average clustering coefficient of 0.685. In the test group, 38 pairs of significant and robust correlations, including 21 pairs of positive correlations and 17 pairs of negative correlations, were identified from 19 genera, with an average degree (AD) of 4.00 (Figure 6B). For the network, the average path length (APL) between nodes was 2.15 edges, with a network diameter (ND) of four edges, and an average clustering coefficient of 0.609.

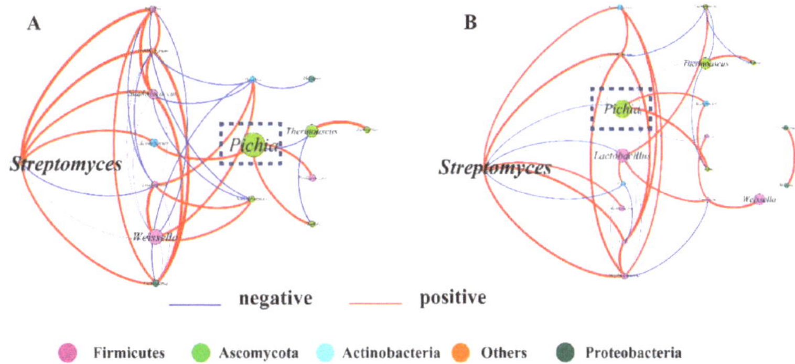

Figure 6. Relationships among microbial communities in the control group (**A**) and test group (**B**). A connection represents a significant ($p < 0.05$) and strong (Spearman's $|\rho| > 0.6$) correlation. The size of each node is in proportion to the relative abundance of each microorganism, and the nodes are colored by phylum (Firmicutes, purple; Ascomycota, green; Actinobacteria, blue; others, orange; Proteobacteria, dark green.). The Spearman's correlation coefficient (ρ) is in proportion to the thickness of each connection (edge) between two nodes. The color of the edges represents the positive (red) or negative (blue) relationship. The *Pichia* circled in the box is the main microorganism that inhibits the growth of *Streptomyces*.

In the test group, the positive interactions between the microbes were significantly enhanced. The correlation between *Pichia* and *Streptomyces* changed from the original indirect promotion to a direct inhibition effect. *Pichia* had a significant negative correlation with *Streptomyces* after inoculation of geosmin-inhibiting microorganisms. These results revealed that inoculation changed the correlation among microbes. In a complex multi-species fermentation system, microbial communities are shaped and stabilized by the interactions between microorganisms, and these interactions greatly determine the composition, dynamics, and functionality of the microbial community [56]. Therefore, our study demonstrated that a shift in microbial interaction played a pivotal role in inhibiting the growth of *Streptomyces* and geosmin production after inoculation. This was also found in a previous study where *S. cerevisiae* inoculation transformed the microbial negative correlations into positive correlations during pomegranate wine fermentation [48]. Therefore, illuminating the microbial interactions is necessary to evaluate the effect of microbial inoculation.

3.4. Volatile Compounds Analysis during Daqu Fermentation

The flavor compounds of the *Daqu* contribute significantly to the flavor of Chinese liquor [4,17], so it is important to maintain the flavor of the *Daqu* stable or enhance the volatile compounds by inoculation with geosmin-inhibiting microorganisms. A total of 30 volatile compounds were determined from the *Daqu* samples, including 8 alcohols, 3 aldehydes, 10 esters, 4 acids, and 5 aromatic compounds (Dataset 2). PLS-DA was performed to analyze the variability in the volatile compounds between the different groups. When two compounds were calculated, the cumulative R^2X, R^2Y, and Q^2 values were 0.666, 0.882, and 0.693, respectively. PLS-DA analysis showed that the volatile compounds in the test group was clearly separated from the control group. Inoculation of geosmin-inhibiting microorganisms changed the volatile compounds in the both the *Daqu* rind and core. To determine which compounds caused segregation, a VIP score (variable importance for the projection) of >1.0 was used. Fourteen compounds were identified for metabolic differences between the two groups, including 4 alcohols (2-methyl-1-propanol, 1-hexanol, 3-methyl-1-pentanol, and heptanol), 4 esters (hexyl acetate, ethyl-2-hydroxyhexanoate, ethyl acetate, and ethyl-2-methylbutyrate), 1 acid (acetic acid), 2 aldehydes (hexanal and 3-methyl-butanal), and 3 aromatics (butylated hydroxytoluene, ethyl phenylacetate, and phenethyl alcohol), as shown in Figure 7. 1-Hexanol and ethyl acetate showed the largest difference between two groups. Inoculation significantly ($p < 0.05$) increased the contents of 1-hexanol, from 0.173 ± 0.104 mg/kg to 0.955 ± 0.008 mg/kg, and ethyl acetate, from 0.262 ± 0.013 mg/kg to 1.338 ± 0.139 mg/kg. 1-Hexanol is a very important compound in Chinese liquor. A previous study found that 1-hexanol was strongly and positively correlated with *Candida* [57]. Ethyl acetate is one of the most abundant esters in Chinese liquor and provides pear-like and banana-like aromas [58]. Ethyl acetate greatly influences the style and quality of the liquor. Previous studies have found that yeasts are the main contributors to ethyl acetate production. For example, co-culture of *S. cerevisiae* and *Wickerhamomyces anomalus* produced a higher level of ethyl acetate during liquor fermentation [59]. A higher level of ethyl acetate in the test group may be attributable to the increase in *Saccharomyces* (Figure 5A).

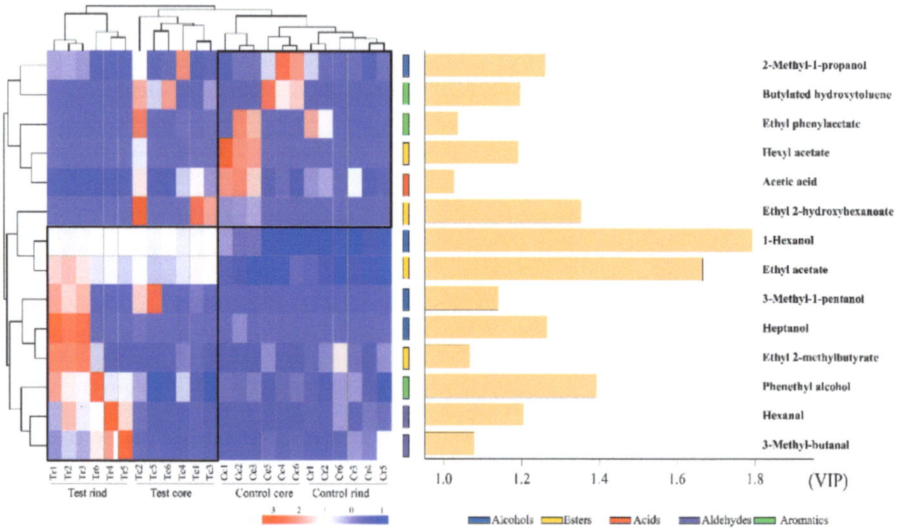

Figure 7. Heatmap of the differential volatile compounds (VIP > 1.0) between the test group and control group. The color scale represents the scaled abundance of each compound, indicated as the Z-score, with red and blue indicating high and low abundances, respectively. The histogram shows the VIP value.

4. Conclusions

Geosmin contamination has always been a serious problem in freshwater [60–62]. However, in recent years, the Chinese liquor industry also has been plagued by geosmin contamination due to its significantly negative effect on the liquor's flavor and quality [2,4]. A musty off-flavor may arouse psychosomatic effects, such as headaches, stress, or stomach upsets [61,63]. In this study, *Streptomyces albus* F5A-1 was selected as a model strain to represent the *Streptomyces* genus, which was the geosmin producer during Chinese liquor production. In turn, a medium-temperature *Daqu* fermentation system was employed as a model to identify the functional strains that can inhibit the growth of *Streptomyces* and thus geosmin production. By combining in vitro antibacterial experiments and in situ inoculation, we obtained six native functional strains (yeasts and *Bacillus* species) as biocontrol agents and achieved the goal of geosmin reduction. In addition, we revealed that microbial interaction significantly affected the growth of *Streptomyces* and thus production of geosmin. This study highlights the potential of biocontrol for controlling off-flavor chemicals at trace levels in complex fermentation systems, and also provides an effective and eco-friendly strategy for controlling geosmin contamination in medium-temperature *Daqu* via regulating microbial communities.

Supplementary Materials: The following supporting information can be downloaded at: https://www.mdpi.com/article/10.3390/fermentation8110588/s1, Figure S1: The schematic diagram of the experimental design and *Daqu* sample collection; Table S1: The geosmin contents in the early stage and the late stage during *Daqu* incubation; Table S2: The physicochemical parameters in the *Daqu*; Table S3: Chao1 index values for fungal and bacterial communities in *Daqu*; Table S4: Shannon index values for fungal and bacterial communities in *Daqu*; Table S5: Microbial community structure during *Daqu* fermentation; Table S6: Table sThe contents of volatile compounds between test group and control group.

Author Contributions: Conceptualization, H.D. and J.W.; methodology, X.Z.; software, J.W.; validation, H.D., X.Z. and J.W.; formal analysis, H.D.; investigation, H.D. and Y.X.; resources, H.D. and Y.X.; data curation, H.D.; writing—original draft preparation, H.D. and J.W.; writing—review and editing, J.W.; visualization, J.W.; supervision, H.D. and Y.X.; project administration, Y.X.; funding acquisition, Y.X. All authors have read and agreed to the published version of the manuscript.

Funding: This work was supported by the National Natural Science Foundation of China (NSFC) (grant 32172176) and Natural Science Foundation of Jiangsu Province of China (grant BK20201341).

Institutional Review Board Statement: Not applicable.

Informed Consent Statement: Not applicable.

Data Availability Statement: Data is presented in the manuscript.

Conflicts of Interest: We declare that we have no known competing financial interest or personal relationships that could have appeared to influence the work reported in this paper.

References

1. Du, H.; Xu, Y. Determination of the microbial origin of geosmin in Chinese liquor. *J. Agric. Food Chem.* **2012**, *60*, 2288–2292. [CrossRef] [PubMed]
2. Zhi, Y.; Wu, Q.; Du, H.; Xu, Y. Biocontrol of geosmin-producing *Streptomyces* spp. by two *Bacillus* strains from Chinese liquor. *Int. J. Food Microbiol.* **2016**, *231*, 1–9. [CrossRef] [PubMed]
3. Gerber, N.N.; Lechevalier, H.A. Geosmin, an earthly-smelling substance isolated from actinomycetes. *Appl. Microbiol.* **1965**, *13*, 935–938. [CrossRef]
4. Du, H.; Fan, W.; Xu, Y. Characterization of geosmin as source of earthy odor in different aroma type Chinese liquors. *J. Agric. Food Chem.* **2011**, *59*, 8331–8337. [CrossRef] [PubMed]
5. Du, H.; Lu, H.; Xu, Y.; Du, X. Community of environmental *Streptomyces* related to geosmin development in Chinese liquors. *J. Agric. Food Chem.* **2013**, *61*, 1343–1348. [CrossRef]
6. McCrummen, S.T.; Wang, Y.F.; Hanson, T.R.; Bott, L.; Liu, S.Y. Culture environment and the odorous volatile compounds present in pond-raised channel catfish (*Ictalurus punctatus*). *Aquacult. Int.* **2018**, *26*, 685–694. [CrossRef]
7. Saadoun, I.M.K.; Schrader, K.K.; Blevins, W.T. Environmental and nutritional factors affecting geosmin synthesis by *Anabaena* sp. *Water Res.* **2001**, *35*, 1209–1218. [CrossRef]
8. Juttner, F.; Watson, S.B. Biochemical and ecological control of geosmin and 2-methylisoborneol in source waters. *Appl. Environ. Microbiol.* **2007**, *73*, 4395–4406. [CrossRef]
9. la Guerche, S.; de Senneville, L.; Blancard, D.; Darriet, P. Impact of the *Botrytis cinerea* strain and metabolism on (-)-geosmin production by *Penicillium expansum* in grape juice. *Anton. Leeuw. Int. J. G.* **2007**, *92*, 331–341. [CrossRef]
10. Morales-Valle, H.; Silva, L.C.; Oliveira, J.M.; Venancio, A.; Lima, N. Microextraction and gas chromatography/mass spectrometry for improved analysis of geosmin and other fungal "off" volatiles in grape juice. *J. Microbiol. Meth.* **2010**, *83*, 48–52. [CrossRef]
11. Omur-Ozbek, P.; Little, J.C.; Dietrich, A.M. Ability of humans to smell geosmin, 2-MIB and nonadienal in indoor air when using contaminated drinking water. *Water Sci. Technol.* **2007**, *55*, 249–256. [CrossRef] [PubMed]
12. Liato, V.; Aider, M. Geosmin as a source of the earthy-musty smell in fruits, vegetables and water: Origins, impact on foods and water, and review of the removing techniques. *Chemosphere* **2017**, *181*, 9–18. [CrossRef]
13. Du, H.; Lu, H.; Xu, Y. Influence of geosmin-producing *Streptomyces* on the growth and volatile metabolites of yeasts during Chinese liquor fermentation. *J. Agric. Food Chem.* **2015**, *63*, 290–296. [CrossRef] [PubMed]
14. Jin, G.Y.; Zhu, Y.; Xu, Y. Mystery behind Chinese liquor fermentation. *Trends Food Sci. Technol.* **2017**, *63*, 18–28. [CrossRef]
15. Song, J.; Tang, H.; Liang, H.; Luo, L.; Lin, W. Effect of bioaugmentation on biochemical characterisation and microbial communities in *Daqu* using *Bacillus*, *Saccharomycopsis* and *Absidia*. *Int. J. Food Sci. Technol.* **2019**, *54*, 2639–2651. [CrossRef]
16. Du, H.; Wang, X.; Zhang, Y.; Xu, Y. Exploring the impacts of raw materials and environments on the microbiota in Chinese *Daqu* starter. *Int. J. Food Microbiol.* **2019**, *297*, 32–40. [CrossRef]
17. He, G.; Dong, Y.; Huang, J.; Wang, X.; Zhang, S.; Wu, C.; Jin, Y.; Zhou, R. Alteration of microbial community for improving flavor character of *Daqu* by inoculation with *Bacillus velezensis* and *Bacillus subtilis*. *LWT-Food Sci.* **2019**, *111*, 1–8. [CrossRef]

18. Zoschke, K.; Engel, C.; Boernick, H.; Worch, E. Adsorption of geosmin and 2-methylisoborneol onto powdered activated carbon at non-equilibrium conditions: Influence of NOM and process modelling. *Water Res.* **2011**, *45*, 4544–4550. [CrossRef]

19. Lin, T.F.; Chang, D.W.; Lien, S.K.; Tseng, Y.S.; Chiu, Y.T.; Wang, Y.S. Effect of chlorination on the cell integrity of two noxious cyanobacteria and their releases of odorants. *J. Water Supply Res. T.* **2009**, *58*, 539–551. [CrossRef]

20. Worley, J.L.; Dietrich, A.M.; Hoehn, R.C. Dechlorination techniques to improve sensory odor testing of geosmin and 2-MIB. *J. AM. Water Works Ass.* **2003**, *95*, 109–117. [CrossRef]

21. Liu, S.; Tang, L.; Wu, M.; Fu, H.; Xu, J.; Chen, W.; Ma, F. Parameters influencing elimination of geosmin and 2-methylisoborneol by K_2FeO_4. *Sep. Purif. Technol.* **2017**, *182*, 128–133. [CrossRef]

22. Behr, M.; Cocco, E.; Lenouvel, A.; Guignard, C.; Evers, D. Earthy and fresh mushroom off-flavors in wine: Optimized remedial treatments. *Am. J. Enol. Viticult.* **2013**, *64*, 545–549. [CrossRef]

23. Fiori, S.; Urgeghe, P.P.; Hammami, W.; Razzu, S.; Jaoua, S.; Migheli, Q. Biocontrol activity of four non- and low-fermenting yeast strains against *Aspergillus carbonarius* and their ability to remove ochratoxin A from grape juice. *Int. J. Food Microbiol.* **2014**, *189*, 45–50. [CrossRef] [PubMed]

24. Kong, Q.; Shan, S.H.; Liu, Q.Z.; Wang, X.D.; Yu, F.T. Biocontrol of *Aspergillus flavus* on peanut kernels by use of a strain of marine *Bacillus megaterium*. *Int. J. Food Microbiol.* **2010**, *139*, 31–35. [CrossRef] [PubMed]

25. Branco, P.; Sabir, F.; Diniz, M.; Carvalho, L.; Albergaria, H.; Prista, C. Biocontrol of *Brettanomyces/Dekkera bruxellensis* in alcoholic fermentations using saccharomycin-overproducing *Saccharomyces cerevisiae* strains. *Appl. Microbiol. Biot.* **2019**, *103*, 3073–3083. [CrossRef] [PubMed]

26. Wang, X.; Du, H.; Zhang, Y.; Xu, Y. Environmental microbiota drives microbial succession and metabolic profiles during Chinese liquor fermentation. *Appl. Environ. Microbiol.* **2018**, *84*, e02369-17. [CrossRef]

27. White, T.J.; Bruns, T.; Lee, S.; Taylor, J. Amplification and direct sequencing of fungal ribosomal RNA genes for phylogenetics. *PCR Protoc. A Guide Methods Appl.* **1990**, *18*, 315–322.

28. Rochelle, P.A.; Fry, J.C.; John Parkes, R.; Weightman, A.J. DNA extraction for 16S rRNA gene analysis to determine genetic diversity in deep sediment communities. *FEMS Microbiol. Lett.* **1992**, *100*, 59–65. [CrossRef]

29. Yang, X.; Jiao, R.; Zhu, X.; Zhao, S.; Liao, G.; Yu, J.; Wang, D. Profiling and characterization of odorous volatile compounds from the industrial fermentation of erythromycin. *Environ. Pollut.* **2019**, *255*, 113130. [CrossRef]

30. Soergel, D.A.; Dey, N.; Knight, R.; Brenner, S.E. Selection of primers for optimal taxonomic classification of environmental 16S rRNA gene sequences. *ISME J.* **2012**, *6*, 1440–1444. [CrossRef]

31. Toju, H.; Tanabe, A.S.; Yamamoto, S.; Sato, H. High-coverage ITS primers for the DNA-based identification of *Ascomycetes* and *Basidiomycetes* in Environmental Samples. *PLoS ONE* **2012**, *7*, e40863. [CrossRef] [PubMed]

32. Caporaso, J.G.; Kuczynski, J.; Stombaugh, J.; Bittinger, K.; Bushman, F.D.; Costello, E.K.; Fierer, N.; Pena, A.G.; Goodrich, J.K.; Gordon, J.I. QIIME allows analysis of high-throughput community sequencing data. *Nat. Methods* **2010**, *7*, 335–336. [CrossRef] [PubMed]

33. Edgar, R.C. Search and clustering orders of magnitude faster than BLAST. *Bioinformatics* **2010**, *26*, 2460–2461. [CrossRef] [PubMed]

34. Liu, Z.; Lozupone, C.; Hamady, M.; Bushman, F.D.; Knight, R. Short pyrosequencing reads suffice for accurate microbial community analysis. *Nucleic Acids Res.* **2007**, *35*, e120. [CrossRef]

35. Bolger, A.M.; Lohse, M.; Usadel, B. Trimmomatic: A flexible trimmer for Illumina sequence data. *Bioinformatics* **2014**, *30*, 2114–2120. [CrossRef]

36. Johnston, D.; Earley, B.; Cormican, P.; Murray, G.; Kenny, D.A.; Waters, S.M.; McGee, M.; Kelly, A.K.; McCabe, M.S. Illumina MiSeq 16S amplicon sequence analysis of bovine respiratory disease associated bacteria in lung and mediastinal lymph node tissue. *BMC Vet. Res.* **2017**, *13*, 118. [CrossRef]

37. Ji, M.; Du, H.; Xu, Y. Structural and metabolic performance of p-cresol producing microbiota in different carbon sources. *Food Res. Int.* **2020**, *132*, 109049. [CrossRef]

38. Hoffman, C.S.; Winston, F. A ten-minute DNA preparation from yeast efficiently releases autonomous plasmids for transformation of *Escherichia coli*. *Gene* **1987**, *57*, 267–272. [CrossRef]

39. Gao, W.J.; Fan, W.L.; Xu, Y. Characterization of the key odorants in light aroma type Chinese liquor by gas chromatography-olfactometry, quantitative measurements, aroma recombination, and omission studies. *J. Agric. Food. Chem.* **2014**, *62*, 5796–5804. [CrossRef]

40. Li, P.; Lin, W.; Liu, X.; Wang, X.; Luo, L. Environmental factors affecting microbiota dynamics during traditional solid-state fermentation of Chinese *Daqu* starter. *Front. Microbiol.* **2016**, *7*, 1237. [CrossRef]

41. Barberan, A.; Bates, S.T.; Casamayor, E.O.; Fierer, N. Using network analysis to explore co-occurrence patterns in soil microbial communities. *ISME J.* **2012**, *6*, 343–351. [CrossRef] [PubMed]

42. Zhang, H.; Wang, L.; Wang, H.; Yang, F.; Chen, L.; Hao, F.; Lv, X.; Du, H.; Xu, Y. Effects of initial temperature on microbial community succession rate and volatile flavors during *Baijiu* fermentation process. *Food Res. Int.* **2020**, *141*, 109887. [CrossRef] [PubMed]

43. Bastian, M.; Heymann, S.; Jacomy, M. Gephi: An open source software for exploring and manipulating networks. *ICWSM* **2009**, *8*, 361–362.

44. Wang, X.; Du, H.; Xu, Y. Source tracking of prokaryotic communities in fermented grain of Chinese strong-flavor liquor. *Int. J. Food Microbiol.* **2017**, *244*, 27–35. [CrossRef]

45. Wei, J.; Du, H.; Zhang, H.; Nie, Y.; Xu, Y. Mannitol and erythritol reduce the ethanol yield during Chinese *Baijiu* production. *Int. J. Food Microbiol.* **2020**, *337*, 108933. [CrossRef]
46. Dimkic, I.; Zivkovic, S.; Beric, T.; Ivanovic, Z.; Gavrilovic, V.; Stankovic, S.; Fira, D. Characterization and evaluation of two *Bacillus* strains, SS-12.6 and SS-13.1, as potential agents for the control of phytopathogenic bacteria and fungi. *Biol. Control* **2013**, *65*, 312–321. [CrossRef]
47. Wu, Q.; Chen, L.; Xu, Y. Yeast community associated with the solid state fermentation of traditional Chinese Maotai-flavor liquor. *Int. J. Food Microbiol.* **2013**, *166*, 323–330. [CrossRef]
48. Wang, X.S.; Ren, X.D.; Shao, Q.Q.; Peng, X.; Zou, W.J.; Sun, Z.G.; Zhang, L.H.; Li, H.H. Transformation of microbial negative correlations into positive correlations by *Saccharomyces cerevisiae* inoculation during pomegranate wine fermentation. *Appl. Environ. Microbiol.* **2020**, *86*, 12. [CrossRef]
49. Liu, W.; Li, H.; Jiang, D.; Zhang, Y.; Zhang, S.; Sun, S. Effect of *Saccharomyces cerevisiae*, *Torulaspora delbrueckii* and malolactic fermentation on fermentation kinetics and sensory property of black raspberry wines. *Food Microbiol.* **2020**, *91*, 103551. [CrossRef]
50. Kong, C.L.; Li, A.H.; Su, J.; Wang, X.C.; Chen, C.Q.; Tao, Y.S. Flavor modification of dry red wine from Chinese spine grape by mixed fermentation with *Pichia fermentans* and *S. cerevisiae*. *LWT-Food Sci. Technol.* **2019**, *109*, 83–92. [CrossRef]
51. Ma, D.; Yan, X.; Wang, Q.; Zhang, Y.; Tao, Y. Performance of selected *P. fermentans* and its excellular enzyme in co-inoculation with *S. cerevisiae* for wine aroma enhancement. *LWT-Food Sci. Technol.* **2017**, *86*, 361–370. [CrossRef]
52. Zhang, H.; Du, H.; Xu, Y. Volatile organic compounds mediated antifungal activity of *Pichia* and its effect on the metabolic profiles of fermentation communities. *Appl. Environ. Microbiol.* **2021**, *87*, e02992-20. [CrossRef] [PubMed]
53. Zhang, H.; Wang, L.; Tan, Y.; Wang, H.; Yang, F.; Chen, L.; Hao, F.; Lv, X.; Du, H.; Xu, Y. Effect of *Pichia* on shaping the fermentation microbial community of sauce-flavor *Baijiu*. *Int J Food Microbiol.* **2021**, *336*, 108898. [CrossRef] [PubMed]
54. Du, H.; Song, Z.W.; Xu, Y. Ethyl carbamate formation regulated by lactic acid bacteria and nonconventional yeasts in solid-state fermentation of Chinese Moutai-flavor liquor. *J. Agric. Food Chem.* **2018**, *66*, 387–392. [CrossRef] [PubMed]
55. Shi, W.; Chai, L.J.; Fang, G.Y.; Mei, J.L.; Lu, Z.M.; Zhang, X.J.; Xiao, C.; Wang, S.T.; Shen, C.H.; Shi, J.S.; et al. Spatial heterogeneity of the microbiome and metabolome profiles of high-temperature Daqu in the same workshop. *Food Res. Int.* **2022**, *156*, 111298. [CrossRef]
56. Lino, F.S.D.; Bajic, D.; Vila, J.C.C.; Sanchez, A.; Sommer, M.O.A. Complex yeast-bacteria interactions affect the yield of industrial ethanol fermentation. *Nat. Commun.* **2021**, *12*, 1498. [CrossRef]
57. He, G.; Huang, J.; Zhou, R.; Wu, C.; Jin, Y. Effect of fortified *Daqu* on the microbial community and flavor in Chinese strong-flavor liquor brewing process. *Front. Microbiol.* **2019**, *10*, 56. [CrossRef]
58. Fan, W.; Shen, H.; Xu, Y. Quantification of volatile compounds in Chinese soy sauce aroma type liquor by stir bar sorptive extraction and gas chromatography-mass spectrometry. *J. Sci. Food Agric.* **2011**, *91*, 1187–1198. [CrossRef]
59. Fan, G.; Teng, C.; Xu, D.; Fu, Z.; Minhazul, K.; Wu, Q.; Liu, P.; Yang, R.; Li, X. Enhanced production of ethyl acetate using co-culture of *Wickerhamomyces anomalus* and *Saccharomyces cerevisiae*. *J. Biosci. Bioeng.* **2019**, *128*, 564–570. [CrossRef]
60. Youn, S.J.; Kim, H.N.; Yu, S.J.; Byeon, M.S. Cyanobacterial occurrence and geosmin dynamics in Paldang Lake watershed, South Korea. *Water Environ. J.* **2020**, *34*, 634–643. [CrossRef]
61. Xue, Q.; Shimizu, K.; Sakharkar, M.K.; Utsumi, M.; Cao, G.; Li, M.; Zhang, Z.; Sugiura, N. Geosmin degradation by seasonal biofilm from a biological treatment facility. *Environ. Sci. Pollut. R.* **2012**, *19*, 700–707. [CrossRef] [PubMed]
62. Zhang, J.; Zhang, H.; Li, L.; Wang, Q.; Yu, J.; Chen, Y. Microbial community analysis and correlation with 2-methylisoborneol occurrence in landscape lakes of Beijing. *Environ. Res.* **2020**, *183*, 109217. [CrossRef] [PubMed]
63. Young, W.F.; Horth, H.; Crane, R.; Ogden, T.; Arnott, M. Taste and odour threshold concentrations of potential potable water contaminants. *Water Res.* **1996**, *30*, 331–340. [CrossRef]

fermentation

Article

Comparison of *Trichoderma longibrachiatum* Xyloglucanase Production Using Tamarind (*Tamarindus indica*) and Jatoba (*Hymenaea courbaril*) Seeds: Factorial Design and Immobilization on Ionic Supports

Alex Graça Contato [1], Ana Claudia Vici [2], Vanessa Elisa Pinheiro [1], Tássio Brito de Oliveira [2], Emanuelle Neiverth de Freitas [1], Guilherme Mauro Aranha [2], Almir Luiz Aparecido Valvassora Junior [2], Carem Gledes Vargas Rechia [3], Marcos Silveira Buckeridge [4] and Maria de Lourdes Teixeira de Moraes Polizeli [1,2,*]

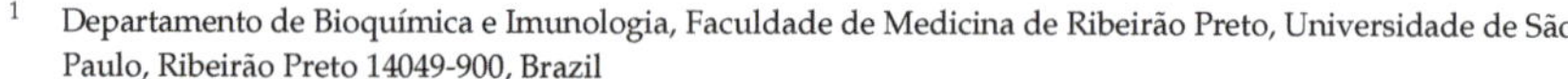

1 Departamento de Bioquímica e Imunologia, Faculdade de Medicina de Ribeirão Preto, Universidade de São Paulo, Ribeirão Preto 14049-900, Brazil
2 Departamento de Biologia, Faculdade de Filosofia, Ciências e Letras de Ribeirão Preto, Universidade de São Paulo, Ribeirão Preto 14050-901, Brazil
3 Departamento de Física e Química, Faculdade de Ciências Farmacêuticas de Ribeirão Preto, Universidade de São Paulo, Ribeirão Preto 14049-900, Brazil
4 Departamento de Botânica, Instituto de Biociências, Universidade de São Paulo, São Paulo 05508-090, Brazil
* Correspondence: polizeli@fflclrp.usp.br; Tel.: +55-(16)-3315-4680

Citation: Contato, A.G.; Vici, A.C.; Pinheiro, V.E.; Oliveira, T.B.d.; Freitas, E.N.d.; Aranha, G.M.; Valvassora Junior, A.L.A.; Rechia, C.G.V.; Buckeridge, M.S.; Polizeli, M.d.L.T.d.M. Comparison of *Trichoderma longibrachiatum* Xyloglucanase Production Using Tamarind (*Tamarindus indica*) and Jatoba (*Hymenaea courbaril*) Seeds: Factorial Design and Immobilization on Ionic Supports. *Fermentation* **2022**, *8*, 510. https://doi.org/10.3390/fermentation8100510

Academic Editors: Xian Zhang and Zhiming Rao

Received: 2 September 2022
Accepted: 27 September 2022
Published: 2 October 2022

Publisher's Note: MDPI stays neutral with regard to jurisdictional claims in published maps and institutional affiliations.

Abstract: Xyloglucan (XG) is the predominant hemicellulose in the primary cell wall of superior plants. It has a fundamental role in controlling the stretching and expansion of the plant cell wall. There are five types of enzymes known to cleave the linear chain of xyloglucan, and the most well-known is xyloglucanase (XEG). The immobilization process can be used to solve problems related to stability, besides the economic benefits brought by the possibility of its repeated use and recovery. Therefore, this study aims at the optimization of the xyloglucanase production of *Trichoderma longibrachiatum* using a central composite rotatable design (CCRD) with tamarind and jatoba seeds as carbon sources, as well as XEG immobilization on ionic supports, such as MANAE (monoamine-*N*-aminoethyl), DEAE (diethylaminoethyl)-cellulose, CM (carboxymethyl)-cellulose, and PEI (polyethyleneimine). High concentrations of carbon sources (1.705%), at a temperature of 30 °C and under agitation for 72 h, were the most favorable conditions for the XEG activity from *T. longibrachiatum* with respect to both carbon sources. However, the tamarind seeds showed 23.5% higher activity compared to the jatoba seeds. Therefore, this carbon source was chosen to continue the experiments. The scaling up from Erlenmeyer flasks to the bioreactor increased the XEG activity 1.27-fold (1.040 ± 0.088 U/mL). Regarding the biochemical characterization of the crude extract, the optimal temperature range was 50–55 °C, and the optimal pH was 5.0. Regarding the stabilities with respect to pH and temperature, XEG was not stable for prolonged periods, which was crucial to immobilizing it on ionic resins. XEG showed the best immobilization efficiency on CM-cellulose and DEAE-cellulose, with activities of 1.16 and 0.89 U/g of the derivative (enzyme plus support), respectively. This study describes, for the first time in the literature, the immobilization of a fungal xyloglucanase using these supports.

Keywords: xyloglucanase; Trichoderma longibrachiatum; Hymenaea courbaril; Tamarindus indica; enzyme immobilization

1. Introduction

Xyloglucan (XG) is the predominant hemicellulose in the primary cell wall of superior plants. This includes all the dicotyledonous and non-gramineous monocotyledonous plants [1]. XG is usually strongly associated with cellulose through hydrogen bonds, forming a tridimensional net of cellulose and xyloglucan [2]. It is probably the second most abundant polymer in nature, after cellulose [3]. It is highly soluble in water, which prevents it from forming crystalline microfibrils such as cellulose [4].

XG is composed of a linear chain of glucose residues linked by β-1,4 bonds, which contain up to 75% of their residues joined to α-D-xylopyranose at the position O-6. Although, some structures of xyloglucan may present a β-D-galactopyranose or an α-L-arabinofuranose linked to the residues of xylose, or even an α-L-fucopyranose connected to residues of galactose [5]. This structural diversity is decurrent from the varied species of plants [6].

Xyloglucan plays a fundamental role during the growth and cellular differentiation of plants, which is related to the control of the stretching and expansion of the cell wall. In some terrestrial plants, XG is the main reserve polysaccharide in the seeds [7]. However, the extraction of XG is not an uncomplicated process, mainly due to the strong hydrogen bonds between cellulose and xyloglucan. In addition, the covalent bonds formed with pectins and xylans make the process even more challenging. Usually, the extraction is performed through an alkaline solution combined with chaotropic agents [8].

The enzymes that cleave XG present great utilities in the degradation and conversion of lignocellulosic biomass, mainly due to its synergistic potential with cellulases at the degradation of the plant cell wall [9–11]. Furthermore, the enzymes that degrade and/or modify xyloglucans can be used in the production of new surfactants of oligoxyloglu-cans [12] in the pharmaceutical [13], textile, and paper industries [14]. Furthermore, the aqueous solutions of XG have a high viscosity, exhibiting Newtonian fluidity, unlike most polysaccharides [15]. That is why they are often used as food additives for increasing viscosity or as stabilizers [16].

Five identified types of enzymes can cleave the linear chain of xyloglucan: endo-β-D-1,4-glucanase, which is specific for xyloglucan, also known as endoxyloglucanase or simply xyloglucanase (XEG) (EC 3.2.1.151); exoxyloglucanase (EC 3.2.1.155); oligoxyloglucan β-glucosidase (EC 3.2.1.120); cellobiohydrolase, which is specific for oligoxyloglucans from the reducing extremities (EC 3.2.1.150); and xyloglucan endotransglucosylase (EC 2.4.1.207) [17].

The immobilization of xyloglucanases can solve problems caused by losses in the stability of free enzymes, which limit their use in large-scale applications. Besides the economic benefits of immobilization, the possibility of its repeated use decreases the general costs of production. This fact is the main reason why, over the last years, science has seen many attempts to obtain immobilized enzymes with high operational and storage stability [18–20].

Due to their constitution, tamarind (*Tamarindus indica* Linn.) and jatoba (*Hymenaea courbaril*) seeds have been utilized for the cultivation of microorganisms or as substrates to produce microbial enzymes [7,21]. In addition, these seeds are rich in xyloglucan, which corresponds to about 40% of their dry mass [22,23].

The ionic adsorption method of enzymatic immobilization is considered simple, cheap, efficient, and reversible [24,25]. In this context, elucidating the enzymatic properties can suggest the vast potential of xyloglucanases for biotechnological applications. Therefore, this study reports the first ever optimization of the production of xyloglucanase from *Trichoderma longibrachiatum* and its immobilization on ionic supports.

2. Material and Methods

2.1. Maintenance of the Fungus and Culture Medium

Trichoderma longibrachiatum LMBC 172 was maintained through the inoculation of its spores in potato dextrose agar medium (PDA) (Sigma-Aldrich, Saint Louis, MO, USA), keeping it through successive transfers in glass tubes containing the same medium, and incubating it at a temperature of 30 °C. Afterward, the test tubes were refrigerated for up to 30 days.

2.2. Submerged Cultivation of the Fungus

The acquirement of xyloglucanases was performed according to Contato et al. [21]. A solution with 10^6–10^7 spores/mL from the fungus was created. The fungus was grown in test tubes and suspended in sterile distilled water, and its spores were counted in a microscope through a Neubauer chamber. The suspension was inoculated into 125 mL Erlenmeyer flasks with 25 mL of Khanna medium (Khanna's salt solution [20×]: NH_4NO_3 (2.0 g), KH_2PO_4 (1.3 g), $MgSO_4.7H_2O$ (0.362 g), KCl (0.098 g), $ZnSO_4.H_2O$ (0.007 g), $MnSO_4.H_2O$ (0.0138 g), $Fe_2(SO_4)_3.6H_2O$ (0.0066 g), $CuSO_4.5H_2O$ (0.0062), distilled water (100 mL) (5.0 mL), yeast extract (0.1 g), carbon source (1.0 g), and distilled water 100 mL) [26]. The media were supplemented with 1% (w/v) of two different carbon sources: tamarind (*Tamarindus indica*, Fabaceae) or jatoba (*Hymenaea courbaril* L., Caesalpinioideae) seeds, which were previously pretreated (boiled in water, dried, and ground to 20 mesh) to secure the sanitary quality of the seeds and avoid the growth of other associated fungi. The Erlenmeyer flasks were incubated at 30 °C under static conditions or were shaken (at 120 rpm), both up to 96 h, with sampling every 24 h. The samples were filtered with a vacuum pump, and the filtrates were used as enzymatic extracts to determine extracellular enzymatic activities, performed in triplicate.

2.3. Optimization of Production through Factorial Design

The central composite rotatable design (CCRD) type "star" was used to evaluate the influence of different variables on the production of xyloglucanases and obtain the best conditions. The design consisted of assays with two independent variables (temperature and concentration of carbon source) in 4 levels (-1.41; -1; $+1$; $+1.41$). The effect of the independent variables was evaluated with respect to the variable response of xyloglucanase activity (mU/mL). The results were adjusted for a second-order polynomial equation. For the construction of the experimental design, the points shown in Table 1 were used, consisting of assays ranging from 24 to 96 h, with samplings every 24 h. The α axial points were chosen due to $\alpha = \sqrt{k}$, where k represents the number of evaluated factors. For $k = 2$, we have the points $\alpha = \pm1.41$ [27].

Table 1. Values used to construct the factorial design for *T. longibrachiatum* LMBC 172.

	-1.41	-1	0	1	1.41
Temperature (°C)	23.95	25.0	30.0	35.0	37.05
Carbon source (%)	0.3	0.5	1.0	1.5	1.7

2.4. Enzymatic Determination

The xyloglucanase activity was measured with xyloglucan (Megazyme®) as substrate [28]. The activity was determined by quantifying the number of reducing sugars using 3,5-dinitrosalicylic acid (DNS), according to the Miller method [29]. The assay mixture consisted of 25 µL of 1% (w/v) substrate solution in distilled water, 10 µL of 50 mM sodium acetate buffer, pH 5.0, and 15 µL of enzyme extract. A blank was carried out for each enzymatic assay by adding the enzyme extract after the assay time had elapsed with 50 µL of DNS. The absorbance was measured at 540 nm, and reducing sugars were quantified using a standard glucose curve (0–1 mg/mL). The detection limit was 3 µg of reducing sugar. The activity unit was defined as the amount of enzyme that releases one µmol of reducing sugar per minute per mL, and it was expressed as milliunit per mL (mU/mL).

2.5. Scaling for Bioreactor

The best cultivation condition was verified through a factorial design. Cultivation was carried out in a 5 L BioFlo 310-New Brunswick® bioreactor (Eppendorf, Hamburg, Germany), containing 3.0 L of workload, and under batch fermentation method to increase enzyme production. The same culture medium previously used for cultivation in Erlenmeyer flasks was sterilized in an autoclave at 121 °C for 30 min and aseptically placed in the reactor. The aeration of 1.0 vvm was performed by continuous injection of filtered, compressed air from a sterile filter. The dissolved oxygen concentration (DO) was controlled, employing a DO probe ranging from 0 to 100%. A volume of 3 mL of antifoam 204 (Sigma® A 6426) was added to the culture medium at the beginning of the process. The fermentation was carried out at 30 °C and for 72 h for *Trichoderma longibrachiatum*, using tamarind seeds as carbon source. Protein concentration and DO were monitored every 24 h through an appropriate collector for the bioreactor, allowing the samples to be taken safely.

The following Equation (1) was used to scale and determine that the agitation speed would be 280 rpm:

$$Ni.tm = 1.54 \, V/Di^3 \tag{1}$$

where:

Ni = stirring speed (1/s);
tm = mixing time constant;
V = volume of medium;
Di = impeller diameter.

2.6. Protein Quantification

The proteins obtained in extracellular solutions were quantified by Bradford's method [30], whereby 40 µL of Bradford's reagent was added to 160 µL of the enzymatic extracts and incubated for 5 min at room temperature. Finally, the absorbance was read on a spectrophotometer (Shimadzu, Kyoto, Japan) at 595 nm, using bovine albumin as standard. The results were expressed in µg of protein/mL.

2.7. Effects of Temperature and pH on the Enzymatic Activity

The effects of temperature and pH on the enzymatic activity were determined for the enzymes produced in the bioreactor. In order to determine the optimal temperature, different temperatures were applied (40 to 80 °C, with intervals of 5 °C between each). To determine thermostability, the enzymes were incubated for up to 24 h at 30 to 70 °C, with intervals of 10 °C between each. In addition, the influence of pH on the enzymatic activity was verified through the solubilization of the substrate in 100 mM of citrate-phosphate buffer (range of pH 3.0–7.0), glycine (range of pH 7.5–9.0), and borate (range of pH 9.5–10.0). All the assays were performed in triplicate with at least three independent experiments.

2.8. Pretreatment of the Crude Extract

Before the immobilization, the crude extracts were pretreated for the removal of the pigments by adsorption/desorption in activated charcoal. For each mL of crude extract, 5 mg of activated charcoal was added under agitation in an ice bath. After 10 min, the mixture was filtered with filter paper and centrifuged at 11.952 g for 10 min, thereby producing the clarified extract.

2.9. Enzymatic Immobilization through Ionic Adsorption

The supports CM (carboxymethyl)-cellulose, DEAE (diethylaminoethyl)-cellulose, MANAE (monoamine-*N*-aminoethyl), and polyethyleneimine (PEI) were used in the immobilization process. The supports were activated in Tris-HCl 10 mM pH 7.0 buffer. The same buffer was added to the extract. The immobilization was performed according to Monteiro et al. [31]. A volume of 10 mL of extract was added to 1 g of the support, which was incubated in a cold chamber under agitation for 48 h. After this timespan, the derivative (support + immobilized enzyme) was filtered and washed with buffer. The immobilization efficiency (*IE*) was defined as the ratio of the amount of enzyme bound to the support over the total amount of enzyme used, according to Equation (2):

$$IE \, (\%) = \frac{enzyme \; immobilized}{enzyme \; loaded} \times 100 \tag{2}$$

2.10. Statistical Analyses

The experimental results were obtained through the average ± standard deviation of three independent extractions. The programs used were GraphPadPrism® (GraphPad Software, San Diego, CA, USA) and Statistica® (StatSoft, Tulsa, OK, USA).

3. Results and Discussion

3.1. Optimization of Cultivation with Tamarind Seeds

The results of the use of these seeds as a carbon source to produce XEG can be seen in Table 2. From the data, it was possible to verify that the best enzymatic activity was obtained at the periods of 48 and 72 h with the usage of 1.705% tamarind seed extract at 30 °C (545.56 and 545.92 mU/mL, respectively) or the zero points (1% tamarind seed at 30 °C) for 96 h in the stationary condition. Regarding the cultivation under agitation, XEG was produced with a more considerable enzymatic activity at 72 h at 30 °C with 1.705% tamarind seed extract, obtaining a value of 817.28 mU/mL.

Table 2. XEG activity in the tests used for the construction of the experimental design for *T. longibrachiatum* cultivated with tamarind seeds.

	Temperature (°C)	Tamarind Seeds (%)	Stationary				Under Agitation			
			24 h	48 h	72 h	96 h	24 h	48 h	72 h	96 h
1	−1 (25)	−1 (0.5)	87.60	66.41	31.34	6.41	19.94	27.12	25.64	7.12
2	1 (35)	−1 (0.5)	27.78	102.82	114.67	138.88	7.48	12.86	135.68	168.80
3	−1 (25)	1 (1.5)	66.24	82.32	120.37	62.32	42.73	214.38	74.78	62.32
4	1 (35)	1(1.5)	55.20	280.97	289.28	415.58	31.52	326.33	451.19	526.69
5	−1.41 (22.95)	0 (1)	71.04	75.85	83.33	48.08	74.78	518.14	430.54	399.56
6	1.41 (37.05)	0 (1)	30.98	61.25	59.47	105.05	56.98	333.68	431.96	632.99
7	0 (30)	−1.41 (0.295)	nd	nd	190.52	411.66	nd	110.04	92.59	145.29
8	0 (30)	1.41 (1.705)	153.13	**545.56**	**545.92**	433.74	99.0	227.91	**817.28**	658.09
9	0 (30)	0 (1)	56.89	312.31	522.42	**537.02**	38.31	148.49	525.04	375.70
10	0 (30)	0 (1)	59.11	314.62	585.38	**628.40**	30.32	142.44	517.80	414.76
11	0 (30)	0 (1)	56.62	335.24	602.90	**633.17**	33.47	215.44	517.81	444.78

Values are expressed in mU/mL. nd = not detected. The maximum values found for each time are in bold.

The best timespan for enabling xyloglucanase activity was 72 h. The ANOVA and the F test (ratio of two variances) were, therefore, calculated only for this interval, where it was verified that the calculated F was 7.93-fold greater than the tabled F when the XEG was produced with tamarind seeds at stationary cultivation and 5.10-fold when produced in cultivation under agitation. Therefore, the null hypothesis was rejected, and there was a significant difference between the groups. According to the graphical representation of the Pareto diagrams and contour graphs (Figure 1), the temperature and the carbon source strongly influence the xyloglucanase activity.

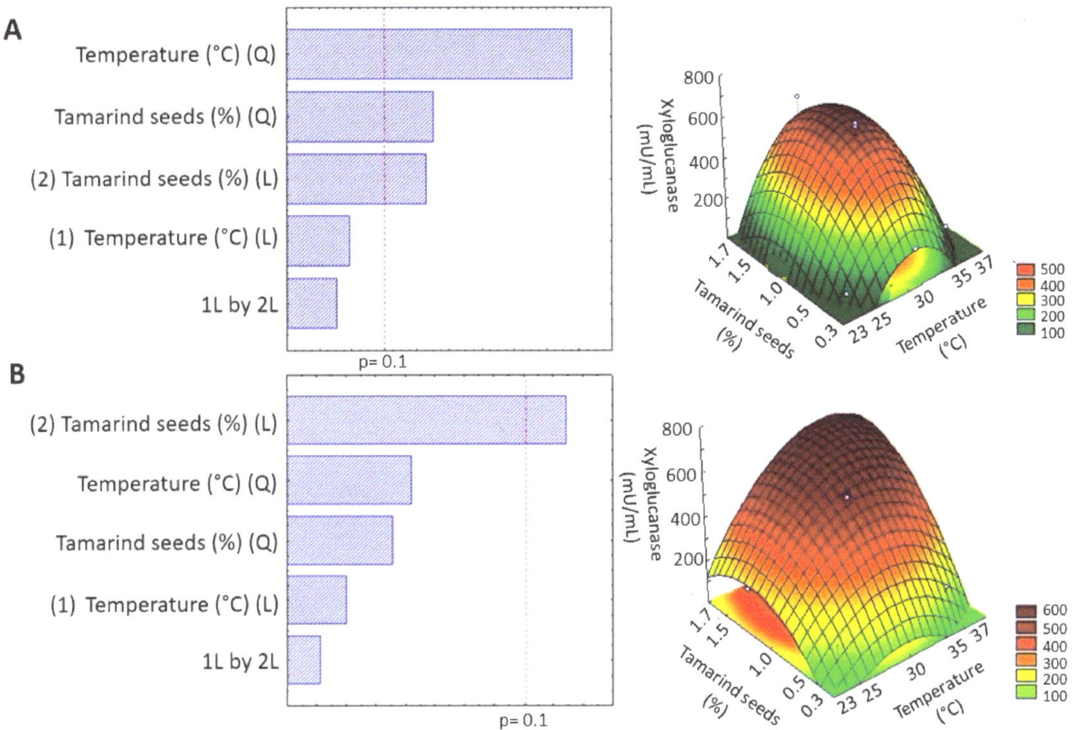

Figure 1. Pareto diagrams and contour graphs of production of *T. longibrachiatum* xyloglucanase with tamarind seeds after 72 h. (**A**)—static cultivation; (**B**)—cultivation under agitation.

3.2. Optimization of Cultivation with Jatoba Seeds

The results of the use of jatoba seeds as a carbon source to produce XEG can be seen in Table 3. It is possible to verify that the most notable enzymatic activity occurred starting from 48 h at the zero points (1% jatoba seed concentration at 30 °C). Still, it doubled 2.3-fold with a 1.705% carbon source and at 30 °C at the 72 h mark for the stationary condition. XEG had a more significant enzymatic activity at 72 h at 30 °C with a 1.705% carbon source for the cultivation under agitation. The approximated values were observed at 48 h (zero points) or 96 h at the same conditions of maximum enzymatic activity (30 °C and 1.705% concentration of the jatoba seed extract).

Via the calculation of the ANOVA and the F test with the time of 72 h (which is also the best timespan for xyloglucanase activity when the jatoba seeds were used as a carbon source), it was verified that the calculated F was 5.25-fold more significant than the tabled F when XEG was produced with jatoba seeds through stationary cultivation, and 2.38-fold greater when it was produced through cultivation under agitation; therefore, the null hypothesis was rejected, with a significant difference between the groups. According to the graphical representation of the Pareto diagrams and contour graphs (Figure 2), the temperature and carbon source strongly influence the xyloglucanase activity when using jatoba seeds as a carbon source.

Table 3. XEG activity in the tests used for the construction of the experimental design for *T. longibrachiatum* cultivated with jatoba seeds.

	Temperature (°C)	Jatoba Seeds (%)	Stationary				Under Agitation			
			24 h	48 h	72 h	96 h	24 h	48 h	72 h	96 h
1	−1 (25)	−1 (0.5)	27.06	14.60	11.40	4.11	4.99	26.71	25.28	8.25
2	1 (35)	−1 (0.5)	21.72	113.89	127.48	117.52	12.82	143.16	305.54	337.59
3	−1 (25)	1 (1.5)	26.35	68.02	87.24	78.24	91.52	100.07	82.62	62.80
4	1 (35)	1(1.5)	70.51	254.26	406.68	138.88	69.09	254.26	305.19	475.41
5	−1.41 (22.95)	0 (1)	89.74	90.81	90.81	49.50	25.64	155.98	205.65	54.49
6	1.41 (37.05)	0 (1)	66.59	55.55	55.55	26.24	72.65	264.23	590.43	559.45
7	0 (30)	−1.41 (0.295)	2.0	113.96	340.44	430.18	7.47	75.5	336.53	450.84
8	0 (30)	1.41 (1.705)	96.5	115.5	**642.07**	319.43	96.15	419.5	**652.40**	540.58
9	0 (30)	0 (1)	67.31	312.31	394.0	449.23	101.49	566.79	621.12	508.82
10	0 (30)	0 (1)	62.41	254.55	398.19	379.26	180.62	593.99	622.0	521.88
11	0 (30)	0 (1)	68.02	256.04	398.87	400.09	137.1	491.79	622.10	501.58

Values are expressed in mU/mL. nd = not detected. Bold values indicate the maximum value found for each time.

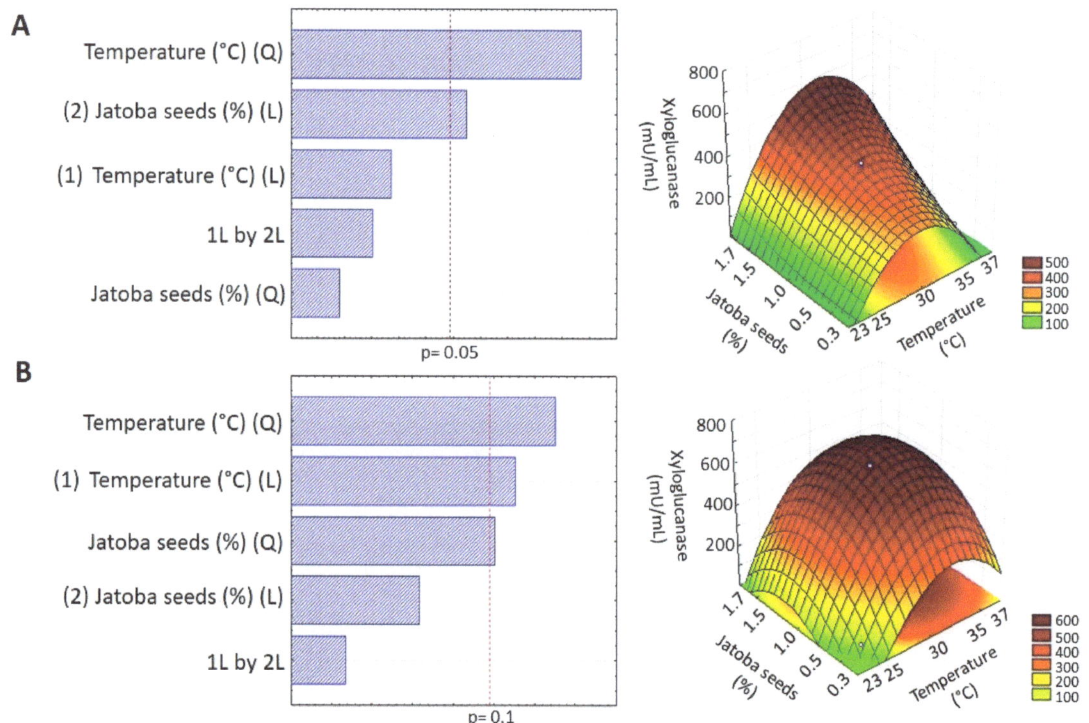

Figure 2. Pareto diagrams and contour graphs detailing the production of *T. longibrachiatum* xyloglucanase with jatoba seeds after 72 h. (**A**)—static cultivation; (**B**)—cultivation under agitation.

When comparing the xyloglucanase levels produced with tamarind seeds or jatoba seeds as carbon sources, the best cultivation condition was 1.705% of tamarind seed extract, at 30 °C, under agitation, and at the time of 72 h.

The experimental design is a statistically convenient technique for planning experiments in the bioprocessing field, including the optimization of enzyme production for biomass hydrolysis [27]. In recent years, several studies have been using optimization strategies to increase the productivity of enzymes [32–34], especially enzymes of the lignocellulolytic complex [35–37]. For example, Tai et al. [38] increased CMCase and xylanase production in *Aspergillus niger* by 124.5 and 78.5%, respectively, using oil palm frond leaves as carbon sources. In comparison, Ezeilo et al. [39] optimized the extracellular cellulases and xylanases produced by *Rhizopus oryzae* using the same carbon source. Another interesting study was conducted by Naidu et al. [40], who optimized the production of laccase, xylanase, and amylase from *Trametes lactinea* and *Pycnoporus sanguineus* using different substrates and evaluated the interactions between them. More recently, Singhal et al. [41] aimed to optimize the cellulase (CMCase) production in *Aspergillus flavus* using wheat straw, an abundantly available lignocellulosic waste, as a substrate. Three parameters, i.e., the nitrogen content (0.25 to 1%), fungal inoculum (0.25 to 1%), and duration (3 to 12 days), were optimized for maximum CMCase production using the Response Surface Methodology. The maximum output of CMCase of 13.89 U/gds was achieved with 0.25% yeast extract, 0.625% fungal inoculum, and a duration of 12 days. There was an almost threefold increase in CMCase production after optimization compared to the screening experiments (4.7 U/gds). All these works demonstrate the importance of a factorial design to optimizing enzyme production.

3.3. Scaling for the Bioreactor to Increase Enzyme Production

Once the best cultivation conditions were verified regarding the temperature, concentration of the carbon source, and scaling of agitation, *T. longibrachiatum* was cultivated in a bioreactor 5 L BioFlo 310-New Brunswick® (Hamburg, Germany) at 30 °C and with 1.705% of tamarind seed extract as a carbon source. A 1.27-fold increase was verified in the XEG activity since a 0.817 ± 0.028 U/mL level of activity was found in the Erlenmeyer flasks (see Table 2), while the activity was 1.040 ± 0.088 U/mL when the bioreactor was used. This result has shown the advantages of up scaling, as it improved the concentration of proteins. Consequently, this fact might have generated a more significant number of enzymes, subsequently favoring immobilization on ionic supports.

3.4. Biochemical Characterization

The crude extract produced using the bioreactor was characterized concerning the optimal temperature and pH, thermal stability, and pH stability. First, the thermostability was determined at 30 to 70 °C, with intervals of 10 °C, while the pH stability was determined at 3.0 to 8.0. Then, the optimal temperature of the crude extract of xyloglucanase was 50 °C, maintaining an excellent activity at 55 °C (Figure 3A). The optimal pH was 5.0 (Figure 3B). Still, a low thermostability (Figure 3C) and low pH stability (Figure 3D) were detected. Finally, the calculation of the T_{50} was performed, which is the temperature at which there was 50% residual activity (Table 4). XEG had a T_{50} of 185 min at the temperature of 50 °C.

Table 4. T_{50} of XEG of *T. longibrachiatum*.

Temperature (°C)	T_{50} (min)
40	138
50	185
60	67
70	66

The values found in this study are consistent with the ones found in the literature, as many studies [1,10,42–45] have performed a biochemical characterization of these parameters for xyloglucanase activity, and it was verified that the best range varies between the temperatures of 30–60 °C. However, there is a minor variation in the literature regarding the optimal pH, with the pH range of 5.0 and 5.5 being those most commonly found.

The low thermostability was a determining factor for the enzymatic immobilization, given that there is proof in the literature that immobilization increases the thermostability and allows for the reuse of the enzyme [46–48].

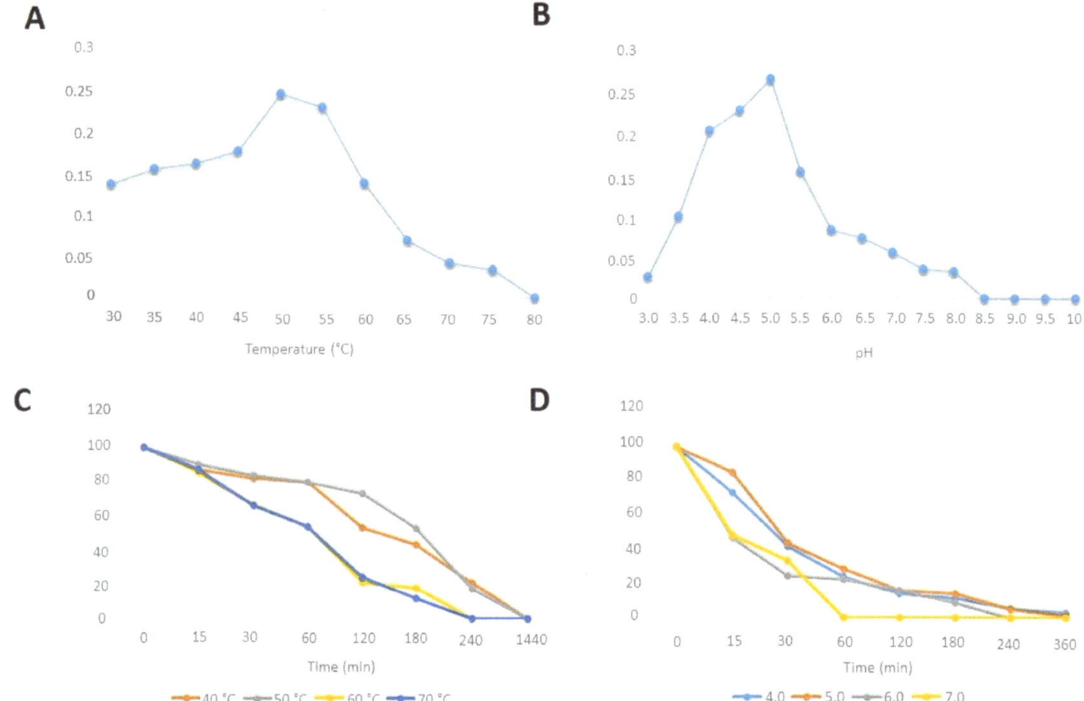

Figure 3. Biochemical characterization of *T. longibrachiatum* xyloglucanase. (**A**)—optimal temperature; (**B**)—optimal pH; (**C**)—thermostability; (**D**)—pH stability.

3.5. Enzymatic Immobilization

The crude enzymatic extract presented a brown-red color, which was a problem for its immobilization. To overcome this problem, the tegument of the tamarind seeds was removed, and it was verified that there was no decrease in the enzymatic activity. Furthermore, the extract was clarified with activated charcoal.

Ionic exchange supports CM-cellulose, DEAE-cellulose, MANAE, and PEI, which efficiently immobilized the xyloglucanase. This is the first time that the immobilization of a xyloglucanase on these supports has been described in the literature. The results, visualized in Table 5, show that the XEG's complete immobilization was obtained in CM-cellulose and DEAE-cellulose, with activities of 1.16 and 0.89 U/g of the derivative (enzyme plus support), respectively, and a hyper-activation of 1.81-fold and 1.39-fold, respectively, which demonstrates an excellent improvement in the XEG activity.

Table 5. Immobilization of XEG of *T. longibrachiatum* on different ionic supports at pH 7.0 after 48 h of immobilization.

Derivative	Total Immobilized Proteins (%)	Derivative Activity (U/g)	Immobilization Efficiency (%)	Hyperactivation
CM-cellulose	18.61	1.16	100	1.81
DEAE-cellulose	15.69	0.89	100	1.39
MANAE	27.63	0.32	49	nd
PEI	28.73	0.01	21	nd

nd = not detected.

The primary means of immobilizing enzymes are to boost the enzyme productivity and operational stability, alongside facilitating the reuse of the enzymes. Notwithstanding the aforementioned benefits, enzyme immobilization promotes high catalytic activity and stability and the convenient handling of enzymes, in addition to their facile separation from reaction mixtures without contaminating the products [49]. Among the few studies in the literature on the immobilization of xyloglucanase, Soares et al. [50] immobilized a recombinant xyloglucanase (XegA) from *Aspergillus niveus* on chitosan-coated ferromagnetic iron oxide nanoparticles functionalized with glutaraldehyde, retained 76.9% of the activity compared to the free enzyme. In comparison, Menon et al. [51] chose to immobilize *Debaromyces hansenii* cells on Ca-alginate beads instead of immobilizing xyloglucanase. The immobilized yeast cells were reused six times at 40 °C with a 100% fermentation efficiency.

4. Conclusions

This study allowed us to conclude that, with the aid of a factorial design, high concentrations of the carbon sources in the culture medium, especially of tamarind seeds, were the most favorable conditions for the highest activity of *T. longibrachiatum* xyloglucanase. Furthermore, the up scaling of the Erlenmeyer flasks for the bioreactor was demonstrated to be an essential strategy for increasing the content of the enzymes secreted. Regarding the biochemical characterization of the crude extract, the optimal temperature range was 50–55 °C, and the optimal pH was 5.0. The pH and temperature stabilities were not stable for prolonged periods, which was crucial to choosing the immobilization process on the ion exchange resins (CM-cellulose, DEAE-cellulose, MANAE, and PEI). This was the first time the immobilization of a xyloglucanase on these chemical supports has been described in the literature.

Author Contributions: Data Curation, A.G.C.; A.C.V.; E.N.d.F.; G.M.A.; A.L.A.V.J.; Methodology, A.G.C.; A.C.V.; V.E.P.; T.B.d.O.; Supervision and responsible for the financial support, M.d.L.T.d.M.P.; Writing—Original Draft, A.G.C.; G.M.A.; Writing—Review & Editing, M.S.B.; C.G.V.R.; M.d.L.T.d.M.P. All authors have read and agreed to the published version of the manuscript.

Funding: The authors thank the Fundação de Amparo à Pesquisa do Estado de São Paulo (FAPESP) for the doctorate grants awarded to A.G. Contato (Processes n° 2017/25862-6 and 2021/07066-3) and E.N Freitas (Process n° 2017/23989-9), and for the post-doctorate grant awarded to T.B. Oliveira (Process n° 2017/09000-4), and the financial support from National Institute of Science and Technology of Bioethanol (Process FAPESP 2014/50884-5), and FAPESP-FCT-Fundação para a Ciência e Tecnologia de Portugal, Process 2018/07522-6. Conselho Nacional de Desenvolvimento Científico (CNPq), process 465319/2014-9. M.L.T.M. Polizeli (process 310340/2021-7), C.G.V. Rechia, and M.S. Buckeridge are Research Fellows of CNPq. G.M. Aranha and A.L.A. Valvassora Jr. had an undergraduate research grant, and V.E. Pinheiro had a doctorate research grant awarded by CNPq. A.C. Vici was a recipient of a post-doctorate scholarship of Coordenação de Aperfeiçoamento de Pessoal de Nível Superior (CAPES) finance code 001.

Institutional Review Board Statement: Not applicable.

Informed Consent Statement: Not applicable.

Acknowledgments: We thank Mauricio de Oliveira for his technical assistance and Mariana Cereia for her English support.

Conflicts of Interest: The authors declare no conflict of interest.

References

1. Vitcosque, G.L.; Ribeiro, L.F.C.; Lucas, R.C.; Silva, T.M.; Ribeiro, L.F.; Damasio, A.R.L.; Farinas, C.S.; Gonçalves, A.Z.L.; Segato, F.; Buckeridge, M.S.; et al. The functional properties of a xyloglucanase (GH12) of *Aspergillus terreus* expressed in *Aspergillus nidulans* may increase performance of biomass degradation. *Appl. Microbiol. Biotechnol.* **2016**, *100*, 9133–9144. [CrossRef] [PubMed]
2. Lima, D.U.; Buckeridge, M.S. Interaction between cellulose and storage xyloglucans: The influence of the degree of galactosylation. *Carbohydr. Polym.* **2001**, *46*, 157–163. [CrossRef]
3. Buckeridge, M.S.; Souza, A.P. *Advances of Basic Science for Second Generation Bioethanol from Sugarcane*, 1st ed.; Springer: Berlin/Heidelberg, Germany, 2017.

4. Kulkarni, A.D.; Joshi, A.A.; Patil, C.L.; Amale, P.D.; Patel, H.M.; Surana, S.J.; Belgamwar, V.S.; Chaudhari, K.S.; Pardeshi, C.V. Xyloglucan: A functional biomacromolecule from drug delivery applications. *Int. J. Biol. Macromol.* **2017**, *104*, 799–812. [CrossRef] [PubMed]

5. Von Schantz, L.; Gullfot, F.; Scheer, S.; Filonova, L.; Gunnarsson, L.C.; Flint, J.E.; Daniel, G.; Nordberg-Karlsson, E.; Brumer, H.; Ohlin, M. Affinity maturation generates greatly improved xyloglucan-specific carbohydrate binding modules. *BMC Biotechnol.* **2009**, *9*, 92. [CrossRef] [PubMed]

6. Fry, S.C.; York, W.S.; Albersheim, P.; Darvill, A.; Hayashi, T.; Joseleau, J.P.; Kato, Y.; Lorences, E.P.; Maclachlan, G.A.; McNeil, M.; et al. An unambiguous nomenclature for xyloglucan-derived oligosaccharides. *Physiol. Plant.* **1993**, *89*, 1–3. [CrossRef]

7. Berezina, O.V.; Herlet, J.; Rykov, S.V.; Kornberger, P.; Zavyalov, A.; Kozlov, D.; Sakhibgaraeva, L.; Krestyanova, I.; Scharwz, W.H.; Zverlov, V.; et al. Thermostable multifunctional GH74 xyloglucanase from *Myceliophthora thermophila*: High-level expression in *Pichia pastoris* and characterization of the recombinant protein. *Appl. Microbiol. Biotechnol.* **2017**, *101*, 5653–5666. [CrossRef]

8. Benkö, Z.; Siika-Aho, M.; Viikari, L.; Réczey, K. Evaluation of the role of xyloglucanase in the enzymatic hydrolysis of lignocellulosic substrates. *Enzym. Microb. Technol.* **2008**, *43*, 109–114. [CrossRef]

9. Song, S.; Tang, Y.; Yang, S.; Yan, Q.; Zhou, P.; Jiang, Z. Characterization of two novel family 12 xyloglucanases from the thermophilic *Rhizomucor miehei*. *Appl. Microbiol. Biotechnol.* **2013**, *97*, 10013–10024. [CrossRef]

10. Damasio, A.R.L.; Rubio, M.V.; Gonçalves, T.A.; Persinoti, G.F.; Segato, F.; Prade, R.A.; Contesini, F.J.; Souza, A.P.; Buckeridge, M.S.; Squina, F.M. Xyloglucan breakdown by endo-xyloglucanase family 74 from *Aspergillus fumigatus*. *Appl. Microbiol. Biotechnol.* **2017**, *101*, 2893–2903. [CrossRef]

11. Brück, S.A.; Contato, A.G.; Gamboa-Trujillo, P.; de Oliveira, T.B.; Cereia, M.; Polizeli, M.L.T.M. Prospection of psychrotrophic filamentous fungi isolated from the High Andean Paramo Region of Northern Ecuador: Enzymatic activity and molecular identification. *Microorganisms* **2022**, *10*, 282. [CrossRef]

12. Greffe, L.; Bessueille, L.; Bulone, V.; Brumer, H. Synthesis, preliminary characterization, and application of novel surfactants from highly branched xyloglucan oligosaccharides. *Glycobiology* **2005**, *15*, 437–445. [CrossRef] [PubMed]

13. Miyazaki, S.; Suisha, F.; Kawasaki, N.; Shirakawa, M.; Yamatoya, K.; Attwood, D. Thermally reversible xyloglucan gels as vehicles for rectal drug delivery. *J. Control. Release* **1998**, *56*, 75–83. [CrossRef]

14. Lima, D.U.; Oliveira, R.C.; Buckeridge, M.S. Seed storage hemicelluloses as wet-end additives in papermaking. *Carbohydr. Polym.* **2003**, *52*, 367–373. [CrossRef]

15. Gidley, D.W.; Nico, J.S.; Skalsey, M. Direct search for two-photon decay modes of orthopositronium. *Phys. Rev. Lett.* **1991**, *66*, 1302–1305. [CrossRef]

16. Ishida, T.; Yaoi, K.; Hiyoshi, A.; Igarashi, K.; Samejima, M. Substrate recognition by glycoside hydrolase family 74 xyloglucanase from the basidiomycete *Phanerochaete chrysosporium*. *FEBS J.* **2007**, *274*, 5727–5736. [CrossRef] [PubMed]

17. Xian, L.; Wang, F.; Yin, X.; Feng, J.X. Identification and characterization of an acidic and acid-stable endoxyloglucanase from *Penicillium oxalicum*. *Int. J. Biol. Macromol.* **2016**, *86*, 512–518. [CrossRef]

18. Brugnari, T.; Pereira, M.G.; Bubna, G.A.; Freitas, E.N.; Contato, A.G.; Corrêa, R.C.G.; Castoldi, R.; Souza, C.G.M.; Polizeli, M.L.T.M.; Bracht, A.; et al. A highly reusable MANAE-agarose-immobilized *Pleurotus ostreatus* laccase for degradation of bisphenol A. *Sci. Total Environ.* **2018**, *634*, 1346–1351. [CrossRef]

19. Wong, H.L.; Hu, N.J.; Juang, T.Y.; Liu, Y.C. Co-Immobilization of xylanase and scaffolding protein onto an immobilized metal ion affinity membrane. *Catalysts* **2020**, *10*, 1408. [CrossRef]

20. Brugnari, T.; Contato, A.G.; Pereira, M.G.; Freitas, E.N.; Bubna, G.A.; Aranha, G.M.; Bracht, A.; Polizeli, M.L.T.M.; Peralta, R.M. Characterisation of free and immobilised laccases from *Ganoderma lucidum*: Application on bisphenol a degradation. *Biocatal. Biotransformation* **2020**, *39*, 71–80. [CrossRef]

21. Contato, A.G.; Oliveira, T.B.; Aranha, G.M.; Freitas, E.N.; Vici, A.C.; Nogueira, K.M.V.; Lucas, R.C.; Scarcella, A.S.A.; Buckeridge, M.S.; Silva, R.N.; et al. Prospection of fungal lignocellulolytic enzymes produced from jatoba (*Hymenaea courbaril*) and tamarind (*Tamarindus indica*) seeds: Scaling for bioreactor and saccharification profile of sugarcane bagasse. *Microorganisms* **2021**, *9*, 533. [CrossRef]

22. Buckeridge, M.S.; Dietrich, S.M.C. Galactomannan from Brazilian legume seeds. *Rev. Bras. Bot.* **1990**, *13*, 109–112.

23. Santos, H.P.; Buckeridge, M.S. The role of the storage carbon of cotyledons in the establishment of seedlings of *Hymenaea courbaril* under different light conditions. *Ann. Bot.* **2004**, *94*, 819–830. [CrossRef] [PubMed]

24. Jesionowski, T.; Zdarta, J.; Krajewska, B. Enzyme immobilization by adsorption: A review. *Adsorption* **2014**, *20*, 801–821. [CrossRef]

25. Andrades, D.; Graebin, N.G.; Kadowaki, M.K.; Ayub, M.A.Z.; Fernandez-Lafuente, R.; Rodrigues, R.C. Immobilization and stabilization of different β-glucosidases using the glutaraldehyde chemistry: Optimal protocol depends on the enzyme. *Int. J. Biol. Macromol.* **2019**, *129*, 672–678. [CrossRef] [PubMed]

26. Khanna, P.; Sundari, S.S.; Kumar, N.J. Production, isolation and partial purification of xylanases from an *Aspergillus* sp. *World J. Microbiol. Biotechnol.* **1995**, *11*, 242–243. [CrossRef] [PubMed]

27. Rodrigues, M.I.; Iemma, A.F. *Planejamento de Experimentos & Otimização de Processos*, 2nd ed.; Editora Cárita: Campinas, Brazil, 2009.

28. Scarcella, A.S.A.; Pasin, T.M.; de Lucas, R.C.; Ferreira-Nozawa, M.S.; de Oliveira, T.B.; Contato, A.G.; Grandis, A.; Buckeridge, M.S.; Polizeli, M.L.T.M. Holocellulase production by filamentous fungi: Potential in the hydrolysis of energy cane and other sugarcane varieties. *Biomass Convers. Biorefining* **2021**, 1–12. [CrossRef]

29. Miller, G.L. Use of dinitrosalicylic acid reagent for determination of reducing sugar. *Anal. Chem.* **1959**, *31*, 426–428. [CrossRef]

30. Bradford, M.M. A rapid and sensitive method for the quantitation of microgram quantities of protein utilizing the principle of protein-dye binding. *Anal. Biochem.* **1976**, *72*, 248–254. [CrossRef]
31. Monteiro, L.M.O.; Pereira, M.G.; Vici, A.C.; Heinen, P.R.; Buckeridge, M.S.; Polizeli, M.L.T.M. Efficient hydrolysis of wine and grape juice anthocyanins by *Malbranchea pulchella* β-glucosidase immobilized on MANAE-agarose and ConA-Sepharose supports. *Int. J. Biol. Macromol.* **2019**, *136*, 1133–1141. [CrossRef]
32. Agrawal, K.; Verma, P. Production optimization of yellow laccase from *Stropharia* sp. ITCC 8422 and enzyme-mediated depolymerization and hydrolysis of lignocellulosic biomass for biorefinery application. *Biomass Convers. Biorefining* **2022**, 1–20. [CrossRef]
33. Contato, A.G.; Inácio, F.D.; Brugnari, T.; de Araújo, C.A.V.; Maciel, G.M.; Haminiuk, C.W.I.; Peralta, R.M.; de Souza, C.G.M. Solid-state fermentation with orange waste: Optimization of Laccase production from *Pleurotus pulmonarius* CCB-20 and decolorization of synthetic dyes. *Acta Sci. Biol. Sci.* **2020**, *42*, e52699. [CrossRef]
34. Contato, A.G.; Inácio, F.D.; Bueno, P.S.A.; Nolli, M.M.; Janeiro, V.; Peralta, R.M.; de Souza, C.G.M. *Pleurotus pulmonarius*: A protease-producing white rot fungus in lignocellulosic residues. *Int. Microbiol.* **2022**, 1–8, *online ahead of print*. [CrossRef] [PubMed]
35. Jana, U.K.; Suryawanshi, R.K.; Prajapati, B.P.; Soni, H.; Kango, N. Production optimization and characterization of mannooligosaccharide generating β-mannanase from *Aspergillus oryzae. Bioresour. Technol.* **2018**, *268*, 308–314. [CrossRef] [PubMed]
36. Neelkant, K.S.; Shankar, K.; Jayalakshmi, S.K.; Sreeramulu, K. Optimization of conditions for the production of lignocellulolytic enzymes by *Sphingobacterium* sp. ksn-11 utilizing agro-wastes under submerged condition. *Prep. Biochem. Biotechnol.* **2019**, *49*, 927–934. [CrossRef]
37. KC, S.; Upadhyaya, J.; Joshi, D.R.; Lekhak, B.; Chaudhary, D.C.; Pant, B.J.; Bajgai, T.R.; Dhital, R.; Khanal, S.; Koirala, N.; et al. Production, characterization, and industrial application of pectinase enzyme isolated from fungal strains. *Fermentation* **2020**, *6*, 59. [CrossRef]
38. Tai, W.Y.; Tan, J.S.; Lim, V.; Lee, C.K. Comprehensive studies on optimization of cellulase and xylanase production by a local indigenous fungus strain via solid state fermentation using oil palm frond as substrate. *Biotechnol. Prog.* **2019**, *35*, e2781. [CrossRef]
39. Ezeilo, U.R.; Wahab, R.A.; Mahat, N.A. Optimization studies on cellulase and xylanase production by *Rhizopus oryzae* UC2 using raw oil palm frond leaves as substrate under solid state fermentation. *Renew. Energy* **2020**, *156*, 1301–1312. [CrossRef]
40. Naidu, Y.; Siddiqui, Y.; Idris, A.S. Comprehensive studies on optimization of ligno-hemicellulolytic enzymes by indigenous white rot hymenomycetes under solid-state cultivation using agro-industrial wastes. *J. Environ. Manag.* **2020**, *259*, 110056. [CrossRef]
41. Singhal, A.; Kumari, N.; Ghosh, P.; Singh, Y.; Garg, S.; Shah, M.P.; Jha, P.K.; Chauhan, D.K. Optimizing cellulase production from *Aspergillus flavus* using response surface methodology and machine learning models. *Environ. Technol. Innov.* **2022**, *27*, 102805. [CrossRef]
42. Master, E.R.; Zheng, Y.; Storms, R.; Tsang, A.; Powlowski, J. A xyloglucan-specific family 12 glycosyl hydrolase from *Aspergillus niger*: Recombinant expression, purification and characterization. *Biochem. J.* **2008**, *411*, 161–170. [CrossRef]
43. McGregor, N.; Morar, M.; Fenger, T.H.; Stogios, P.; Lenfant, N.; Yin, V.; Xu, X.; Evdokimova, E.; Cui, H.; Henrissat, B.; et al. Structure-function analysis of a mixed-linkage β-glucanase/xyloglucanase from the key ruminal bacteroidetes *Prevotella bryantii* B14. *J. Biol. Chem.* **2016**, *291*, 1175–1197. [CrossRef] [PubMed]
44. Han, Y.; Ban, Q.; Li, H.; Hou, Y.; Jin, M.; Han, S.; Rao, J. DkXTH8, a novel xyloglucan endotransglucosylase/hydrolase in persimmon, alters cell wall structure and promotes leaf senescence and fruit postharvest softening. *Sci. Rep.* **2016**, *6*, 39155. [CrossRef] [PubMed]
45. Morales-Quintana, L.; Beltrán, D.; Mendez-Yañez, Á.; Valenzuela-Riffo, F.; Herrera, R.; Moya-León, M.A. Characterization of FcXTH2, a novel xyloglucan endotransglycosylase/hydrolase enzyme of Chilean strawberry with hydrolase activity. *Int. J. Mol. Sci.* **2020**, *21*, 3380. [CrossRef]
46. Mateo, C.; Palomo, J.M.; Fernandez-Lorente, G.; Guisan, J.M.; Fernandez-Lafuente, R. Improvement of enzyme activity, stability and selectivity via immobilization techniques. *Enzym. Microb. Technol.* **2007**, *40*, 1451–1463. [CrossRef]
47. Zdarta, J.; Meyer, A.S.; Jesionowski, T.; Pinelo, M. A general overview of support materials for enzyme immobilization: Characteristics, properties, practical utility. *Catalysts* **2018**, *8*, 92. [CrossRef]
48. Han, P.; Zhou, X.; You, C. Efficient multi-enzymes immobilized on porous microspheres for producing inositol from starch. *Front. Bioeng. Biotechnol.* **2020**, *8*, 380. [CrossRef]
49. Wahab, R.A.; Elias, N.; Abdullah, F.; Ghoshal, S.K. On the taught new tricks of enzymes immobilization: An all-inclusive overview. *React. Funct. Polym.* **2020**, *152*, 104613. [CrossRef]
50. Soares, J.M.; Carneiro, L.A.; Barreto, M.Q.; Ward, R.J. Co-immobilization of multiple enzymes on ferromagnetic nanoparticles for the depolymerization of xyloglucan. *Biofuels Bioprod. Biorefining* **2022**. [CrossRef]
51. Menon, V.; Prakash, G.; Rao, M. Enzymatic hydrolysis and ethanol production using xyloglucanase and *Debaromyces hansenii* from tamarind kernel powder: Galactoxyloglucan predominant hemicellulose. *J. Biotechnol.* **2010**, *148*, 233–239. [CrossRef]

 *fermentation*

Article

Heterologous Expression of Thermotolerant α-Glucosidase in *Bacillus subtilis* 168 and Improving Its Thermal Stability by Constructing Cyclized Proteins

Zhi Wang, Mengkai Hu, Ming Fang, Qiang Wang, Ruiqi Lu, Hengwei Zhang, Meijuan Xu, Xian Zhang * and Zhiming Rao *

The Key Laboratory of Industrial Biotechnology of Ministry of Education, School of Biotechnology, Jiangnan University, Wuxi 214122, China
* Correspondence: zx@jiangnan.edu.cn (X.Z.); raozhm@jiangnan.edu.cn (Z.R.); Tel.: +86-137-7140-1977 (X.Z.); +86-139-2113-5816 (Z.R.)

Abstract: α-glucosidase is an essential enzyme for the production of isomaltooligosaccharides (IMOs). Allowing α-glucosidase to operate at higher temperatures (above 60 °C) has many advantages, including reducing the viscosity of the reaction solution, enhancing the catalytic reaction rate, and achieving continuous production of IMOs. In the present study, the thermal stability of α-glucosidase was significantly improved by constructing cyclized proteins. We screened a thermotolerant α-glucosidase (AGL) with high transglycosylation activity from *Thermoanaerobacter ethanolicus* JW200 and heterologously expressed it in *Bacillus subtilis* 168. After forming the cyclized α-glucosidase by different isopeptide bonds (SpyTag/SpyCatcher, SnoopTag/SnoopCatcher, SdyTag/SdyCatcher, RIAD/RIDD), we determined the enzymatic properties of cyclized AGL. The optimal temperature of all cyclized AGL was increased by 5 °C, and their thermal stability was generally improved, with SpyTag-AGL-SpyCatcher having a 1.74-fold increase compared to the wild-type. The results of molecular dynamics simulations showed that the RMSF values of cyclized AGL decreased, indicating that the rigidity of the cyclized protein increased. This study provides an efficient method for improving the thermal stability of α-glucosidase.

Keywords: α-glucosidase; isomaltooligosaccharides; isopeptide bonds; thermal stability

Citation: Wang, Z.; Hu, M.; Fang, M.; Wang, Q.; Lu, R.; Zhang, H.; Xu, M.; Zhang, X.; Rao, Z. Heterologous Expression of Thermotolerant α-Glucosidase in *Bacillus subtilis* 168 and Improving Its Thermal Stability by Constructing Cyclized Proteins. *Fermentation* **2022**, *8*, 498. https://doi.org/10.3390/fermentation8100498

Academic Editor: Fabrizio Beltrametti

Received: 2 September 2022
Accepted: 26 September 2022
Published: 29 September 2022

Publisher's Note: MDPI stays neutral with regard to jurisdictional claims in published maps and institutional affiliations.

1. Introduction

Isomaltooligosaccharides (IMOs) are functional oligosaccharides consisting of 2 to 10 glucosyl saccharide units linked by α-1, 6-glycosidic bonds [1]. The prebiotic component of IMOs mainly include isomaltose, panose and isomaltotriose [2]. IMOs are also known as "Bifidus factors" due to their ability to effectively promote the growth and multiplication of *Bifidobacterium* in the human body [3]. As a common prebiotic with water-soluble dietary fiber function, low caloric value [4], and anti-cavity properties [5], IMOs are widely used as food additives and feed ingredients in the food industries [6] and animal husbandry [7].

IMOs are mainly produced in three ways (Figure 1): 1. Utilizing the retrosynthesis of glucoamylase, glucose is synthesized into oligosaccharides such as isomaltose and maltose in high glucose concentration solutions [8]. However, the method of producing IMOs has the disadvantages of low yield, complex products, and long production cycles, which makes it difficult to be applied industrially. 2. Sucrose and maltose are converted by dextransucrase, and the maltose in the mixture reacts with the glucose produced by sucrose hydrolysis to produce IMOs [9]. 3. IMOs are industrially produced mainly by α-glucosidase catalyzed maltose. The method consists of three main steps: (1) Starch is liquefied by heat-resistant α-amylase to produce oligosaccharides and dextrins. (2) Further saccharification of oligosaccharides and dextrins by glycosylases occurs, such as β-amylase and pullulanase. (3) α-glucosidase (EC 3.2.1.20) transfers a glucosyl residue from the donor substrate to the 6-OH group of the non-reducing glucose unit and produces IMOs [10]. The research and application of amylase and pullulanase have been well established. With the increasing requirements of industrial production, α-glucosidase has become a key enzyme for the industrial production of IMOs.

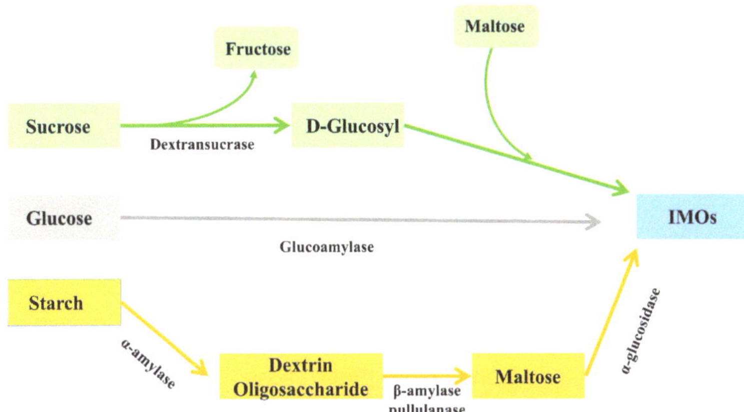

Figure 1. Three main ways for the production of IMOs: Utilizing the retrosynthesis of glucoamylase (grey routine), sucrose and maltose are converted by dextransucrase (green routine), maltose catalyzed by α-glucosidase (yellow routine).

Besides hydrolysis activity, some α-glucosidases also have transglycosylation activity [11]. The transglycosylation reaction mediated by α-glucosidase consists of two sequential steps [12]: 1. Nucleophilic residues (Asp or Glu) attack the glycoside and form a covalent bond with the split glycosidic bond. 2. Another (acid/base) catalytic residue (Asp or Glu) mediates the transfer of the glycosyl portion to the receptor molecule (the glycoside substrate), completing the glycosylation and eventually forming IMOs. Most industrially applied α-glucosidases are derived from *Aspergillus niger*, generally more stable below 50 °C. It quickly becomes inactivated when the temperature is above 60 °C [13]. However, it is beneficial for reducing the viscosity of the reaction solution, improving the catalytic efficiency, and increasing the production of IMOs by adequately increasing the temperature of the transglycosylation reaction [14]. Poor thermal stability has become a limiting factor for the commercial application of α-glucosidase. Therefore, the screening and modification of α-glucosidase with good thermal stability are of great significance for the efficient production of IMOs [15,16].

Currently, the thermal stability of α-glucosidase is mainly improved by two methods. The most common method is to obtain heat-resistant α-glucosidase by cloning the α-glucosidase of thermophilic microorganisms. Hung et al. isolated a strain of *Geobacillus* sp. expressing α-glucosidase from an extreme environment [17]. Zhang et al. heterologously expressed α-glucosidase (GSJ) from this source in *Escherichia coli* (*E. coli*) [14]. The optimal temperature of the recombinant α-glucosidase was 65 °C, and the $t_{1/2}$ (Half-life) was 84 h at 60 °C. Another method to improve the thermal stability of α-glucosidase is site-directed mutagenesis. Zhou et al. mutated four sites (Leu152, Asn208, Lys285, and Thr430) in *T. tengcongensis* MB4-derived α-glucosidase (TtGluA). As a result, the T_{50} (the temperature when the enzyme activity is reduced by half) of mutant K285P was increased by 10.5 °C [18].

Isopeptide bond is the irreversible covalent bond formed spontaneously by the side chains of lysine (Lys) and asparagine/aspartate (Asn/Asp) residues, which can effectively mediate the cyclization of proteins [19–21]. The Tag/Catcher system consist of two short polypeptide tags; the Catcher is able to specifically bind the Tag peptide and catalyze the formation of an isopeptide bond between two amino acid side chains. The thermal stability of the protein can be effectively enhanced by the formation of cyclized proteins.

SpyTag/SpyCatcher [22] and SnoopTag/SnoopCatcher [21] have been widely used to construct cyclized proteins, while SdyTag/SdyCatcher [23] was less used. Chen et al. [24] used both site-directed mutagenesis and cyclization of isopeptide bonds to improve the thermal stability of trehalose synthase (Tres), and the final results showed that the $t_{1/2}$ of all four cyclized Tres increased by 2–3 times at 55 °C, and the T_{50} increased by 7.5–15.5 °C. Further, the improved thermal stability of cyclized Tres was much better than that of the fixed-point mutation. Wang et al. [25] used SpyTag/SpyCatcher to mediate the cyclization of lichenase, and the cyclized protein showed significantly higher activity than the linear protein after heat treatment. The same strategy was also applied to the construction of cyclized phytase [26] and firefly luciferase [27].

This research aims to construct α-glucosidases with heat resistance and potential industrial applications. We heterologous expressed α-glucosidase from *Thermoanaerobacter ethanolicus* JW200 [28] and *Geobacillus* sp. Strain HTA-462. Three traditional isopeptide bonds (SpyTag/SpyCatcher, SnoopTag/SnooopCatcher, SdyTag/SdyCatcher) and a novel pair of short peptide tags [29] (RIAD/RIDD), which were considered to be effective in forming cyclized proteins and thus improving the thermal stability of the enzyme, were applied in the construction of cyclized α-glucosidase. Furthermore, the enzymatic properties of cyclized α-glucosidase were compared with wild-type α-glucosidase, and molecular dynamics simulations were performed to analyze the reasons for the improved thermal stability of cyclized α-glucosidase. Finally, we successfully constructed cyclized α-glucoside with significantly improved thermal stability.

2. Materials and Methods

2.1. Materials

The strains and plasmids used to construct the recombinant α-glucosidase are shown in Table S1. *E. coli* JM109 was used for gene cloning and *Bacillus subtilis* 168 was used for gene expression. PrimeStar and restriction endonucleases *Eco*R I, *Hin*d III, *Nde* I, and *Mlu* I, were purchased from Nanjing Novozymes (Nanjing, China). LB medium (peptone 10 g/L yeast powder 5 g/L NaCl 10 g/L) was used to culture the organisms. Maltose, kanamycin, and glucopyranoside were purchased from Shanghai Biotech (Shanghai, China). The Ni-NTA purification column used for protein purification was purchased from Thousand Pure Co. (Wuxi, China). The IMOs standards for HPLC were purchased from Shanghai Yuanye Biotechnology Co. (Shanghai, China).

2.2. Methods

2.2.1. Heterologous Expression of α-Glucosidase in *B. subtilis* 168

The nucleotide sequence of α-glucosidase from *Thermoanaerobacter ethanolicus* JW200 (*agl*, EF635970.1) and *Geobacillus* sp. strain HTA-462 (*gsj*, AB15481) were obtained by NCBI online search (https://www.ncbi.nlm.nih.gov, accessed on 5 July 2022), and the gene fragment with *Eco*R I and *Hin*d III digestion sites was synthesized by Anznta Biotechnology (Suzhou, China). Meanwhile, 6×Histag was connected to the N-terminal terminus of recombinant α-glucosidase for protein purification. The *Eco*R I and *Hin*d III digested gene fragments were ligated with pMA5 plasmid and transferred into *E. coli* JM109 for cloning, colony PCR was performed for initial verification of plasmids. The verified positive transformants were picked and inoculated into a 10 mL LB medium containing 0.1 mg/mL kanamycin, incubated overnight for 12 h, and then the recombinant plasmids were extracted and sent to GENEWIZ Biotechnology (Suzhou, China) for sequencing. The recombinant plasmid was transferred into *B. subtilis* 168 for heterologous expression of α-glucosidase according to the method of You et al. [30]. Recombinant α-glucosidases are called AGL and GSJ, which are encoded by the *agl* and *gsj* genes, respectively.

2.2.2. Expression Analysis and Purification of Recombinant α-Glucosidase

The recombinant strains were inoculated into a 50 mL LB medium and cultured at 37 °C, 200 rpm for 20 h. The cells in the culture medium were harvested by centrifuging at $8000\times g$ for 5 min, and the supernatant was discarded. The cells were washed twice with 10 mL of phosphate-buffered saline (PBS) buffer (pH 7.4) and resuspended with 5 mL buffer. Cells were disrupted by sonication (power was set to 90 W) at 4 °C for 20 min, and the mixture solution was centrifuged at $12,000\times g$ for 20 min to remove cell debris. The supernatant was loaded onto Ni-NTA resin pre-equilibrated with 50 mM PBS buffer (pH 7.4) containing 500 mM NaCl; the targeted proteins were eluted with 50 mM PBS buffer (pH 7.4) containing 500 mM NaCl and 200 mM imidazole. The expression of recombinant α-glucosidase was analyzed by SDS-polyacrylamide gel electrophoresis (SDS-PAGE), and the concentration was determined by the Bradford method.

2.2.3. Determination of Recombinant α-Glucosidase Activity

α-glucosidase was obtained and purified according to Section 2.2.3, and purified α-glucosidase was used for the determination of enzyme activity. The enzymatic activity of recombinant α-glucosidase was determined by the pNPG method [31]. The reaction mixture (200 µL) contained 20 mM pNPG, 50mM PBS buffer (pH 7.0), and 50 uL of α-glucosidase. The reaction was performed at 60 °C for 20 min and terminated by adding 600 µL 1.0 M Na_2CO_3. The amount of pNP was characterized by the absorbance value at 410 nm. One unit (1 U) of enzyme activity was defined as the amount of the enzyme producing 1 µmol pNP per min, and specific activity was defined as units per mg protein.

2.2.4. Determination of Transglycosylation Activity

The mixture (1 mL) containing 50 mM PBS buffer (pH 7.0), 300 mg maltose, and 50 µL enzyme solution was incubated at 60 °C for different times. The content of transglycosylation products during the reaction were detected by HPLC (Agilent 1260, Santa Clara, CA, USA) on an XBridge NH_2-column (4.6 mm$\times$250 mm, 3.5 µm). The mobile phase consisted of 70% (v/v) acetonitrile and 30% (v/v) water; a RID detector (Agilent Co., Santa Clara, CA, USA) was used for detection. The temperature of the column and detector is kept at 40 °C.

2.2.5. Construction of the Cyclized α-Glucosidase

The nucleotide sequences of SpyTag/SpyCatcher, SnoopTag/SnoopCatcher, Sdy-Tag/SnoopCatcher, and RIDD/RIAD with linker (3$\times$GGGS) were synthesized by Anznta Biotechnology (Suzhou, China). RIAD, SpyTag, SnoopTag, and SdyTag were connected to the N-terminus of α-glucosidase, while RIDD, SpyCatcher, SnoopCatcher, and Sdy-Catcher were connected to the C-terminus. With the expression of recombinant vectors in *B. subtilis* 168, four cyclized α-glucosidase variants (SpyTag-AGL-SpyCatcher, SnoopTag-AGL-SnoopCatcher, SdyTag-AGL-SdyCatcher, and RIAD-AGL-RIDD) were obtained. Using SpyTag-AGL-SpyCatcher as an example, to verify the formation site of the isopeptide bond, we mutated the key site (K11) of SpyTag-AGL-SpyCatcher, and the primers used for mutation are shown in Table S2. If the mutated SpyTag-AGL-SpyCatcher is changed to a linear protein, the mutated site is proved to be the site for the formation of the isopeptide bond. The changes in molecular weight of the cyclized proteins were analyzed by SDS-PAGE and combined with the 3D structure of cyclized AGL to determine whether cyclization occurred.

2.2.6. Characterization of the Enzymatic Properties

The optimal temperature of recombinant α-glucosidase was determined by measuring the enzyme activity at different reaction temperatures for 20 min at pH 7.0.

The thermal stability of recombinant α-glucosidase was measured by incubating the purified enzyme at 65 °C for 42 h in a substrate-free PBS buffer (pH 7.0), and samples were taken every 6 h to determine the residual enzyme activity.

The optimal pH of recombinant α-glucosidase was determined by measuring the enzyme activity at different pH in different buffers (acetate buffer, pH 4.0–6.0; sodium phosphate buffer, pH 6.0–8.0; glycine-NaOH buffer, pH 8.0–10.0) at 60 °C.

The pH stability of recombinant α-glucosidase was determined by incubating the purified enzyme in various buffers (acetate buffer, pH 5.0–6.0; sodium phosphate buffer, pH 7.0–8.0; glycine-NaOH buffer, pH 9.0) at 4 °C for 24 h and the residual enzyme activity was determined at 60 °C.

Kinetic parameters were determined by measuring the enzymatic activity at pH 7.0, 65°C, with various concentrations (10–50 mM) of pNPG as substrate. After the reaction, the experimental data were analyzed using GraphPad Prism 8.0 to determine the V_{max}, K_{cat}, and K_m values.

2.2.7. Molecular Dynamics Simulation of Cyclized α-Glucosidase

Due to the lack of suitable crystal templates, the amino acid sequence (Table S3) of wild-type and cyclized α-glucosidase were submitted to AlphaFold 2 software to obtain the three-dimensional structure of cyclized α-glucosidase. The 3D structure of cyclized α-glucosidase was analyzed by pymol, and Gromacs (http://www.gromacs.org/, accessed on 5 July 2022) simulated the molecular dynamics of recombinant α-glucosidase at 338K. Specifically, the thermal fluctuations of wild-type and cyclized AGL were analyzed using an OPLS-AA force field. The enzyme was surrounded by H_2O containing 0.15 M NaCl with pH 7.0 in a dodecahedron box, and the distance between the protein and the edge of the box was set to 1.2 nm. The water molecules were described by the simple point charge (SPC) explicit solvent model. Due to the lack of isopeptide bond parameters in the classical force field, we modified the force field by defining the non-standard residues of LYT, ASQ and ADN manually based on LYS, ASP and ASN. According to the method of Chen et al. [24], we modified the aminoacids.rtp and aminoacids.hdb files and updated the specbond.dat file. The indexes of "peptide tags", "linkers" and "AGL" were generated by the command "gmx make_ndx" (Gromacs). The stability of the system is assessed by the root mean square deviation (RMSD). Root mean square fluctuations (RMSF) were used to assess the stability of protein residues.

3. Result and Discussion

3.1. Heterologous Expression of Thermotolerant α-Glucosidase in B. Subtilis 168

As a critical enzyme for the production of IMOs, the poor thermal stability of α-glucosidase is a significant factor limiting its application. Therefore, we searched the BRENDA (https://www.brenda-enzymes.org//, accessed on 5 July 2022) enzyme database and selected α-glucosidase with good heat resistance and industrial application potential. In the present study, we constructed recombinant plasmids by amplifying codon-optimized gene fragments and ligating them into the pMA5 vector. The recombinant strains *B. Subtilis* 168/pMA5-*gsj* and *B. Subtilis* 168/pMA5-*agl* were obtained after transferring the recombinant plasmids into *B. subtilis* 168. The results of SDS-PAGE (Figure 2) indicated that both *gsj* (code a 59 kDa protein) and *agl* (code a 96 kDa protein) genes could be expressed normally in *B. subtilis* 168. The enzyme activity of recombinant α-glucosidase was determined by the pNPG method. The results showed that the enzyme activity of GSJ was about 3.7 U·mL^{-1}, and AGL was about 2.5 U·mL^{-1}.

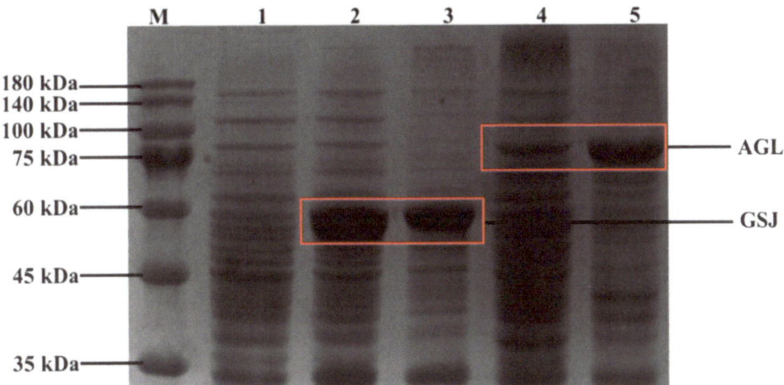

Figure 2. SDS-PAGE analysis of recombinant α-glucosidase. Lane M, protein marker; Lane 1, Control (*B. Subtilis* 168/pMA5 crude enzyme); Lane 2, *B. Subtilis* 168/pMA5-*gsj* crude enzyme; Lane 3, *B. Subtilis* 168/pMA5-*gsj* broken cell precipitate; Lane 4, *B. Subtilis* 168/pMA5-*agl* crude enzyme; Lane 5 *B. Subtilis* 168/pMA5-*agl* broken cell precipitate.

3.2. Transglucosylation Activity of Recombinant α-Glucosidase

α-glucosidase has both transglycosylation and hydrolysis activities. At high concentrations of maltose (G2) solution, α-glucosidase initially synthesizes panose (PN) and maltotriose (G3). With the increasing amount of glucose (Glc) delivered from the glycosylation step, α-glucosidase starts to synthesize isomaltose (IG2) and isomaltotriose (IG3) [7]. The transglycosylation activity of α-glucosidase is essential for the production of IMOs. Therefore, to determine the transglycosylation activity of recombinant α-glucosidase, we added an appropriate amount of recombinant α-glucosidase to a high concentration of maltose solution and detected the changes of each component during the reaction by HPLC. As shown in Figure 3a, after 12 h of transglycosylation reaction, only 17.2% of maltose was converted to IMOs in the mixture containing GSJ, while 72.8% of maltose was hydrolyzed to glucose. In the AGL mixture, 54% of maltose was converted to IMOs, and 33.3% was hydrolyzed to glucose (Figure 3b). The results show that the hydrolysis activity of GSJ is much higher than the transglycosylation activity, which implies that GSJ is unsuitable for producing IMOs. In contrast, AGL has a high transglycosylation activity and can be applied to the production of IMOs. The α-glucosidase derived from *Aspergillus niger* is widely used in producing α-glucosidase due to its good transglycosylation activity. In addition, the α-glucosidase from *Aspergillus awamori* and *Aspergillus carbonarious* also showed transglycosylation activity [32]. In suitable conditions, *Aspergillus niger*-derived α-glucosidase can convert about 55% of maltose to IMOs [33]; this is similar to the transglycosylation activity of the recombinant α-glucosidase in this study, which also indicates the potential of recombinant α-glucosidase to be used for the production of IMOs.

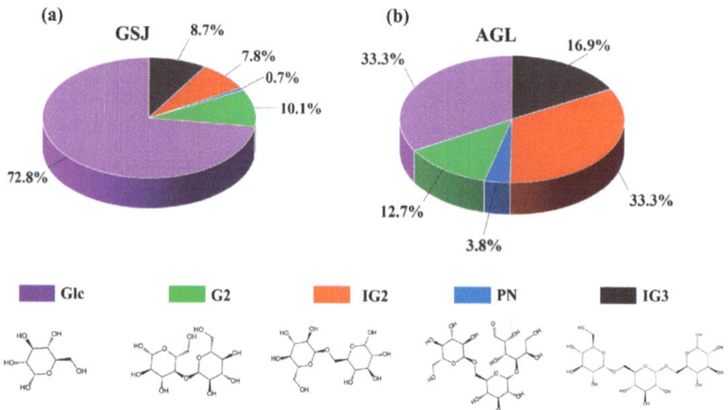

Figure 3. Analysis of transglycosylation products. The transglycosylation products of maltose by GSJ (**a**) and AGL (**b**).

3.3. Construction of Cyclized AGL to Improve the Thermal Stability of Recombinant α-Glucosidase

The cyclic peptide is a type of naturally occurring polypeptide molecule with a circular structure [34]. Due to its good stability, genetic engineering, protein engineering, other methods have developed a series of molecular cyclization techniques, such as protein trans-splicing (PTS) [35], expressed protein ligation (EPL) mediated by intrinsic peptide [36], and sortagging mediated by Sortase [37]. Although these cyclization methods can improve the stability of proteins, achieving the above linkage methods requires complex reaction conditions or the involvement of specific catalysts, and the reaction efficiency is low [38]. Isopeptide bond has been widely used due to its specificity, stability, and rapid reaction. We used three traditional isopeptide bonds (SpyTag/SpyCatcher, SnoopTag/SnoopCatcher, SdyTag/SnoopCatcher) and a novel short peptide tag (RIAD/RIDD) to form cyclized AGL. According to the method in Section 2.2.5 and Figure S1, we constructed recombinant plasmids and expressed them in *B. subtilis* 168. As shown in Figure 4, the results of SDS-PAGE indicated that the recombinant proteins RIAD-AGL-RIDD (~103 kDa), SpyTag-AGL-SpyCatcher (~103 kDa), SnoopTag-AGL-SnoopCatcher (~104 kDa), and SdyTag-AGL-SdyCatcher (~103 kDa) were successfully expressed, and the molecular weights were increased. To verify whether the isopeptide bond was formed, we mutated the key site of SpyTag-AGL-SpyCatcher. It was found that the band was located between cyclized AGL and wild-type AGL, indicating that AGL was successfully cyclized by isopeptide bonds. After purifying the cyclized protein, we assayed the enzyme activity of cyclized AGL (Table 1). We found that the activity of SpyTag-AGL-SpyCatcher, SnoopTag-AGL-SnoopCatcher, and SdyTag-AGL-SdyCatcher had no significant change. In contrast, the enzyme activity of RIAD-AGL-RIDD decreased by 56.5%. Therefore, we performed further enzymatic property characterization of SpyTag-AGL-SpyCatcher, SnoopTag-AGL-SnoopCatcher, and SdyTag-AGL-SdyCatcher.

Table 1. Specific enzyme activities of wild-type and cyclized AGL.

Enzyme	Specific Enzyme Activity (U·mg^{-1})	Relative Enzyme Activity (100%)
WT	27.91 ± 0.5	100.0 ± 1.7
RIAD-AGL-RIDD	12.01 ± 0.7	43.5 ± 2.5
SpyTag-AGL-SpyCatcher	26.43 ± 0.3	94.7 ± 1.1
SnoopTag-AGL-SnoopCatcher	28.55 ± 0.6	102.3 ± 2.1
SdyTag-AGL-SdyCatcher	27.18 ± 0.3	97.4 ± 1.1

All assays are performed in triplicate and the standard deviation of biological replicates is expressed as a numerical error.

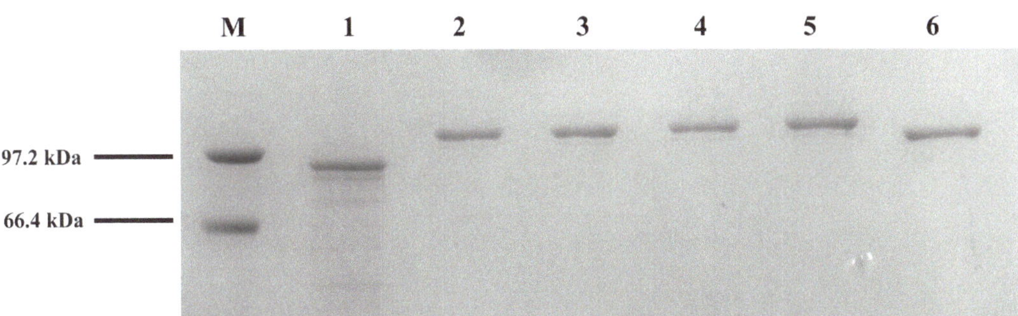

Figure 4. SDS-PAGE analysis of cyclized AGL. Lane M, protein marker; Lane 1, Control (wild type AGL); Lane 2, RIAD-AGL-RIDD; Lane 3, SpyTag-AGL-SpyCatcher; Lane 4, SnoopTag-AGL-SnoopCatcher; Lane 5, SdyTag-AGL-SdyCatcher; Lane 6, linear AGL.

3.4. Optimal Temperature and Thermal Stability of Cyclized AGL

To visually investigate the changes in enzyme activity at different temperatures, the optimal temperature of wild-type and cyclized AGL was characterized by the relative enzyme activity at different temperatures (40–75 °C). As shown in Figure 5a, the optimal temperature of all cyclized AGL was increased by 5 °C compared to the wild-type AGL. In addition, the relative enzyme activity of all cyclized AGL was higher than wild-type AGL at temperatures above 60 °C.

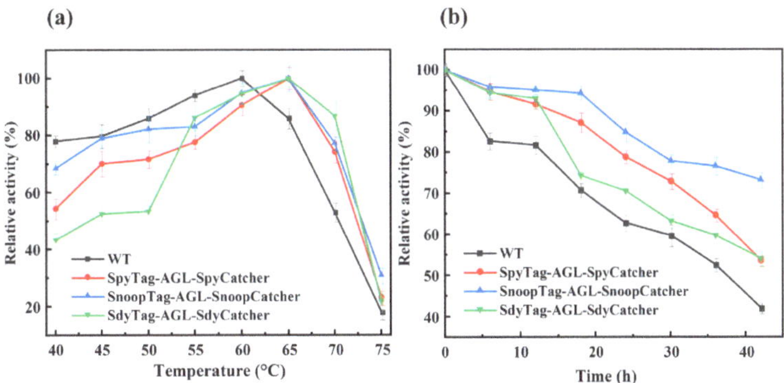

Figure 5. Effects of temperature on the activity and stability of wild-type and cyclized AGL: (**a**) optimal temperature, (**b**) thermal stability at 65 °C. All assays were performed in triplicate, and the standard deviations of biological replicates are indicated by error bars.

For the thermal stability experiments, we incubated the recombinant α-glucosidase at 65 °C for 42 h and measured the enzyme activity. The results are shown in Figure 5b. Compared to wild-type AGL, all cyclized AGL thermal stability was improved. After incubation at 65 °C for 42 h, the residual enzyme activities of SpyTag-AGL-SpyCatcher and SdyTag-AGL-SdyCatcher were 53.7% and 54.2%, respectively, while the wild-type AGL was 42.1%. In contrast, the thermal stability of SnoopTag-AGL-SnoopCatcher was significantly improved, and the residual enzyme activity after 42 h remained 73.2%, which was 1.74-fold higher than the wild type.

This result was similar to previous studies on cyclized proteins' optimal temperature and thermal stability. While the cyclized protein increased thermal stability, the optimal temperature was also increased. Chen et al. [24] found that the optimal temperature of cyclized Tres increased by 5–10 °C after using various isopeptide bonds. Wang et al. [25] used SpyCatcher/SpyTag to cause spontaneous cyclization of lichenase, and the optimal temperature of cyclized lichenase was increased by 5 °C. Likewise, Si et al. [27] constructed cyclized luciferase by SpyCatcher/SpyTag, which eventually increased the optimal temperature of luciferase by 5 °C.

3.5. Optimal pH and pH Stability of Cyclized AGL

We determined the enzyme activity of recombinant α-glucosidase at different pH according to method Section 2.2.5. As shown in Figure 6a, the optimal pH value was 6.0 for wild-type AGL and cyclized AGL, and the relative activity of all cyclized AGL was slightly higher than that of wild-type AGL at pH values in the range of 4.0–10.0. The residual enzyme activities of wild-type AGL and cyclized AGL were similar after 24 h incubation in different pH buffers (Figure 6b).

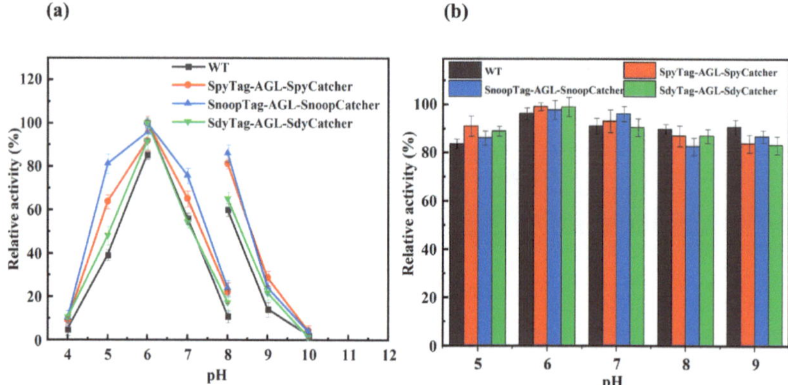

Figure 6. Effects of pH on the activity and stability of wild-type and cyclized AGL: (**a**) optimal pH, (**b**) pH stability. All assays were performed in triplicate, and the standard deviations of biological replicates are indicated by error bars.

3.6. Kinetic Analysis of Recombinant α-Glucosidase

The kinetic parameters of recombinant α-glucosidase were determined at pH 7.0 and 65 °C with pNPG as the substrate. The K_m, V_{max} and K_{cat} values of cyclized AGL were similar to the wild-type AGL (Table 2), which indicated that the cyclization reaction does not affect the catalytic efficiency of recombinant α-glucosidase. However, there is no consistent conclusion on whether the insertion of isopeptide bonds affects the catalytic efficiency of the protein. Previous studies have shown no effect of cyclization on the K_m values of luciferase [27] and β-lactamase [26], while the kinetic parameters of xylanase [39] and lichenase [25] showed significant differences in polysaccharide before and after cyclization.

Table 2. Kinetic parameters of wild-type AGL and cyclized AGL.

Enzyme	K_m (mM)	V_{max} (U mg^{-1})	K_{cat} (S^{-1})
WT	1.83 ± 0.06	36.42 ± 0.12	282.31 ± 2.1
SpyTag-AGL-SpyCatcher	1.86 ± 0.03	35.94 ± 0.07	276.38 ± 7.2
SnoopTag-AGL-SnoopCatcher	1.82 ± 0.05	36.12 ± 0.15	278.92 ± 6.8
SdyTag-AGL-SdyCatcher	1.86 ± 0.06	36.31 ± 0.19	281.59 ± 6.5

All assays are performed in triplicate and the standard deviation of biological replicates is expressed as a numerical error.

3.7. Molecular Dynamics Simulation Analysis of Cyclized AGL

The three-dimensional models of cyclized AGL were simulated to explore the molecular mechanism of increased thermal stability of cyclized AGL. Furthermore, MD simulation was performed at 338 K by GROMACS. Due to the lack of a suitable crystal structure as a template, we conducted homology modeling by ALPHA-FOLD to obtain the 3D structure of cyclized AGL. As shown in Figure 7, all three isopeptide bonds successfully cyclized the recombinant protein, and cyclization had no effect on the active sites of AGL compared to wild-type AGL (Figure S2). In addition, there was no significant change in the distance between the terminals in the wild-type and cyclized versions (Figure S3), indicating that the isopeptide bond did not distort the enzyme. The structural alignment revealed that cyclization did not cause conformational changes in AGL (Figure S4).

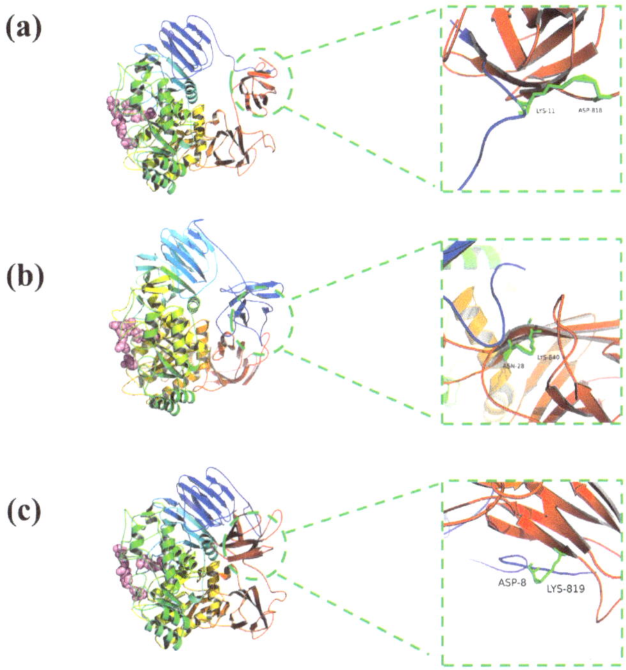

Figure 7. 3D structure of cyclized AGL (the pink spheres highlight the enzyme active sites). (**a**) SpyTag-AGL-SpyCatcher, (**b**) SnoopTag-AGL-SnoopCatcher, (**c**) SdyTag-AGL-SdyCatcher.

In the present study, we analyzed the dynamic behavior of the cyclized AGL systems by performing 30-ns MD simulations, and the RMSD value characterized the stability of the systems. As shown in Figure 8a, the WT system fluctuate between 0–0.4 nm until it reaches equilibrium at 5 ns. SpyTag-AGL-SpyCatcher, SnoopTag-AGL-SpyCatcher, and SdyTag-AGL-SdyCatcher systems fluctuate from 0 to 0.48 nm and maintain equilibrium between 5 and 30 ns. SnoopTag-AGL-SpyCatcher system has the highest RMSD value (0.48 nm), while the four systems were maintaining equilibrium. These results indicated that the four systems remained stable between 5 and 30 ns, and the 5 to 30 ns simulation trajectories could be used for further analysis, including Rg and RMSF.

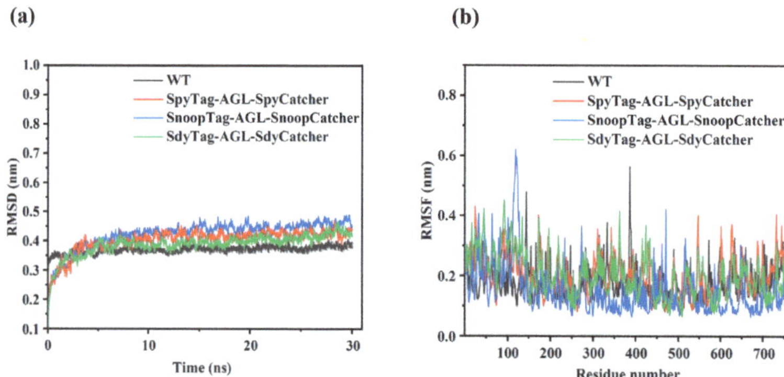

Figure 8. MD simulations of wild-type AGL and cyclized AGL. (**a**) RMSD values, (**b**) RMSF values.

As shown in Figure 8b, the fluctuations of most residues were similar in the four systems. The WT system's RMSF values were higher than the cyclized AGL system at residues 150–200, 320–350, 380–400, 560–590, and 700–740. This result shows that the rigidity of the cyclized AGL increased in these regions, contributing to the increased thermal stability of the cyclized AGL. Moreover, the RMSF values of the SnoopTag-AGL-SnoopCatcher system are lower than WT, SpyTag-AGL-SpyCatcher, and SdyTag-AGL-SdyCatcher systems at residues 300–400, 600–700, which explained the better thermal stability of SnoopTag-AGL-SnoopCatcher. In summary, the results of MD simulations are consistent with the experimental data.

4. Conclusions

In the present study, we heterologously expressed α-glucosidase with industrial application potential in *B. subtilis* 168. based on the molecular cyclization of isopeptide bonds, we constructed cyclized proteins to improve the thermal stability of recombinant α-glucosidase. Among them, SpyTag/SpyCatcher can significantly improve the thermal stability of the recombinant protein. MD simulations showed that the rigidity of the cyclized proteins was increased, indicating that improving the thermal stability of α-glucosidase by isopeptide bonds is an effective method. Due to its good thermal stability and high transglycosylation activity, cyclized AGL can be used to produce IMOs continuously. Therefore, in future studies, we will try to immobilize the enzyme with different materials and determine the optimal conditions for immobilization. The immobilized enzyme is used in the continuous production of IMOs. In addition, we will select suitable mutation sites for targeted mutagenesis by rational design to further improve the thermal stability and transglycosylation activity of recombinant α-glucosidase.

Supplementary Materials: The following supporting information can be downloaded at: https://www.mdpi.com/article/10.3390/fermentation8100498/s1, Figure S1: Construction of recombinant plasmids for cyclized AGL; Figure S2: Three-dimensional structure of wild-type AGL; Figure S3: The distance between the terminals; Figure S4: Structural alignment of cyclized AGL with wild-type; Table S1: strains and plasmids used in this study; Table S2: Primers used to construct the cyclized proteins; Table S3: Amino acid sequence of cyclized protein.

Author Contributions: Conceptualization, Z.R. and X.Z.; methodology, Z.W.; software, H.Z.; validation, Q.W., M.F. and R.L.; formal analysis, M.X.; investigation, M.H.; resources, M.H.; data curation, Z.W.; writing—original draft preparation, Z.W.; writing—review and editing, Z.W.; visualization, M.H.; supervision, X.Z.; project administration, X.Z.; funding acquisition, Z.R. All authors have read and agreed to the published version of the manuscript.

Funding: This research was funded by National Key Research and Development Program of China, (No. 2021YFC2100900), National Natural Science Foundation of China (No. 32171471, No. 32071470), Key Research and Development Project of Shandong Province, China (2019JZZY020605), the Project Funded by the Priority Academic Program Development of Jiangsu Higher Education Institutions, Top-notch Academic Programs Project of Jiangsu Higher Education Institutions.

Institutional Review Board Statement: Not applicable.

Informed Consent Statement: Not applicable.

Data Availability Statement: The data that support the findings of this study are available from the corresponding author upon reasonable request.

References

1. Huang, Z.; Li, Z.; Su, Y.; Zhu, Y.; Zeng, W.; Chen, G.; Liang, Z. Continuous Production of Isomalto-oligosaccharides by Thermo-inactivated Cells of Aspergillus niger J2 with Coarse Perlite as an Immobilizing Material. *Appl. Biochem. Biotechnol.* **2018**, *185*, 1088–1099. [CrossRef] [PubMed]
2. Shi, Q.; Hou, Y.; Juvonen, M.; Tuomainen, P.; Kajala, I.; Shukla, S.; Goyal, A.; Maaheimo, H.; Katina, K.; Tenkanen, M. Optimization of Isomaltooligosaccharide Size Distribution by Acceptor Reaction of Weissella confusa Dextransucrase and Characterization of Novel alpha-(1 -> 2)-Branched Isomaltooligosaccharides. *J. Agric. Food Chem.* **2016**, *64*, 3276–3286. [CrossRef] [PubMed]
3. Ojha, S.; Mishra, S.; Chand, S. Production of isomalto-oligosaccharides by cell bound alpha-glucosidase of *Microbacterium* sp. *Lwt-Food Sci. Technol.* **2015**, *60*, 486–494. [CrossRef]
4. Soto-Maldonado, C.; Concha-Olmos, J.; Caceres-Escobar, G.; Menses-Gomez, P. Sensory evaluation and glycaemic index of a food developed with flour from whole (pulp and peel) overripe banana (Musa cavendishii) discards. *Lwt-Food Sci. Technol.* **2018**, *92*, 569–575. [CrossRef]
5. Wu, Q.; Pi, X.e.; Liu, W.; Chen, H.; Yin, Y.; Yu, H.D.; Wang, X.; Zhu, L. Fermentation properties of isomaltooligosaccharides are affected by human fecal enterotypes. *Anaerobe* **2017**, *48*, 206–214. [CrossRef]
6. Madsen, L.R., II; Stanley, S.; Swann, P.; Oswald, J. A Survey of Commercially Available Isomaltooligosaccharide-Based Food Ingredients. *J. Food Sci.* **2017**, *82*, 401–408. [CrossRef]
7. Gu, X.L.; Li, H.; Song, Z.H.; Ding, Y.N.; He, X.; Fan, Z.Y. Effects of isomaltooligosaccharide and Bacillus supplementation on sow performance, serum metabolites, and serum and placental oxidative status. *Anim. Reprod. Sci.* **2019**, *207*, 52–60. [CrossRef]
8. Nakakuki, T. Present status and future of functional oligosaccharide development in Japan. *Pure Appl. Chem.* **2002**, *74*, 1245–1251. [CrossRef]
9. Bivolarski, V.; Vasileva, T.; Bozov, P.; Iliev, I. INFLUENCE OF DIFFERENT ACCEPTORS ON SYNTHESIS OF GLUCOOLIGOSAC-CHARIDES BY PURIFIED DEXTRANSUCRASE FROM LEUCONOSTOC MESENTEROIDES URE 13. *Comptes Rendus De L Acad. Bulg. Des. Sci.* **2013**, *66*, 1405–1412.
10. Sorndech, W.; Sagnelli, D.; Blennow, A.; Tongta, S. Combination of amylase and transferase catalysis to improve IMO compositions and productivity. *Lwt-Food Sci. Technol.* **2017**, *79*, 479–486. [CrossRef]
11. Rudeekulthamrong, P.; Sawasdee, K.; Kaulpiboon, J. Production of long-chain isomaltooligosaccharides from maltotriose using the thermostable amylomaltase and transglucosidase enzymes. *Biotechnol. Bioprocess Eng.* **2013**, *18*, 778–786. [CrossRef]
12. Ma, M.; Okuyama, M.; Sato, M.; Tagami, T.; Klahan, P.; Kumagai, Y.; Mori, H.; Kimura, A. Effects of mutation of Asn694 in Aspergillus niger alpha-glucosidase on hydrolysis and transglucosylation. *Appl. Microbiol. Biotechnol.* **2017**, *101*, 6399–6408. [CrossRef] [PubMed]
13. Chen, D.-L.; Tong, X.; Chen, S.-W.; Chen, S.; Wu, D.; Fang, S.-G.; Wu, J.; Chen, J. Heterologous Expression and Biochemical Characterization of alpha-Glucosidase from Aspergillus niger by Pichia pastroris. *J. Agric. Food Chem.* **2010**, *58*, 4819–4824. [CrossRef] [PubMed]
14. Zhang, F.; Wang, W.; Bah, F.B.M.; Song, C.; Zhou, Y.; Ji, L.; Yuan, Y. Heterologous Expression of a Thermostable alpha-Glucosidase from *Geobacillus* sp. Strain HTA-462 by Escherichia coli and Its Potential Application for Isomaltose-Oligosaccharide Synthesis. *Molecules* **2019**, *24*, 1413. [CrossRef] [PubMed]
15. Zhou, C.; Xue, Y.; Ma, Y. Evaluation and directed evolution for thermostability improvement of a GH 13 thermostable alpha-glucosidase from Thermus thermophilus TC11. *BMC Biotechnol.* **2015**, *15*, 97. [CrossRef] [PubMed]
16. Park, J.-E.; Park, S.R.; Woo, J.Y.; Hwang, H.S.; Cha, J.; Lee, H. Enzymatic Properties of a Thermostable alpha-Glucosidase from Acidothermophilic Crenarchaeon Sulfolobus tokodaii Strain 7. *J. Microbiol. Biotechnol.* **2013**, *23*, 56–63. [CrossRef] [PubMed]
17. Hung, V.S.; Hatada, Y.; Goda, S.; Lu, J.; Hidaka, Y.; Li, Z.J.; Akita, M.; Ohta, Y.; Watanabe, K.; Matsui, H.; et al. alpha-Glucosidase from a strain of deep-sea Geobacillus: A potential enzyme for the biosynthesis of complex carbohydrates. *Appl. Microbiol. Biotechnol.* **2005**, *68*, 757–765. [CrossRef]
18. Zhou, C.; Xue, Y.; Ma, Y. Enhancing the thermostability of alpha-glucosidase from Thermoanaerobacter tengcongensis MB4 by single proline substitution. *J. Biosci. Bioeng.* **2010**, *110*, 12–17. [CrossRef]

19. Zakeri, B.; Howarth, M. Spontaneous Intermolecular Amide Bond Formation between Side Chains for Irreversible Peptide Targeting. *J. Am. Chem. Soc.* **2010**, *132*. [CrossRef]
20. Zakeri, B.; Fierer, J.O.; Celik, E.; Chittock, E.C.; Schwarz-Linek, U.; Moy, V.T.; Howarth, M. Peptide tag forming a rapid covalent bond to a protein, through engineering a bacterial adhesin. *Proc. Natl. Acad. Sci. USA* **2012**, *109*, E690–E697. [CrossRef]
21. Veggiani, G.; Nakamura, T.; Brenner, M.D.; Gayet, R.V.; Yan, J.; Robinson, C.V.; Howarth, M. Programmable polyproteams built using twin peptide superglues. *Proc. Natl. Acad. Sci. USA* **2016**, *113*, 1202–1207. [CrossRef]
22. Fierer, J.O.; Veggiani, G.; Howarth, M. SpyLigase peptide-peptide ligation polymerizes affibodies to enhance magnetic cancer cell capture. *Proc. Natl. Acad. Sci. USA* **2014**, *111*, E1176–E1181. [CrossRef]
23. Wong, J.X.; Gonzalez-Miro, M.; Sutherland-Smith, A.J.; Rehm, B.H.A. Covalent Functionalization of Bioengineered Polyhydroxyalkanoate Spheres Directed by Specific Protein-Protein Interactions. *Front. Bioeng. Biotechnol.* **2020**, *8*, 44. [CrossRef] [PubMed]
24. Chen, Y.; Zhao, Y.; Zhou, X.; Liu, N.; Ming, D.; Zhu, L.; Jiang, L. Improving the thermostability of trehalose synthase from Thermomonospora curvata by covalent cyclization using peptide tags and investigation of the underlying molecular mechanism. *Int. J. Biol. Macromol.* **2021**, *168*, 13–21. [CrossRef] [PubMed]
25. Wang, J.; Wang, Y.; Wang, X.; Zhang, D.; Wu, S.; Zhang, G. Enhanced thermal stability of lichenase from *Bacillus subtilis* 168 by SpyTag/SpyCatcher-mediated spontaneous cyclization. *Biotechnol. Biofuels* **2016**, *9*, 79. [CrossRef]
26. Schoene, C.; Bennett, S.P.; Howarth, M. SpyRing interrogation: Analyzing how enzyme resilience can be achieved with phytase and distinct cyclization chemistries. *Sci. Rep.* **2016**, *6*, 21151. [CrossRef] [PubMed]
27. Si, M.; Xu, Q.; Jiang, L.; Huang, H. SpyTag/SpyCatcher Cyclization Enhances the Thermostability of Firefly Luciferase. *PLoS ONE* **2016**, *11*, e0162318. [CrossRef]
28. Wang, Y.-H.; Jiang, Y.; Duan, Z.-Y.; Shao, W.-L.; Li, H.-Z. Expression and characterization of an alpha-glucosidase from Thermoanaerobacter ethanolicus JW200 with potential for industrial application. *Biologia* **2009**, *64*, 1053–1057. [CrossRef]
29. Kang, W.; Ma, T.; Liu, M.; Qu, J.; Liu, Z.; Zhang, H.; Shi, B.; Fu, S.; Ma, J.; Lai, L.T.F.; et al. Modular enzyme assembly for enhanced cascade biocatalysis and metabolic flux. *Nat. Commun.* **2019**, *10*, 4248. [CrossRef]
30. You, J.; Yang, C.; Pan, X.; Hu, M.; Du, Y.; Osire, T.; Yang, T.; Rao, Z.J.B.T. Metabolic engineering of *Bacillus subtilis* for enhancing riboflavin production by alleviating dissolved oxygen limitation. *Bioresour. Technol.* **2021**, *333*, 125228. [CrossRef]
31. Jung, J.H.; Seo, D.H.; Holden, J.F.; Kim, H.S.; Baik, M.Y.; Park, C.S. Broad substrate specificity of a hyperthermophilic alpha-glucosidase from Pyrobaculum arsenaticum. *Food Sci. Biotechnol.* **2016**, *25*, 1665–1669. [CrossRef] [PubMed]
32. Anindyawati, T.; Ann, Y.-G.; Ito, K.; Iizuka, M.; Minamiura, N. Two kinds of novel α-glucosidases from Aspergillus awamori KT-11: Their purifications, properties and specificities. *J. Ferment. Bioeng.* **1998**, *85*, 465–469. [CrossRef]
33. Shimba, N.; Shinagawa, M.; Hoshino, W.; Yamaguchi, H.; Yamada, N.; Suzuki, E.-i. Monitoring the hydrolysis and transglycosylation activity of alpha-glucosidase from Aspergillus niger by nuclear magnetic resonance spectroscopy and mass spectrometry. *Anal. Biochem.* **2009**, *393*, 23–28. [CrossRef] [PubMed]
34. Cascales, L.; Craik, D.J. Naturally occurring circular proteins: Distribution, biosynthesis and evolution. *Org. Biomol. Chem.* **2010**, *8*, 5035–5047. [CrossRef]
35. Boecker, J.K.; Friedel, K.; Matern, J.C.J.; Bachmann, A.-L.; Mootz, H.D. Generation of a Genetically Encoded, Photoactivatable Intein for the Controlled Production of Cyclic Peptides. *Angew. Chem. Int. Ed.* **2015**, *54*, 2116–2120. [CrossRef]
36. De Rosa, L.; Cortajarena, A.L.; Romanelli, A.; Regan, L.; D'Andrea, L.D. Site-specific protein double labeling by expressed protein ligation: Applications to repeat proteins. *Org. Biomol. Chem.* **2012**, *10*, 273–280. [CrossRef]
37. Agwa, A.J.; Craik, D.J.; Schroeder, C.I. Cyclizing Disulfide-Rich Peptides Using Sortase A. *Methods Mol. Biol. (Clifton N.J.)* **2019**, *2012*, 29–41. [CrossRef]
38. Minato, Y.; Ueda, T.; Machiyama, A.; Shimada, I.; Iwai, H. Segmental isotopic labeling of a 140 kDa dimeric multi-domain protein CheA from Escherichia coli by expressed protein ligation and protein trans-splicing. *J. Biomol. NMR* **2012**, *53*, 191–207. [CrossRef]
39. Lin, Y.; Jin, W.; Wang, J.; Cai, Z.; Wu, S.; Zhang, G. A novel method for simultaneous purification and immobilization of a xylanase-lichenase chimera via SpyTag/SpyCatcher spontaneous reaction. *Enzym. Microb. Technol.* **2018**, *115*, 29–36. [CrossRef]

fermentation

Article

Optimized Recombinant Expression and Characterization of Collagenase in *Bacillus subtilis* WB600

Yaqing Zhu [1], Linlin Wang [2], Kaixuan Zheng [1], Ping Liu [3], Wenkang Li [1], Jian Lin [1], Wenjing Liu [4], Shoushui Shan [1,*], Liqin Sun [1,*] and Hailing Zhang [1,*]

[1]   Department of Biological Engineering, College of Life Sciences, Yantai University, Yantai 264005, China
[2]   Department of Food Engineering, Shandong Business Institute, Yantai 264670, China
[3]   Central Innovation Biotechnology (Shandong) Co., Ltd., Yantai 264005, China
[4]   School of Medicine, Yunnan University, Kunming 650091, China
*   Correspondence: xinshibio@163.com (S.S.); sliqin2005@163.com (L.S.); hailing1203@hotmail.com (H.Z.); Tel.: +86-535-6902638 (H.Z.)

Abstract: Background: The collagenase encoding gene *col* was cloned into a pP43NMK vector and amplified in *Escherichia coli* JM109 cells. The shuttle vector pP43NMK was used to sub-clone the *col* gene to obtain the vector pP43NMK-*col* for the expression of collagenase in *Bacillus subtilis* WB600. The enzyme was characterized and the composition of the expression medium and culture conditions were optimized. Methods: The expressed recombinant enzyme was purified by ammonium sulfate, ultrafiltration, and through a nickel column. The purified collagenase had an activity of 9405.54 U/mg. Results: The recombinant enzyme exhibited optimal activity at pH 9.0 and 50 °C. Catalytic efficiency of the recombinant collagenase was inhibited by Fe^{3+} and Cu^{2+}, but stimulated by Co^{2+}, Ca^{2+}, Zn^{2+}, and Mg^{2+}. The optimal conditions for its growth were at pH 7.0 and 35 °C, using 15 g/L of fructose and 36 g/L of yeast powder and peptone mixture (2:1) at 260 rpm with 11% inoculation. The maximal extracellular activity of the recombinant collagenase reached 2746.7 U/mL after optimization of culture conditions, which was 2.4-fold higher than that before optimization. Conclusions: This study is a first attempt to recombinantly express collagenase in *B. subtilis* WB600 and optimize its expression conditions, its production conditions, and possible scale-up.

Keywords: collagenase; *Bacillus subtilis* WB600; protein purification; characterization; recombinant protein expression

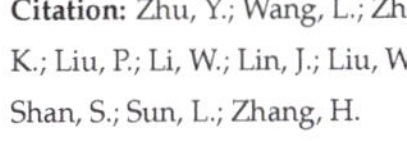
check for updates

Citation: Zhu, Y.; Wang, L.; Zheng, K.; Liu, P.; Li, W.; Lin, J.; Liu, W.; Shan, S.; Sun, L.; Zhang, H. Optimized Recombinant Expression and Characterization of Collagenase in *Bacillus subtilis* WB600. *Fermentation* **2022**, *8*, 449. https://doi.org/10.3390/fermentation8090449

Academic Editors: Xian Zhang and Zhiming Rao

Received: 16 August 2022
Accepted: 7 September 2022
Published: 9 September 2022

Publisher's Note: MDPI stays neutral with regard to jurisdictional claims in published maps and institutional affiliations.

1. Introduction

Collagenases can be classified as metallocollagenases and serine collagenases that specifically hydrolyze the three-dimensional spiral structure of natural collagen [1,2]. They are involved in the degradation of the extracellular matrices of animal cells [3]. Enzymes catalyzing collagen hydrolysis include matrix metalloproteinases (MMPs), which are zinc-containing enzymes [4]; they usually require calcium for their optimum activity and stability [5]. Collagenases are found in several bacterial species such as *Bacillus cereus* [6,7], *Clostridium histolyticum* [8], *Bacillus subtilis* [9], and *Actinomyces* [10]. However, activity efficiency of most bacterial collagenases does not meet the needs of industrial processes, which are often carried out under conditions not optimal for the stability and activity of collagenases. Nevertheless, collagenases have several industrial, biotechnological, pharmacological, and medicinal applications. They are used to treat cardiovascular diseases, neurodegenerative diseases, cancer, arthritis [10–14], and liver fibrosis [15].

Collagenase genes of different origins were heterologously expressed in *Escherichia coli* (*E. coli*) [6–8]. However, *E. coli* is not a suitable host for enzyme production in the food industry due to plasmid instability and the presence of endotoxins. *Bacillus subtilis* (*B. subtilis*) is a non-pathogenic and bacteriophage-resistant host, which is generally regarded as safe [16] compared to *E. coli*. Moreover, *B. subtilis* can use various signal peptides to non-specifically secrete recombinant proteins for the production of industrial enzymes [17] and act as microbial cell factories [18]. *B. subtilis* WB600 lacks six protease genes; therefore, secreted proteins are not digested [19].

The present study describes the successful expression of collagenase in *B. subtilis* WB600 by using a modified shuttle vector pP43NMK. Furthermore, the culture conditions, including the key media components and the fermentation conditions, were optimized. The results obtained here may be useful for the industrial production of collagenase. The production of recombinant collagenase in *Bacillus subtilis* can simplify the purification process for its secretory expression, and the enzymes can be safely and widely used in the food industry due to their non-pathogenicity, which avoids the effects of harmful virulence factors [20].

2. Materials and Methods

2.1. Media Composition and Culture Conditions

The *Bacillus* strains were cultured in Luria Bertani (LB) medium containing peptone (10 g/L), yeast extract (5 g/L), and sodium chloride (10 g/L) at 37 °C and shaken at 200 rpm. Terrific Broth (TB), comprising peptone (12 g/L), yeast extract (24 g/L), glycerol (0.4% v/v), KH_2PO_4 (0.23 g/L), and K_2HPO_4 (1.25 g/L), was used as the fermentation medium for collagenase expression. Kanamycin (Kan; 50 µg/mL) was added to the LB and TB media. Collagenase was expressed in 100 mL of growth media in 300 mL conical flasks at 37 °C for 48 h.

Plasmid extraction kit: Tiangen Biotech (Beijing) Co., Ltd., Beijing, China. DNA Marker: Takara Bio Co., Ltd., Beijing, China. Ultra-micro spectrophotometer: NP80 Mobile, Implen International Trading Co., Ltd., Beijing, China. Shaking incubator: ZQTY-70V, Shanghai Zhichu Instrument Co., Ltd., Shanghai, China. Gel imaging system: 4600SF, Shanghai Tianneng Technology Co., Ltd., Shanghai, China. Autoclave pot: D-1 autoclave, Beijing Faen Science and Trade Co., Ltd., Beijing, China. Vertical electrophoresis instrument: Mini-protean, Bio-Rad Laboratories, Inc., Shanghai, China. Agarose level electrophoresis instrument: DYCP-31D, Beijing Liuyi Biotechnology Co., Ltd., Beijing, China. UV/VIS spectrophotometer: TU-1810, Beijing General Instrument Co., Ltd., Beijing, China. High-speed refrigerated centrifuge: ALLegra, Beckman Coulter, Inc., California, USA. Digital display constant temperature water bath pot: HH-4A, Guohua Instrument Manufacturing Co., Ltd., Changzhou, China. Thermal cycler: T100, Bio-Rad Laboratories, Inc., Shanghai, China.

2.2. Gene Optimization Methodology

Codon usage bias of the optimization sequence was adjusted against a proprietary reference codon bias established by multi-omics data analysis to fit the profile of highly expressed genes in the target host and optimized by Sangon Biotech (Shanghai) Co., Ltd., Shanghai, China, with their homemade software (http://192.168.19.50:8080/cool/, accessed on 6 May 2022). Various parameters were meticulously evaluated, including codon usage bias, codon context, GC profile, negative CpG islands, splicing sites, negative cis-acting elements, hidden stop codon, mRNA secondary structure, RNA instability, repeat sequences, restriction sites, and any undesired motif. Artificial gene synthesis was carried out by PCR, digestion, ligation, cell transformation, and other technologies.

2.3. Plasmid Construction and Transformation for Collagenase Expression

B. amyloliquefaciens has been proven able to synthesize collagenase and secrete it extracellularly, and the dermal collagen could be degraded by its fermentation broth, which could be applied to leather production [21]. The application of collagenase in the food industry has attracted much attention, and it is necessary to find a collagenase gene with stable enzymatic properties, accompanied by the potential for mass production in the microbiology industry. According to the GenBank database, the sequence identity of the collagenase encoding gene from *B. amyloliquefaciens* and *Bacillus velezensis* is 100%;

therefore, the gene synthesis was carried out according to the sequence published by *Bacillus velezensis* in this study, whose GenBank code is CP011686.1 with the locus tag AB13_2557. The collagenase gene sequence was extracted from GenBank (*col*, GenBank ID CP011686.1, locus tag: AB13_2557) and restriction enzyme sites' sequences for *Pst* I and *Hind* III were synthesized and ligated into the plasmid pP43NMK purchased from Sangon Biotech (Shanghai) Co., Ltd., Shanghai, China. The ligated plasmid was used to transform *E. coli* JM109 competent cells. Colonies carrying the pP43NMK-*col* plasmid were confirmed by colony PCR and double digestion by the restriction enzymes *Pst* I and *Hind* III. Following amplification in *E. coli* JM109, the expression plasmids were extracted and were used to transform *B. subtilis* WB600 [1].

2.4. Recombinant Expression and Purification of Collagenase

The *B. subtilis* cells were grown for 14–16 h in 50 mL LB media at 37 °C with stirring at 220 rpm. An aliquot of the overnight culture (2% *v/v*) was inoculated into 100 mL of fermentation medium in a 300 mL conical flask and incubated for 48 h and harvested by centrifugation at 10,000 rpm at 4 °C for 20 min.

Saturated ammonium sulfate solution was carefully added to the supernatant at different concentrations (20%, 30%, 40%, 50%, 60%, 70%, 80%, 90%) and incubated for 14–16 h at 4 °C. The precipitates were dissolved in 10 mM Tris-HCl buffer (pH 7.5) and dialyzed against the same buffer to completely remove the ammonium sulfate. The enzyme solution was concentrated using a 10 kDa Amicon filter (Merck, Munich, Germany) and purified using a His-Tag Purification Resin (Beyotime Biotechnology, Shanghai, China). The purified recombinant collagenase was then analyzed by sodium dodecyl sulfate–polyacrylamide gel electrophoresis (SDS-PAGE).

2.5. Sodium Dodecyl Sulphate–Polyacrylamide Gel Electrophoresis

SDS-PAGE was carried out to determine the successful purification of the recombinant protein and its molecular mass. Coomassie Brilliant Blue R-250 was used for staining the gel, and the molecular mass of collagenase was estimated on a 12% polyacrylamide gel based on protein standards [22].

2.6. Protein Quantification and Enzyme Assay of Collagenase

The protein concentration was estimated using the Bradford method; the absorbance was measured at 595 nm and bovine serum albumin was used as the protein standard [23].

Collagenase activity was determined by the colorimetric method. Briefly, 1 mL of 1 mg/mL gelatin solution was mixed with 500 μL of Tris-HCl (pH 7.5) and 100 μL of the enzyme solution and incubated for 40 min at 37 °C. Subsequently, the reaction was stopped by adding 500 μL of trichloroacetic acid (10%, *m/v*), 900 μL of acetic acid buffer (pH 5.4), and 1 mL of indigohydrone to the mixture, followed by incubation in boiling water for 10 min, then immediately cooling down in ice water. Finally, 4 mL 60% ethanol was added to the mixture, and the absorbance was read at 570 nm.

One unit (U) of enzyme activity was defined as the amount of enzyme catalyzing the formation of 1.0 μg glycine per minute.

2.7. Effects of pH, Temperature, and Metal Ions on Collagenase Activity and Stability

Collagenase activity was assayed under standard conditions at various temperatures (30–80 °C) to determine the optimum temperature of the recombinant enzyme. Thermal stability of the enzyme was measured for any residual activity at different temperatures for 2 h. All experiments were performed in triplicate and repeated three times. The residual activity was expressed as a percentage of the control sample activity, based on the assumption that the activity of control sample (in the absence of any additives) was 100%.

For the pH dependence assay, the collagenase enzymatic activity was measured using buffers at different pH levels, including sodium citrate buffer (pH 5.0), phosphate buffer (pH 6.0–8.0), and glycine–NaOH buffer (pH 9.0–11.0). To determine the pH stability of the recombinant collagenase, the enzyme was incubated in the above buffers for 2 h, followed by the measurement of the residual enzymatic activity under standard assay conditions.

To analyze the effect of metal ions on the enzyme activity, the enzyme solution was incubated in different metal ion solutions, including Fe^{3+}, Zn^{2+}, Co^{2+}, Ca^{2+}, Mg^{2+}, and Cu^{2+} (5 mM) for 2 h. The residual activity of the enzyme in each solution was measured individually and compared.

2.8. Optimization of the Fermentation Conditions for the Recombinant Expression of Collagenase

The effects of media composition and culture conditions on the recombinant expression of collagenase in *B. subtilis* WB600 were examined.

The carbon sources (glucose, mannitol, maltose, fructose, glycerin, and sucrose) and nitrogen sources (beef paste, ammonium sulfate, urea, gelatin, peptone, and yeast extract) were varied with an initial total nitrogen concentration of 10.0 g/L. The initial concentrations of carbon source and nitrogen source analyzed were 5.0, 10.0, 15.0, 20.0, 25.0, and 30.0 g/L and 6.0, 12.0, 18.0, 24.0, 30.0, 36.0, and 42.0 g/L, respectively. The effect of metal ions (Zn^{2+}, Mg^{2+}, Na^+, and Ca^{2+}) on collagenase production was also investigated.

The optimized medium was adjusted at different initial pH levels (5.0, 6.0, 7.0, 8.0, 9.0), stirring speeds (200, 220, 240, 260, 280 rpm), temperatures (25, 30, 35, 40, 45 °C), and inoculation concentrations (5%, 7%, 9%, 11%, 13%, 15%). All the values of enzyme activities were averaged from three replicates with standard deviations.

3. Results and Discussion

3.1. Gene Optimization

Gene expression is regulated and influenced by various factors such as codon usage bias, ribosome binding, and mRNA structure. The gene optimization process takes into consideration as many factors as possible, resulting in gene sequences that can optimally express proteins. There are numerous possibilities of mRNA sequences coding for the same protein. However, an advanced algorithm was used to screen and generate an optimal sequence from tens of thousands of candidate sequences.

The 936 bp collagenase sequence was optimized for optimal protein expression in *B. subtilis*, as shown in Figures 1 and 2. Codon usage bias was adjusted to fit the highest expression profile of the target host; CAI (Codon Adaptation Index) was upgraded from 0.81 to 0.96 (a CAI of 0.8–1.0 is regarded as good for high expression). Average GC content was adjusted from 47.1% to 40.3% and unfavorable peaks were removed. Repeated sequences in the original sequence were removed to avoid the formation of stem-loop structures in the mRNA and to facilitate the protein synthesis process. Undesired motifs, including restriction enzyme sites for use in sub-cloning and negative cis-acting sites, were modified. The whole sequence was fine-tuned to increase the translation efficiency and prolong the half-life of mRNA.

3.2. Expression of Recombinant Collagenase

DNA fragments around 936 bp were obtained, encoding for a protein comprising 312 amino acids. The appropriate recombinant plasmid was confirmed by double digestion with the restriction endonucleases *Pst* I and *Hind* III (Figure 3a) and was subsequently transformed into competent *B. subtilis* cells.

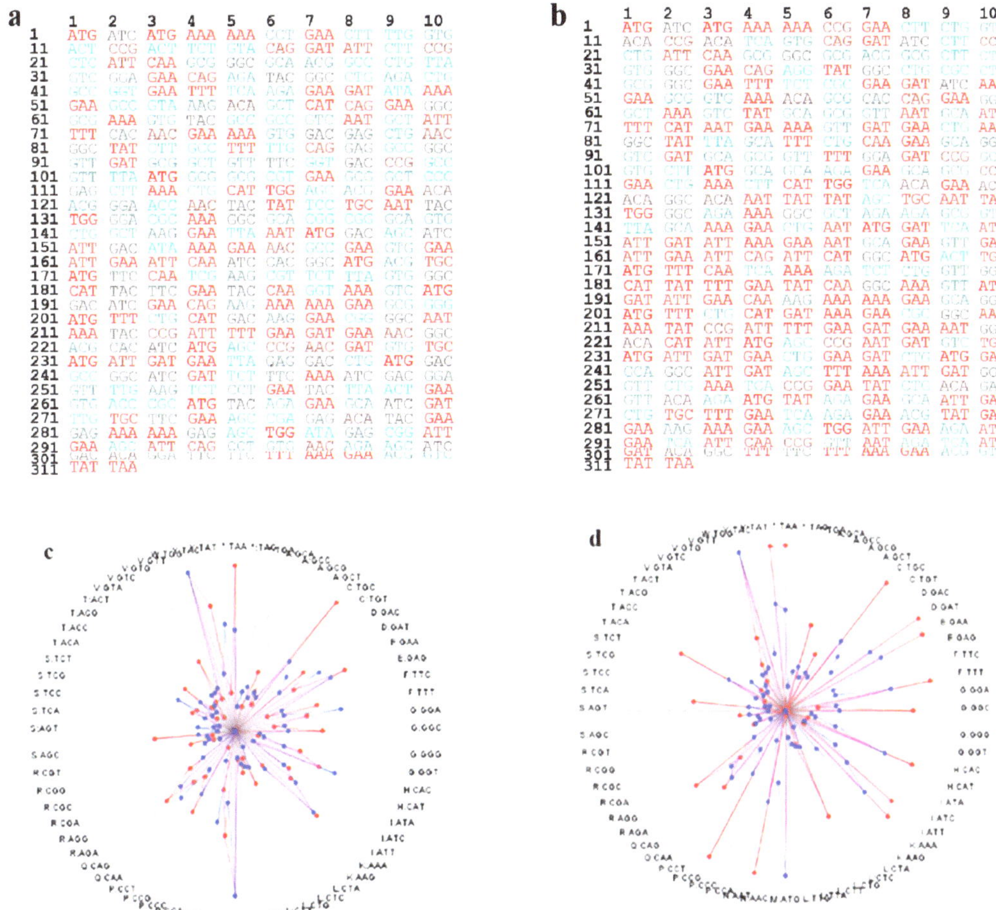

Figure 1. (**a**) Original sequence. (**b**) Optimized sequence. Relative codon frequency distribution. Color of codons indicates the frequency of that codon with respect to the host. Rarely used codons are shown in cyan and frequently used codons are shown in red. A redder codon indicates higher frequency. Redder codons compared with the other codons mean that they are better suited for the host codon bias, resulting in high expression. (**c**) Original sequence. (**d**) Optimized sequence. Codon relative frequency radar plot. The relative codon frequency distribution shows the frequency of each individual codon and this radar plot shows the suitability of the codon usage profile between the optimized sequence (show in red) and host (shown in blue). A better curve match means it is more adequate.

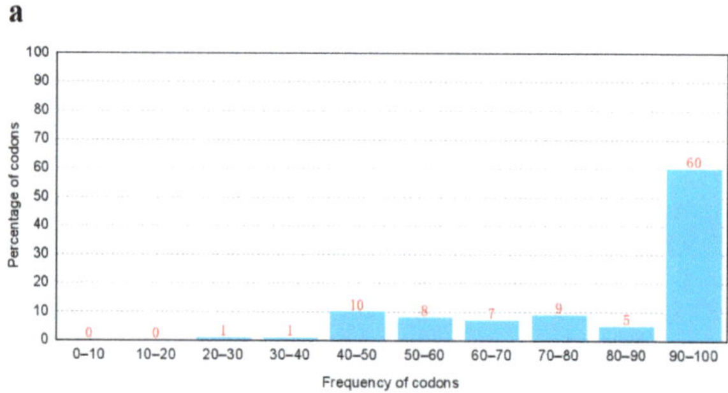

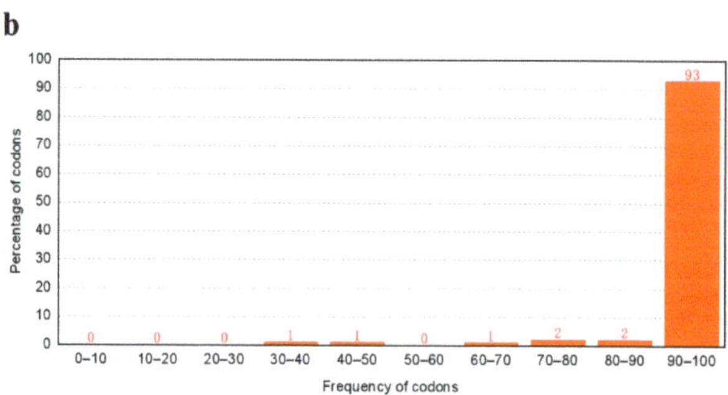

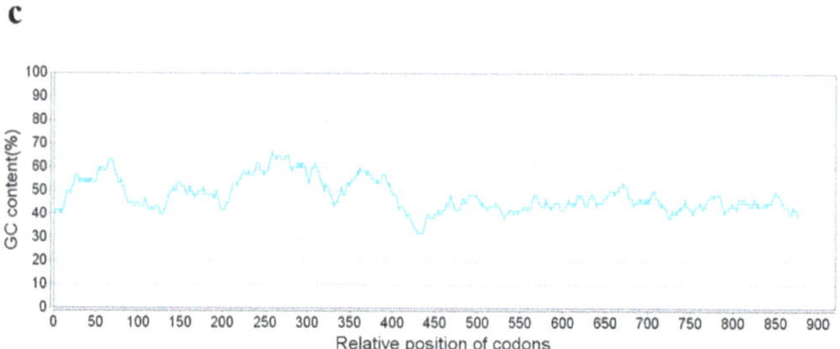

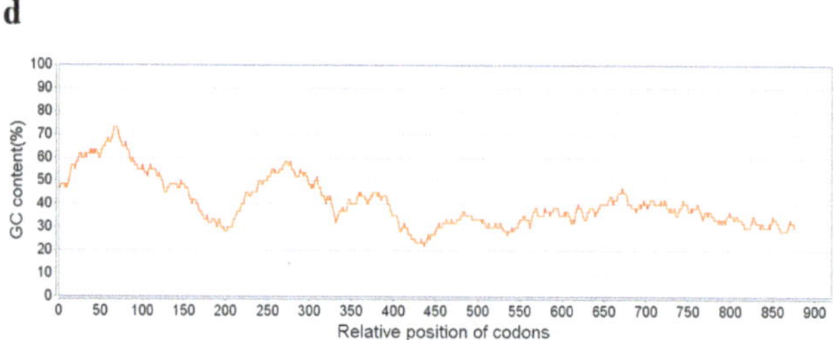

Figure 2. (**a**) Original sequence. (**b**) Optimized sequence. Frequency of optimal codons. Shows the percentage distribution of codons; codons with a higher percentage are more frequently used. (**c**) Original Sequence (47.1%). (**d**) Optimized sequence (40.3%). GC content adjustment. The ideal percentage range of GC content is between 30% and 70%, ideally 40–60%.

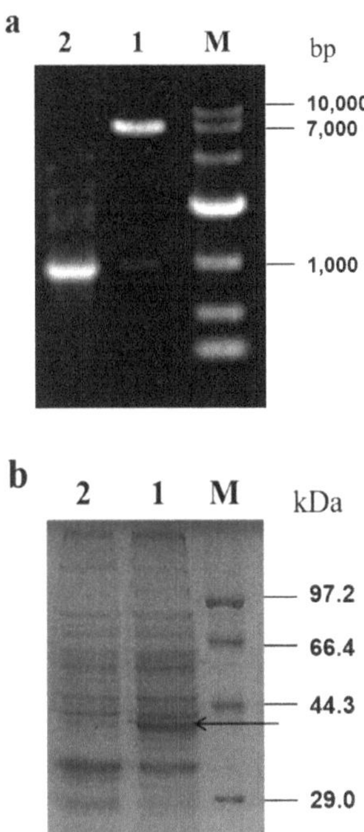

Figure 3. Cloning and expression of collagenase. (**a**) Electrophoresis gel of the double-digested plasmid from *B. subtilis* WB600 to show the presence of the *col* gene. M, DNA ladder; lane 1, recombinant plasmid digestion; lane 2, recombinant plasmid PCR. (**b**) The recombinant collagenase was successfully expressed. M, molecular weight marker; lane 1, *B. subtilis* WB600/pP43NMK-*col* fermentation supernatant, the arrow points to the target protein; lane 2, *B. subtilis* WB600/pP43NMK fermentation supernatant.

The molecular mass of the recombinant protein was predicted to be 35.4 kDa using the ExPASy program (http://web.expasy.org/protparam/ accessed on 6 May 2022). Meanwhile, recombinant collagenase from other bacteria such as *Actinomycestes* [10], *Pseudoalteromonas agarivorans* [24], *Bacillus cereus* [7], *Vibrio alginolyticus* [8], and *Lucilia Sericata* [25] were reported as 150 kDa, 52.5 kDa, 55 kDa, 90 kDa, and 52 kDa, respectively.

B. subtilis WB600/pP43NMK-*col* was cultured in the fermentation medium at 220 rpm and 37 °C for the expression of collagenase. Under such expression conditions, the extracellular activity of collagenase was 1145.16 U/mL. SDS-PAGE analyses of the supernatant of the culture broth indicated a distinctive protein band at around 35.4 kDa, confirming the expression of collagenase. This band was not present in the negative control samples (Figure 3b), consistent with the enzyme activity assay results. In summary, recombinant collagenase was successfully expressed in *B. subtilis* WB600.

3.3. Purification of the Recombinantly Expressed Collagenase

The supernatant of the bulk-cultured recombinant collagenase was concentrated with 70% ammonium sulfate, and the precipitate was dissolved in Tris-HCl (pH 7.5) buffer. The solution was then ultrafiltered against the Tris-HCl (pH 7.5) buffer. Subsequently, the ultrafiltrate was passed through a nickel column. The fractions containing the collagenolytic activity were pooled, concentrated, and stored at −20 °C. The results of the collagenase purification using ammonium sulfate precipitation are summarized in Table 1. The optimal concentration of ammonium sulfate for purification was 70%, resulting in a specific activity of 6156.67 U/mg. Using a three-step procedure, the specific activity of the enzyme was increased from 1998.73 U/mg (crude enzyme) to 9405.54 U/mg (purified enzyme), and the enzyme was purified 4.71-fold with a yield of 2.14% from the crude extract (Table 2). The native bacteria *B. subtilis* DB104 expressing *col*H and its specific activity is estimated to be 1210 U/mg [26]. Cloned from *Grimontia (Vibrio) hollisae* 1706B and expressed by the *Brevi bacillus* system, the purified recombinant enzyme had a specific activity of 5314 U/mg [27]. The strain SM1988^T, which was a Gram-negative unipolar flagellar-shaped bacterium, expressed the collagenase enzyme with a specific activity of 384.14 U/mg [28]. Compared with them, the purified enzyme was higher, which may be due to the different measurement methods of enzyme activity, the different adaptation substrates, and the adequate purification effect of this study, which maintains the vitality of collagenase to a certain extent.

Table 1. Specific activity of the enzyme at different concentrations of ammonium sulfate precipitation.

Ammonium Sulfate Concentration (%)	Specific Activity (U/mg)
20	4498.19
30	2925.32
40	2165.25
50	2177.78
60	4921.74
70	6156.67
80	5509.75
90	4000.21

Table 2. Purification of collagenase from *B. subtillis* WB600/pP43NMK-*col*.

Steps	Total Activity (U)	Total Protein (mg)	Specific Activity (U/mg)	Purification (Fold)	Yield (%)
Cultivate supernatant	114,516.78	57.29	1998.73	1	100
Ammonium sulfate precipitation	58,034.55	9.42	6156.67	3.08	50.67
Ultrafiltration	8329.20	0.94	8860.85	4.43	7.27
Nickel column purification	2445.44	0.26	9405.54	4.71	2.14

As observed from the SDS-PAGE, the recombinant collagenase was efficiently purified. Its relative molecular mass was estimated to be 35.4 kDa (Figure 4).

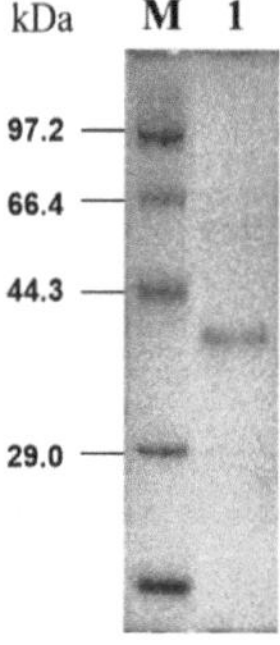

Figure 4. SDS-PAGE of the purified collagenase expressed by *B. subtilis* WB600/pP43NMK-*col*. M, molecular weight markers; lane 1, purified collagenase.

3.4. Effect of pH, Temperature, and Metal Ions on the Stability of Collagenase

As presented in Table 3, the maximum collagenase activity was observed at pH 9.0. The enzyme was relatively stable during the pH treatment for 60 min, with approximately 60% of its activity retained in the pH range of 8–10 for 90 min (Figure 5a). This result is consistent with the collagenase activity from *Streptomyces parvulus* [29].

Table 3. Effect of temperature, pH, and metal ions on recombination collagenase activity.

Parameters	Enzyme Activity (U/mL)
Temperature (°C)	
30	1449.39 ± 50
40	1600.96 ± 74
50	2055.65 ± 61
60	1202.49 ± 68
70	1035.44 ± 71
80	716.02 ± 53
pH	
5	1644.96 ± 56
6	850.47 ± 82
7	877.36 ± 53
8	1419.24 ± 76
9	2005.94 ± 69
10	1173.15 ± 64
11	680.16 ± 78
Metal ions	
Control	1599.33 ± 25
Fe^{3+}	1235.90 ± 34
Zn^{2+}	1816.08 ± 51
Co^{2+}	2343.30 ± 39
Ca^{2+}	2542.94 ± 43
Mg^{2+}	1908.97 ± 37
Cu^{2+}	174.95 ± 45

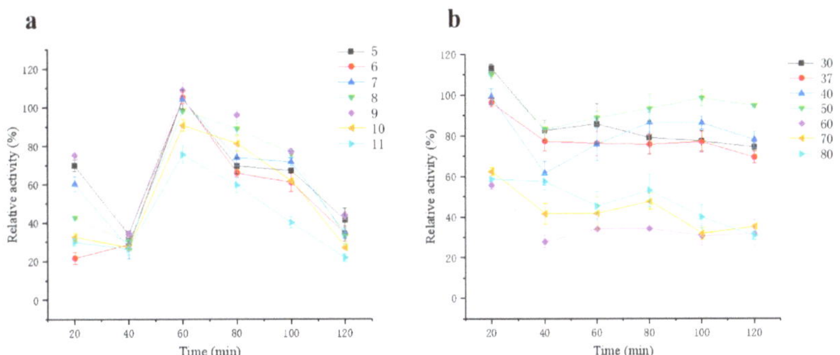

Figure 5. (**a**) Effect of pH on the stability of recombinant collagenase. (**b**) Effect of temperature on the stability of recombinant collagenase.

Table 3 and Figure 5b illustrate the temperature effect on the collagenase activity. The optimum activity of the enzyme was observed at 50 °C, and about 75% of its activity was retained between 30 and 50 °C. The enzyme activity declined sharply when the temperature exceeded 50 °C. After 2 h of pre-incubation at different temperatures, collagenase activity decreased markedly after 40 min of incubation but showed stability in a temperature range of 30–50 °C, with more than 60% of its activity retained. However, pre-incubation above 60 °C substantially abrogated the enzyme activity, with less than 30% of its activity retained after incubation at 80 °C for 2 h. A similar result was observed with a thermophilic collagenase, which showed stability across a range of pH levels (7.0–8.5) and temperatures (40–60 °C), with an optimal pH of 8.0 and temperature of 60 °C [10]. The optimum collagenase activity of *B. cereus* was reported to be at a temperature of 45 °C and had a wide working ranges of pH values and temperatures (7.2–11.0 and 25–50 °C, respectively) [30]. Moreover, more than 50% of the collagenase activity remained after 10 min of incubation at 60 °C [31].

The effect of a few metal ions such as Fe^{3+}, Zn^{2+}, Co^{2+}, Ca^{2+}, Mg^{2+}, and Cu^{2+} on the enzyme activity was evaluated at a final concentration of 5 mM (Table 3). The catalytic efficiency of the recombinant collagenase was inhibited by Fe^{3+} and Cu^{2+}, but stimulated by Zn^{2+}, Co^{2+}, Mg^{2+}, and Ca^{2+}. The most notable inhibition was observed in the presence of Cu^{2+}, while the biggest stimulation was observed in the presence of Ca^{2+}. A similar result was observed in a *Pseudoalteromonas agarivorans* strain where the enzyme activity was strongly activated by Ca^{2+} [24], similar to that of a *Bacillus pumilus* Col-J [31]. This can be attributed to the collagenase structure (Figure 6) that shows high stability in the presence of Ca^{2+} owing to the formation of the collagen–Ca^{2+}–collagenase complex [24].

3.5. Optimization of Carbon Sources, Nitrogen Sources, Metal Ions for the Production of Collagenase by Recombinant B. subtilis WB600

Ten percent of different carbon sources were added to the fermentation medium, while the other components were not changed. After incubation by stirring for 48 h in the fermentation culture, the enzyme activity and mass were determined. As presented in Figure 7a, the best carbon source for collagenase production by *B. subtilis* was fructose, with a collagenase activity at 2237.78 U/mL. The recombinant *B. subtilis* grew better with the addition of glucose [32], showing that it could absorb glucose better and consume nutrients quickly, but with less enzyme production.

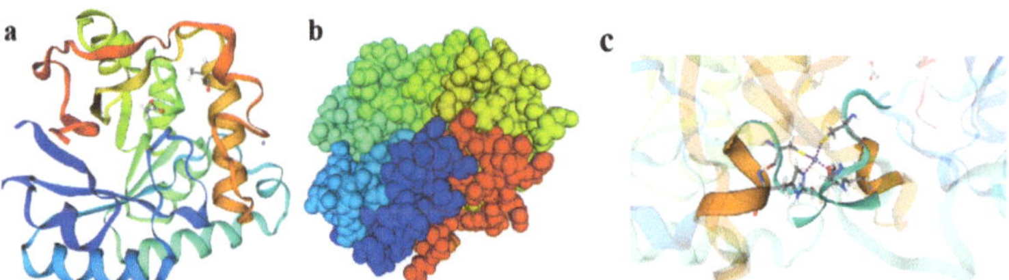

Figure 6. (**a**) Cartoon representation of the collagenase obtained from the SWISS-MODEL (https: //swissmodel.expasy.org/ accessed on 6 May 2022). The gray ball represents zinc ion. Prediction and interpretation the zinc ion and acetate ion were through QM/MM optimization. Both the polarization by the catalytic site Zn301 and the interaction with the acetate moiety of Glu227 promote the elongation of the O-H bond of the Zn-ligated nucleophilic water [33]. (**b**) Spacefill representation of the collagenase (**c**) SWISS-MODEL matched template (SMTL ID: 5d88.1 The Structure of the U32 Peptidase Mk0906), U32 collagenase family. The interaction between zinc ions and residues is shown. According to the results of the experiment, it is also proved that Ca^{2+} and Zn^{2+} bind, which promotes the occurrence of enzymatic lysis and the improvement of enzymatic activity [24,34,35].

Determination of the optimum concentration of fructose was investigated by assaying various fructose concentrations ranging between 5 and 30 g/L at a pH of 7.5, temperature of 37 °C, and with shaking at 220 rpm (Figure 7b). The optimum fructose concentration was found to be 15 g/L, where collagenase activity was at a maximum (2158 U/mL). However, the cell growth was increased with an increasing starch concentration, reaching a maximum at 25 g/L.

In summary, in view to optimize both the collagenase activity and bacterial growth, the optimal concentration of fructose was selected at 15 g/L, and was used for subsequent studies.

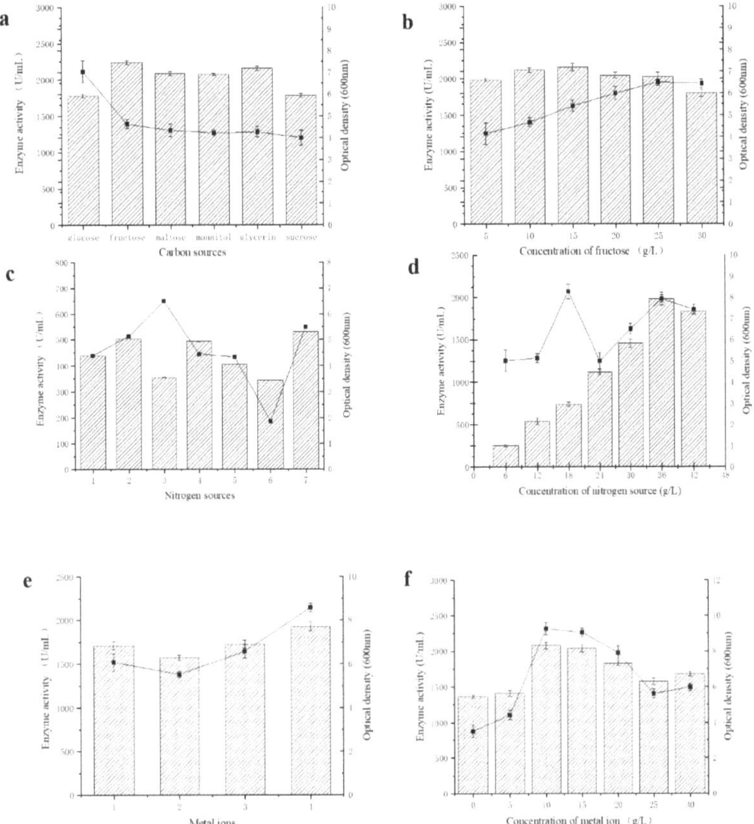

Figure 7. Optimization of the growth media for recombinant expression of collagenase in *B. subtilis* WB600. The histogram represents the activity of recombinant collagenase and the line represents the bacterial growth. (**a**) Effect of carbon sources on the enzyme activity and cell growth. (**b**) Effect of concentration of fructose on the enzyme activity and cell growth. (**c**) Effect of nitrogen sources on the enzyme activity and cell growth. 1, beef paste; 2, ammonium sulphate; 3, yeast powder; 4, peptone; 5, urea; 6, gelatin; 7, mixture of peptone and yeast powder. (**d**) Effect of the concentration of nitrogen source on the enzyme activity and bacterial growth. (**e**) Effect of metal ions on the enzyme activity and cell growth. 1, Zn^{2+}; 2, Mg^{2+}; 3, Na^{2+}; 4, Ca^{2+}. (**f**) Effect of concentration of the Ca^{2+} metal ion on enzyme activity and bacterial growth. The data are presented as mean $\pm$ standard deviation from three independent experiments.

To investigate the influence of nitrogen sources on the expression of recombinant collagenase by *B. subtilis*, the maximum collagenase activity (531.86 U/mL) was observed when peptone was used together with yeast powder; the bacterial growth increased slightly. As shown in Figure 7c, when yeast powder was added alone, the enzyme activity was relatively low, but the cell growth was high. This could have been due to the yeast powder being conducive to the growth of *B. subtilis* but less promotive of enzyme production.

The yeast powder and peptone were mixed in a ratio of 2:1, and Figure 7d shows the effects of different yeast and peptone concentrations (6–42 g/L) on the expression of collagenase by *B. subtilis*. Bacterial growth was increased at a concentration of 18 g/L. The best mixture concentration for optimal collagenase activity was 36 g/L.

Figure 7e shows that recombinant *B. subtilis* had the highest bacterial growth and collagenase enzyme activity (1926.9 U/mL) when Ca^{2+} was added, which shows that

Ca^{2+} can promote the growth of bacteria and retain the optimal activity of collagenase. Different concentrations (1–30 g/L) of Ca^{2+} were added to the fermentation medium; the recombinant *B. subtilis* grew well and expressed collagenase at its highest when the concentration of Ca^{2+} was 10 g/L. This further confirms that Ca^{2+} affects the cell growth and expression potential of *B. subtilis* (Figure 7f).

3.6. Factors Affecting the Recombinant Production of Collagenase by B. subtilis WB600

The initial pH, rotational speed, temperature, and percentage of inoculation are process variables that affect the growth of bacteria and enzyme production. These variables were separately studied in order to evaluate their effects on collagenase production (Figure 8).

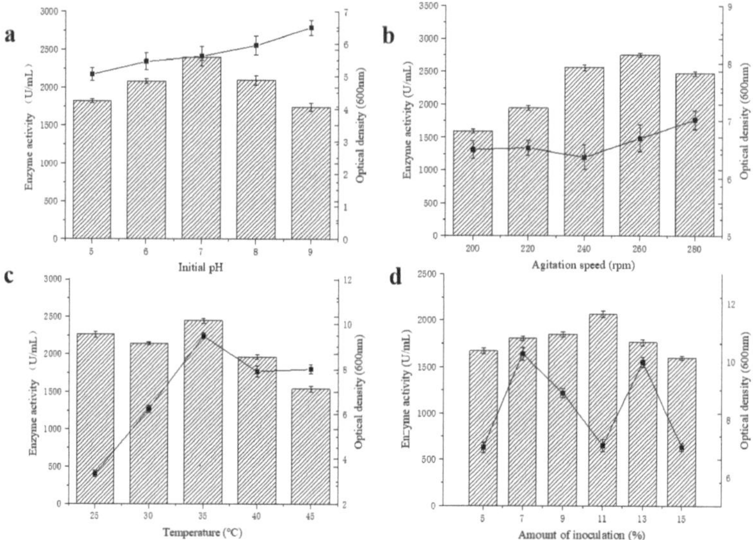

Figure 8. Optimization of the expression conditions for producing collagenase in *B. subtilis* WB600. Histogram, the activity of recombinant collagenase. Line, bacterial cell growth. (**a**) Effect of initial pH on the collagenase activity and bacterial cell growth. (**b**) Effect of agitation speed. (**c**) Effect of expression temperature. (**d**) Effect of inoculation percentage. The data are presented as mean ± standard deviation from three independent experiments.

3.6.1. Effect of Initial pH

The pH affects the growth of microbial cells and the protein expression yield [36]. The effect of the initial pH on the production of collagenase by *B. subtilis* was studied at 37 °C in the presence of 15.0 g/L of fructose and 36 g/L of a yeast and peptone mixture, and shaken at 220 rpm. Different initial pH levels ranging from 5.0 to 9.0 were studied. As indicated in Figure 8a, the cell growth was maximized at a pH of 9.0. However, the maximum enzyme activity (2391.4 U/mL) was obtained at pH 7.0. This is in agreement with the predicted pH for maximum collagenolytic activity at pH 7.21 [37]. In this context, pH 7.0 was selected for further experiments (Table 3).

3.6.2. Effect of Agitation Speed

Agitation speed is important for microorganism growth and the expression of collagenase during fermentation. Oxygen is required by bacteria to grow well and, therefore, to be able to express collagenase. Agitation of the medium ensures that the medium is well aerated. The effects of different agitation speeds (200–280 rpm) on *B. subtilis* growth and collagenase production were examined at 37 °C in 15.0 g/L of fructose and a 36 g/L of peptone and yeast mixture at a pH of 7.0. As depicted in Figure 8b, the highest collagenase activity (2746.7 U/mL) was reached when the agitation speed was 260 rpm. However, the highest cell growth of recombinant *B. subtilis* was reached at 280 rpm. These results suggest that a high agitation speed promoted fast growth of the recombinant *B. subtilis*, but the higher shear force caused by the violent shaking and the over accumulation of metabolites could also negatively affect enzyme production [35,38]. The agitation speed for future experiments was taken at 260 rpm because maximum collagenase activity was achieved at that speed.

3.6.3. Effect of Temperature on the Cell Growth and Expression of Collagenase

Temperature is one of the most important factors to be considered because it affects the enzyme activity, cell growth, and expression of the enzyme [39]. Different temperatures (25–45 °C) were evaluated for their effects on collagenase activity and bacterial cell growth in 15.0 g/L of fructose and 36 g/L of a peptone and yeast mixture at pH 7.0, with shaking at 260 rpm. The maximum collagenase activity (2444.3 U/mL) and cell growth were reached at 35 °C (Figure 8c). Moreover, the bacterial growth increased significantly with the increase in temperature, particularly from 25 °C to 35 °C. However, with a further increase in temperature, the collagenase activity markedly decreased. A possible reason for the decrease in collagenase activity at a high temperature could be attributed to a decrease in the transport of intermediate metabolites to the cells [40]. Another reason could be that the enzyme was denatured at temperatures higher than 35 °C.

3.6.4. Effect of the Percentage of Inoculation

The percentage of inoculation has a significant effect on the yield of recombinant protein expression by bacteria. A small amount of inoculation will affect the cell growth of recombinant strains and will thus affect the enzyme production during the logarithmic growth period of the bacteria. A small percentage of inoculation will increase the consumption of nutrients, and it will be difficult to meet the needs of the growing recombinant strain in a short period of time.

To evaluate the effect of inoculation amount on the activity of collagenase, six different inoculation amounts ranging from 5% to 15% (v/v) were set up under the experimental situations of 35 °C, pH 7.0, and 260 rpm. As shown in Figure 8d, it can be concluded that different inoculation percentages have different effects on the recombinant expression of collagenase. The enzyme activity is the highest at 11% inoculation, but the cell growth is low, which may be due to the large number of bacteria but limited nutrients, affecting the cell growth but maximizing the use of the nutrients for collagenase production.

4. Conclusions

In this study, the collagenase encoding gene *col* was successfully expressed in *B. subtilis* WB600 under the pP43NMK vector. The collagenase sequence contains an open reading frame of 936 bp, which encodes a protein of 312 amino acid residues with a predicted molecular mass of 35.4 kDa. The recombinant enzyme was purified by ammonium sulfate, ultrafiltration, and nickel column with a collagenase activity of 9405.54 U/mg. The enzyme exhibited maximal activity at pH 9.0 and 50 °C. Catalytic efficiency of the recombinant collagenase was inhibited by Fe^{3+} and Cu^{2+}, but stimulated by Co^{2+}, Ca^{2+}, Zn^{2+}, and Mg^{2+}. The optimal conditions for expression were at pH 7.0 and 35 °C, using 15 g/L of fructose and 36 g/L of yeast powder and peptone mixture (2:1) at 260 rpm with 11% inoculation. The maximal extracellular activity of collagenase reached 2746.7 U/mL after culturing in the optimization culture conditions, which was 2.4-fold higher than before optimization. This study is a first attempt to express recombinant collagenase and to optimize its production for possible scale-up.

Author Contributions: Conceptualization, Y.Z. and H.Z.; methodology, Y.Z. and L.W.; software, Y.Z.; validation, P.L. and K.Z.; formal analysis, Y.Z.; investigation, Y.Z. and S.S.; resources, W.L. (Wenkang Li) and W.L. (Wenjing Liu); data curation, P.L. and W.L. (Wenkang Li); writing—original draft preparation, Y.Z. and L.W.; writing—review and editing, H.Z. and L.S.; visualization, Y.Z.; supervision, H.Z. and S.S.; project administration, J.L. and L.S.; funding acquisition, H.Z. and L.S. All authors have read and agreed to the published version of the manuscript.

Funding: This work was supported by the National Natural Science Foundation of China (no. 32001836), the Natural Science Foundation of Shandong Province (ZR201911180224), the Natural Science Foundation of Shandong Province (ZR2020MC043), the Outstanding Youth Project of Yunnan Provincial Department of Science and Technology (no. 2019F1019), and the Yantai Science and Technology Development Plan (SK21H266).

Institutional Review Board Statement: Not applicable.

Informed Consent Statement: Not applicable.

Conflicts of Interest: The authors declare no conflict of interest.

References

1. Taga, Y.; Tanaka, K.; Hattori, S.; Mizuno, K. In-depth correlation analysis demonstrates that 4-hydroxyproline at the Yaa position of Gly-Xaa-Yaa repeats dominantly stabilizes collagen triple helix. *Matrix Biol. Plus.* **2021**, *10*, 100067. [CrossRef] [PubMed]
2. Bertini, I.; Fragai, M.; Luchinat, C.; Melikian, M.; Toccafondi, M.; Lauer, J.L.; Fields, G.B. Structural basis for matrix metalloproteinase 1-catalyzed collagenolysis. *J. Am. Chem. Soc.* **2012**, *134*, 2100–2110. [CrossRef] [PubMed]
3. Tanaka, K.; Okitsu, T.; Teramura, N.; Iijima, K.; Hayashida, O.; Teramae, H.; Hattori, S. Recombinant collagenase from *Grimontia hollisae* as a tissue dissociation enzyme for isolating primary cells. *Sci. Rep.* **2020**, *10*, 3927. [CrossRef]
4. Varghese, A.; Chaturvedi, S.S.; Fields, G.B.; Karabencheva-Christova, T.G. A synergy between the catalytic and structural Zn(II) ions and the enzyme and substrate dynamics underlies the structure-function relationships of matrix metalloproteinase collagenolysis. *J. Biol. Inorg. Chem.* **2021**, *26*, 583–597. [CrossRef] [PubMed]
5. Laronha, H.; Caldeira, J. Structure and Function of Human Matrix Metalloproteinases. *Cells* **2020**, *9*, 1076. [CrossRef] [PubMed]
6. Song, Y.; Fu, Y.; Huang, S.; Liao, L.; Wu, Q.; Wang, Y.; Ge, F.; Fang, B. Identification and antioxidant activity of bovine bone collagen-derived novel peptides prepared by recombinant collagenase from *Bacillus cereus*. *Food Chem.* **2021**, *349*, 129143. [CrossRef]
7. Cheng, X.; Abfalter, C.M.; Schönauer, E.; Ponnuraj, K.; Huemer, M.; Gadermaier, G.; Regl, C.; Briza, P.; Ferreira, F.; Huber, C.G.; et al. Cloning, Purification and Characterization of the Collagenase ColA Expressed by *Bacillus cereus* ATCC 14579. *PLoS ONE* **2016**, *11*, e0162433. [CrossRef]
8. Ducka, P.; Eckhard, U.; Schonauer, E.; Kofler, S.; Gottschalk, G.; Brandstetter, H.; Nuss, D. A universal strategy for high-yield production of soluble and functional clostridial collagenases in *E. coli*. *Appl. Microbiol. Biotechnol.* **2009**, *83*, 1055–1065. [CrossRef]
9. Rochima, E.; Sekar, N.; Buwono, I.D.; Afrianto, E.; Pratama, R.I. Isolation and Characterization of Collagenase from *Bacillus Subtilis* (Ehrenberg, 1835); ATCC 6633 for Degrading Fish Skin Collagen Waste from Cirata Reservoir, Indonesia. *Aquat. Procedia* **2016**, *7*, 76–84. [CrossRef]
10. Abood, A.; Salman, A.M.M.; Abdelfattah, A.M.; El-Hakim, A.E.; Abdel-Aty, A.M.; Hashem, A.M. Purification and characterization of a new thermophilic collagenase from *Nocardiopsis dassonvillei* NRC2aza and its application in wound healing. *Int. J. Biol. Macromol.* **2018**, *116*, 801–810. [CrossRef]
11. Behl, T.; Kaur, G.; Sehgal, A.; Bhardwaj, S.; Singh, S.; Buhas, C.; Judea-Pusta, C.; Uivarosan, D.; Munteanu, M.A.; Bungau, S. Multifaceted Role of Matrix Metalloproteinases in Neurodegenerative Diseases: Pathophysiological and Therapeutic Perspectives. *Int. J. Mol. Sci.* **2021**, *22*, 1413. [CrossRef] [PubMed]
12. Folgueras, A.R.; Pendas, A.M.; Sanchez, L.M.; Lopez-Otin, C. Matrix metalloproteinases in cancer: From new functions to improved inhibition strategies. *Int. J. Dev. Biol.* **2004**, *48*, 411–424. [CrossRef] [PubMed]
13. Lauer-Fields, J.L.; Juska, D.; Fields, G.B. Matrix metalloproteinases and collagen catabolism. *Biopolymers* **2002**, *66*, 19–32. [CrossRef] [PubMed]
14. Spinale, F.G. Matrix metalloproteinases: Regulation and dysregulation in the failing heart. *Circ. Res.* **2002**, *90*, 520–530. [CrossRef] [PubMed]
15. Luo, J.; Zhang, Z.; Zeng, Y.; Dong, Y.; Ma, L. Co-encapsulation of collagenase type I and silibinin in chondroitin sulfate coated multilayered nanoparticles for targeted treatment of liver fibrosis. *Carbohydr. Polym.* **2021**, *263*, 117964. [CrossRef]
16. Yang, H.; Qu, J.; Zou, W.; Shen, W.; Chen, X. An overview and future prospects of recombinant protein production in *Bacillus subtilis*. *Appl. Microbiol. Biotechnol.* **2021**, *105*, 6607–6626. [CrossRef]
17. Cui, W.; Han, L.; Suo, F.; Liu, Z.; Zhou, L.; Zhou, Z. Exploitation of *Bacillus subtilis* as a robust workhorse for production of heterologous proteins and beyond. *World J. Microbiol. Biotechnol.* **2018**, *34*, 145. [CrossRef]
18. Gu, Y.; Xu, X.; Wu, Y.; Niu, T.; Liu, Y.; Li, J.; Du, G.; Liu, L. Advances and prospects of *Bacillus subtilis* cellular factories: From rational design to industrial applications. *Metab. Eng.* **2018**, *50*, 109–121. [CrossRef]

19. Wu, X.C.; Lee, W.; Tran, L.; Wong, S.L. Engineering a *Bacillus subtilis* expression-secretion system with a strain deficient in six extracellular proteases. *J. Bacteriol.* **1991**, *173*, 4952–4958. [CrossRef] [PubMed]
20. Duarte, A.S.; Correia, A.; Esteves, A.C. Bacterial collagenases—A review. *Crit. Rev. Microbiol.* **2016**, *42*, 106–126. [CrossRef]
21. Cao, S.; Li, D.; Ma, X.; Xin, Q.; Song, J.; Lu, F.; Li, Y. A novel unhairing enzyme produced by heterologous expression of keratinase gene (kerT) in *Bacillus subtilis*. *World J. Microbiol. Biotechnol.* **2019**, *35*, 122. [CrossRef] [PubMed]
22. Matsushita, O.; Yoshihara, K.; Katayama, S.; Minami, J.; Okabe, A. Purification and characterization of Clostridium perfringens 120-kilodalton collagenase and nucleotide sequence of the corresponding gene. *J. Bacteriol.* **1994**, *176*, 149–156. [CrossRef] [PubMed]
23. Bradford, M.M. A rapid and sensitive method for the quantitation of microgram quantities of protein utilizing the principle of protein-dye binding. *Anal. Biochem.* **1976**, *72*, 248–254. [CrossRef]
24. Bhattacharya, S.; Bhattacharya, S.; Gachhui, R.; Hazra, S.; Mukherjee, J. U32 collagenase from *Pseudoalteromonas agarivorans* NW4327: Activity, structure, substrate interactions and molecular dynamics simulations. *Int. J. Biol. Macromol.* **2019**, *124*, 635–650. [CrossRef] [PubMed]
25. Alipour, H.; Raz, A.; Dinparast Djadid, N.; Zakeri, S. Expression of a New Recombinant Collagenase Protein of *Lucilia Sericata* in SF9 Insect Cell as a Potential Method for Wound Healing. *Iran. J. Biotechnol.* **2019**, *17*, e2429. [CrossRef]
26. Jung, C.M.; Matsushita, O.; Katayama, S.; Minami, J.; Ohhira, I.; Okabe, A. Expression of the colH gene encoding Clostridium histolyticum collagenase in *Bacillus subtilis* and its application to enzyme purification. *Microbiol. Immunol.* **1996**, *40*, 923–929. [CrossRef] [PubMed]
27. Teramura, N.; Tanaka, K.; Iijima, K.; Hayashida, O.; Suzuki, K.; Hattori, S.; Irie, S. Cloning of a novel collagenase gene from the gram-negative bacterium *Grimontia* (*Vibrio*) *hollisae* 1706B and its efficient expression in *Brevibacillus choshinensis*. *J. Bacteriol.* **2011**, *193*, 3049–3056. [CrossRef]
28. Li, J.; Cheng, J.H.; Teng, Z.J.; Sun, Z.Z.; He, X.Y.; Wang, P.; Shi, M.; Song, X.Y.; Chen, X.L.; Zhang, Y.Z.; et al. Taxonomic and Enzymatic Characterization of *Flocculibacter collagenilyticus* gen. nov., sp. nov., a Novel Gammaproteobacterium with High Collagenase Production. *Front. Microbiol.* **2021**, *12*, 621161. [CrossRef]
29. Sakurai, Y.; Inoue, H.; Nishii, W.; Takahashi, T.; Iino, Y.; Yamamoto, M.; Takahashi, K. Purification and characterization of a major collagenase from Streptomyces parvulus. *Biosci. Biotechnol. Biochem.* **2009**, *73*, 21–28. [CrossRef]
30. Pequeno, A.C.L.; Arruda, A.A.; Silva, D.F.; Duarte Neto, J.M.W.; Silveira Filho, V.M.; Converti, A.; Marques, D.A.V.; Porto, A.L.F.; Lima, C.A. Production and characterization of collagenase from a new Amazonian *Bacillus cereus* strain. *Prep. Biochem. Biotechnol.* **2019**, *49*, 501–509. [CrossRef]
31. Wu, Q.; Li, C.; Li, C.; Chen, H.; Shuliang, L. Purification and characterization of a novel collagenase from *Bacillus pumilus* Col-J. *Appl. Biochem. Biotechnol.* **2010**, *160*, 129–139. [CrossRef]
32. Klausmann, P.; Hennemann, K.; Hoffmann, M.; Treinen, C.; Aschern, M.; Lilge, L.; Morabbi Heravi, K.; Henkel, M.; Hausmann, R. *Bacillus subtilis* High Cell Density Fermentation Using a Sporulation-Deficient Strain for the Production of Surfactin. *Appl. Microbiol. Biotechnol.* **2021**, *105*, 4141–4151. [CrossRef] [PubMed]
33. Chen, B.; Kang, Z.; Zheng, E.; Liu, Y.; Gauld, J.W.; Wang, Q. Hydrolysis Mechanism of the Linkers by Matrix Metalloproteinase-9 Using QM/MM Calculations. *J. Chem. Inf. Model.* **2021**, *61*, 5203–5211. [CrossRef] [PubMed]
34. Itoh, T.; Nakagawa, E.; Yoda, M.; Nakaichi, A.; Hibi, T.; Kimoto, H. Structural and biochemical characterisation of a novel alginate lyase from *Paenibacillus* sp. str. FPU-7. *Sci. Rep.* **2019**, *9*, 14870. [CrossRef] [PubMed]
35. Singh, W.; Fields, G.B.; Christov, C.Z.; Karabencheva-Christova, T.G. Effects of Mutations on Structure-Function Relationships of Matrix Metalloproteinase-1. *Int. J. Mol. Sci.* **2016**, *17*, 1727. [CrossRef]
36. Sarchami, T.; Johnson, E.; Rehmann, L. Optimization of fermentation condition favoring butanol production from glycerol by *Clostridium pasteurianum* DSM 525. *Bioresour. Technol.* **2016**, *208*, 73–80. [CrossRef]
37. Lima, C.A.; Marques, D.A.; Barros Neto, B.; Lima Filho, J.L.; Carneiro-da-Cunha, M.G.; Porto, A.L. Fermentation medium for collagenase production by Penicillium aurantiogriseum URM4622. *Biotechnol Prog.* **2011**, *27*, 1470–1477. [CrossRef]
38. Wang, Y.; Chen, S.; Zhao, X.; Zhang, Y.; Wang, X.; Nie, Y.; Xu, Y. Enhancement of the production of *Bacillus naganoensis* pullulanase in recombinant *Bacillus subtilis* by integrative expression. *Protein Expr. Purif.* **2019**, *159*, 42–48. [CrossRef]
39. Yang, X.; Xiao, X.; Liu, D.; Wu, R.; Wu, C.; Zhang, J.; Huang, J.; Liao, B.; He, H. Optimization of Collagenase Production by *Pseudoalteromonas* sp. SJN2 and Application of Collagenases in the Preparation of Antioxidative Hydrolysates. *Mar. Drugs* **2017**, *15*, 377. [CrossRef]
40. Amorim, C.; Silverio, S.C.; Goncalves, R.F.S.; Pinheiro, A.C.; Silva, S.; Coelho, E.; Coimbra, M.A.; Prather, K.L.J.; Rodrigues, L.R. Downscale fermentation for xylooligosaccharides production by recombinant Bacillus subtilis 3610. *Carbohydr. Polym.* **2019**, *205*, 176–183. [CrossRef]

Article

A Novel Salt-Tolerant L-Glutaminase: Efficient Functional Expression, Computer-Aided Design, and Application

Hengwei Zhang, Mengkai Hu, Qing Wang, Fei Liu, Meijuan Xu, Xian Zhang * and Zhiming Rao *

Key Laboratory of Industrial Biotechnology of the Ministry of Education, Laboratory of Applied Microorganisms and Metabolic Engineering, School of Biotechnology, Jiangnan University, Wuxi 214122, China
* Correspondence: zx@jiangnan.edu.cn (X.Z.); raozhm@jiangnan.edu.cn (Z.R.)

Abstract: The low productivity in long fermentation duration and high-salt working conditions limit the application of L-glutaminase in soy sauce brewing. In this study, a novel L-glutaminase (LreuglsA) with eminent salt tolerance was mined and achieved more than 70% activity with 30% NaCl. To improve the robustness of the enzyme at different fermentation strategies, mutation LreuglsA^{H105K} was built by a computer-aided design, and the recombinant protein expression level, an essential parameter in industrial applications, was increased 5.61-fold with the synthetic biology strategy by improving the mRNA stability. Finally, the LreuglsA^{H105K} functional expression box was contributed to *Bacillus subtilis* 168 by auxotrophic complementation, and the production in a 5-L bioreactor was improved to 2516.78 $\pm$ 20.83 U mL^{-1}, the highest production ever reported. When the immobilized cells were applied to high-salt dilute-state soy sauce brewing, the L-glutamate level was increased by 45.9%. This work provides insight into the salt-tolerant enzyme for improving the efficiency of industrial applications.

Keywords: L-glutaminase; computer-aided design; expression level; 5-L bioreactor; high-salt dilute-state soy sauce

Citation: Zhang, H.; Hu, M.; Wang, Q.; Liu, F.; Xu, M.; Zhang, X.; Rao, Z. A Novel Salt-Tolerant L-Glutaminase: Efficient Functional Expression, Computer-Aided Design, and Application. *Fermentation* 2022, 8, 444. https://doi.org/10.3390/fermentation8090444

Academic Editor: Michael Kupetz

Received: 7 August 2022
Accepted: 1 September 2022
Published: 6 September 2022

Publisher's Note: MDPI stays neutral with regard to jurisdictional claims in published maps and institutional affiliations.

1. Introduction

The L-glutamate generated by the reaction is the primary substance responsible for the umami taste of food. Approximately 46% of the L-glutamate in brewed soy sauce is produced from L-glutamine during soy sauce fermentation [1]. Currently, most soy sauces enhance their umami taste by adding monosodium glutamate (MSG), but this often leads to an excessive intake of sodium and thus increases the risk of cardiovascular and cerebrovascular diseases [2,3]. Therefore, the natural enzymatic method can effectively solve this problem by increasing the glutamate content in soy sauce.

L-Glutaminase (EC 3.5.1.2) produces L-glutamate by hydrolyzing the amino group of the amide group in L-glutaminase and is widely distributed in microorganisms such as yeast, bacteria, and fungi [4]. This enzyme has significant potential for application in food and medicine [5]. However, during traditional soy sauce brewing, L-glutaminase activity in *Aspergillus oryzae*, one of the most critical microorganisms in soy sauce Koji, is inhibited because of the high-salt environment, leading to the production of higher levels of the flavorless pyroglutamine rather than the flavor-presenting substance glutamate [6]. Therefore, it is necessary to develop an enzymatic method that can maintain a high stability and enzymatic activity in high-salt soy sauce brewing, which can effectively increase the content of naturally produced L-glutamate in soy sauce products.

According to the current research, the existing L-glutaminase has defects in enzyme activity, stability, and salt tolerance that are not conducive to large-scale industrial production. Kumar et al. found that the L-glutaminase from *Bacillus* sp. LKG-01 (MTCC 10401), isolated from the Gangotri District of Uttarakhand in the Himalayas, is also stable, with 80% relative enzyme activity in the presence of a 25% salt concentration [7]. The

L-glutaminase from *Micrococcus luteus* K-3 showed the maximum enzyme activity in the presence of 1.71 M NaCl, and it showed >90% activity in the presence of 3.08 M NaCl (about an 18% salt concentration) [8,9]. However, the enzyme source is related to food safety, and genetic sources that do not meet the food safety standards are often considered potentially unsafe, and the activity and thermal stability of the enzymes did not meet the industrial requirements. A membrane-bound salt-tolerant and thermally stable L-glutaminase from *Lactobacillus rhamnosus* was screened. The activity of L-glutaminase increased approximately two-fold in the presence of 5% salt, and 90% activity remained in the presence of 15% salt [10]. According to the sodium dodecyl sulfate-polyacrylamide gel electrophoresis (SDS-PAGE) results, one L-glutaminase bound to the cell membrane in *L. reuteri* KCTC3594 was isolated and characterized, and the sizes of the two primary brands of the enzyme were 70 and 50 kDa [11]. Although *L. rhamnosus* and *L. reuteri* are food safe probiotics, the membrane-bound nature of these enzymes increases the complexity of the food brewing process and limits the industrial application of these L-glutaminases (Table 1) [12]. On the other hand, some L-glutaminases also synthesize theanine through glutamyl hydrolysis or the transfer reaction, inhibiting caffeine formation and improving tea broth's flavor [13,14]. In addition to its applications in the food industry, L-glutaminase is used in enzyme therapy for cancer, especially for acute lymphocytic leukemia, because tumor cells have no mechanism to synthesize L-glutamine and take it up as an exogenous resource [15]. The cytotoxic effect of an L-glutaminase from *Bacillus subtilis* OHEM11 (MK389501) indicated significant safety in Vero cells, with high anticancer activity against the NFS-60, HepG-2, and MCF-7 cancer cell lines [16]. L-Glutaminase from *Streptomyces canarius* FR (KC460654) exerted inhibitory effects against Hep-G2 cells, HeLa cells, HCT-116 cells, and RAW 264.7 cells [17].

Table 1. Review of the different sources of glutaminase.

Source	Optimum pH	Optimum Temp (°C)	Specific Gravity (U/mg)	Thermal Stability	Salt Tolerant	References
A. oryzae	9	41	40.12	NR	+	[6]
Bacillus sp. LKG-01	11	70	584.2	+++	+++	[7]
M. luteus K-3	8	50	190	+	+++	[8]
L. rhamnosus	7	50	NR	++	+++	[10]
L. reuteri KCTC3594	7.5	40	16.4	+	++	[11]

NR: No Reported; +: intensity.

Therefore, due to the limitations of a low yield and low enzyme activity of glutaminase, a heterologous high-efficiency expression can solve the problems of high production costs and challenging applications. *B. subtilis*, a Gram-positive model bacterium, has long been used as a microbial cell factory because of its low codon preference, high productivity, and low requirements for fermentation media. In addition, due to its generally regarded as safe (GRAS) status, *B. subtilis* has been used in nutritional food production [18]. Synthetic biology strategies based on *B. subtilis* for enzyme or biochemical production have been developed to render *B. subtilis* more suitable for industrial production [19].

Recently, no publication has reported that a glutaminase can simultaneously possess excellent industrial properties with high enzymatic activity, stability, and salt tolerance. In this study, we mined a nonmembrane-bound L-glutaminase (LreuglsA) from *L. reuteri* DSM20016. LreuglsA had significant specific enzyme activity and exhibited a salt tolerance. Based on a computer-aided rational design, the thermal stability of the enzyme was increased to make it more adaptable for soy sauce production. Subsequently, we replaced the antibiotic resistance gene based on the *B. subtilis*-pMA5 expression system to eliminate the potential hazard to food safety. Moreover, we further increased the expression of LreuglsA in *B. subtilis* by designing a hairpin loop for the ribosome-binding site (RBS) region. Finally, we performed scale-up experiments of the recombinant strain in a 5-L bioreactor and applied the cells in soy sauce brewing after immobilization.

2. Materials and Methods

2.1. Gene Source, Plasmids, and Strains

The host strains for gene cloning and expression were *Escherichia coli* JM109 and *B. subtilis* 168. The shuttle expression plasmid pMA5 was employed for the expression and mutagenesis studies. The DNA fragments containing the L-glutaminase structural gene from *L. reuteri* DSM20016 (NCBI accession number: ABQ83511) were synthesized by GENEWIZ after the codon optimization of *B. subtilis* 168. The C112-02 ClonExpress II One Step Cloning Kit was purchased from Vazyme (Nanjing, China). PCR amplification of the L-glutaminase gene was performed with an appropriate primer pair. Restriction enzymes and PCR reagents were purchased from TaKaRa (Dalian, China). The linearized pMA5 plasmid and glsA genes were assembled to form the recombinant plasmid pMA5-glsA. The TIANprep Mini Plasmid Kit and TIANgel Purification Kit were purchased from Tiangen Biotech (Beijing, China). All experiments were performed according to the product manuals. All other chemicals were purchased from Sinopharm or Merck.

The plasmid was chemically transformed into competent *E. coli* JM109 cells. After extraction and purification from an agarose gel, the plasmid was transformed into *B. subtilis*, and the positive transformants were picked and sent for sequencing to verify the correct sequence. The original strains and plasmids were sourced from our laboratory.

2.2. Bioinformatics Analysis and Screening for Positive Variants

AlphaFold2, an advanced tool for predicting a protein structure based on machine learning, exhibited the best performance in CASP14 [20]. The structural model of LreuglsA was acquired by modeling using AlphaFold2. We validated the multimerization profile of LreuglsA using active page and modeled homologous protein complexes using AlphaFold2-Multimer [21]. The program PyMOL was used to analyze variations in the molecular tertiary structure. The programs Clustal, ESPript 3.0, and Schrödinger Maestro 12.8 were used to determine the active center residues based on multiple sequence alignment (MSA) and ligand docking.

The position-specific scoring matrix (PSSM), representing the conservation of residues, was generated using psiblast in NCBI BLAST 2.9 with uniref90 (https://www.uniprot.org/ (accessed on 3 May 2022)) by setting the number of iterations to 3 and the E-value to 0.01. By analyzing the conservation of these protein residues and evolutionary information from the PSSM, all positions in the protein were examined for the evolvability of the L-glutaminase family and other homologous proteins. The conserved residues were assigned higher scores at the corresponding positions. Based on the PSSM score, we screened evolvable residues to allow mutations. FoldX predicted the overall stability of virtual saturation mutation at all sites and predicted protein thermostability by comparing the changes in Gibbs energy (ddG and $\Delta\Delta G$) after mutation [22]. The conserved residues analyzed from the PSSM were removed from the candidate list. Adverse interactions within the proteins might lead to increased instability of the protein structure. Molecular dynamics (MD) simulations were used to characterize the microscopic evolution of systems at the atomic level by calculating the atomic motion of the protein in the solvent and to visually observe the mechanism and principle of the experimental observation. Based on the calculations obtained from the MD simulation, root mean square fluctuation (RMSF) values further refining the B-factors of the residues were used to indicate the flexible variability at the amino acid sites [23]. In this study, the MD simulation of the qualified models was performed using GROMACS 2019.6 [24].

2.3. Construction of a Food Safe Recombinant B. Subtilis 168 Strain

D-alanine is an essential component of the *B. subtilis* 168 cell wall and is produced by the D-alanine racemase encoded by the alrA gene (Gene ID: 939942). Microorganisms carrying antibiotic resistance genes are usually not allowed in the food industry [25]. The alrA gene was knocked down to obtain a food-grade L-glutaminase-harboring recombinant *B. subtilis* 168 without the antibiotic resistance gene that would be suitable for application in the food industry. The alrA gene on the replicative plasmids might complement the chromosomal deletion of the alrA gene and provide selective pressure for maintaining the plasmids. The recombinant plasmid pMA5-ΔkanR-alrA-glsA was transformed into *B. subtilis* 168/ΔalrA to obtain strain BSW1.

The alrA gene was knocked out using the Cre/loxP site-specific gene operating system, based on the principle that Cre recombinase specifically recognizes loxP sites and catalyzes the deletion, inversion, and exchange of fragments between two lox sites.

2.4. Overexpression through Synthetic Biology Components

mRNAs play an essential role in posttranscriptional regulation. The secondary structure at the P-terminus of a mRNA is not conducive to ribosome binding and translation initiation. However, some studies have found that a reasonable hairpin structure before the initiation codon ATG will increase, not decrease, the expression of structural genes in eukaryotic cells [26,27]. Viegas and Xiao revealed the feasibility of this strategy in prokaryotic cells of *E. coli* and Bacillus licheniformis, respectively, by assessing the effect of a hairpin structure on the expression [28,29]. In the present study, a hairpin structure was designed directly in front of the open reading frame (ORF), preventing mRNA degradation by the RNA enzyme. The folding free energy of the hairpin rings was calculated using RNAfold [30]. The optimized Shine–Dalgarno (SD) sequence was designed as a single loop after and on the hairpin ring to balance the mRNA–ribosome-binding efficiency and stability. The designed 5′-UTR sequence was inserted in front of the glsA gene by PCR, and pMA5 replaced the original 5′-UTR sequence. Then, chemical transformation was used to construct the *B. subtilis* 168 strain BSW2-BSW9 for overexpression.

For the RT-qPCR analysis of mRNA stability, cells cultured to the logarithmic phase in LB medium were harvested for mRNA extraction. After adding 1 mM rifampicin, the culture was sampled and placed in liquid nitrogen at different times to prevent mRNA degradation. Total RNA was purified using the RNAprep Pure Bacteria Kit (Tiangen Biotech, Beijing, China). The cDNA templates were synthesized with the HiScript II Q RT SuperMix for the qPCR kit (Vazyme Biotech, Nanjing, China) according to the manufacturer's instructions. RT-qPCR was performed using a StepOnePlus instrument (Applied Biosystems, Waltham, MA, USA). The relative expression of the 16S rRNA gene was determined as the internal standard. The primer sequences qPCR-F (ACCAACGACAA-GAAAGCCGA) and qPCR-R (AGACGTTCGATGGCGTACAG) were used for RT-qPCR. The expression levels of LreuglsA were characterized using the 2-ΔΔt method. Each assay was repeated three times.

2.5. L-Glutaminase Expression and Purification

The recombinant strains were cultured in 10 mL of LB medium containing 20 μg mL^{-1} kanamycin for 12 h at 37 °C. Then, a 5% inoculum was transferred to 50 mL of LB medium containing 20 μg mL^{-1} kanamycin, and the cells were cultured for 30 h at 30 °C. The strains were centrifugally separated and washed with phosphate-buffered saline (PBS, pH 7.4). Then, the bacterial suspension was sonicated for 30 min (for 2 s at 5-s intervals) after mixing with 5 mL of PBS containing 20 μL of lysozyme (200 mg/mL). Finally, the mixture was centrifuged at 12,000 rpm for 30 min to separate the cell debris. After centrifugation, the resulting supernatant was a LreuglsA crude enzyme solution and was used to measure the L-glutaminase activity.

A His tag was inserted at the N-terminus of LreuglsA. After SDS-PAGE verification of the expression level, the crude enzyme was purified using Ni-NTA affinity chromatography according to the manufacturer's protocol (GE Healthcare Bio-Sciences, Uppsala, Sweden). The purified LreuglsA and crude enzyme were analyzed using SDS-PAGE analysis.

2.6. Enzyme Activity Assay

The L-glutaminase activity was calculated by testing the L-glutamate content produced. The reaction was performed at a specific temperature for 5 min and terminated by adding 100 μL of 15% trichloroacetic acid (TCA). The reaction mixture (1 mL) contained 880 μL of 200 mM L-glutamine and 20 μL of L-glutaminase. After centrifugation, the L-glutamate concentration in the supernatant was measured using a biosensor analyzer (Institute of Biology, Shandong Academy of Sciences). One unit (U) of enzymatic activity of L-glutaminase was defined as the amount of enzyme required to produce 1 μmol of L-glutamate per minute. The Bradford Protein Assay Kit (Sangon Biotech (Shanghai) Co., Ltd., Shanghai, China) was used to determine the protein concentration according to the manufacturer's instructions.

2.7. Fermentation and Application of L-Glutaminase in Soy Sauce Brewing

The production capacity of the bioreactor shows the industrial application potential of the recombinant strain. The recombinant bacterial strain was cultured by two-stage seed expansion, and 100 mL of the seed culture were transferred into a 5-L bioreactor (Dibi'er Bio-Engineering Co., Ltd., Shanghai, China) containing 2 L of fermentation medium. DO-stat fed-batch fermentation strategies were used for fermentation. The pH was maintained at 7.0 using NH_4OH (30% v/v) and the feed medium (sucrose, 500 g L^{-1}; K_2HPO_4, 2.612 g L^{-1}; KH_2PO_4, 2.041 g L^{-1}; $MgSO_4 \cdot 7H_2O$, 1.845 g L^{-1}; and NaCl, 5 g L^{-1}).

Studies have shown that brewed soy sauce is rich in nutrients and contains a variety of physiologically active substances. High-salt diluted-state (HSDS)-brewed soy sauce has a rich and full-bodied aroma and taste [6]. The soybeans were cleaned and separated, soaked for 8 h, and then steamed at 125 °C for 15 min. The cooked soybeans were naturally cooled to 40 °C and then mixed (soybean:flour = 4:1, 0.05% spores of a species) to obtain a 0-h sample of Koji, which was then placed in an incubator for culture. Then, the soybeans were placed in the incubator to produce the Koji. The Koji was turned in a timely manner during the incubation process to control the temperature of the product, and the temperature was not allowed to exceed 28–30 °C to prevent burning. After 44 h, the finished product was obtained. The mash was mixed with brine (24° Be/20 °C) at a ratio of 1:2.2 and placed in an incubator at 37 °C for fermentation. After 120 days, the soy sauce was sampled and filtered to obtain raw soy sauce and, finally, heat-treated at 90 °C for 30 min to obtain sterilized soy sauce. The free amino acid content in the finished soy sauce was determined using HPLC.

3. Results and Discussion

3.1. A Novel L-Glutaminase Displays Significant Salt Tolerance

As a GRAS strain, the application value of *B. subtilis* in the food industry is self-evident [19]. A food safe strain without antibiotic resistance must be constructed. The Food and Agriculture Organization of the United Nations (FAO) and the World Health Organization (WHO) have stressed that antibiotic resistance genes that are resistant to clinically used antibiotics should not appear in food [31]. The food-grade strain BSW1 lacking the alrA gene and transformed with pMA5-ΔkanR-alrA-glsA was successfully constructed to obtain the recombinant L-glutaminase-expressing strain for food industry applications (Figure 1A). After 30 h of cultivation in LB medium, the intracellular L-glutaminase activity was 16.32 ± 0.56 U mL^{-1}, and the expression of LreuglsA was analyzed using SDS-PAGE (Figure 1B). A 6His tag was added to the N-terminus of LreuglsA, and Ni^{2+} affinity chromatography was used to purify LreuglsA. Separation on a 12% SDS-PAGE gel indicated that the protein was purified and that the molecular mass was approximately 30 kDa, as previously predicted (Figure 1C).

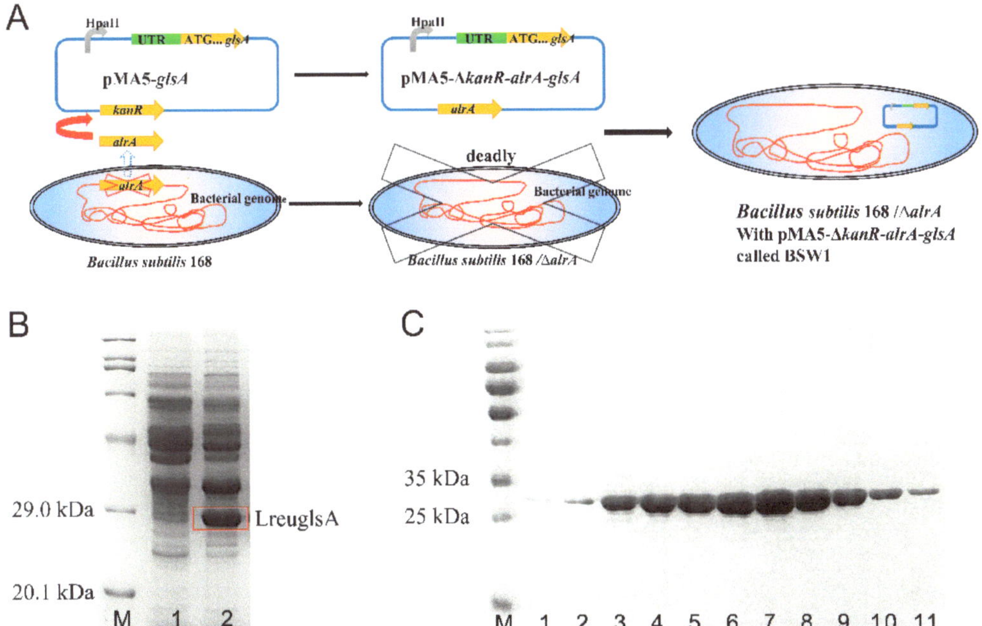

Figure 1. Construction of food-grade *B. subtilis*. (**A**) *B. subtilis* 168, in which the *alrA* gene was knocked down, did not grow in basic medium. D-Alanine was added when screening the auxotrophic strain *B. subtilis* 168/ΔalrA. The recombinant plasmid pMA5-ΔkanR-*alrA*-glsA, in which the alrA gene was used to replace the pMA5 recombinant plasmid containing the Kan resistance gene, was transformed into *B. subtilis* 168/ΔalrA. (**B**) SDS-PAGE analysis of LreuglsA expression in BSW1. The three lanes show the marker (Lane M), pMA5-ΔkanR-alrA expression in BSW0 (Lane 1), and pMA5-ΔkanR-*alrA*-glsA expression in BSW0 (Lane 2). The protein size was approximately 30 kDa, according to the marker. (**C**) SDS-PAGE analysis of purified LreuglsA. Ni2+-affinity chromatography was used to purify the protein. The gradient elution process of protein samples during purification is reflected in the image (Lanes 1–11).

The enzymatic properties play a decisive role in the industrial application of enzyme preparations. The enzymatic reaction was performed under different conditions to investigate the properties of LreuglsA. The maximum reaction rate was achieved at a temperature of 50 °C and pH 7.5 (50 mM Tris-HCl buffer, Figure 2A,B). The specific activity of the purified enzyme was 1048.14 ± 7.83 U mg^{-1}. The enzyme remained stable (more than 60% activity remained) after an incubation for 216 h at 4 °C in pH 6–7 buffer without any protective agent. In terms of temperature stability, the enzyme reached a T1/2 of approximately 44 h at an industrial working temperature of 37 °C; however, the activity reduced to less than 50% after 30 min of incubation at 50 °C (Figure 2C). The enzyme kinetic constant Km was measured as 51.24 ± 4.39 mM, and the Vmax was 7.84 ± 0.33 mM min^{-1} under the optimum reaction conditions (Figure 2D). We investigated the effect of adding 1 mM metal ions on the enzyme activity. Na$^+$ activated the enzyme, and the other metal ions inhibited (less than 30%) the enzyme to some extent (Figure 2E).

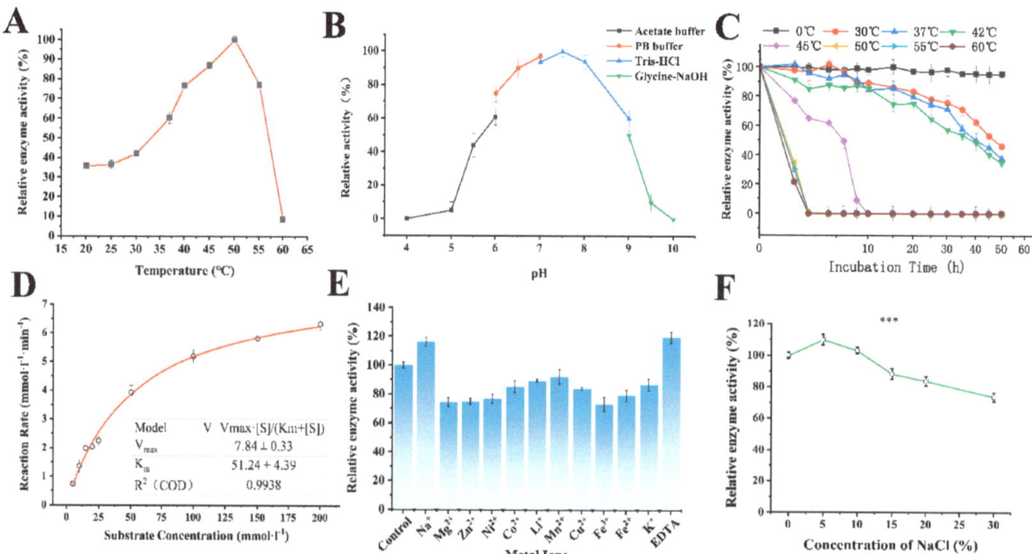

Figure 2. Enzymatic properties of LreuglsA. The activities are presented as relative values observed in various ranges, as described below. (**A**) Effect of temperature on LreuglsA activity. (**B**) Effect of pH on LreuglsA activity. (**C**) Effect of stability at different temperatures on LreuglsA activity. (**D**) Enzyme kinetic fitting. The Michaelis–Menten equation was applied using the Origin nonlinear fitted Hill function (n = 1). (**E**) Effect of metal ions on LreuglsA activity. (**F**) Effects of salt concentrations on the glutaminase activity of LreuglsA. Each experimental data point represents the average of three independent experiments, and the error bars indicate standard deviations. *** $p < 0.005$.

The most significant feature for the industrial value of LreuglsA was its salt tolerance. The high-salt normothermic conditions in brewing processes such as those of soy sauce limit the application of food enzyme preparations. The LreuglsA enzyme activity peaked at a 5% salt concentration, reaching 110.03% of the activity measured under blank conditions. At a near-saturated NaCl concentration (30%), 73.85% of the activity under blank conditions was maintained (Figure 2F). Moreover, in 20% NaCl solution, the thermal stability of the enzyme increased by more than 70% at 50 °C. A potential explanation for this result may be that, at low salt concentrations, the ions in the solvent enhance the interaction force of hydrophobic residues inside the protein and increase the stability. However, at high salt concentrations, the weakening of the hydrated layer formed by the charged residues on the protein surface and the increasing difficulty of the active pocket to take up substrates from the solvent lead to a decrease in enzyme activity. Based on these data, LreuglsA has a good potential for industrialization. Its ability to maintain the necessary activity under high-salt conditions indicates that it can be used in high-salt industrial brewing processes such as those used for soy sauce.

3.2. Structural Prediction and Identification of the Active Center of LreuglsA

AlphaFold2 is a valuable tool to obtain deeper insights into the structure of the LreuglsA protein. The active page shows a comparable molecular weight of active LreuglsA to the bovine serum protein (66.43 kDa), suggesting that, in the reactivated state, LreuglsA is present in a homodimeric form. We obtained a dimer model with an "iptm + ptm" score of 0.93 and a residue pLDDT score greater than 80 using AlphaFold2-Multimer for homologous dimer modeling.

Thus far, we have identified the possible active sites by performing a sequence alignment between LreuglsA and glutaminases from *Bacillus amyloliquefaciens* (BalglsA),

B. subtilis (BsglsA), *B. licheniformis* (BliglsA), *E. coli* (EcoglsA), *Geobacillus stearothermophilus* (GbtglsA), and *Heyndrickxia sporothermodurans* (HstglsA) to identify possible LreuglsA active sites ([32] Figure 3A), combined with the structural analysis of BsglsA and EcoglsA and alignment of highly conserved sequences. In the catalytic pocket, we identified a catalytic triplet, S60-V61-S62-K63, which is the β-lactamase motif (S-X-X-K) and is considered essential for catalysis [33]. We screened a suitable docking result using Glide version 91117 (mmshare version 54117) in Maestro 12.8 software for molecular docking, with the following docking values: Best Emodel = −42.29, E = −34.31, Eint = 7.15, and GlideScore = −4.56. The amino acid residues E59, S60, E157, and Y188 form hydrogen bonds with L-glutamine (Figure 3B). Based on Schrödinger Glide Docking, we mutated the residues into alanine to verify the reliability of the molecular docking results. After mutating each of the four residues mentioned above that form hydrogen bonds with glutamine, the glutaminase was inactive. These results allowed us to determine the active center of LreuglsA.

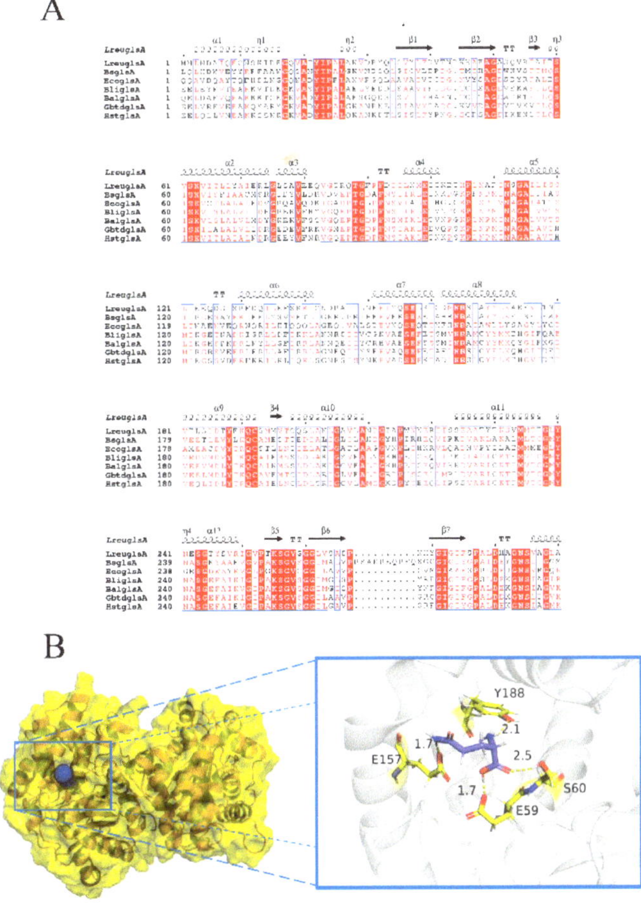

Figure 3. Structural prediction and identification of the active center. (**A**) Analysis of conserved residues required for enzyme function by a sequence alignment of known L-glutaminases. The color scheme option was normal in ESPript 3 (http://espript.ibcp.fr (accessed on 3 May 2022)). (**B**) The PDB structure generated by AlphaFold2-Multimer and molecular docking via Schrödinger. The images were generated using PyMOL. The blue part is the substrate entry and exit channel.

3.3. A H105K Mutant Is Designed Using Computer-Aided Mutation Prediction to Increase Thermostability

In industrial applications, temperature stability has always been an issue of concern. Greater temperature stability may broaden the prospective applications of LreuglsA. We identified regions of LreuglsA with high RMSF values by performing an MD simulation and screened for energy differences using FoldX for virtual saturation mutations in the entire sequence to increase the temperature stability. Mutations in residues near the active pocket were excluded, as these residues may be critical for the reaction (Figure 4A). Meanwhile, evolutionary information was employed by PSSM to identify positions that might be mutated into conserved residues related to heat resistance in L-glutaminase (Figure 4B). Based on the computer-aided mutation prediction, eight mutant sites were selected after a visual inspection, and the mutants were expressed in *B. subtilis* 168.

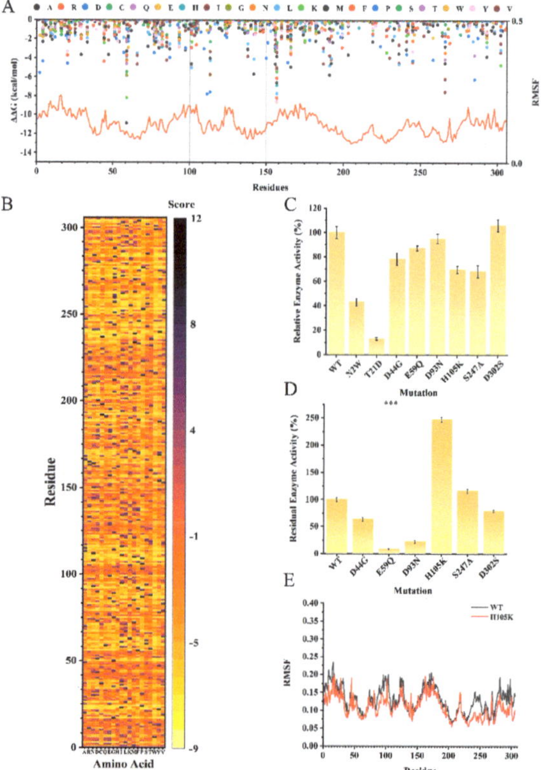

Figure 4. Computer-aided mutation prediction for increasing the thermostability. (**A**) The ddG value of the virtual saturation mutation and the RMSF of LreuglsA. The colored circles represent the energy change after virtual mutation, and the part where ddG is positive is not shown. The red line is the RMSF of LreuglsA, which indicates the side-chain flexibility of LreuglsA. (**B**) Evolvable residues based on the PSSM score to allow mutations. The more conserved amino acids scored higher at this position. (**C**) Enzyme activity of mutants reacted at 50 °C for 5 min. The activities are presented as relative values observed for the mutants. (**D**) Enzyme activity of mutants after 72 h of incubation at 37 °C. *** $p < 0.005$. (**E**) RMSF values of the side-chain atoms for the WT and H105K systems. The black line represents the outcome for the WT system, and the red line represents the outcome for the H105K system. The activities are presented as relative values observed for the mutants. Each experimental data point is the average of three independent experiments, and the error bars indicate standard deviations.

As the solvent environment of crude enzymes is closer to that of industrial catalytic microbial cell factories, crude enzymes are used to screen mutants with improved stability, and according to the report, the change of the folding free energy of different mutants will affect the protein expression level. The thermodynamic stabilities of the wild-type and mutant enzymes were compared at 37 °C. Crude enzyme activity was determined at 50 °C to exclude variants with decreased activity compared to wild-type LreuglsA (Figure 4C). After an incubation at 37 °C for 72 h, the residual enzyme activity of L-glutaminase was measured to compare the thermostabilities of the wild-type and mutant proteins. The results of the preliminary screen suggested that the enzyme activity of the N2W and T21D mutants was 50% lower than that of the wild-type enzyme (Figure 4C). No follow-up studies were conducted for these two mutation sites. After 72 h of incubation, the relative enzyme activity of the H105K mutant was 247.36% of that of the wild-type enzyme. A comparison of the RMSF values of the H105K mutant and the wild-type enzyme by MD simulations of 30 ns at 37 °C showed that the H105K mutant was less structurally flexible than the wild-type protein, resulting in the excellent thermal stability of the H105K mutant (Figure 4D). Additionally, an energy-based ddG of -1.93 kJ mol^{-1} confirmed this possibility. Next, we verified the salt tolerance of the H105K mutant, and the relative enzyme activity in the presence of 20% NaCl reached more than 85% of that measured in the absence of NaCl, indicating that the mutation did not affect the salt tolerance of LreuglsA.

3.4. Improving Enzyme Production with a Portable 5′-UTR

The secondary hairpin structure of the 5′-UTR has been reported to increase the stability of mRNA, thereby increasing the level of structural gene expression. A series of particular secondary structures based on the RBS of pMA5 were developed and inserted into the 5′-UTR as a stabilizer to increase the translation efficiency (Figure 5A). The sequence of the 5′-UTR used in this study is listed in Figure 5A. We included the completely unpaired SD sequence within or after the hairpin loop to adjust the distance between the hairpin structure and ATG. Driven by BSW3 (with UTR3), the L-glutaminase activity increased 5.61-fold from 16.32 ± 0.56 U mL^{-1} to 91.56 ± 2.43 U mL^{-1} (Figure 5B). The bound ribosome and hairpin ring secondary structure protected the mRNA from digestion, as described in previous research. RT-qPCR was performed to determine the stability of the mRNA after the addition of the UTR element. The RT-qPCR results showed that the half-life of the mRNA containing UTR3 was 5.2-fold higher than that of the control group (Figure 5C). Thus, in *B. subtilis*, synthetic biological 5′-UTR components represent a new strategy to increase the protein expression. As shown by the RNAfold results, the original RBS region of pMA5 also formed a hairpin loop structure. However, the SD sequence was present in paired bases, which was unfavorable for ribosome recognition and binding.

In contrast, for the artificially designed RBS sequence, the SD sequence located in the loop did not affect ribosome binding. However, the effect of different minimum free energies of hairpin loops on the expression was noticeable. The finer optimization of the minimum free energy for hairpin ring uncoupling warrants further investigation.

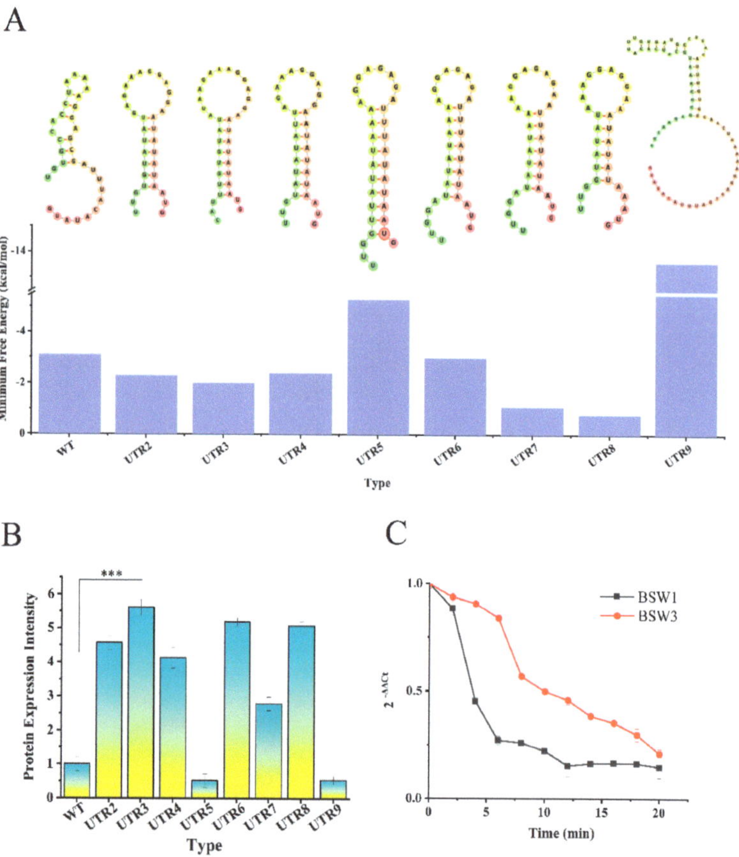

Figure 5. Protein overexpression with a portable 5′-UTR. (**A**) MFE plain structures of the 5′-UTR and its corresponding minimum free energy values are presented in the bar chart below. (**B**) Protein expression levels from constructs with different 5′-UTRs located before the ATG start codon. *** $p < 0.005$. (**C**) Intracellular LreuglsA mRNA levels in BSW1 and BSW3 at different times after the addition of rifampicin. Each experimental data point is the average of three independent experiments, and the error bars indicate standard deviations.

3.5. Immobilized BSW3-H105K Cells Significantly Increase Glutamate Production during Soy Sauce Brewing

Generally, because stirred-tank bioreactors have a more automated control system, they provide a better production environment for microbial cell factories. Therefore, the process for the production of LreuglsA by BSW3 was transferred to a 5-L bioreactor to evaluate the kinetic parameters during scale-up. We selected the most suitable fermentation medium for LreuglsA by BSW3-H105K cells using speed–DO-coupled replenishment (Figure 6A). During fermentation, with DO controlled at 35% and the stirring rate down-off, feeding was started at 15 mL/h when the DO content was higher than 40% in the logarithmic growth phase of fermentation, and the feed flow rate was increased when the DO content started increasing again. Eventually, the L-glutaminase activity reached 2516.78 ± 20.83 U mL^{-1}, and the OD600 reached 93.90 ± 2.01. Regarding enzyme production, by exploring the bacterial growth curve and enzyme production kinetics, L-glutaminase was identified as a growth-associated product in the bioreactor culture, with an average yield of 26.79 U mL^{-1} per unit OD600 (Figure 6B).

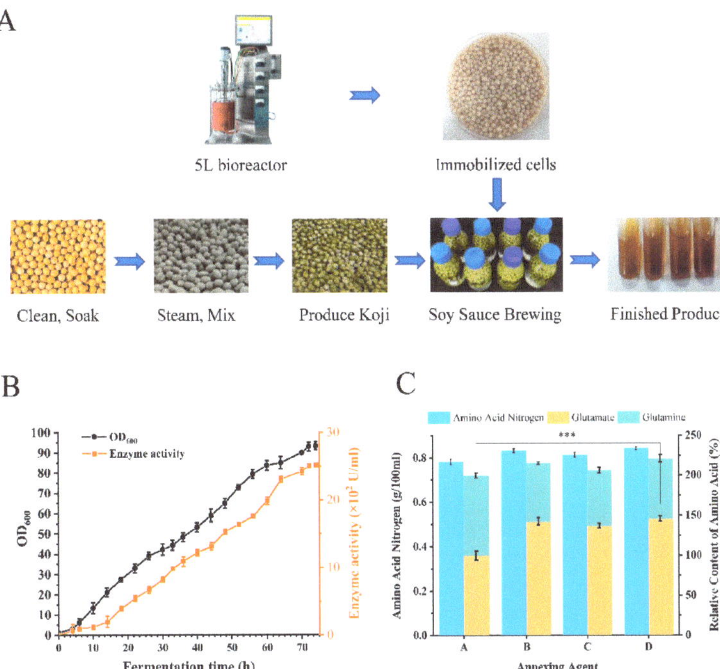

Figure 6. Fermentation, immobilization, and soy sauce brewing. (**A**) BSW3 was fermented in a 5-L bioreactor, and 500 mL of cells were immobilized after fermentation; the resulting immobilized cells were added to the HSDS soy sauce brew in six portions. (**B**) The fermentation process parameters of the recombinant strain BSW3 in 5-L fed-batch fermentation. (**C**) Levels of amino acid nitrogen, glutamate, and glutamine in the produced soy sauce. Group A was the control group, the cell-disrupting crude enzyme was added to group B, washed whole cells were added to group C, and immobilized cells were added to group D. Each experimental data point represents the average of three independent experiments, and the error bars indicate standard deviations. *** $p < 0.005$.

The addition of LreuglsA to the soy sauce brewing process revealed the potential applicability of this enzyme in the brewing industry. The levels of amino acid nitrogen and free amino acids are the characteristic indexes used to determine the degree of fermentation in soy sauce brewing. Cell immobilization can effectively prolong the stability of the enzyme [34], and the addition of immobilized cells in the soy sauce brewing process might effectively prevent the cell contents and the enzyme itself from influencing the flavor of the soy sauce (Figure 6A). Cells harvested from 500 mL of fermentation culture were washed with PBS and mixed with an equal volume of 2.5% sodium alginate. The mixture was dropped through a syringe into a 10% CaCl$_2$ solution to form immobilized capsules of 3–5 mm. In soy sauce brewing, crude enzyme liquid, whole cells, and immobilized cells (prepared separately from the same batch of 500 mL of fermentation broth) were divided into six batches and added to the fermentation solution every 20 days. The formaldehyde method was used to determine the amino acid nitrogen content, and HPLC was used to determine the free amino acid content. According to the Chinese hygiene standards for soy sauce (GB/T 18186-2000), the soy sauce from the control group without additives reached the first-class standard with a 0.783 $\pm$ 0.003 g per 100 mL amino acid nitrogen content, and soy sauce fermented with immobilized BSW3 reached the superfine standard with a final amino acid nitrogen content of 0.841 $\pm$ 0.002 g per 100 mL. Regarding the glutamate content, the addition of all three forms of glutaminase increased the glutamate content, with the best-immobilized BSW3 increasing it by 45.9% from 5.0152 $\pm$ 0.048 g L^{-1}

to 7.319 ± 0.068 g L^{-1} (Figure 6C). Thus, the addition of salt-tolerant LreuglsA to the soy sauce brewing process increased the L-glutamate content in the soy sauce and improved its quality. Laboratory-level soy sauce brewing may reveal the potential of this strategy for soy sauce production. The application of immobilized cells in soy sauce potentially improves the enzyme stability and multibatch utilization and effectively reduces the alteration of soy sauce flavor due to the presence of food additives. Although cell immobilization may provide more favorable working conditions for the enzyme, high-salt conditions are not conducive to the maintenance of the stability of recombinant bacteria. These results will be valuable for investigating the directed evolution of salt tolerance of host bacteria and the immobilization of enzymes.

4. Conclusions

In this work, a food-grade recombinant bacterium was constructed with a salt-tolerant L-glutaminase based on a synthetic biology strategy. First, we mined for a novel L-glutaminase with significant salt tolerance and enzymatic activity, and based on AlphaFold2 modeling and computational biology, we screened for mutation sites with enhanced temperature stability. By optimizing the free energy of the 5′-UTR and synthetic biological application of the *alrA* gene, the constructed food-grade BSW3 achieved an enzyme activity of 2516.78 ± 20.83 U/mL in a 5-L bioreactor, which is the highest production reported to date. Finally, the immobilized BSW3 was effective in soy sauce brewing, improving the glutamic acid, the important flavoring substance in soy sauce, by 45.9%. This study provided reference for further improving the quality of soy sauce, and the study of salt tolerance of the enzyme provided valuable material.

Author Contributions: Conceptualization, Z.R. and X.Z.; methodology, H.Z., F.L., M.X. and M.H.; software, H.Z.; validation, H.Z.; formal analysis, M.X.; investigation, M.H. and Q.W.; resources, M.H.; data curation, H.Z.; writing—original draft preparation, H.Z.; writing—review and editing, H.Z.; visualization, M.H.; supervision, H.Z.; project administration, Z.R. and X.Z.; and funding acquisition, Z.R. and X.Z. All authors have read and agreed to the published version of the manuscript.

Funding: This work was supported by grants from the National Key Research and Development Program of China (No. 2021YFC2100900), National Natural Science Foundation of China (No. 32171471 and No. 32071470), and a Project funded by the Priority Academic Program Development of Jiangsu Higher Education Institutions, Top-notch Academic Programs Project of Jiangsu Higher Education Institutions.

Institutional Review Board Statement: Not applicable.

Informed Consent Statement: Not applicable.

Data Availability Statement: The data that support the findings of this study are available from the corresponding author upon reasonable request.

Conflicts of Interest: The authors declare no conflict of interest.

References

1. Halpern, B.P. Glutamate and the flavor of foods. *J. Nutr.* **2000**, *130*, 910S–914S. [CrossRef] [PubMed]
2. Geha, R.S.; Beiser, A.; Ren, C.; Patterson, R.; Greenberger, P.A.; Grammer, L.C.; Ditto, A.M.; Harris, K.E.; Shaughnessy, M.A.; Yarnold, P.R.; et al. Review of alleged reaction to monosodium glutamate and outcome of a multicenter double-blind placebo-controlled study. *J. Nutr.* **2000**, *130*, 1058S–1062S. [CrossRef] [PubMed]
3. Graudal, N.A.; Hubeck-Graudal, T.; Jurgens, G. Effects of low sodium diet versus high sodium diet on blood pressure, renin, aldosterone, catecholamines, cholesterol, and triglyceride. *Cochrane Database Syst. Rev.* **2020**, *12*, CD004022. [CrossRef] [PubMed]
4. Binod, P.; Sindhu, R.; Madhavan, A.; Abraham, A.; Mathew, A.K.; Beevi, U.S.; Sukumaran, R.K.; Singh, S.P.; Pandey, A. Recent developments in l-glutaminase production and applications—An overview. *Bioresour. Technol.* **2017**, *245*, 1766–1774. [CrossRef]
5. Nandakumar, R.; Yoshimune, K.; Wakayama, M.; Moriguchi, M. Microbial glutaminase: Biochemistry, molecular approaches and applications in the food industry. *J. Mol. Catal. B Enzym.* **2003**, *23*, 87–100. [CrossRef]
6. Tadanobu Nakadai, S.N. Use of Glutaminase for Soy Sauce Made by Koji or a Preparation of Proteases from Aspergillus oryzae. *J. Ferment. Bioeng.* **1989**, *67*, 158–162. [CrossRef]

7. Kumar, L.; Singh, B.; Adhikari, D.K.; Mukherjee, J.; Ghosh, D. A temperature and salt-tolerant L-glutaminase from gangotri region of uttarakhand himalaya: Enzyme purification and characterization. *Appl. Biochem. Biotechnol.* **2012**, *166*, 1723–1735. [CrossRef]
8. Nandakumar, R.; Wakayama, M.; Nagano, Y.; Kawamura, T.; Sakai, K.; Moriguchi, M. Overexpression of salt-tolerant glutaminase from Micrococcus luteus K-3 in Escherichia coli and its purification. *Protein Expr. Purif.* **1999**, *15*, 155–161. [CrossRef]
9. Moriguchi, M.; Sakai, K.; Tateyama, R.; Furuta, Y.; Wakayama, M. Isolation and Characterization of Salt-Tolerant Glutaminases from Marine Micrococcus-Luteus K-3. *J. Ferment. Bioeng.* **1994**, *77*, 621–625. [CrossRef]
10. Weingand-Ziade, A.; Gerber-Decombaz, C.; Affolter, M. Functional characterization of a salt- and thermotolerant glutaminase from Lactobacillus rhamnosus. *Enzym. Microb. Technol.* **2003**, *32*, 862–867. [CrossRef]
11. Jeon, J.M.; Lee, H.I.; Han, S.H.; Chang, C.S.; So, J.S. Partial purification and characterization of glutaminase from Lactobacillus reuteri KCTC3594. *Appl. Biochem. Biotechnol.* **2010**, *162*, 146–154. [CrossRef]
12. Amobonye, A.; Singh, S.; Pillai, S. Recent advances in microbial glutaminase production and applications-a concise review. *Crit. Rev. Biotechnol.* **2019**, *39*, 944–963. [CrossRef] [PubMed]
13. Pu, H.; Wang, Q.; Zhu, F.; Cao, X.; Xin, Y.; Luo, L.; Yin, Z. Cloning, expression of glutaminase fromPseudomonas nitroreducensand application to theanine synthesis. *Biocatal. Biotransform.* **2012**, *31*, 1–7. [CrossRef]
14. Sakhaei, M.; Alemzadeh, I. Enzymatic Synthesis of Theanine in the Presence of L-glutaminase Produced by Trichoderma koningii. *Appl. Food Biotechnol.* **2017**, *4*, 113–121. [CrossRef]
15. Kroemer, G.; Pouyssegur, J. Tumor cell metabolism: Cancer's Achilles' heel. *Cancer Cell* **2008**, *13*, 472–482. [CrossRef] [PubMed]
16. Orabi, H.; El-Fakharany, E.; Abdelkhalek, E.; Sidkey, N. Production, optimization, purification, characterization, and anti-cancer application of extracellular L-glutaminase produced from the marine bacterial isolate. *Prep. Biochem. Biotechnol.* **2020**, *50*, 408–418. [CrossRef]
17. Reda, F.M. Kinetic properties of Streptomyces canarius L- Glutaminase and its anticancer efficiency. *Braz. J. Microbiol.* **2015**, *46*, 957–968. [CrossRef]
18. Liu, Y.; Su, A.; Li, J.; Ledesma-Amaro, R.; Xu, P.; Du, G.; Liu, L. Towards next-generation model microorganism chassis for biomanufacturing. *Appl. Microbiol. Biotechnol.* **2020**, *104*, 9095–9108. [CrossRef]
19. Liu, Y.; Liu, L.; Li, J.; Du, G.; Chen, J. Synthetic Biology Toolbox and Chassis Development in Bacillus subtilis. *Trends Biotechnol.* **2019**, *37*, 548–562. [CrossRef]
20. Jumper, J.; Evans, R.; Pritzel, A.; Green, T.; Figurnov, M.; Ronneberger, O.; Tunyasuvunakool, K.; Bates, R.; Zidek, A.; Potapenko, A.; et al. Highly accurate protein structure prediction with AlphaFold. *Nature* **2021**, *596*, 583–589. [CrossRef]
21. Evans, R.; O'Neill, M.; Pritzel, A.; Antropova, N.; Senior, A.; Green, T.; Žídek, A.; Bates, R.; Blackwell, S.; Yim, J.; et al. Protein complex prediction with AlphaFold-Multimer. *BioRxiv* **2022**. [CrossRef]
22. Schymkowitz, J.; Borg, J.; Stricher, F.; Nys, R.; Rousseau, F.; Serrano, L. The FoldX web server: An online force field. *Nucleic Acids Res.* **2005**, *33*, W382–W388. [CrossRef] [PubMed]
23. Sun, Z.; Liu, Q.; Qu, G.; Feng, Y.; Reetz, M.T. Utility of B-Factors in Protein Science: Interpreting Rigidity, Flexibility, and Internal Motion and Engineering Thermostability. *Chem. Rev.* **2019**, *119*, 1626–1665. [CrossRef] [PubMed]
24. Abraham, M.J.; Murtola, T.; Schulz, R.; Páll, S.; Smith, J.C.; Hess, B.; Lindahl, E. GROMACS: High performance molecular simulations through multi-level parallelism from laptops to supercomputers. *SoftwareX* **2015**, *1–2*, 19–25. [CrossRef]
25. Xia, Y.; Chen, W.; Zhao, J.X.; Tian, F.W.; Zhang, H.; Ding, X.L. Construction of a new food-grade expression system for *Bacillus subtilis* based on theta replication plasmids and auxotrophic complementation. *Appl. Microbiol. Biotechnol.* **2007**, *76*, 643–650. [CrossRef]
26. Schwanhausser, B.; Busse, D.; Li, N.; Dittmar, G.; Schuchhardt, J.; Wolf, J.; Chen, W.; Selbach, M. Global quantification of mammalian gene expression control. *Nature* **2011**, *473*, 337–342, Erratum in *Nature* **2013**, *495*, 126–127. [CrossRef]
27. Kozak, M. Influences of mRNA secondary structure on initiation by eukaryotic ribosomes. *Proc. Natl. Acad. Sci. USA* **1986**, *83*, 2850–2854. [CrossRef]
28. Viegas, S.C.; Apura, P.; Martinez-Garcia, E.; de Lorenzo, V.; Arraiano, C.M. Modulating Heterologous Gene Expression with Portable mRNA-Stabilizing 5′-UTR Sequences. *ACS Synth. Biol.* **2018**, *7*, 2177–2188. [CrossRef]
29. Xiao, J.; Peng, B.; Su, Z.; Liu, A.; Hu, Y.; Nomura, C.T.; Chen, S.; Wang, Q. Facilitating Protein Expression with PorTable 5′-UTR Secondary Structures in Bacillus licheniformis. *ACS Synth. Biol.* **2020**, *9*, 1051–1058. [CrossRef] [PubMed]
30. Hofacker, I.L.; Stadler, P.F. Memory efficient folding algorithms for circular RNA secondary structures. *Bioinformatics* **2006**, *22*, 1172–1176. [CrossRef]
31. Codex Alimentarius Commission. *Guideline for the Conduct of Food Safety Assessment of Foods Derived from Recombinant-DNA Plants*; Codex Alimentarius Commission: Rome, Italy, 2003.
32. Robert, X.; Gouet, P. Deciphering key features in protein structures with the new ENDscript server. *Nucleic Acids Res.* **2014**, *42*, W320–W324. [CrossRef] [PubMed]
33. Oliva, M.; Dideberg, O.; Field, M.J. Understanding the acylation mechanisms of active-site serine penicillin-recognizing proteins: A molecular dynamics simulation study. *Proteins* **2003**, *53*, 88–100. [CrossRef] [PubMed]
34. Shin, K.C.; Sim, D.H.; Seo, M.J.; Oh, D.K. Increased Production of Food-Grade d-Tagatose from d-Galactose by Permeabilized and Immobilized Cells of Corynebacterium glutamicum, a GRAS Host, Expressing d-Galactose Isomerase from Geobacillus thermodenitrificans. *J. Agric. Food Chem.* **2016**, *64*, 8146–8153. [CrossRef] [PubMed]

 *fermentation*

Article

Design of 5′-UTR to Enhance Keratinase Activity in *Bacillus subtilis*

Jun Fang [1,2], Guanyu Zhou [1,2], Xiaomei Ji [1,2], Guoqiang Zhang [1,2], Zheng Peng [1,2,*] and Juan Zhang [1,2,*]

[1] Key Laboratory of Industrial Biotechnology, Ministry of Education, School of Biotechnology, Jiangnan University, 1800 Lihu Road, Wuxi 214122, China
[2] Science Center for Future Foods, Jiangnan University, 1800 Lihu Road, Wuxi 214122, China
* Correspondence: zhengpeng@jiangnan.edu.cn (Z.P.); zhangj@jiangnan.edu.cn (J.Z.)

Abstract: Keratinase is an important industrial enzyme, but its application performance is limited by its low activity. A rational design of 5′-UTRs that increases translation efficiency is an important approach to enhance protein expression. Herein, we optimized the 5′-UTR of the recombinant keratinase KerZ1 expression element to enhance its secretory activity in *Bacillus subtilis* WB600 through Spacer design, RBS screening, and sequence simplification. First, the A/U content in Spacer was increased by the site-directed saturation mutation of G/C bases, and the activity of keratinase secreted by mutant strain *B. subtilis* WB600-SP was 7.94 times higher than that of KerZ1. Subsequently, the keratinase activity secreted by the mutant strain *B. subtilis* WB600-SP-R was further increased to 13.45 times that of KerZ1 based on the prediction of RBS translation efficiency and the multi-site saturation mutation screening. Finally, the keratinase activity secreted by the mutant strain *B. subtilis* WB600-SP-R-D reached 204.44 KU mL^{-1} by reducing the length of the 5′ end of the 5′-UTR, which was 19.70 times that of KerZ1. In a 5 L fermenter, the keratinase activity secreted by *B. subtilis* WB600-SP-R-D after 25 h fermentation was 797.05 KU mL^{-1}, which indicated its high production intensity. Overall, the strategy of this study and the obtained keratinase mutants will provide a good reference for the expression regulation of keratinase and other industrial enzymes.

Keywords: keratinase; 5′-UTR; spacer; RBS; sequence simplification; Gibbs free energy

Citation: Fang, J.; Zhou, G.; Ji, X.; Zhang, G.; Peng, Z.; Zhang, J. Design of 5′-UTR to Enhance Keratinase Activity in *Bacillus subtilis*. *Fermentation* **2022**, *8*, 426. https:// doi.org/10.3390/fermentation8090426

Academic Editors: Zhiming Rao and Xian Zhang

Received: 13 August 2022
Accepted: 19 August 2022
Published: 27 August 2022

Publisher's Note: MDPI stays neutral with regard to jurisdictional claims in published maps and institutional affiliations.

1. Introduction

Keratin is an insoluble protein waste widely distributed in the epidermis and its appendages of animals, such as hair, feathers, nails, claws, carapace, horns and beaks [1–3]. Due to the richness of cysteine and glycine, and the cross-linked disulfide bonds formed between cysteine residues, keratin has a dense structure and low solubility in water [1]. The current chemical and physical methods for extracting keratin inevitably involve processes such as high temperature, microwave, and strong acid or alkali, which will not only cause product damage but also huge energy consumption and environmental burden [4].

Keratinase is a hydrolase with the ability to specifically degrade keratin and is considered to be an important biological enzyme that can improve existing processes in many industries such as tanning, cosmetics, detergents and feed [5–10]. Keratinases are mainly secreted by bacteria, fungi and actinomycetes found in soil, water or various sources rich in keratin substances [11–13]. The native keratinase gene has been extensively recombinantly expressed in conventional engineered strains [14,15]. Compared with the higher misfolding rate of *Escherichia coli* and the long-term fermentation of *Pichia*, *Bacillus subtilis* has the advantages of a short cycle and strong secretion capacity [16–20]. Strategies such as promoter optimization, signal peptide screening and pro-peptide engineering have all been used to regulate the expression of keratinase in *B. subtilis* [15,17,18]. However, the lower activity and expression capacity remain challenges for keratinase toward application.

The 5′-untranslated region (5′-UTR) of prokaryotic mRNAs contains the initiation codon, the Shine–Dalgarno (SD) sequence and the translational enhancer sequence, which play an important role in mRNA stability and translation initiation [21–23]. Since translation initiation is the rate-limiting step in gene expression, 5′-UTRs are often engineered to regulate protein expression levels in various biotechnological applications, including biosensor development, metabolic engineering and gene circuits [24–27]. Salis et al. established the "RBS calculator" to aid in the design of RBS sequences, which can increase the target translation initiation rate by 100,000-fold in *E. coli* [28]. Josh et al. constructed a random 5′-UTR library and trained a convolutional neural network on activity data obtained from the library in high-throughput parallel growth experiments to accurately predict *Saccharomyces cerevisiae* 5′-UTR elements with potentially high expression capacity [29]. In addition, although it is difficult to form a complete system, the modification of other structures of the mRNA 5′-UTR beyond the ribosome binding site is still widely reported. Xiao et al. developed a portable 5′-UTR sequence for enhancing the protein export of the industrial strain *Bacillus licheniformis* DW2. The optimized SD sequence is presented in single-stranded form on the hairpin loop for better ribosome recognition and recruitment. By optimizing the free energy of folding, this 5′-element can effectively enhance the expression of eGFP by about 50-fold [30].

In a previous study, we have successfully expressed the recombinant keratinase KerZ1 in *B. subtilis* WB600 with the P43 promoter [18]. Herein, we will optimize the 5′-UTR of the recombinant keratinase KerZ1 expression element to enhance its activity in *B. subtilis* WB600 by Spacer design, RBS screening and sequence simplification. This study will be a paradigm for enhancing protein expression by designing 5′-UTRs in *B. subtilis*.

2. Materials and Methods

2.1. Gene, Plasmids and Strains

In previous studies, we have expressed the keratinase gene from *Bacillus licheniformis* in *B. subtilis* WB600 to obtain the recombinant keratinase KerZ1 [18]. Strains, plasmids and primers used in this study are listed in Table 1. All plasmids were derived from the backbone vector pP43NMK-*Ker*. *E. coli* JM109 was used for plasmid cloning and enrichment. All mutants achieved secretory expression in *B. subtilis* WB600. In site-directed saturation mutagenesis, the entire plasmid was amplified using mutant primers containing the degenerate base N using plasmid pP43NMK-*ker* as a template. Subsequently, the circularization of all linearized plasmids was done using Gibson assembly. All plasmids were chemically transformed into competent cells of *E. coli* JM109 and *B. subtilis* WB600.

Table 1. List of strains, plasmids and primers used in this study.

Name	Sequence (5′-3′)
Strains	
JM109	*Escherichia coli*
WB600	*Bacillus subtilis* 168 derivate, missing *nprE aprE epr bpr mpr nprB*
WB600-*Ker*	*Bacillus subtilis* WB600 contains plasmid pP43NMK-*Ker*
Plasmids	
pP43NMK	Amp^r, Km^r, *E. coli–B. subtilis* shuttle vector
pP43NMK-*Ker*	pP43NMK derivate with *B. licheniformis Ker* gene under the control of the promoter P$_{43}$
Primers	
Spacer	
Spacer-3N-F	TTATAGGTAAGAGAGGAAT<u>N</u>TA<u>N</u>A<u>N</u>ATGATGAGGAAAAAGAGTTTTTGGCTTGG
Spacer-3N-R	ATTCCTCTCTTACCTATAATGGTACCGCTAT
RBS	
RBS-6N-F	TAGCGGTACCATTATAGG<u>NNNNNN</u>AGGAATGTATAGATGATGAGGAAAAAGAGTTTTTG
RBS-6N-R	CCTATAATGGTACCGCTATCACTTTATATTTTACATAATCG
5′ end	
Sim1-F	TAAAATATAAAGTGATAGCGTACCATTATAGGTATTGGAGGAATGTACAC
Sim2-F	TAAAATATAAAGTGATAGGTACCATTATAGGTATTGGAGGAATGTACAC
Sim3-F	TAAAATATAAAGTGATAGTACCATTATAGGTATTGGAGGAATGTACAC
Sim4-F	ATTATGTAAAATATAAAGTATAGTACCATTATAGGTATTGGAGGAATGTACAC
Sim5-F	ATTATGTAAAATATAAATATAGTACCATTATAGGTATTGGAGGAATGTACAC

Table 1. *Cont.*

Name	Sequence (5′-3′)
Sim6-F	ATGTAAAATATAAAGTGATGCGGTACCATTATAGGTATTGGAGGAATGTACAC
Sim7-F	ATGTAAAATATAAAGTGAGCGGTACCATTATAGGTATTGGAGGAATGTACAC
Sim1C-F	TAAAATATAAAGTGATAGCGTAACATTATAGGTATTGGAGGAATGTACAC
Sim2C-F	TAAAATATAAAGTGATAGGTAACATTATAGGTATTGGAGGAATGTACAC
Sim3C-F	TAAAATATAAAGTGATAGTAACATTATAGGTATTGGAGGAATGTACAC
Sim4C-F	ATTATGTAAAATATAAAGTATAGTAACATTATAGGTATTGGAGGAATGTACAC
Sim6C-F	ATGTAAAATATAAAGTGATGCGGTAACATTATAGGTATTGGAGGAATGTACAC
Sim7C-F	ATGTAAAATATAAAGTGAGCGGTAACATTATAGGTATTGGAGGAATGTACAC
Sim8-F	ATGTAAAATATAAAGTGGCGGTACCATTATAGGTATTGGAGGA
Sim9-F	GATTATGTAAAATATAAAGGGCGGTACCATTATAGGTATTGGAGGA
Sim10-F	ATTATGTAAAATATAAACGCGGTACCATTATAGGTATTGGAGGA
Sim11-F	ATTATGTAAAATATAAAGTACCATTATAGGTATTGGAGGAATGTACAC
Sim123-R	TATCACTTTATATTTTACATAATCGCGCGCTTTTTTTC
Sim4591011-R	TTTATATTTTACATAATCGCGCGCTTTTTTTCACG
Sim678-R	CACTTTATATTTTACATAATCGCGCGCTTTTTTTC

2.2. Medium and Culture Conditions

Escherichia coli JM109 was cultured in Luria–Bertani medium (yeast powder 5, peptone 10, sodium chloride 10) g L^{-1} supplemented with 100 μg mL^{-1} ampicillin at 37 °C for 12 h. *B. subtilis* WB600 cells carrying recombinant plasmids were cultured in fermentation medium (glucose 30 g L^{-1}, yeast extract 5.72 g L^{-1}, soybean meal 40 g L^{-1}, $Na_2HPO_4 \cdot 12H_2O$ 3 g L^{-1}, KH_2PO_4 1.5 g L^{-1}, $MgSO_4 \cdot 7H_2O$ 0.3 g L^{-1}) supplemented with 50 μg mL^{-1} kanamycin at 37 °C for 24 h. For site-directed saturation mutagenesis, 96-well plates were used for the culturing and screening of recombinant strains. Transformants of *B. subtilis* WB600 were picked into 96-well plates containing 1 mL of fermentation medium and cultured with shaking at 37 °C for 24 h.

2.3. Keratinase Activity

A reaction system containing 150 μL of 50 mM Gly/NaOH buffer (pH 9.0), 100 μL of 2.5% soluble keratin (CAS RN: 69430-36-0) and 50 μL of appropriately diluted enzyme was incubated at 60 °C for 20 min. Then, 200 μL of 0.5 mol L^{-1} trichloroacetic acid (TCA) was added to stop the reaction, and the system was centrifuged at 12000× g for 2 min. 200 μL of the supernatant was added to 1 mL of 4% Na_2CO_3, followed by 200 μL of Folin–Ciocalteu reagent, and the chromogenic system was mixed and incubated at 50 °C for 10 min. Keratinase activity was calculated from absorbance at 660 nm and a tyrosine standard curve. All experiments were repeated three times, and the control group was mixed with trichloroacetic acid before adding the enzyme solution.

2.4. Analysis of 5′-UTR Secondary Structure

The translation initiation efficiency after the mutation of the RBS sequence in the *Bacillus subtilis* recombinant keratinase KerZ1 expression system was predicted using RBS Calculator v2.0 (https://salislab.net/software/forward, accessed on 4 March 2022) [28]. The 5′-UTR sequence of the keratinase expression element was uploaded to the online server mfold (http://www.unafold.org/mfold/applications/rna-folding-form.php, accessed on 20 June 2022) for secondary structure prediction and Gibbs free energy calculation [31].

2.5. Fermentation Performance Validation in Fermenter

Culture validation was performed using a 5 L fermenter (T&J Bio-engineering Co., Ltd., Shanghai, China) containing 3.0 L fermentation medium. The initial fermentation temperature was set to 37 °C, and the pH of the system was maintained at 7 by automatically pumping in ammonia. During fermentation, the aeration rate was set at 0.5 vvm and the dissolved oxygen (DO) was maintained at 20−30% by correlating the DO with the stirring speed. Glucose with a concentration of 720 g L^{-1} was replenished in two stages at 40 mL h^{-1} and 30 mL h^{-1}.

3. Results and Discussion

3.1. Replacement of Spacer Sequence C/G to A/T

The Spacer sequence is located between the RBS and the target gene. Studies have shown that the activation of translation by A/U-rich Spacer sequences is independent of SD sequences, initiation codons and prior cistron translation [32]. A/U-rich sequences may improve translation efficiency by enhancing interaction with ribosomal protein S1 [33]. To improve the translation efficiency of keratinase, we performed saturation mutations on the C/G bases in the Spacer sequence of the 5′-UTR element of keratinase KerZ1 (Figure 1a).

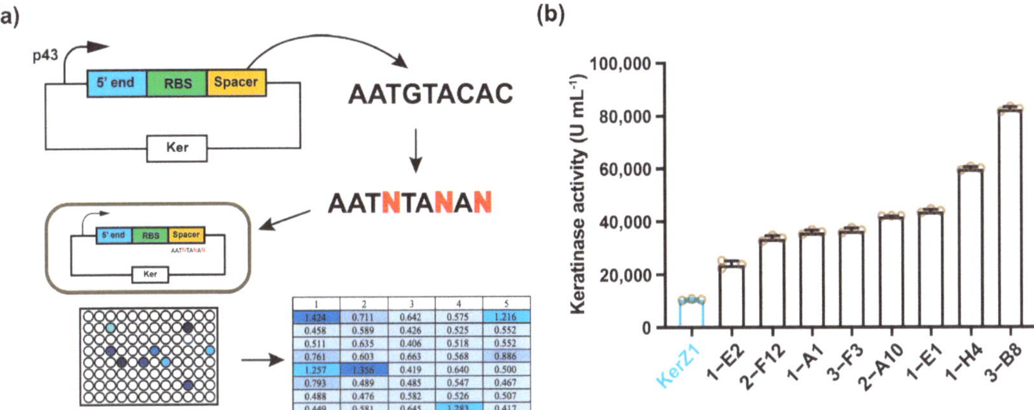

Figure 1. Base substitution of Spacer to enhance keratinase activity. (**a**) Schematic diagram of base substitution design of Spacer and screening process of mutant strains. (**b**) Fermentation in shake flasks to rescreen Spacer mutant strains. KerZ1 was secreted by the original strain *B. subtilis* WB600 pP43NMK-*Ker*.

High-throughput screening was performed for the Spacer sequence by measuring the keratinase activity secreted by the mutant strains. After preliminary screening, eight Spacer sequence mutant strains with improved keratinase activity in 96-well plate fermentation were obtained, namely 1-A1, 1-E1, 1-E2, 1-H4, 2-A10, 2-F12, 3-B8 and 3-F3. Subsequently, rescreening was performed at the shake flask level. The activity assay showed that the keratinase activity secreted by the mutant strains obtained after preliminary screening was significantly higher than that of the original strain (KerZ1, 10380 U mL^{-1}). Among them, the keratinase activities secreted by strains 1-H4 and 3-B8 were significantly higher than others, reaching 59,845 and 82,435 U mL^{-1}, which were 5.76 and 7.94 times that of KerZ1, respectively (Figure 1b).

Sequencing showed that the C/G bases in the Spacer sequence of the screened strains were multi-mutated to A/T bases, and the corresponding mRNA bases were A/U bases (Table 2). Similar results were shown in a study by Ilya A. et al.; A/U-rich enhancers derived from highly expressed phoP genes increased monocistronic mRNA expression nearly five-fold [32]. However, in the mutant strain 3-B8 with the highest keratinase activity, the −3 site of the Spacer sequence was mutated from C to T, and the −1 site was mutated from C to G, which was not a complete A/U substitution. Perhaps the presence of G bases in the Spacer sequence may also improve the expression efficiency of the protein, which remains to be verified in the future. Finally, we named 3-B8 as *B. subtilis* WB600-SP and carried out the next transformation on this basis.

Table 2. Spacer base mutants and their corresponding keratinase activities.

Strains	Spacer (5′-3′)	Activity (U mL^{-1})
KerZ1	AATGTACAC	10,380
1-A1	AATTTACAT	35,935
1-E1	AATTTACAT	43,800
1-E2	AATTTACAC	23,745
1-H4	AATTTATAA	59,845
2-A10	AATTTACAT	41,915
2-F12	AATTTACAA	33,615
3-B8	AATGTATAG	82,435
3-F3	AATTTACAT	36,565

3.2. Screening of Ribosome Binding Sites (RBS)

RBS sequence is the core of 5 'UTR, usually composed of 4–9 nucleotides and rich in a and G bases [34]. In prokaryotes, the RBS sequence also has the SD sequence, which can complementarily pair with the 3′ end of ribosomal 16S rRNA to facilitate ribosome binding to mRNA [35]. The easier the ribosome binds to RBS, the more stable the complex formed and the higher the translation initiation efficiency. To screen for RBS sequences with higher translation initiation efficiency, we performed saturation mutations on the original RBS sequences (Figure 2a). Based on the original RBS sequence (GTAAGAGAGG) of plasmid p43NMK, the 6-base group gradually moved closer to the translation initiation site (NNNNNNGAGG, GNNNNNNAGG until GTAANNNNNN) to predict the effect of RBS sequence mutation on the translation initiation rate [28]. The results show that the generation of mutations closer to the translation initiation site has less effect on the translation initiation rate. Positions 2–7 of RBS have the highest possible translation initiation efficiency when predicted for mutations (Figure 2b). Therefore, primers were designed to perform saturation mutation on positions 2–6 of RBS, that is, the −13 to −18 region of the total sequence, to obtain a strain with enhanced keratinase secreting activity.

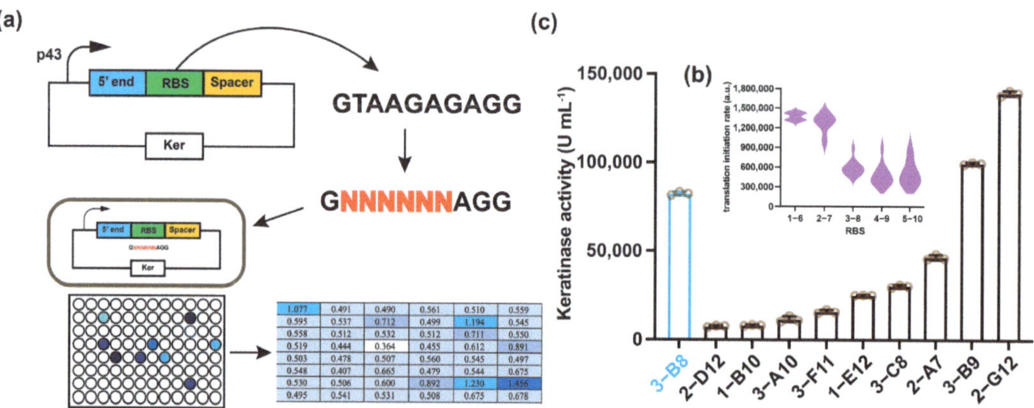

Figure 2. Base optimization of RBS to increase translation initiation rate to enhance keratinase activity. (**a**) Schematic diagram of base optimization design of RBS and screening process of mutant strains. (**b**) Predicted transcription initiation rates corresponding to mutations in different regions of the RBS sequence. (**c**) Fermentation in shake flasks to rescreen RBS mutant strains.

The theoretical number of transformants were selected three times for screening to achieve 95% coverage. The primary screening in the 96-well plate showed that the keratinase activity secreted by nine strains was higher than that of the control strain *B. subtilis* WB600-SP, namely 1-B10, 1-E12, 2-A7, 2-D12, 2-G12, 3-A10, 3-B9, 3-C8, 3-F11. After rescreening in the shake flask, only the keratinase activities secreted by 2-G12 and 3-B9 were significantly higher than those of the control, reaching 139,650 and 99,860 U mL^{-1}, which were 13.45 and 9.62 times that of the original strain (KerZ1, 10380 U mL^{-1}), respectively (Figure 2c). Although the translation initiation efficiency prediction results were different from the actual expression, the translation initiation efficiency of the 2-G11 and 3-B9 strains with increased activity in the shake flask was significantly higher than that of the original strain (Table 3), which indicates that it is effective to screen RBS sequences based on the prediction of translation initiation efficiency. We named 2-G12 as *B. subtilis* WB600-SP-R.

Table 3. RBS base mutants and their corresponding keratinase activities.

Strains	RBS (5′-3′)	Activity (U mL^{-1})	Translation Initiation Rate (au)
SP	GTAAGAGAGG	82,435	72.08
1-B10	GCTGCACAGG	7695	11.63
1-E12	GCTTGCGAGG	25,350	26.81
2-A7	GGGAAGTAGG	45,740	362.40
2-D12	GATGGTAAGG	7510	37.82
2-G12	GTATTGGAGG	139,650	85.58
3-A10	GAAAGACAGG	11,060	23.71
3-B8	GGACGGAAGG	99,860	127.08
3-C8	GTGTTGCAGG	30,395	29.45
3-F11	GGGGGCTAGG	15,710	121.96

3.3. Simplification of the 5′ End Sequence

There is a 9 bp single chain sequence at the 5′-end of the 5′-UTR stem-loop structure, which has no additional function other than carrying the initiation site for transcription. Studies have shown that the simplification of the single-stranded sequence is helpful to weaken the influence of the downstream expressed gene sequence on the 5′-UTR region and enhance the protein expression [30]. In addition, simplifying the single-chain sequence can adjust the proportion of each base in the 5′-UTR region [36]. To this end, we used the 5′-UTR region of KerZ1 as a template to adjust the base ratio and prevent additional stem-loop structures by deleting a/T or C/G bases one by one. At the same time, for the change of stem ring structure caused by the introduction of mutation, a base mutation was introduced to maintain the original stem ring structure, as shown in Table 4.

Table 4. Mutant sequences for simplified design of the 5′-UTR region.

Mutants	5′-UTR Sequence (5′-3′)
2-G12	GTGATAGCGGTACCATTATAGGTATTGGAGGAATGTACAC
Sim1	GTGATAGC–GTACCATTATAGGTATTGGAGGAATGTACAC
Sim1C	GTGATAGC–GTAACATTATAGGTATTGGAGGAATGTACAC
Sim2	GTGATAG–--GTACCATTATAGGTATTGGAGGAATGTACAC
Sim2C	GTGATAG–--GTAACATTATAGGTATTGGAGGAATGTACAC
Sim3	GTGATA——GTACCATTATAGGTATTGGAGGAATGTACAC
Sim3C	GTGATA——GTAACATTATAGGTATTGGAGGAATGTACAC
Sim4	GT–ATA——GTACCATTATAGGTATTGGAGGAATGTACAC
Sim4C	GT–ATA——GTAACATTATAGGTATTGGAGGAATGTACAC
Sim5	–T–ATA——GTACCATTATAGGTATTGGAGGAATGTACAC
Sim6	GTGAT–GCGGTACCATTATAGGTATTGGAGGAATGTACAC
Sim6C	GTGAT–GCGGTAACATTATAGGTATTGGAGGAATGTACAC
Sim7	GTGA—-GCGGTACCATTATAGGTATTGGAGGAATGTACAC
Sim7C	GTGA—-GCGGTAACATTATAGGTATTGGAGGAATGTACAC
Sim8	GTG——GCGGTACCATTATAGGTATTGGAGGAATGTACAC
Sim9	G–G——GCGGTACCATTATAGGTATTGGAGGAATGTACAC
Sim10	————GCGGTACCATTATAGGTATTGGAGGAATGTACAC
Sim11	—————GTACCATTATAGGTATTGGAGGAATGTACAC

Note: – is the deleted base; the underlined base is the modified base to maintain the stem-loop structure.

A keratinase activity test showed that the activity of mutant Sim3C reached 204.44 KU mL^{-1} (Figure 3a), which was 19.70 times that of keratinase KerZ1 (10,380 U mL^{-1}), and was named *B. subtilis* WB600-SP-R-D. SDS-PAGE was used to verify the keratinases secreted by the optimal mutant strains obtained under different strategies (Figure 3d). The results showed that the expression of keratinase increased gradually. In addition, we also obtained several mutants with increased keratinase activity to varying degrees. Using mfold to predict the secondary structure of mRNA, we found that 5′ deletions altered the mRNA secondary structure (Figures 3b,c). The SD sequence is located on the stem-loop to form a single chain, which is more conducive to its recognition and to the recruitment of ribosomes. In addition, the Gibbs free energy of Sim3C (−1.60) was lower than that of 2-G12 (−7.70), indicating that Sim3C mRNA is more easily bound to ribosomes and has a higher translation initiation rate.

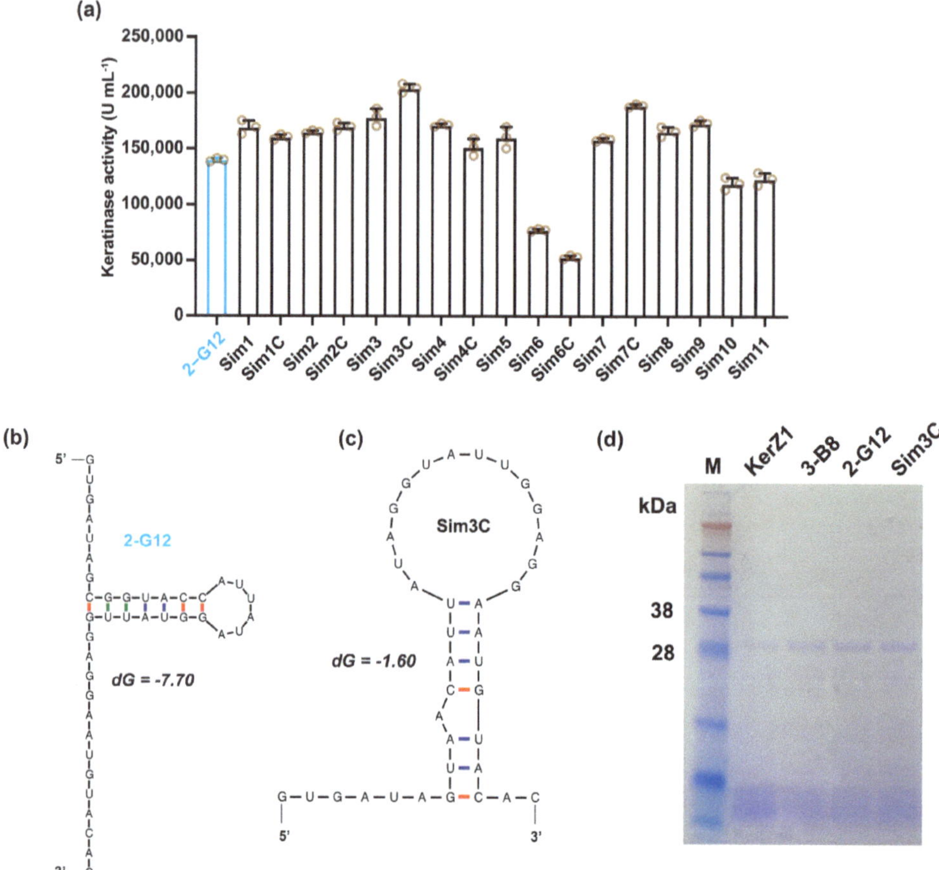

Figure 3. Deletion of the 5′ end of the 5′ UTR to enhance keratinase activity. (**a**) Keratinase activity of mutant strains with deletions at the 5′ end of the 5′ UTR. (**b**) mRNA secondary structure of the original sequence at the 5′ end of the 5′ UTR. (**c**) mRNA secondary structure of mutant Sim3C at the 5′ end of the 5′ UTR. (**d**) SDS-PAGE verified the keratinase secreted by the optimal mutant strains obtained under different strategies.

3.4. Fermentation Performance of B. subtilis WB600-SP-R-D

Exploring the enzyme-producing properties of mutant strains in larger systems is not only the basis for the development of industrial strains but also the verification of production strength. To this end, we fermented the mutant strain *B. subtilis* WB600-SP-R-D in a 5 L fermenter to test its ability and strength to secrete keratinase. Previous studies have shown that excess glucose causes recombinant *Bacillus subtilis* to produce large amounts of lactic acid as a by-product, which inhibits bacterial growth [37]. While it is possible to automatically pump ammonia to adjust the pH of the environment, this reduces the ability of cells to produce keratinase [15,38]. Therefore, 720 g L^{-1} of glucose was continuously fed for 10–22 h to keep the glucose concentration at 1–30 g L^{-1} during fermentation. The feeding was stopped after the 22nd hour, so that the glucose was almost exhausted at the end of the fermentation, which was more conducive to the subsequent separation of keratinase. After 27 h of fermentation, the mutant strain *B. subtilis* WB600-SP-R-D reached the peak activity of 797.05 KU mL^{-1} at 25th hour of fermentation, which was 76.79 times that of the original strain (Figure 4). Therefore, the mutant *B. subtilis* WB600-SP-R-D after 5′-UTR optimization showed excellent and stable enzyme production ability.

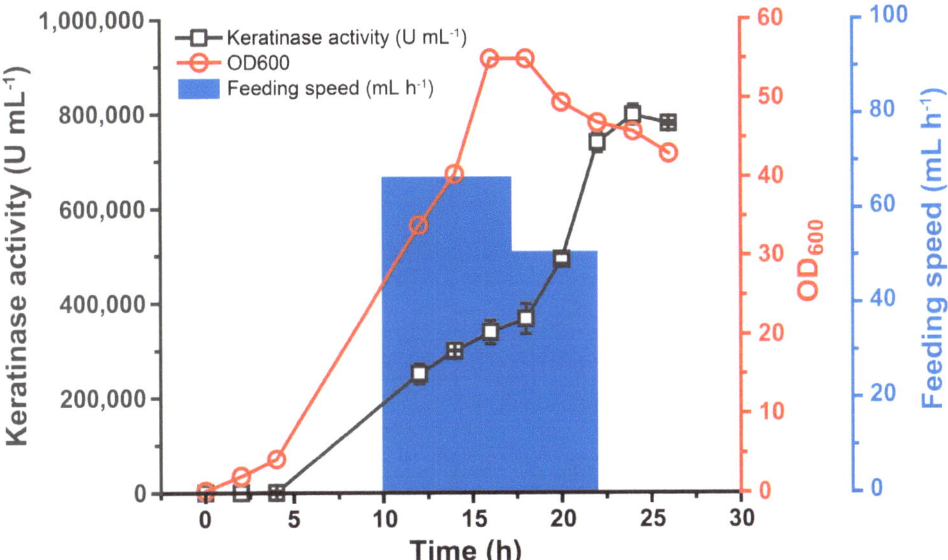

Figure 4. Fermentation performance of mutant strain *B. subtilis* WB600-SP-R-D. Fed-batch fermentation of strain *B. subtilis* WB600-SP-R-D in a 5 L fermenter. The glucose concentration was maintained within 1–30 g L^{-1} by adding 720 g L^{-1} glucose. Then the feed was stopped after the 22nd hour, and the glucose was almost exhausted by the end of the fermentation.

4. Conclusions

Keratinase is a promising keratin treatment, but the low activity is still the resistance to use. In this study, we obtained several mutant strains with increased keratinase secretion through the modification of 5′-UTR. A/U base substitution in Spacer region, RBS sequence optimization and the deletion of a single-stranded sequence at the 5′ end made the keratinase activity secreted by mutant *B. subtilis* WB600-SP-R-D reach 204.44 KU mL^{-1}, which was 19.70 times that of keratinase KerZ1 (10,380 U mL^{-1}). In addition, *B. subtilis* WB600-SP-R-D had the highest activity in the 5 L fermenter at 797.05 KU mL^{-1}, showing excellent enzyme production capacity and stability. In conclusion, the rational design of 5′-UTR can significantly improve the expression activity of keratinase in *B. subtilis*, which will be a typical example of the regulation of recombinant protein expression.

Author Contributions: Conceptualization, J.F. and Z.P.; methodology, J.F., G.Z. (Guanyu Zhou) and Z.P.; validation, J.F., G.Z. (Guanyu Zhou) and X.J.; formal analysis, J.F., G.Z. and Z.P. (Guanyu Zhou); data curation, J.F., G.Z. (Guanyu Zhou) and Z.P.; writing—original draft preparation, J.F. and Z.P.;

writing—review and editing, Z.P., G.Z. (Guoqiang Zhang) and J.Z.; supervision, Z.P., G.Z. (Guoqiang Zhang) and J.Z. All authors have read and agreed to the published version of the manuscript.

Funding: This work was supported by the National Key Research and Development Program of China (2021YFC2104000) and a grant from the Key Technologies R & D Program of Jiangsu Province (BE2021624).

Informed Consent Statement: Not applicable.

Data Availability Statement: Not applicable.

Conflicts of Interest: The authors declare no conflict of interest.

References

1. Wang, B.; Yang, W.; McKittrick, J.; Meyers, M.A. Keratin: Structure, mechanical properties, occurrence in biological organisms, and efforts at bioinspiration. *Prog. Mater. Sci.* **2016**, *76*, 229–318. [CrossRef]
2. Meyers, M.A.; Chen, P.-Y.; Lin, A.Y.-M.; Seki, Y. Biological materials: Structure and mechanical properties. *Prog. Mater. Sci.* **2008**, *53*, 1–206. [CrossRef]
3. Donato, R.K.; Mija, A. Keratin Associations with Synthetic, Biosynthetic and Natural Polymers: An Extensive Review. *Polymers* **2019**, *12*, 32. [CrossRef] [PubMed]
4. Callegaro, K.; Brandelli, A.; Daroit, D.J. Beyond plucking: Feathers bioprocessing into valuable protein hydrolysates. *Waste Manag.* **2019**, *95*, 399–415. [CrossRef]
5. Brandelli, A.; Daroit, D.J.; Riffel, A. Biochemical features of microbial keratinases and their production and applications. *Appl. Microbiol. Biotechnol.* **2010**, *85*, 1735–1750. [CrossRef]
6. Fang, Z.; Yong, Y.-C.; Zhang, J.; Du, G.; Chen, J. Keratinolytic protease: A green biocatalyst for leather industry. *Appl. Microbiol. Biotechnol.* **2017**, *101*, 7771–7779. [CrossRef]
7. Zhang, Z.; Li, D.; Zhang, X. Enzymatic decolorization of melanoidins from molasses wastewater by immobilized keratinase. *Bioresour. Technol.* **2019**, *280*, 165–172. [CrossRef]
8. Peng, Z.; Mao, X.; Mu, W.; Du, G.; Chen, J.; Zhang, J. Modifying the Substrate Specificity of Keratinase for Industrial Dehairing to Replace Lime-Sulfide. *ACS Sustain. Chem. Eng.* **2022**, *10*, 6863–6870. [CrossRef]
9. Gupta, R.; Sharma, R.; Beg, Q.K. Revisiting microbial keratinases: Next generation proteases for sustainable biotechnology. *Crit. Rev. Biotechnol.* **2013**, *33*, 216–228. [CrossRef]
10. Verma, A.; Singh, H.; Anwar, S.; Chattopadhyay, A.; Tiwari, K.K.; Kaur, S.; Dhilon, G.S. Microbial keratinases: Industrial enzymes with waste management potential. *Crit. Rev. Biotechnol.* **2017**, *37*, 476–491. [CrossRef]
11. Qiu, J.; Wilkens, C.; Barrett, K.; Meyer, A.S. Microbial enzymes catalyzing keratin degradation: Classification, structure, function. *Biotechnol. Adv.* **2020**, *44*, 107607. [CrossRef] [PubMed]
12. Su, C.; Gong, J.-S.; Qin, J.; Li, H.; Li, H.; Xu, Z.-H.; Shi, J.-S. The tale of a versatile enzyme: Molecular insights into keratinase for its industrial dissemination. *Biotechnol. Adv.* **2020**, *45*, 107655. [CrossRef] [PubMed]
13. Daroit, D.J.; Brandelli, A. A current assessment on the production of bacterial keratinases. *Crit. Rev. Biotechnol.* **2014**, *34*, 372–384. [CrossRef]
14. Prakash, P.; Jayalakshmi, S.K.; Sreeramulu, K. Purification and characterization of extreme alkaline, thermostable keratinase, and keratin disulfide reductase produced by *Bacillus halodurans* PPKS-2. *Appl. Microbiol. Biotechnol.* **2010**, *87*, 625–633. [CrossRef] [PubMed]
15. Peng, Z.; Zhang, J.; Song, Y.; Guo, R.; Du, G.; Chen, J. Engineered pro-peptide enhances the catalytic activity of keratinase to improve the conversion ability of feather waste. *Biotechnol. Bioeng.* **2021**, *118*, 2559–2571. [CrossRef] [PubMed]
16. Hu, H.; Gao, J.; He, J.; Yu, B.; Zheng, P.; Huang, Z.; Mao, X.; Yu, J.; Han, G.; Chen, D. Codon Optimization Significantly Improves the Expression Level of a Keratinase Gene in Pichia pastoris. *PLoS ONE* **2013**, *8*, e58393. [CrossRef] [PubMed]
17. Su, C.; Gong, J.-S.; Sun, Y.-X.; Qin, J.; Zhai, S.; Li, H.; Li, H.; Lu, Z.-M.; Xu, Z.-H.; Shi, J.-S. Combining Pro-peptide Engineering and Multisite Saturation Mutagenesis To Improve the Catalytic Potential of Keratinase. *ACS Synth. Biol.* **2019**, *8*, 425–433. [CrossRef]
18. Peng, Z.; Mao, X.; Zhang, J.; Du, G.; Chen, J. Biotransformation of keratin waste to amino acids and active peptides based on cell-free catalysis. *Biotechnol. Biofuels* **2020**, *13*, 61. [CrossRef]
19. Fang, Z.; Zhang, J.; Liu, B.; Jiang, L.; Du, G.; Chen, J. Cloning, heterologous expression and characterization of two keratinases from *Stenotrophomonas maltophilia* BBE11-1. *Process Biochem.* **2014**, *49*, 647–654. [CrossRef]
20. Cui, W.; Han, L.; Suo, F.; Liu, Z.; Zhou, L.; Zhou, Z. Exploitation of Bacillus subtilis as a robust workhorse for production of heterologous proteins and beyond. *World J. Microbiol. Biotechnol.* **2018**, *34*, 145. [CrossRef]
21. Gingold, H.; Pilpel, Y. Determinants of translation efficiency and accuracy. *Mol. Syst. Biol.* **2011**, *7*, 481. [CrossRef] [PubMed]
22. Espah Borujeni, A.; Salis, H.M. Translation Initiation is Controlled by RNA Folding Kinetics via a Ribosome Drafting Mechanism. *J. Am. Chem. Soc.* **2016**, *138*, 7016–7023. [CrossRef] [PubMed]
23. Andreeva, I.; Belardinelli, R.; Rodnina, M.V. Translation initiation in bacterial polysomes through ribosome loading on a standby site on a highly translated mRNA. *Proc. Natl. Acad. Sci. USA* **2018**, *115*, 4411–4416. [CrossRef] [PubMed]

24. Yu, Q.; Li, Y.; Ma, A.; Liu, W.; Wang, H.; Zhuang, G. An efficient design strategy for a whole-cell biosensor based on engineered ribosome binding sequences. *Anal. Bioanal. Chem.* **2011**, *401*, 2891–2898. [CrossRef] [PubMed]
25. Zhou, S.; Ding, R.; Chen, J.; Du, G.; Li, H.; Zhou, J. Obtaining a Panel of Cascade Promoter-5′-UTR Complexes in *Escherichia coli*. *ACS Synth. Biol.* **2017**, *6*, 1065–1075. [CrossRef]
26. Zhang, D.; Zhou, C.Y.; Busby, K.N.; Alexander, S.C.; Devaraj, N.K. Light-Activated Control of Translation by Enzymatic Covalent mRNA Labeling. *Angew. Chem. Int. Ed. Engl.* **2018**, *57*, 2822–2826. [CrossRef]
27. Viegas, S.C.; Apura, P.; Martinez-García, E.; de Lorenzo, V.; Arraiano, C.M. Modulating Heterologous Gene Expression with Portable mRNA-Stabilizing 5′-UTR Sequences. *ACS Synth. Biol.* **2018**, *7*, 2177–2188. [CrossRef]
28. Salis, H.M.; Mirsky, E.A.; Voigt, C.A. Automated design of synthetic ribosome binding sites to control protein expression. *Nat. Biotechnol.* **2009**, *27*, 946–950. [CrossRef]
29. Cuperus, J.T.; Groves, B.; Kuchina, A.; Rosenberg, A.B.; Jojic, N.; Fields, S.; Seelig, G. Deep learning of the regulatory grammar of yeast 5′ untranslated regions from 500,000 random sequences. *Genome Res.* **2017**, *27*, 2015–2024. [CrossRef]
30. Xiao, J.; Peng, B.; Su, Z.; Liu, A.; Hu, Y.; Nomura, C.T.; Chen, S.; Wang, Q. Facilitating Protein Expression with Portable 5′-UTR Secondary Structures in *Bacillus licheniformis*. *ACS Synth. Biol.* **2020**, *9*, 1051–1058. [CrossRef]
31. Zuker, M. Mfold web server for nucleic acid folding and hybridization prediction. *Nucleic Acids Res.* **2003**, *31*, 3406–3415. [CrossRef]
32. Osterman, I.A.; Evfratov, S.A.; Sergiev, P.V.; Dontsova, O.A. Comparison of mRNA features affecting translation initiation and reinitiation. *Nucleic Acids Res.* **2013**, *41*, 474–486. [CrossRef]
33. Lee, H.-M.; Ren, J.; Kim, W.Y.; Vo, P.N.L.; Eyun, S.-I.; Na, D. Introduction of an AU-rich Element into the 5′ UTR of mRNAs Enhances Protein Expression in Escherichia coli by S1 Protein and Hfq Protein. *Biotechnol. Bioprocess Eng.* **2021**, *26*, 749–757. [CrossRef]
34. Zelcbuch, L.; Antonovsky, N.; Bar-Even, A.; Levin-Karp, A.; Barenholz, U.; Dayagi, M.; Liebermeister, W.; Flamholz, A.; Noor, E.; Amram, S.; et al. Spanning high-dimensional expression space using ribosome-binding site combinatorics. *Nucleic Acids Res.* **2013**, *41*, e98. [CrossRef] [PubMed]
35. Amin, M.R.; Yurovsky, A.; Chen, Y.; Skiena, S.; Futcher, B. Re-annotation of 12,495 prokaryotic 16S rRNA 3′ ends and analysis of Shine-Dalgarno and anti-Shine-Dalgarno sequences. *PLoS ONE* **2018**, *13*, e0202767. [CrossRef] [PubMed]
36. Komarova, E.S.; Chervontseva, Z.S.; Osterman, I.A.; Evfratov, S.A.; Rubtsova, M.P.; Zatsepin, T.S.; Semashko, T.A.; Kostryukova, E.S.; Bogdanov, A.A.; Gelfand, M.S.; et al. Influence of the spacer region between the Shine–Dalgarno box and the start codon for fine-tuning of the translation efficiency in *Escherichia coli*. *Microb. Biotechnol.* **2020**, *13*, 1254–1261. [CrossRef] [PubMed]
37. Liu, Y.; Liu, L.; Shin, H.-D.; Chen, R.R.; Li, J.; Du, G.; Chen, J. Pathway engineering of *Bacillus subtilis* for microbial production of N-acetylglucosamine. *Metab. Eng.* **2013**, *19*, 107–115. [CrossRef] [PubMed]
38. Gu, Y.; Lv, X.; Liu, Y.; Li, J.; Du, G.; Chen, J.; Rodrigo, L.-A.; Liu, L. Synthetic redesign of central carbon and redox metabolism for high yield production of N-acetylglucosamine in *Bacillus subtilis*. *Metab. Eng.* **2019**, *51*, 59–69. [CrossRef]

 *fermentation*

Article

A High-Throughput Absolute Abundance Quantification Method for the Characterisation of Daqu Core Fungal Communities

Hai Du [†], Jia Sun [†], Tianci Zhou and Yan Xu *

Laboratory of Brewing Microbiology and Applied Enzymology, Key Laboratory of Industrial Biotechnology of Ministry of Education School of Biotechnology, Jiangnan University, Wuxi 214100, China; duhai88@126.com (H.D.); sunjia0605@163.com (J.S.); 6180201098@stu.jiangnan.edu.cn (T.Z.)
* Correspondence: yxu@jiangnan.edu.cn; Tel.: +86-510-85964112; Fax: +86-510-85918201
† These authors contributed equally to this work.

Abstract: An inherent issue in high-throughput sequencing applications is that they provide compositional data for relative abundance. This often obscures the true biomass and potential functions of fungi in the community. Therefore, we presented a high-throughput absolute quantification (HAQ) method to quantitatively estimate the fungal abundance in Daqu. In this study, five internal standard plasmids (ISPs) were designed for the fungal ITS2 subregion with high length variations. Five ISPs were then utilised to establish standard curves with a quantitative concentration range of 10^3–10^7 cells/g, and this was used to quantify the core fungi, including Basidiomycota, Ascomycota, and Mucoromycota. Using three types of mature Daqu from different regions, we demonstrated that the HAQ method yielded community profiles substantially different from those derived using relative abundances. Then, the HAQ method was applied to the Daqu during fermentation. The initial formation of the Daqu surface occurred in the fourth stage, which was mainly driven by moisture. The key fungi that caused the initial formation of the Daqu surface included *Hyphopichia burtonii*, *Saccharomycopsis fibuligera*, and *Pichia kudriavzevii*. The initial formation of the Daqu core occurred in the fifth stage, which was mainly affected by moisture and reducing the sugar content. The key fungi that cause the initial formation of the Daqu core included *S. fibuligera* and *Paecilomyces verrucosus*. We conclude that the HAQ method, when applied to ITS2 gene fungal community profiling, is quantitative and that its use will greatly improve our understanding of the fungal ecosystem in Daqu.

Keywords: high-throughput sequencing; internal standard plasmids; standard curves; quantitative Daqu core fungi

Citation: Du, H.; Sun, J.; Zhou, T.; Xu, Y. A High-Throughput Absolute Abundance Quantification Method for the Characterisation of Daqu Core Fungal Communities. *Fermentation* **2022**, *8*, 345. https://doi.org/10.3390/fermentation8080345

Academic Editors: Fabrizio Beltrametti and Johannes Wöstemeyer

Received: 28 May 2022
Accepted: 14 July 2022
Published: 22 July 2022

Publisher's Note: MDPI stays neutral with regard to jurisdictional claims in published maps and institutional affiliations.

1. Introduction

Chinese liquor, called Baijiu, is a traditional fermented alcoholic drink originating in China, one of the oldest distilled liquors in the world. It has a profound influence on the global liquor industry [1]. Most of the famous liquors, such as Moutai liquor, Wuliangye, and Fen liquor, are brewed using the traditional Daqu method. According to the temperature of the production, traditional Chinese Daqu can be divided into medium-temperature Daqu (45~50 °C), medium-high-temperature Daqu (50~59 °C), and high-temperature Daqu (Above 60 °C) [2]. The brewing of Chinese liquor is inseparable from Daqu [3–5]. It can directly transfer the abundant microbial strains that are useful to liquor brewing to the fermented grains [6]. At the same time, Daqu also provides fermented grains with rich metabolites, mainly the decomposition products of protein and starch and their transformed substances [7], which play an important role in the taste and flavour of Chinese liquor [8]. Daqu contains abundant fungi, including Mucoromycota, Ascomycota, and Basidiomycota [4,5,9,10]. These fungi are important saccharifying and fermenting agents [11] that provide various enzymes [1] essential for the production of Chinese liquor. Daqu

contributes 61–80% of the fungi present during Baijiu fermentation and is closely related to the yield and flavour of fresh Baijiu [12]. Therefore, an in-depth analysis of fungal community succession rates during the Daqu-making process is of great importance for providing a dynamic perspective to optimise the Daqu-making process, improving the quality of Daqu and liquor production. The fungal community has multiple components that are normally abundant and are typically investigated using high-throughput sequencing technology [13]. Nevertheless, the relative abundance determined using high-throughput sequencing cannot reflect the number of fungal communities or dissimilarity among the samples [13,14]. When the total number of microorganisms in different samples is inconsistent, the relative abundance comparison may lead to erroneous conclusions [14,15]. Therefore, to unravel the complexity of fungi in different samples, the fungal communities must be characterised quantitatively. However, the research on the absolute quantification of Daqu fungal communities has not yet been carried out.

Currently, various methods are used for the absolute quantification of microorganisms based on high-throughput sequencing technology [16–22]. High-throughput sequencing is combined with microbial quantification techniques, such as fluorescence staining and flow cytometry (FCM) [19–23], quantitative PCR (qPCR) [22,24], or microbial biomass carbon (MBC) [25], to obtain the absolute abundance of microbial communities. However, a combination of multiple methods increases the study time and may affect the accuracy of the final results. In contrast, the absolute quantification of microorganisms using the internal standard method has been identified as more reliable [13,22,26,27]. This method involves spiking an internal standard to the sample, followed by performing high-throughput sequencing consistent with the target strains. This brings the experimental error of the internal standard close to that of the target strains. For example, previously, internal DNA standards were spiked into samples to quantitatively estimate the microbial abundances per unit volume of filtered seawater, which yielded more accurate results [14]. Despite the advantages of the internal standard method, this method is commonly used for the quantification of bacteria and still faces many challenges when applied to the quantification of fungi. The main constraints are as follows: First, unlike the concentrated distribution of the bacterial 16S sequence, the fungal ITS2 sequence is highly variable in length. It causes the ITS lengths of different fungi to be quite different, and the difference in sequence lengths can cause deviations in high-throughput sequencing [28]. This makes the quantitative analysis of fungi more challenging than that of bacteria. Second, the production process of Daqu is time-sequential, and microorganisms experience growth, reproduction, and death during the production process. Therefore, the absolute content of fungi in Daqu production is dynamic, but the description of absolute quantitative data of fungi is lacking.

To overcome the existing deficits in the fungal quantification method, we selected mature Daqu as the model ecosystem to establish the high-throughput absolute quantification (HAQ) method. Mature Daqu fungal communities usually form under controlled conditions, and many replicate communities can be generated [9,29]. Moreover, these fungal communities can be reproducibly cultured using known media. In this study, we improved the absolute quantification method based on the internal standard. First, three representative aroma characteristics of Chinese liquors were selected: light-aroma type, strong-aroma type, and sauce-aroma type. Based on the experimental data of our group, analysing the data of Daqu and fermented grains, the core fungi were screened. Next, based on the screened fungal ITS2 information, five ISPs were constructed to increase the accuracy of the quantification. Then, high-throughput sequencing and qPCR were used to determine the added concentration of the mixed ISPs and genome extraction efficiency. Standard curves were used to quantify the core fungi. Lastly, we verified the feasibility of this method for three types of mature Daqu. It was also applied to the fermentation process of medium-high-temperature Daqu for quantitative analysis of the formation law of the spatial structure of fungi during fermentation. This study indicates a novel direction for quantitative fungi profiling and provides new insights into the functions of Daqu core fungi.

2. Materials and Methods

2.1. Design and Verification of ISF and ISP

The primers ITS3: 5′-GCATCGATGAAGAACGCAGC-3′ [30] and ITS4: 5′-TCCTCCGC TTATTGATATGC-3′ [30] were selected at both ends of the internal standard fragments (ISFs). The 116-bp sequence (Table S3) was cited as a specific DNA fragment in the ISF [15]. Other sequences were designed using random DNA generation tools (http: //www.novopro.cn/tools/random_dna.html, accessed on 1 June 2020). The obtained ISF was compared in the NCBI (https://www.ncbi.nlm.nih.gov/, accessed on 8 July 2020), UNITE (https://mothur.org/wiki/unite_its_database/, accessed on 8 July 2020), and ITS2 (http://its2.bioapps.biozentrum.uni-wuerzburg.de/, accessed on 8 July 2020) databases, and homologous sequences were observed. The ISF was synthesised by Genewiz Company (Suzhou, China), and *BamHI* (3′) and *SalI* (5′) restriction sites were added at both ends.

Plasmid pET-28a (Table S3), digested by *BamHI* and *SalI*, and ISF were ligated using an In-Fusion HD cloning kit, and the ligation product, pET-28a-ISF, was transformed into *Escherichia coli* JM109 (Table S3) for plasmid cloning. Subsequently, the desired transformants were directly screened on LB plates containing antibiotics (50 µg/mL kanamycin) and were verified using PCR. The preparation of *E. coli* JM109-competent cells and gene fragments and the plasmid transformation were all performed according to the methods described by Zhang [31]. After measuring the ISP concentration using a Thermo Scientific NanoDrop 8000 UV–Vis Spectrophotometer (NanoDrop Technologies, Wilmington, DE, USA), the plasmid copy number was calculated according to the equation described by Dhanasekaran [32]. After dilution with sterile deionised water, the gradient concentrations of ISP in the order of 1×10^{11} to 1×10^{6} copies/mL were stored at $-20\ ^{\circ}$C for later use.

2.2. Screening and Fungi

The mature Daqu and Jiupei of the Chinese soy sauce aroma type, Chinese strong aroma type, and Chinese light aroma type were analysed to screen for fungi. For our study, we put forth three requirements as follows: (1) average abundance > 1% in a single sample, (2) the frequency of occurrence > 50% in all samples, and (3) the absolute value of Spearman's coefficient (| R |) > 0.5.

2.3. Sample Collection

Mature Daqu were collected from traditional Chinese liquor production industries. Among them, high-temperature mature Daqu, medium-high-temperature mature Daqu, and medium-temperature mature Daqu were used in the production of Chinese soy sauce aroma wine, Chinese strong aroma wine, and Chinese light aroma liquor, respectively. Three mature Daqu samples were collected from different production teams. Lastly, all nine samples were transferred to the lab on ice and stored at $-80\ ^{\circ}$C until DNA extraction.

The samples during Daqu fermentation were collected from a typical medium-high-temperature Daqu in a winery in Anhui Province. Choose 3 Qufang to collect the samples. Take 500 g Daqu surface and Daqu core in each Qufang and then transfer to a $-80\ ^{\circ}$C refrigerator for storage. The samples were collected at the end of each stage of the medium- and high-temperature Daqu, namely on 1, 3, 10, 18, 24, and 31 days respectively, for a total of 18 samplings.

2.4. Construction of the HAQ Method

2.4.1. Selection of ISPs Addition Concentration

To explore the optimal added concentration of ISPs and genome extraction efficiency, we designed a simple method, using ISP II for the experiment. The above experiments included 24 treatments as follows: 7 g of high-temperature mature Daqu was spiked with 1 mL of ISP II at seven different concentrations (10^{11}, 10^{10}, 10^{9}, 10^{8}, 10^{7}, 10^{6}, and 0 copies/mL corresponding to treatments T11, T10, T9, T8, T7, T6, and Control, respectively). qPCR was performed using these genomes as templates and SpecF and SpecR as primers to plot the standard curve of the ISPII extracted from Daqu. Equation (1) calculates the extraction efficiency of the ISP in Daqu. Then, 1 mL of mixed fungi at five concentrations (10^3–10^7 cells/mL each) were spiked with 1 mL of the optimal ISP II concentration, corresponding to treatments M3, M4, M5, M6, and M7, respectively. Moreover, the medium-high-temperature mature Daqu (7 g) and medium-temperature mature Daqu (7 g) were spiked with 1 mL ISP II at six concentrations (10^{11}, 10^{10}, 10^9, 10^8, 10^7, and 10^6 copies/mL) to compare the extraction efficiency of ISP II in different mature Daqu.

2.4.2. Establishment of the Core Fungal Standard Curve

One fungus was selected per genus for absolute quantification based on the ease of culture, and a total of 16 core fungi were cultured. Absolute quantitative fungi were isolated from different mature types of Daqu and Jiupei. After culturing in the YEPD medium, a hemocytometer was used to quantify each core fungus. Here, molds were counted in units of spores (germ cells). Lastly, all fungi were mixed at five concentration gradients (each fungus was 10^3–10^7 cells/mL) for later use. First, 1 mL of mixed fungi (5 mL) was spiked with 1 mL of mixed ISPs; then, the genome was extracted and HTS performed in triplicate. The obtained HTS data were analysed and organised, using Equation (2) to calculate the concentration of a certain fungus in the mix, and the correspondence between the ISP and the fungus should be noted. Taking the LG value of the concentration of fungus A (copies/g) as the X-axis, and the LG value of the corresponding fungal A concentration (cells/g) as the Y-axis, a standard curve y = ax + b was established, where x refers to the concentration of fungi (copies/g), and y refers to the concentration of fungi (cells/g).

2.5. Application of HAQ Method

High-temperature mature Daqu (7 g), medium-high-temperature mature Daqu (7 g), and medium-temperature mature Daqu (7 g) were spiked with 1 mL of five mixed ISPs (HQ, MHQ, and MQ). Medium-high-temperature mature Daqu were spiked with 1 mL of sterile water (MHC). Eighteen medium-high-temperature Daqu samples during Daqu fermentation were spiked with 1 mL of five mixed ISPs. All samples were analysed in triplicate.

All samples were mixed thoroughly, and genomic DNA was extracted according to the methods described by Song [33]. The extracted DNA was stored at $-80\ ^\circ$C before qPCR.

2.6. Quantitative PCR

qPCR was performed using a StepOnePlus instrument (Applied Biosystems, Foster City, CA, USA) and a commercial kit (AceQ Universal SYBR qPCR MasterMix. Vazyme, Nanjing, China). We selected a pair of primers, SpecF (5′-GCGGTAAGGTGAAGAGTG-3′) and SpecR 5′-GGCTAACGAGACAACTGC-3′), to detect the copy numbers of the Spec gene of the ISP.

The standard curve was generated using the 10-fold serial dilution of ISP II. The 20-μL qPCR reaction mixture contained 10 μL of SYBR mix (Vazyme), 0.4 μL of each primer (10 mmol/L), 1 μL of DNA template, and 8.2 μL of sterile and DNA-free water. The reaction was executed under the following thermocycler conditions: 95 $^\circ$C for 5 min and 40 cycles of 95 $^\circ$C for 10 s and 60 $^\circ$C for 30 s. Then, the specificity of the PCR products was determined using a melting curve analysis [34]. All qPCR reactions were conducted in triplicate.

2.7. Amplification and Sequencing

The fungal ITS region was amplified using primers ITS3 and ITS4 [30]. The PCR products were purified and carefully evaluated using a Thermo Scientific NanoDrop 8000 UV–Vis Spectrophotometer (NanoDrop Technologies, Wilmington, DE, USA). The high-throughputs were then sequenced using an Illumina MiSeq platform (Illumina, San Diego, CA, USA) at AuwiGene Technology Co., Ltd., Beijing, China.

All raw sequences generated via Qiime (v1.8.0) were processed [35]. In short, the raw sequences were filtered for quality determination, and only those over 200 bp were selected for further analysis. Then, sequences with ambiguous bases ('N') were removed using

Trimmomatic (version 0.32) [36]. Chimera sequences were removed using the UCHIME algorithm [37]. The QIIME UPARSE pipeline was used to cluster (97% sequence similarity) the high-quality sequences into OTUs. In addition, OTUs of ITS high-throughput sequences were mapped to Unite (version 7.0) [38]. For further accurate verification of species information, the fungal OTUs were manually corrected using the CBS database (http://www.westerdijkinstitute.nl, accessed on 11 June 2021).

2.8. Statistical Analysis

The extraction efficiency of the ISP was calculated using the following equation:

$$E = \frac{Rb}{Ra} \tag{1}$$

where E is the extraction efficiency of the ISP II; Ra and Rb are the concentrations of the ISP II (copies/g); and, after genome extraction, the concentration of ISP II (copies/g) in the sample, respectively.

The semi-quantitative fungal data from samples were collected using the following equation:

$$Aa = \frac{Ra(1 - Rr) \times Ar}{Rr \times E} \tag{2}$$

where Aa is semi-quantitative measurements of the fungi (copies/g), Ar is the relative abundance of the fungi, and Rr is the relative abundances of ISP. Different fungi corresponded to the different ISP. Then, using Equations (1) and (2), the semi-quantitative results in the mixed fungi were calculated with different concentrations. Finally, using the semi-quantitative result of the fungus as the X-axis and the cell concentration of the fungus by the haemocytometer as the Y-axis, standard curves were established based on the mixed ISPs.

The alpha diversity was calculated by analysing the Shannon index using QIIME after rarefying all samples to the same sequencing depth (46,339 reads). Statistical significance of the differences between alpha diversity was investigated using one-way analysis of variance, followed by Tukey's post hoc test using IBM® SPSS® Amos™ 22 (Arbuckle, 2013). PCoA was performed to examine the fungal community dissimilarity of different samples based on Bray–Curtis distances. A hierarchical clustering analysis was performed using the SIMCAP+ software package (v.13.0. Umetrics, Umea, Sweden) to illustrate the differences in fungal community compositions among the samples.

2.9. Data Availability

All sequences generated were submitted to the NCBI database under the accession number PRJNA649523.

3. Results

3.1. Sequence Distribution of Fungal ITS2 and Construction of ISP

By analysing the fungal ITS2 sequence information, we found that its length in mature Daqu was usually 200–440 bp (Table S1). Among them, the sequences of 320–360 bp accounted for 49.94%, those of 260–320 bp accounted for 36.23%, those of 200–260 bp accounted for 12.06%, and those of 360–380 bp accounted for 1.04%. The sequences of other lengths accounted for less than 1% (Figure 1A). These data demonstrate that the length of the fungal ITS2 sequence varies greatly. To improve the quantitative accuracy, we designed five internal standard fragments (ISFs) (Table 1) based on the lengths and GC contents of 30 screened fungal ITS2 sequences (Table S2). Then, we used five ISFs (Figure 1B) to obtain five internal standard plasmids (ISPs) (Table 2). Enzyme digestion and ISP PCR were used to successfully verify the five ISPs (Figure S1).

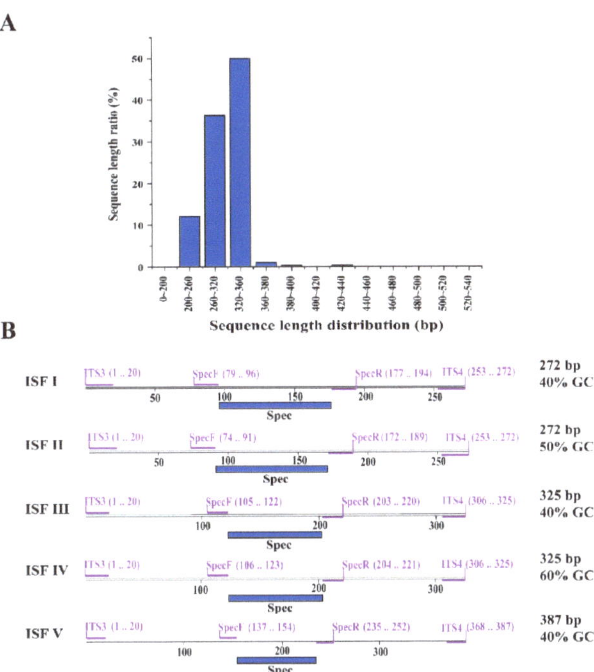

Figure 1. Length distribution and five ISFs. (**A**) Fungal ITS2 sequence length distribution. (**B**) The sequence lengths and GC contents of ISF I, II, III, IV, and V were 272 bp, 40% GC; 272 bp, 50% GC; 325 bp, 40% GC; 325 bp, 60% GC, and 387 bp, 40% GC, respectively. Spec indicates a specific fragment of 116 bp.

Table 1. Screened fungi and five ISFs [a].

Group	ITS2 Length	ITS2 GC Content	ISF Length	ISF GC Content	Core Fungi
1	231~300 bp	39~42% GC	272 bp	40% GC	*Kodamaea ohmeri, Hyphopichia burtonii, Wickerhamomyces anomalus*
2	231~300 bp	45~52% GC	272 bp	50% GC	*Candida versatilis, Candida metapsilosis, Kazachstania humilis, Pichia kudriavzevii.*
3	300~350 bp	36~50% GC	325 bp	40% GC	*Saccharomyces* sp., *Rhizopus arrhizus, Saccharomycopsis fibuligera, Rhizopus microsporus, Rhodotorula mucilaginosa, Rhizomucor pusillus.*
4	300~350 bp	56~63% GC	325 bp	60% GC	*Candida athensensis, Pichia sporocuriosa, Paecilomyces verrucosus, Aspergillus amstelodami, Aspergillus* sp., *Thermoascus crustaceus, Thermomyces lanuginosus, Rasamsonia composticola, Monascus purpureus, Thermoascus aurantiacus, Aspergillus flavus, Leiotheciume llipsoideum, Aspergillus costiformis.*
5	350~413 bp	31~44% GC	387 bp	40% GC	*Saccharomyces cerevisiae, Kazachstania bulderi, Lichtheimiaceae ramosa, Schizosaccharomyces pombe.*

[a] Based on 30 screened fungal ITS2 lengths and GC contents, which were divided into five groups as follows: sequence length < 300 bp, divided into 38–45% GC, 46–56% GC; sequence length 300–350 bp, divided into 36~50% GC, 56~63% GC; sequence length > 350 bp, divided into 31~49% GC. The average sequence length and GC content in each group were used as the sequence length and GC content of the corresponding ISF.

Table 2. Five ISPs and 16 core fungi.

ISP [a]	Core Fungi	Source [b]
I	*Hyphopichia burtonii*	LBMAE
	Wickerhamomyces anomalus	LBMAE
II	*Candida metapsilosis*	LBMAE
	Pichia kudriavzevii	LBMAE
III	*Saccharomycopsis fibuligera*	LBMAE
	Rhizopus microsporus	LBMAE
	Rhodotorula mucilaginosa	LBMAE
IV	*Paecilomyces verrucosus*	LBMAE
	Thermomyces lanuginosus	LBMAE
	Rasamsonia composticola	LBMAE
	Monascus purpureus	LBMAE
	Aspergillus flavus	LBMAE
V	*Saccharomyces cerevisiae*	LBMAE
	Kazachstania bulderi	LBMAE
	Lichtheimia ramosa	LBMAE
	Schizosaccharomyces pombe	LBMAE

[a] ISP with ISF I, ISF II, ISF III, ISF IV, and ISF V were named ISP I, ISP II, ISP III, ISP IV, and ISP V, respectively.
[b] LBMAE, Lab of Brewing Microbiology and Applied Enzymology at Jiangnan University (the strains were isolated from fermented grains; all strains are available to the public for free).

3.2. Selection of ISP Concentrations and Application Verification

To determine the optimal added concentration of ISPs and the efficiency of genome extraction, a convenient method was developed based on ISP II (Table 2). Here, we used the extraction efficiency of ISP II to reflect that of the sample genome. Using Equation (1), the extraction efficiency of ISP II at different concentrations in high-temperature mature Daqu was determined to be 17.42~18.71% (Figure 2b). Meanwhile, we selected the data with CT values between 15 and 25 for further analysis. Accordingly, 1×10^7–1×10^9 copies/mL was the initial concentration of the mixed ISPs that was added (Figure 2a).

To further determine the suitable concentrations of the ISPs, the high-temperature mature Daqu genome DNA containing different concentrations of ISP II was subjected to high-throughput sequencing (Figure 2c). The results showed that the T10 and T9 experimental groups significantly changed the abundance of fungal communities in the high-temperature mature Daqu compared to that in the control. The relative abundance of *Aspergillus ruber* in the T8 experimental group was reduced by approximately 10% compared to that in the control. The abundance of fungal communities in the T7 experimental group was most similar to that of the control. Therefore, a concentration of 1×10^7 copies/mL of the mixed ISPs was selected as the optimal concentration of the mixed ISPs.

Moreover, there existed a linear relationship ($R^2 > 0.99$) between the relative abundance of the fungal communities and the ISP II that was added to the high-temperature mature Daqu (Figure 3a). This indicated that the added ISP II could be correspondingly reflected in high-throughput sequencing so that the abundance of other fungi could be accurately determined. Meanwhile, the melting curve revealed a consistent melting temperature (Figure 3b), indicating that the amplified fragment had higher specificity. In addition, we compared the extraction efficiency of ISP II in three types of mature Daqu (Figure 3c) and determined it to be 17.42~19.89%, which indicated that ISP II was stable and could be used in different types of mature Daqu.

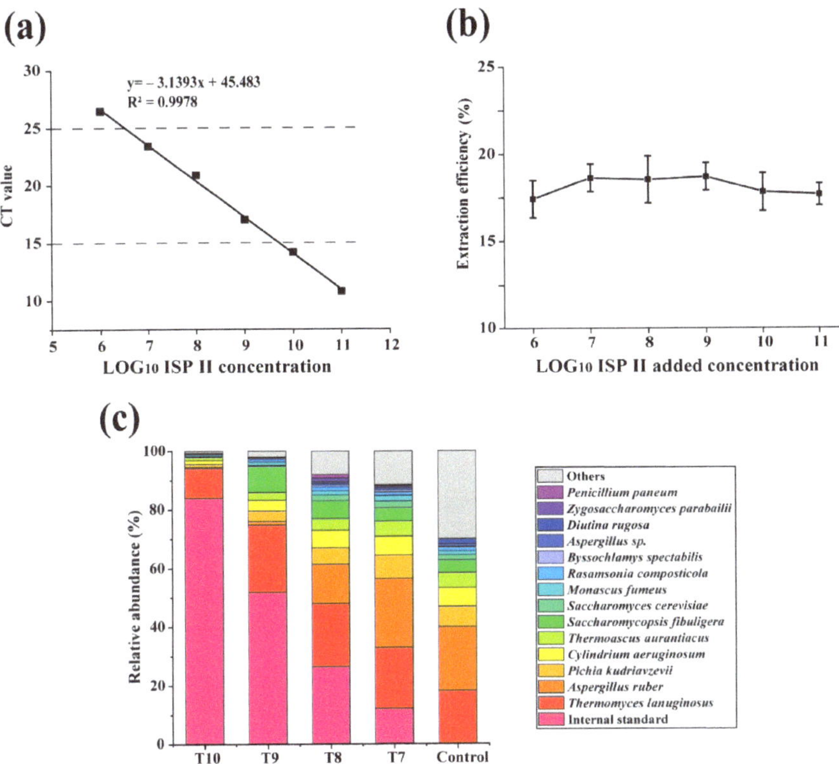

Figure 2. Selection of mixed ISP-added concentrations. (**a**) Standard curves of ISP II. (**b**) The extraction efficiency of ISP II in high-temperature mature Daqu. (**c**) Relative abundance of the high-temperature mature Daqu with different concentrations of ISP II added at the species level.

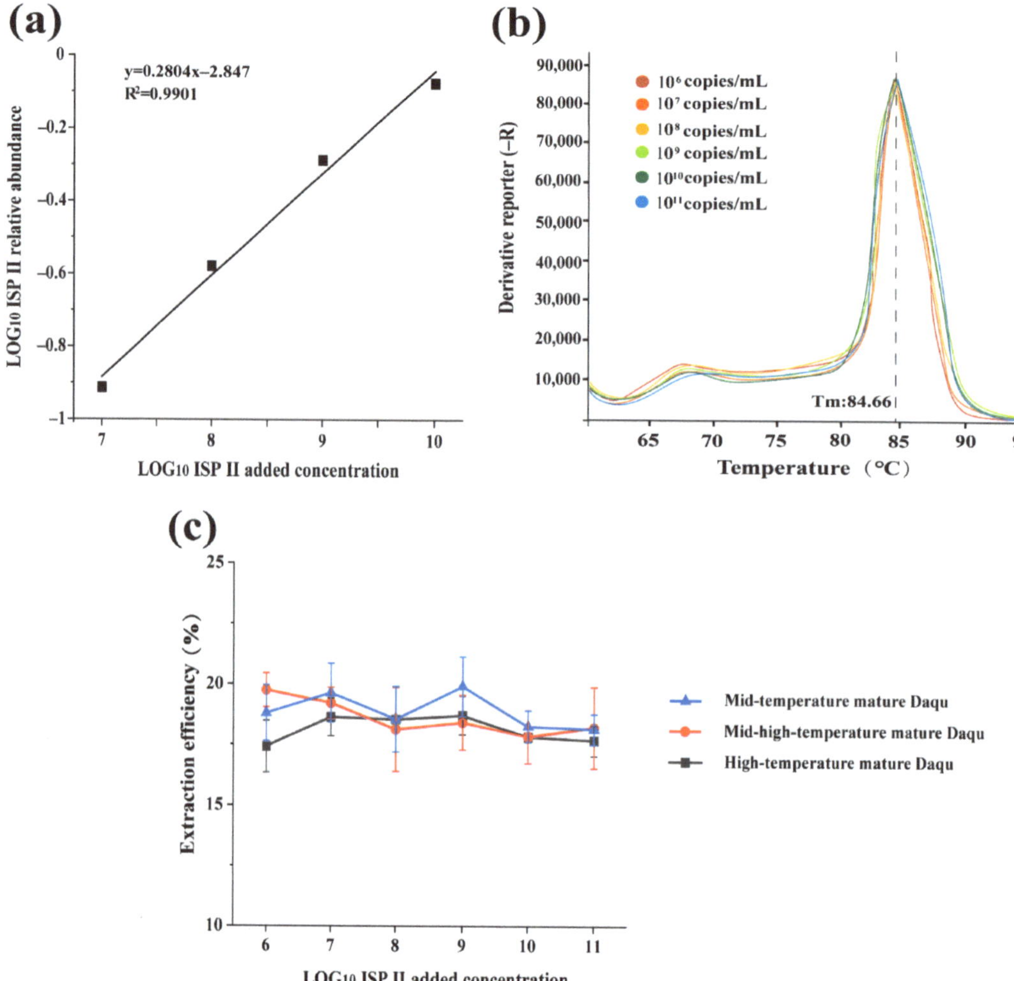

Figure 3. Verification of mixed ISP-added concentrations. (**a**) The regression equation of ISP II in high-temperature mature Daqu. (**b**) Melting curve of ISP II in high-temperature mature Daqu. (**c**) The extraction efficiency of ISP II in different mature Daqu.

3.3. Construction of the HAQ Method

First, 16 core fungi (Table 2), including Ascomycota, Basidiomycota, and Mucoromycota, were selected for cultivation from 30 screened fungi (Table 1) and mixed at five different concentration gradients ($10^3 \sim 10^7$ cells/mL for each fungal solution). Using Equation (1), the extraction efficiency of ISP II in mixed fungi was found to be 19.24~19.58% (Figure S2). The average extraction efficiency (19.42%) of ISP II in mixed fungi was used in the calculation. Similarly, we established 16 standard curves for the core fungi using Equation (2). Figure 4 shows that the HAQ method was accurate ($R^2 > 0.99$), and the concentration range of fungal quantification was $10^3 \sim 10^7$ cells/g. Finally, we calculated the semi-quantitative results of 16 core fungi in mature Daqu using Equation (2), and the results were substituted into the standard curve to achieve absolute quantification of the target fungi.

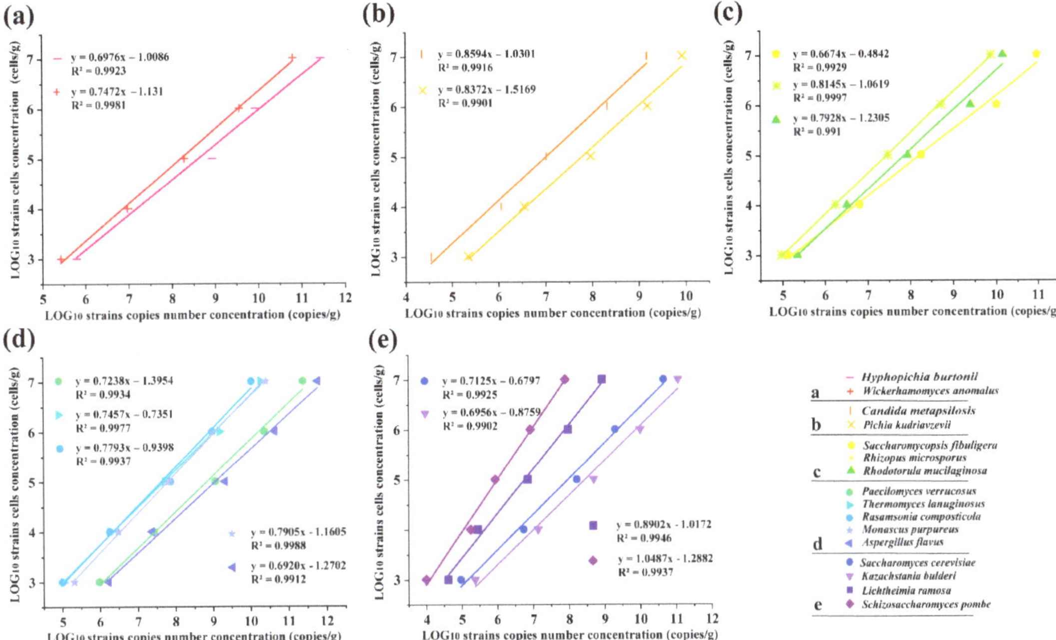

Figure 4. Standard curves of 16 core fungi. (**a**) Based on ISP I, used for the absolute quantification of *Hyphopichia burtonii* and *Wickerhamomyces anomalus*. (**b**) Based on ISP II, used for the absolute quantification of *Candida metapsilosis* and *Pichia kudriavzevii*. (**c**) Based on ISP III, used for the absolute quantification of *Saccharomycopsis fibuligera*, *Rhizopus microspores*, and *Rhodotorula mucilaginosa*. (**d**) Based on ISP IV, used for the absolute quantification of *Paecilomyces verrucosus*, *Thermomyces lanuginosus*, *Rasamsonia composticola*, *Monascus purpureus*, and *Aspergillus flavus*. (**e**) Based on ISP V, used for the absolute quantification of *Saccharomyces cerevisiae*, *Kazachstania bulderi*, *Lichtheimia ramose*, and *Schizosaccharomyces pombe*.

3.4. Application I: Case Studies of Different Mature Daqu

Figure 5 showed that a total of 128 species of fungi were detected in the medium-temperature Daqu samples, a total of 157 species of fungi were detected in the medium-high-temperature Daqu samples, and a total of 164 species of fungi were detected in the high-temperature Daqu samples. Among them, 121 fungi were detected in the middle-temperature Daqu and the high-temperature Daqu, and the shared fungal rate reached 73.78%. One hundred and fifteen fungi were detected in both medium-temperature Daqu and medium-high-temperature Daqu, and the shared fungal rate reached 73.25%. One hundred and thirty fungi were detected in both medium-high-temperature Daqu and high-temperature Daqu, and the shared fungal rate reached 79.27%. A total of 108 fungi were detected in the three types of Daqu, and the shared fungal rate reached 65.85%. The above conclusions showed that the three different Daqu had small differences in the fungal community structure of Daqu. Therefore, we speculated that the difference in fungal biomass was an important factor that caused the differences in the three types of Daqu.

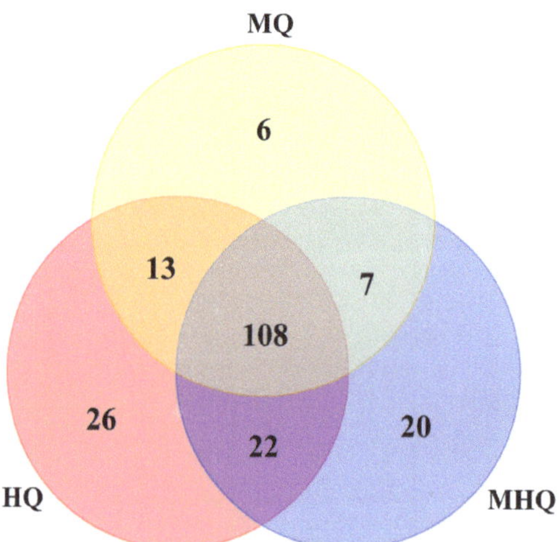

Figure 5. Venn diagram of the fungal community in different mature Daqu. MQ: Medium-temperature Daqu, MHQ: Medium-high-temperature Daqu, and HQ: High-temperature Daqu.

High-throughput sequencing was used to characterise the fungal community structures in different types of mature Daqu. A total of 556,068 high-quality reads from the internal transcribed spacer (ITS) region were obtained from all 12 samples. The rarefaction curves of the fungal communities approached the saturation plateau, which illustrated that the fungal communities were well-represented at the sequencing depth (Figure S2A). The alpha diversity of mature Daqu was determined using the Shannon index (Figure S2B). We found that there was no significant difference in the fungal diversity between the MHQ (medium-high-temperature mature Daqu were spiked with 1 mL of five mixed ISPs) and MHC (medium-high-temperature mature Daqu was spiked with 1 mL of sterile water), which indicated that the addition of mixed ISPs had little effect on fungal diversity. A principal coordinate analysis (PCoA) of the fungal community was carried out based on the Bray–Curtis distances (Figure 6a). Overall, these findings explained 67.5% of the total variation in the differences between fungal communities. The results showed that the fungal community structure of MHQ was very close to that of MHC. This further proved that the addition of mixed ISPs had little effect on the fungal community structures. These were also verified by HCA, which demonstrated a clustering pattern similar to that of PCoA (Figure 6b).

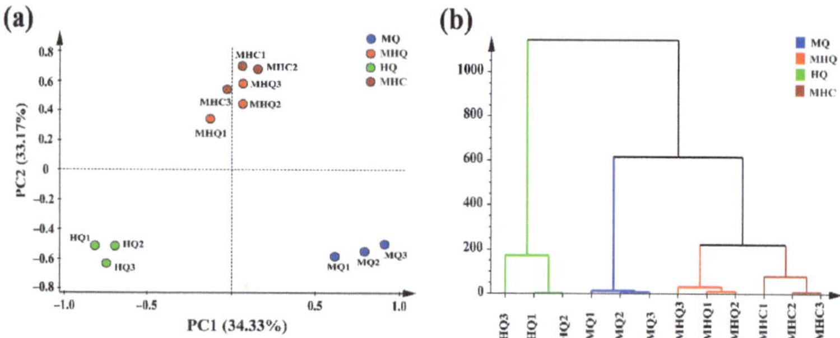

Figure 6. Fungal community structure of different mature Daqu. (**a**) Principal coordinate analysis (PCoA) of the fungal communities based on Bray-Curtis distances. (**b**) Hierarchical clustering analysis (HCA) of fungal communities. Triplicate samples are shown as '#1' to '#3'.

To more accurately characterise the fungal community structures in three types of mature Daqu, 16 core fungi were quantified using the HAQ methods. Figure 7a shows that all 16 core fungi had extremely high abundance in MQ (medium-temperature mature Daqu were spiked with 1 mL of five mixed ISPs), MHQ, and HQ (high-temperature mature Daqu were spiked with 1 mL of five mixed ISPs), occupying 93.52 ± 2.45%, 81.03 ± 1.80%, and 77.59 ± 1.08% respectively. These findings indicated that the abundances of the 16 selected fungi could be used as reference standards for analysing mature Daqu fungal community structures. After absolute quantification, we found that the absolute abundance of the 16 core fungi in the three types of mature Daqu varied, and the total amounts of the 16 core fungi were also dissimilar (Figure 7b). Table 3 shows that the total abundance of the 16 core fungi in MHQ ($1.76 \times 10^7 \pm 4.36 \times 10^5$ cells/g) was the highest, followed by that in MQ ($4.36 \times 10^6 \pm 6.12 \times 10^5$ cells/g) and HQ ($2.86 \times 10^5 \pm 5.09 \times 10^4$ cells/g). Moreover, only the absolute abundance of *Paecilomyces verrucosus* was of the same order of magnitude (10^3 cells/g) in the three types of mature Daqu (Figure S4). The absolute abundances of the remaining 15 fungi in HQ were all lower than those in MQ and MHQ. The absolute abundances of *Thermomyces lanuginosus* and *Wickerhamomyces anomalus* in MQ, MHQ, and HQ were 10^6, 10^5, and 10^4 cells/g, respectively, whereas that of *Lichtheimia ramosa* in MQ, MHQ, and HQ was 10^6, 10^5, and 10^3 cells/g, respectively. Therefore, this could be a criterion for identifying different types of mature Daqu.

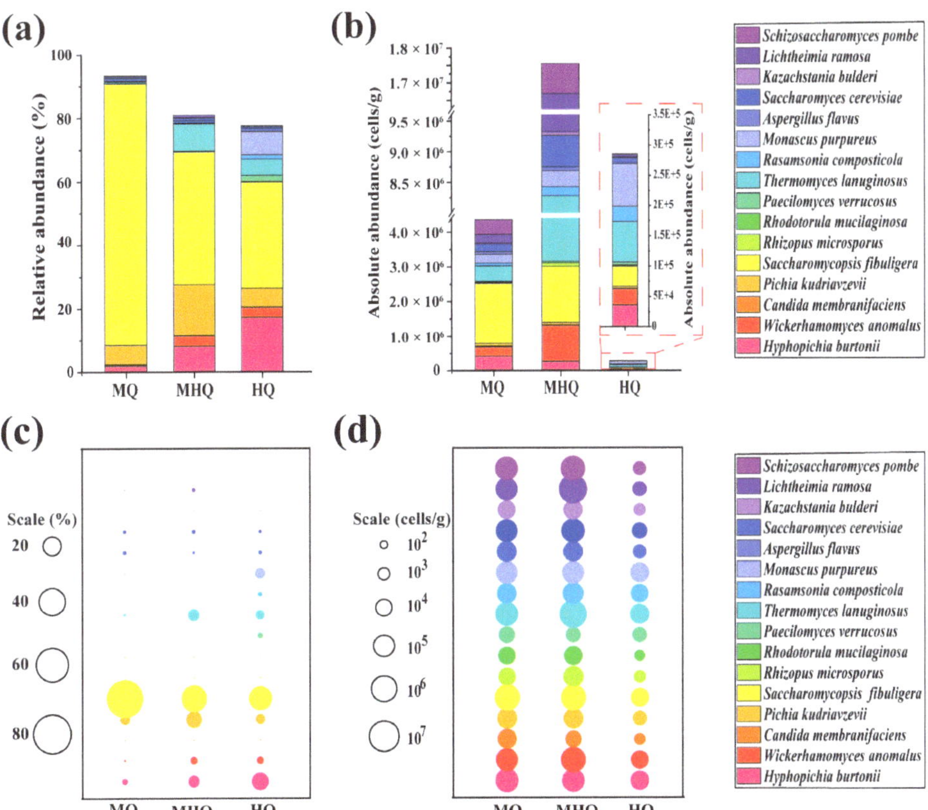

Figure 7. Comparison of the relative and absolute abundance of 16 core fungi. (**a,c**) Relative abundance of 16 core fungal communities in mature Daqu at the species level. (**b,d**) Absolute abundance of 16 core fungal communities in mature Daqu at the species level.

Table 3. Comparison of the relative and absolute abundance of 16 core fungi.

Fungi	Relative Abundance (%)			Absolute Abundance (Cells/g)		
	Medium-Temperature Daqu	Medium-High-Temperature Daqu	High-Temperature Daqu	Medium-Temperature Daqu	Medium-High-Temperature Daqu	High-Temperature Daqu
Hyphopichia burtonii	2.02 ± 0.21	8.15 ± 0.10	17.31 ± 0.52	$4.30 \times 10^5 \pm 1.16 \times 10^5$	$2.66 \times 10^5 \pm 2.47 \times 10^4$	$3.62 \times 10^4 \pm 7.14 \times 10^3$
Wickerhamomyces anomalus	0.26 ± 0.02	3.33 ± 0.53	3.13 ± 0.28	$2.57 \times 10^5 \pm 7.80 \times 10^4$	$1.05 \times 10^6 \pm 2.51 \times 10^5$	$2.65 \times 10^4 \pm 1.75 \times 10^3$
Candida membranifaciens	0.23 ± 0.02	0.12 ± 0.01	0.12 ± 0.01	$2.23 \times 10^4 \pm 3.60 \times 10^3$	$1.12 \times 10^4 \pm 6.58 \times 10^3$	$3.86 \times 10^2 \pm 3.66 \times 10^1$
Pichia kudriavzevii	6.13 ± 1.09	16.07 ± 0.71	5.91 ± 0.53	$8.11 \times 10^4 \pm 4.01 \times 10^5$	$7.28 \times 10^4 \pm 3.13 \times 10^4$	$1.94 \times 10^3 \pm 2.41 \times 10^2$
Saccharomyces fibuligera	82.33 ± 2.75	41.79 ± 4.74	33.41 ± 1.59	$1.74 \times 10^6 \pm 2.25 \times 10^3$	$1.62 \times 10^6 \pm 2.57 \times 10^5$	$3.35 \times 10^4 \pm 1.38 \times 10^3$
Rhizopus microsporus	0.03 ± 0.01	0.10 ± 0.03	0.12 ± 0.06	$1.97 \times 10^4 \pm 7.20 \times 10^3$	$9.68 \times 10^4 \pm 1.43 \times 10^4$	$1.12 \times 10^3 \pm 6.01 \times 10^2$
Rhodotorula mucilaginosa	0.06 ± 0.01	0.08 ± 0.00	0.09 ± 0.01	$1.81 \times 10^4 \pm 4.15 \times 10^3$	$3.54 \times 10^4 \pm 4.19 \times 10^3$	$4.69 \times 10^2 \pm 7.16 \times 10^1$
Paecilomyces verrucosus	0.03 ± 0.01	0.01 ± 0.01	1.90 ± 0.08	$9.89 \times 10^3 \pm 2.87 \times 10^4$	$5.68 \times 10^3 \pm 1.14 \times 10^3$	$4.87 \times 10^3 \pm 1.01 \times 10^3$
Thermomyces lanuginosus	0.43 ± 0.06	8.61 ± 1.90	5.19 ± 0.93	$4.56 \times 10^5 \pm 3.39 \times 10^4$	$5.13 \times 10^6 \pm 7.67 \times 10^3$	$6.72 \times 10^4 \pm 1.07 \times 10^4$
Rasamsonia composticola	0.03 ± 0.02	0.06 ± 0.03	1.3 ± 0.57	$7.70 \times 10^4 \pm 5.51 \times 10^4$	$1.44 \times 10^5 \pm 4.80 \times 10^4$	$2.54 \times 10^4 \pm 1.93 \times 10^4$
Monascus purpureus	0.22 ± 0.03	0.19 ± 0.15	7.27 ± 0.67	$2.48 \times 10^5 \pm 1.16 \times 10^5$	$2.60 \times 10^4 \pm 4.62 \times 10^4$	$7.04 \times 10^4 \pm 2.46 \times 10^4$
Aspergillus flavus	0.94 ± 0.61	0.52 ± 0.11	0.97 ± 0.03	$7.96 \times 10^4 \pm 6.25 \times 10^4$	$6.23 \times 10^4 \pm 4.58 \times 10^4$	$2.43 \times 10^3 \pm 5.61 \times 10^2$
Saccharomyces cerevisiae	0.67 ± 0.09	1.05 ± 0.18	0.66 ± 0.10	$2.18 \times 10^5 \pm 6.96 \times 10^4$	$5.16 \times 10^5 \pm 9.12 \times 10^4$	$7.14 \times 10^3 \pm 1.22 \times 10^3$
Kazachstania bulderi	0.11 ± 0.01	0.13 ± 0.01	0.17 ± 0.01	$2.90 \times 10^4 \pm 8.87 \times 10^3$	$5.66 \times 10^4 \pm 7.90 \times 10^3$	$1.38 \times 10^3 \pm 6.60 \times 10^1$
Lichtheimia ramosa	0.04 ± 0.02	0.80 ± 0.13	0.04 ± 0.01	$2.46 \times 10^5 \pm 2.28 \times 10^5$	$7.37 \times 10^6 \pm 2.33 \times 10^5$	$3.48 \times 10^3 \pm 5.67 \times 10^2$
Schizosaccharomyces pombe	0.01 ± 0.00	0.01 ± 0.00	0.01 ± 0.00	$4.26 \times 10^5 \pm 5.21 \times 10^4$	$8.72 \times 10^5 \pm 5.93 \times 10^5$	$2.02 \times 10^3 \pm 1.45 \times 10^3$
Total abundance	93.52 ± 2.45	81.03 ± 1.80	77.59 ± 1.08	$4.36 \times 10^6 \pm 6.12 \times 10^5$	$1.76 \times 10^7 \pm 4.36 \times 10^5$	$2.86 \times 10^5 \pm 5.09 \times 10^4$

Figure 7c,d show that the relative abundance of the 16 core fungi in MQ was dominated by *Saccharomycopsis fibuligera* (82.33%), followed by *Pichia kudriavzevii* (6.13%) and *Hyphopichia burtonii* (2.02%). The fungi with the highest absolute abundance were *S. fibuligera* ($1.74 \times 10^6 \pm 2.25 \times 10^3$ cells/g), *T. lanuginosus* ($4.56 \times 10^5 \pm 3.39 \times 10^4$ cells/g), and *H. burtonii* ($4.30 \times 10^5 \pm 1.16 \times 10^5$ cells/g). The relative abundances of the 16 core fungi in MHQ were dominated by *S. fibuligera* (41.78%), followed by *P. kudriavzevii* (16.07%) and *T. lanuginosus* (8.61%). The fungi with the highest absolute abundance were *Lichtheimia ramosa* ($7.37 \times 10^6 \pm 2.33 \times 10^5$ cells/g), *T. lanuginosus* ($5.13 \times 10^6 \pm 7.67 \times 10^3$ cells/g), and *S. fibuligera* ($1.62 \times 10^6 \pm 2.57 \times 10^5$ cells/g). The relative abundances of the 16 core fungi in HQ were dominated by *S. fibuligera* (33.41%), followed by *H. burtonii* (17.31%) and *Monascus purpureus* (7.27%). Fungi with the highest absolute abundance were *M. purpureus* ($7.04 \times 10^4 \pm 2.46 \times 10^4$ cells/g), *T. lanuginosus* ($6.72 \times 10^4 \pm 1.07 \times 10^4$ cells/g), and *H. burtonii* ($3.62 \times 10^4 \pm 7.14 \times 10^3$ cells/g). These results indicated that the relative and absolute abundances may be inconsistent even in the same sample.

In addition, we found (Figure S6) that the abundance of *H. burtonii* was underestimated by 7.71% in MQ and overestimated in MHQ (8.54%) and HQ (9.64%) compared to the absolute abundance. The abundances of *W. anomalus*, *T. lanuginosus*, *Rasamsonia composticola*, *M. purpureus*, *Saccharomyces cerevisiae*, *L. ramosa*, and *Schizosaccharomyces pombe* were all underestimated in MQ, MHQ, and HQ. Among them, the fungus that was most underestimated in the three types of Daqu was *T. lanuginosus*, which, in MQ, MHQ, and HQ, was underestimated by 10.0%, 18.6%, and 16.8%, respectively. The abundances of *P. kudriavzevii* and *Saccharomyces fibuligera* in MQ, MHQ, and HQ were overestimated; in particular, the abundance of *S. fibuligera* in MQ, MHQ, and HQ was overestimated by 48.1%, 42.3%, and 31.4%, respectively. The abundances of the remaining six core fungi had almost no deviation (absolute value of abundance deviation < 1%).

3.5. Application II: Case Studies of Medium-High Temperature Daqu during Fermentation

ITS amplicon technology has accelerated the study of Daqu fungi, deepening the understanding of the fungal community structure of Daqu. However, the formation principle of the space structure of fungi in medium-high-temperature Daqu is unclear.

During the whole process of Daqu fermentation, a total of 81 fungi were detected in the Daqu surface sample, and a total of 84 fungi were detected in the Daqu core sample. Among them, 78 species of fungi were common in the Daqu surface samples and Daqu core samples. The shared fungi rate was over 90% (Figure 8). This showed that there was almost no difference in the structure of the fungal community during Daqu fermentation. Therefore, we speculated that the gap in the fungal biomass was an important factor causing the difference between the Daqu surface and Daqu core.

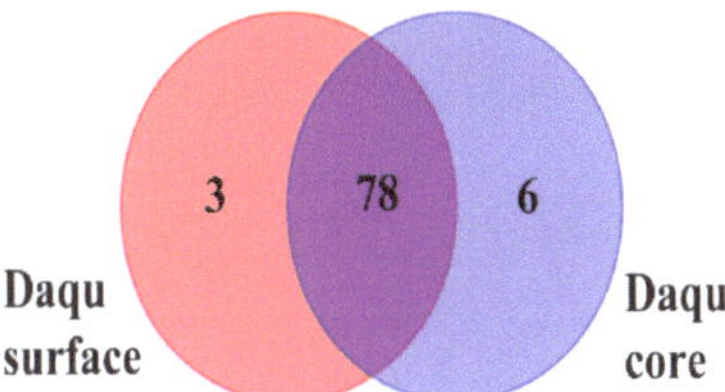

Figure 8. Veen diagram of the fungal community on Daqu surface and Daqu core during fermentation.

To find the key time points for the formation of the medium-high-temperature Daqu surface and Daqu core, the PCoA was used to characterise the fungal community during Daqu fermentation (Figure 9). Figure 9A,B showed that these findings, respectively, explained 57.82% and 56.76% of the total variation in the differences between fungal communities. From the perspective of the spatial sequence level, 4~6 stages during Daqu fermentation were the important stages for the initial formation of the Daqu surface and Daqu core. Among them, Daqu surface fungi had undergone significant changes compared with Daqu core from the fourth stage, and the fungal community of Daqu core became stable, indicating that Daqu core had initially formed in the fourth phase. The clusters of fungal community in the 5~6 stages showed that the Daqu core had initially formed at the fifth stage.

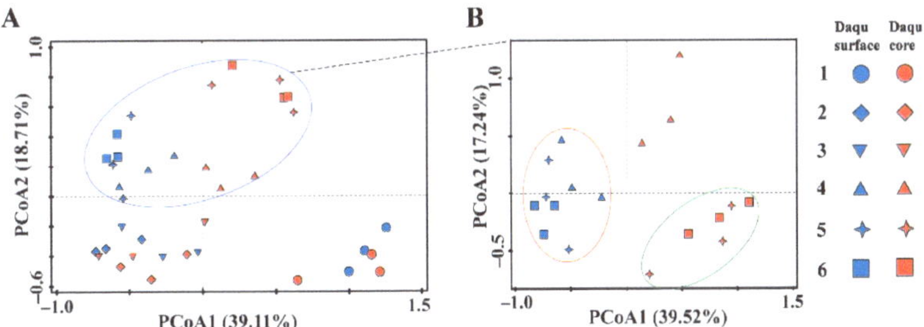

Figure 9. Principal coordinate analysis of fungi during Daqu fermentation. (**A**) Based on 1~6 stages during Daqu fermentation. (**B**) Based on 4~6 stages during Daqu fermentation.

A variety of fungi were involved when the Daqu surface formed in the fourth stage. However, the importance of different fungi to Daqu surface formation would vary. SIMCA software was used to perform partial least squares regression (PLS) to evaluate the importance of fungi and Daqu surface formation. In the fourth stage, there are four important fungi that differ from the Daqu surface and Daqu core, namely *H. burtonii*, *W. anomalus*, *P. kudriavzevii*, and *S. fibuligera*. These four kinds of yeasts play an important role in the formation of Daqu surface. Among them, there was a significant difference ($p < 0.05$) between *P. kudriavzevii* and *S. fibuligera* in the Daqu surface and these in the Daqu core. There was a very significant difference in *H. burtonii* ($p < 0.01$). These three yeasts were key fungi in the formation of the Daqu surface.

To analyse the driving factors of the initial formation about the Daqu surface during Daqu fermentation, a redundant analysis (RDA) was used to analyse the influence of pH, moisture, and reducing the sugar content on the fungal community. Overall, these findings explained 93.01% of the total variation in the differences between fungal communities. Combining Figure 10 and Table 4, it was found that the initial formation of the Daqu surface during Daqu fermentation was mainly driven by moisture (51.8%). *H. burtonii*, *S. fibuligera*, and *P. kudriavzevii* were the key fungi for the initial formation of the Daqu surface during Daqu fermentation. *W. anomalus* also made an important contribution to the initial formation of the Daqu surface.

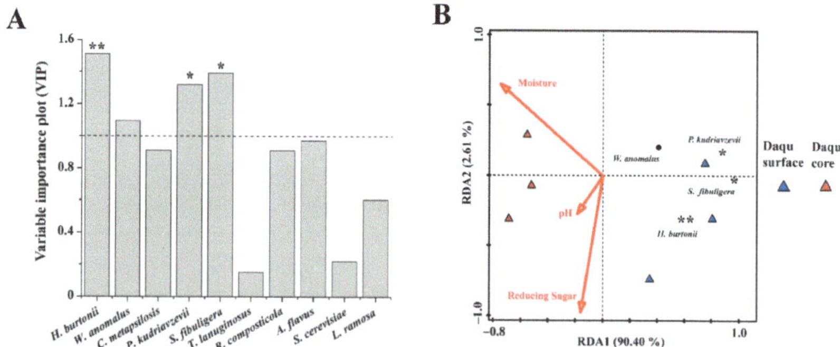

Figure 10. The initial formation of the Daqu surface during fermentation. (**A**) The results of partial least squares regression between the Daqu surface and Daqu core in the fourth stage during fermentation. (**B**) Redundant analysis of the fungi and physical and chemical parameters in the fourth stage. *: $p < 0.05$ and **: $p < 0.01$.

Table 4. The relationship between the fungal community and physicochemicals in the fourth stage.

Name	Explains %	Contribution %	Psedo-F	*p*
Moisture	51.8	55.5	4.3	0.136
Reducing sugar	33.5	35.9	6.8	0.042
pH	8.1	8.7	2.4	0.218

In order to explore the reason for the initial formation of the Daqu core during Daqu fermentation, we compared the differences in the fungal community between the fourth stage and the fifth~sixth stages. Meanwhile, use SIMCA software to find fungi with a VIP value greater than 1.0. As shown in Figure 11, there are six important differences between the fourth stage of the Daqu core and the fifth~sixth stages of the Daqu core. Among them, three kinds of yeasts are *C. metapsilosis*, *P. kudriavzevii*, and *S. fibuligera*. Three kinds of molds are *P. verrucosus*, *T. lanuginosus*, and *A. flavus*. These six fungi played an important role in the initial formation of the Daqu core during Daqu fermentation. There was a significant difference ($p < 0.05$) between *S. fibuligera* and *P. verrucosus* in the fourth stage of the Daqu core and those in the fifth~sixth stages of the Daqu core. They were the key fungi that caused the initial formation of the Daqu core.

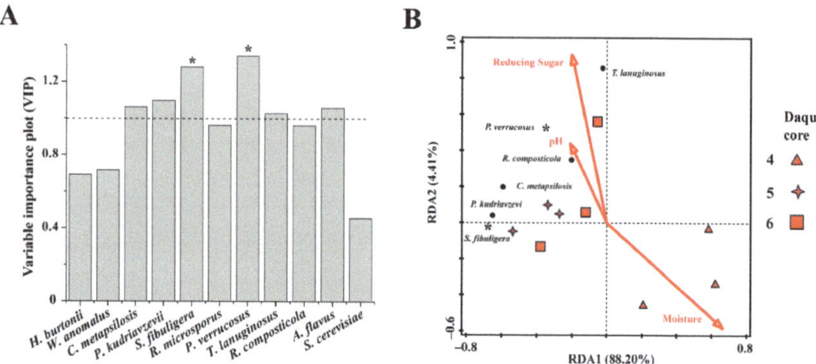

Figure 11. The initial formation of the Daqu core during fermentation. (**A**) The results of partial least squares regression between the Daqu surface and Daqu core. *: $p < 0.05$. (**B**) The results of partial least squares regression between the Daqu core in the fourth stage and Daqu core in the fifth~sixth stages.

To analyse the driving factors of the initial formation of the Daqu core during Daqu fermentation, RDA was used to analyse the influence of pH, moisture, and reducing sugar on the Daqu core fungal community. Overall, these findings explained 92.61% of the total variation in the differences between fungal communities. Combining Figure 11 and Table 5, it could be found that the initial formation of the Daqu core was affected by moisture and reducing sugar. *S. fibuligera* and *P. verrucosus* were the key fungi for the initial formation of the Daqu core during Daqu fermentation. *T. lanuginosus*, *R. composticola*, *C. metapsilosis*, and *P. kudriavzevii* also made an important contribution to the initial formation of the Daqu core.

Table 5. The relationship between fungal community and physicochemical in the Daqu core for 4~6 stages.

Name	Explains %	Contribution %	Pseudo-F	*p*
Moisture	19.1	43.9	1.7	0.194
Reducing sugar	15.4	35.5	1.4	0.268
pH	9	20.6	0.8	0.404

In this study, the microbial community interaction network was used to analyse the relationship between Daqu fungus microorganisms. Figure 12 showed that there were 13 positive correlations and 12 negative correlations between the four key fungi and other fungi. Among them, *H. burtonii* had a significant positive correlation with *W. anomalus*, *P. kudriavzevii*, *S. fibuligera*, and *A. flavus* and a significant negative correlation with *P. verrucosus* and *M. purpureus*. *P. kudriavzevii* had a significant positive correlation with *H. burtonii*, *W. anomalus*, and *S. fibuligera* and a significant negative correlation with *R. mucilaginosa*, *P. verrucosus*, *T. lanuginosus*, *R. composticola*, *M. purpureus*, and *A. flavus*. *S. fibuligera* had a significant positive correlation with *H. burtonii*, *W. anomalus*, *P. kudriavzevii*, and *A. flavus* and a significant negative correlation with *P. verrucosus*, *T. lanuginosus*, *R. composticola*, and *M. purpureus*. *P. verrucosus* had a significant positive correlation with *C. metapsilosis*, *T. lanuginosus*, *R. composticola*, and *M. purpureus* and a significant negative correlation with *H. burtonii*, *W. anomalus*, *P. kudriavzevii*, and *S. fibuligera*.

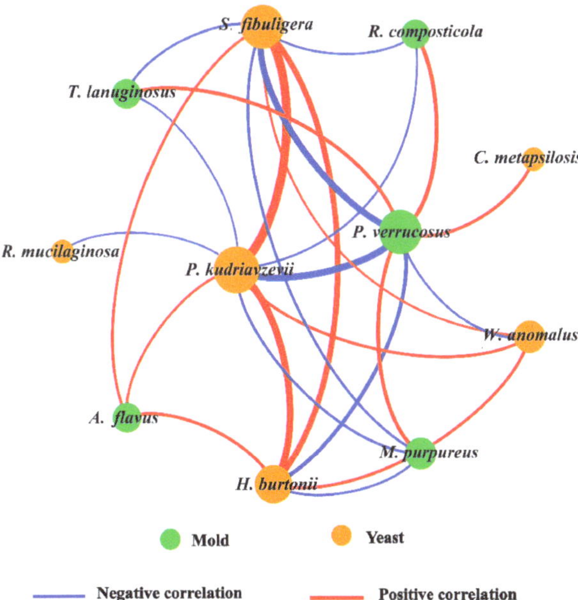

Figure 12. Network analysis of the key fungi in Daqu.

4. Discussion

Absolute quantitation of the microbiota is essential for all aspects of microbial ecology [13]. The relative abundance obtained by high-throughput sequencing can partially characterise the fungal community in the sample, while the absolute abundance based on quantitative methods can more objectively describe the actual abundance of the fungi. To quantify the core fungal community, we designed a HAQ method in which the structurally stable and highly reproducible mature Daqu was selected as a model ecosystem. Unlike the direct use of sequencing data to quantify microorganisms, this method quantifies fungi based on the construction of a standard curve.

Despite the advantages of the internal standard method, this method is commonly used for the quantification of bacteria and still faces many challenges when applied to the quantification of fungi. Figure 1A shows that the fungal ITS2 sequence is highly variable in length. Moreover, the difference in sequence lengths can cause deviations in high-throughput sequencing [28]. In the face of the absolute quantification of fungi, the corresponding internal standard sequence length should be designed according to the ITS sequence length of different fungi. Therefore, we constructed five ISPs based on the sequence lengths and GC contents of 30 screened fungal ITS2. The 16 core fungi from the screening culture were divided by five ISPs as a way to compensate for the lack of accuracy in quantifying all fungi using one internal standard substance (Figure 1B and Table S2). Recently, several studies have used one internal standard to attempt an absolute quantification of all fungi [13–15]. For example, a synthetic plasmid was used as the internal standard for the absolute quantitation of fungal abundance in environmental samples, which is easier to operate [13]. However, owing to the high length variations of the fungal ITS2 sequence, this method may reduce the accuracy of quantification [15,28]. Therefore, it would be more accurate to quantify core fungi using five internal standards compared to merely using one internal standard to quantify all fungi.

It is strongly recommended to determine the optimal level of ISP addition for a given sample. Since spiking the internal standard to the sample before DNA isolation can obtain the most accurate results [13], our method too followed this principle. Figure 1B shows that the five ISFs have a similar basic composition. Furthermore, the most common sequence of fungal ITS is 272 bp [39], and 50% GC is the regular GC content (same sequence length and GC content as ISF II) [28]. Therefore, we used ISP II to determine the added concentration of mixed ISPs. Since the CT values between 15 and 25 had a higher accuracy for a given target concentration [40,41], combined with the results of high-throughput sequencing, we selected 10^7 copies/mL as the final added concentration of the mixed ISPs (Figure 2a,c). When the concentration of ISP II was high, the fungal community structures showed obvious changes (Figure 2c). However, a concentration of the mixed ISPs hardly altered the measured structure of the fungal community in the three types of mature Daqu (Figures 6 and S2B). The linear relationship (Figure 3b) demonstrated a strong association between the relative abundance of ISP II and its added concentration in high-temperature mature Daqu. Similar results were reported by Yang et al. [15] and Lou et al. [24], wherein the slopes of the internal standard output and input amounts were close to 1. This indicated that the ISP II concentration of the changes could be linearly reflected in the high-throughput sequencing results. Therefore, the constructed mixed ISPs could be used as internal standards for the determination of the absolute abundance of the fungal community.

In studies of the high-throughput absolute quantification of bacteria, it has been shown that, once the internal standard strain is added to the sample to be tested, it is difficult to extract it completely, and the losses are significant [15]. This also reflects that the same problem exists for the extraction of genomes from samples to be tested. Therefore, the extraction efficiency of the sample to be tested should be taken into account when quantifying the microbial community by the internal standard method. The extraction efficiency of the sample genome was calculated using ISP II with a 116-bp specific fragment. The accuracy of this method mainly depends on the linearity of the standard curves [42]. The R^2 (>0.99) of the standard curves indicated the accuracy of the qPCR (Figure 2a) [41].

Meanwhile, the consistent melting temperature showed that the amplified 116-bp fragment had a strong specificity (Figure 3b). Therefore, the calculation of the extraction efficiency of the sample genome based on ISP II was reliable. Furthermore, the extraction efficiency of ISP II in the three types of mature Daqu was 17.42~19.89% (Figure 3c), which was similar to the results obtained by Yang [28]. We also used this method of qPCR to characterise the extraction efficiency of mixed fungi, the results of which were within the above range (Figure S3).

Presently, the absolute quantitative method is mainly used to quantify microbial communities in soil. In this study, we used the culturable technique to establish standard curves of 16 core fungi to achieve absolute abundance quantification of core fungal communities and applied them to different mature Daqu. The results showed that the differences in the fungal community structure of the three mature Daqu were small (Figure 5). Therefore, we inferred that the difference in fungal biomass was an important factor in the difference between the three mature Daqu. Using the HTS technique, we found that the 16 core fungi used for absolute quantification all occupied high abundance in three mature Daqu (Figure 7a), indicating that the 16 core fungi have important reference values for resolving the fungal community structure of the mature Daqu. Meanwhile, we also found that the relative and absolute abundances have opposite trends in different samples [15,24–26]. Notably, we found similar results in the same sample. For example, the fungus with highest relative abundance among the 16 core fungi in MHQ was *S. fibuligera* (41.78%), whereas, in terms of their absolute abundance, *L. ramosa ramosa* (41.95%) was the most dominant, and *S. fibuligera* only accounted for 9.22% (Figure S5). Moreover, we revealed the deviation in abundance of the 16 core fungi (Figure S6). Compared to their absolute abundance, we found that two fungi were overestimated in the three types of mature Daqu. Among them, the overestimation of *S. fibuligera* (36% GC) may be caused by the low GC content of the ITS2 sequence. All seven fungi were underestimated in the three types of mature Daqu. Among them, the underestimation of *T. lanuginosus* (59% GC), *R. composticola* (62% GC), and *M. purpureus* (60% GC) may result from the high GC content of the ITS2 sequence. The underestimation of *S. cerevisiae* (381 bp, 43% GC), *L. ramosa* (399 bp, 39% GC), and *S. pombe* (421 bp, 31% GC) may result from the long sequence length and low GC content of the ITS2 sequence. The above results indicate that the use of multiple internal standards can help obtain accurate quantitative results.

In this study, the changes of the physical and chemical indexes of the Daqu surface and Daqu core were tracked during the Daqu fermentation, and the driving factors of the initial formation about Daqu surface and Daqu core were quantitatively analysed by means of RDA in combination with the HAQ. The figure (Figure 8) showed that the fungal community structure of the Daqu surface and Daqu core fungi was similar. The difference in fungal biomass was an important factor in the formation of the Daqu surface and Daqu core. To find the key time points for the formation of the medium-high-temperature Daqu surface and Daqu core, the fungal community during Daqu fermentation characterised at the OUT level based on PCoA. Figure 9 showed that, from the spatial structure level, 4–6 stages during Daqu fermentation were important stages for the initial formation of the Daqu surface and Daqu core. Among them, *H. burtonii, S. fibuligera, P. kudriavzevii,* and *P. verrucosus* were the key fungi that caused the formation of Daqu surface and Daqu core. To further analyse the correlations between the four key fungi and other core fungi, we used a microbial interaction network analysis to resolve the relationship between Daqu fungus microorganisms, and the results showed a total of 13 positive and 12 negative correlations between the four key fungi and other fungi. *H. burtonii, S. fibuligera,* and *P. kudriavzevii* have been proven to be important functional fungi in Chinese liquor fermentation [43]. *H. burtonii* could produce various important flavour substances, such as ethyl acetate and 4-hydroxy-2-butanone [44,45]. *P. kudriavzevii* could produce metabolites such as phenethyl alcohol that had an important contribution to liquor flavour [46]. It could also ensure the stability of the microbial community and the fermentation process [47]. S. fibuligera could increase the activity of saccharified starch and acid protease and the rate of ethanol synthesis in the early stage of fermentation [48]. It could also improve the aroma of wine and produce many pleasant aroma compounds, such as ethyl acetate and ethyl butyrate [49].

5. Conclusions

In conclusion, the relative abundance determined using high-throughput sequencing can be partially helpful in characterising fungal communities. The actual abundance of fungi can be described more objectively based on the absolute abundance of the quantitative method. In this study, five ISPs were constructed to quantify fungi because of the length variations of the fungal ITS2 sequence. The HAQ method can absolutely quantify 16 fungi and more truly characterise the core fungal community structures. Furthermore, we found that fungal ITS2 sequences with lower GC contents may be easily overestimated or vice versa. The Daqu surface during fermentation initially formed in the fourth stage, which was mainly driven by the moisture. Among them, *H. burtonii*, *S. fibuligera*, and *P. kudriavzevii* were the key fungi. The Daqu core during fermentation initially formed in the fifth stage, which was mainly affected by the moisture and reducing sugar. Among them, *S. fibuligera* and *P. verrucosus* were the key fungi. The HAQ method can be used for the quantification of core fungi in different samples and serve as a valuable reference for studying fungal interactions, potential functions, and energy metabolism.

Supplementary Materials: The following supporting information can be downloaded at: https://www.mdpi.com/article/10.3390/fermentation8080345/s1. Figure S1. Verification of ISFs and ISPs. (A) Enzyme digestion of pET-28a plasmid and ISFs. (B) Plasmid map of ISP. (C) PCR verification. Figure S2. Rarefaction curves and box plots. (A) Rarefaction curves of the fungal ITS2 region sequences of all samples. (B) Box plots showing the Shannon index values of fungal communities among MQ, MHQ, HQ, and MHC. Sample groups with different letters and colours indicate significant differences ($p < 0.05$), as determined using one-way analysis of variance and Tukey's post hoc test. Figure S3. Extraction efficiency of ISP II in mixed fungi with different concentrations. Figure S4. Absolute abundance of 16 core fungi in different mature Daqu. Figure S5. Relative and absolute abundance of 16 core fungi. (A) The total relative abundance of the 16 core fungi was regarded as '1'. (B) The total absolute abundance of the 16 core fungi was regarded as '1'. Figure S6. The abundance deviation of 16 core fungi. The abundance deviation was calculated according to the following equation: $As = \frac{Ar}{R} - \frac{Aa}{A}$. Where As is the abundance deviation of a fungus (%); Ar and Aa are the relative and absolute abundance of a fungus, respectively; and R and A are the total relative and absolute abundances of the 16 core fungi, respectively. The positive value of abundance deviation represented overestimation, and the negative value represented underestimation. (A–C) were in MQ, MHQ, and HQ, respectively. Table S1. High-quality sequence distribution statistics. Table S2. ITS2 sequence and GC content of screened core fungi. Table S3. Strain, plasmid, and sequence used in this study.

Author Contributions: Conceptualisation, Writing—original draft, Formal analysis, Investigation, Data curation, H.D. and J.S.; Conceptualisation, Formal analysis, Investigation, Data curation, Writing—review and editing, T.Z.; and Conceptualisation, Writing—review and editing, Y.X. All authors have read and agreed to the published version of the manuscript.

Funding: This research was funded by the National Natural Science Foundation of China (NSFC) (grant 32172176), and the Natural Science Foundation of Jiangsu Province of China (grant BK20201341).

Institutional Review Board Statement: Not applicable.

Informed Consent Statement: Not applicable.

Data Availability Statement: Not applicable.

Conflicts of Interest: The authors declare no conflict of interest.

References

1. Jin, G.; Zhu, Y.; Xu, Y. Mystery behind Chinese liquor fermentation. *Trends Food Sci. Technol.* **2017**, *63*, 18–28. [CrossRef]
2. Xing, G.; Zonghua, A.O.; Wang, S.; Deng, B.; Wang, X.; Dong, Z. Analysis of the Change in Physiochemical Indexes during the Production Process of Daqu of Different Temperature. *Liquor-Mak. Sci. Technol.* **2014**, *11*, 592421.
3. Zou, W.; Zhao, C.Q.; Luo, H.B. Diversity and Function of Microbial Community in Chinese Strong-Flavor Baijiu Ecosystem: A Review. *Front. Microbiol.* **2018**, *9*, 671. [CrossRef] [PubMed]
4. Du, H.; Wang, X.S.; Zhang, Y.H.; Xu, Y. Exploring the impacts of raw materials and environments on the microbiota in Chinese Daqu starter. *Int. J. Food Microbiol.* **2019**, *297*, 32–40. [CrossRef]
5. Fan, G.S.; Fu, Z.L.; Teng, C.; Liu, P.X.; Wu, Q.H.; Rahman, M.K.R.; Li, X.T. Effects of aging on the quality of roasted sesame-like flavor Daqu. *Bmc Microbiol.* **2020**, *20*, 67. [CrossRef]
6. Pan, L.; Lin, W.; Xiong, L.; Wang, X.; Luo, L. Environmental Factors Affecting Microbiota Dynamics during Traditional Solid-state Fermentation of Chinese Daqu Starter. *Front. Microbiol.* **2016**, *7*, 1237.
7. Li, P.; Sha, L.; Cheng, L.; Luo, L. Analyzing the relation between the microbial diversity of DaQu and the turbidity spoilage of traditional Chinese vinegar. *Appl. Microbiol. Biotechnol.* **2014**, *98*, 6073. [CrossRef]
8. Pang, X.N.; Han, B.Z.; Huang, X.N.; Zhang, X.; Hou, L.F.; Cao, M.; Gao, L.J.; Hu, G.H.; Chen, J.Y. Effect of the environment microbiota on the flavour of light-flavour Baijiu during spontaneous fermentation. *Sci. Rep.* **2018**, *8*, 3396. [CrossRef]
9. Li, P.; Lin, W.F.; Liu, X.; Wang, X.W.; Gan, X.; Luo, L.X.; Lin, W.T. Effect of bioaugmented inoculation on microbiota dynamics during solid-state fermentation of Daqu starter using autochthonous of Bacillus, Pediococcus, Wickerhamomyces and Saccharomycopsis. *Food Microbiol.* **2017**, *61*, 83–92. [CrossRef]
10. Wang, X.S.; Du, H.; Zhang, Y.; Xu, Y. Environmental Microbiota Drives Microbial Succession and Metabolic Profiles during Chinese Liquor Fermentation. *Appl. Environ. Microbiol.* **2018**, *84*, e02369-17. [CrossRef]
11. Zheng, X.W.; Yan, Z.; Nout, M.J.R.; Smid, E.J.; Zwietering, M.H.; Boekhout, T.; Han, J.S.; Han, B.Z. Microbiota dynamics related to environmental conditions during the fermentative production of Fen-Daqu, a Chinese industrial fermentation starter. *Int. J. Food Microbiol.* **2014**, *182*, 57–62. [CrossRef] [PubMed]
12. Fan, G.S.; Fu, Z.L.; Teng, C.; Wu, Q.H.; Liu, P.X.; Yang, R.; Minhazul, K.; Li, X.T. Comprehensive analysis of different grades of roasted-sesame-like flavored Daqu. *Int. J. Food Prop.* **2019**, *22*, 1205–1222. [CrossRef]
13. Tkacz, A.; Hortala, M.; Poole, P.S. Absolute quantitation of microbiota abundance in environmental samples. *Microbiome* **2018**, *6*, 110. [CrossRef] [PubMed]
14. Lin, Y.J.; Gifford, S.; Ducklow, H.; Schofield, O.; Cassar, N. Towards Quantitative Microbiome Community Profiling Using Internal Standards. *Appl. Environ. Microbiol.* **2019**, *85*, e02634-18. [CrossRef] [PubMed]
15. Yang, L.; Lou, J.; Wang, H.Z.; Wu, L.S.; Xu, J.M. Use of an improved high-throughput absolute abundance quantification method to characterize soil bacterial community and dynamics. *Sci. Total Environ.* **2018**, *633*, 360–371. [CrossRef]
16. Dannemiller, K.C.; Lang-Yona, N.; Yamamoto, N.; Rudich, Y.; Peccia, J. Combining real-time PCR and next-generation DNA sequencing to provide quantitative comparisons of fungal aerosol populations. *Atmos. Environ.* **2014**, *84*, 113–121. [CrossRef]
17. Nishino, S.; Okahashi, N.; Matsuda, F.; Shimizu, H. Absolute quantitation of glycolytic intermediates reveals thermodynamic shifts in Saccharomyces cerevisiae strains lacking PFK1 or ZWF1 genes. *J. Biosci. Bioeng.* **2015**, *120*, 280–286. [CrossRef]
18. Nayfach, S.; Pollard, K.S. Toward Accurate and Quantitative Comparative Metagenomics. *Cell* **2016**, *166*, 1103–1116. [CrossRef]
19. Props, R.; Kerckhof, F.M.; Rubbens, P.; De Vrieze, J.; Sanabria, E.H.; Waegeman, W.; Monsieurs, P.; Hammes, F.; Boon, N. Absolute quantification of microbial taxon abundances. *ISME J.* **2017**, *11*, 584–587. [CrossRef]
20. Liu, X.R.; Li, J.; Yu, L.; Pan, H.; Liu, H.Y.; Liu, Y.M.; Di, H.J.; Li, Y.; Xu, J.M. Simultaneous measurement of bacterial abundance and composition in response to biochar in soybean field soil using 16S rRNA gene sequencing. *Land Degrad. Dev.* **2018**, *29*, 2172–2182. [CrossRef]
21. Jiang, S.Q.; Yu, Y.N.; Gao, R.W.; Wang, H.; Zhang, J.; Li, R.; Long, X.H.; Shen, Q.R.; Chen, W.; Cai, F. High-throughput absolute quantification sequencing reveals the effect of different fertilizer applications on bacterial community in a tomato cultivated coastal saline soil. *Sci. Total Environ.* **2019**, *687*, 601–609. [CrossRef] [PubMed]
22. Du, R.B.; Wu, Q.; Xu, Y. Chinese Liquor Fermentation: Identification of Key Flavor-Producing Lactobacillus spp. by Quantitative Profiling with Indigenous Internal Standards. *Appl. Environ. Microbiol.* **2020**, *86*, e00456-20. [CrossRef] [PubMed]
23. Vandeputte, D.; Kathagen, G.; D'Hoe, K.; Vieira-Silva, S.; Valles-Colomer, M.; Sabino, J.; Wang, J.; Tito, R.Y.; De Commer, L.; Darzi, Y.; et al. Quantitative microbiome profiling links gut community variation to microbial load. *Nature* **2017**, *551*, 507–511. [CrossRef]
24. Lou, J.; Yang, L.; Wang, H.Z.; Wu, L.S.; Xu, J.M. Assessing soil bacterial community and dynamics by integrated high-throughput absolute abundance quantification. *PeerJ* **2018**, *6*, e4514. [CrossRef] [PubMed]
25. Zhang, Z.J.; Qu, Y.Y.; Li, S.Z.; Feng, K.; Wang, S.; Cai, W.W.; Liang, Y.T.; Li, H.; Xu, M.Y.; Yin, H.Q.; et al. Soil bacterial quantification approaches coupling with relative abundances reflecting the changes of taxa. *Sci. Rep.* **2017**, *7*, 4837. [CrossRef]
26. Smets, W.; Leff, J.W.; Bradford, M.A.; McCulley, R.L.; Lebeer, S.; Fierer, N. A method for simultaneous measurement of soil bacterial abundances and community composition via 16S rRNA gene sequencing. *Soil Biol. Biochem.* **2016**, *96*, 145–151. [CrossRef]
27. Tourlousse, D.M.; Yoshiike, S.; Ohashi, A.; Matsukura, S.; Noda, N.; Sekiguchi, Y. Synthetic spike-in standards for high-throughput 16S rRNA gene amplicon sequencing. *Nucleic Acids Res.* **2017**, *45*, e23. [CrossRef]
28. Nilsson, R.H.; Anslan, S.; Bahram, M.; Wurzbacher, C.; Baldrian, P.; Tedersoo, L. Mycobiome diversity: High-throughput sequencing and identification of fungi. *Nat. Rev. Microbiol.* **2019**, *17*, 95–109. [CrossRef]

29. Ban, S.B.; Chen, L.N.; Fu, S.X.; Wu, Q.; Xu, Y. Modelling and predicting population of core fungi through processing parameters in spontaneous starter (Daqu) fermentation. *Int. J. Food Microbiol.* **2022**, *363*, 109493. [CrossRef]
30. Rama, T.; Davey, M.; Norden, J.; Halvorsen, R.; Blaalid, R.; Mathiassen, G.; Alsos, I.; Kauserud, H. Fungi Sailing the Arctic Ocean: Speciose Communities in North Atlantic Driftwood as Revealed by High-Throughput Amplicon Sequencing. *Microb. Ecol.* **2016**, *72*, 295–304. [CrossRef]
31. Zhang, L.J.; Cao, Y.L.; Tong, J.N.; Xu, Y. An Alkylpyrazine Synthesis Mechanism Involving L-Threonine-3-Dehydrogenase Describes the Production of 2,5-Dimethylpyrazine and 2,3,5-Trimethylpyrazine by Bacillus subtilis. *Appl. Environ. Microbiol.* **2019**, *85*, e01807-19. [CrossRef] [PubMed]
32. Dhanasekaran, S.; Doherty, T.M.; Kenneth, J.; Grp, T.B.T.S. Comparison of different standards for real-time PCR-based absolute quantification. *J. Immunol. Methods* **2010**, *354*, 34–39. [CrossRef] [PubMed]
33. Song, Z.W.; Du, H.; Zhang, Y.; Xu, Y. Unraveling Core Functional Microbiota in Traditional Solid-State Fermentation by High-Throughput Amplicons and Metatranscriptomics Sequencing. *Front. Microbiol.* **2017**, *8*, 1294. [CrossRef]
34. Xu, W.; Huang, Z.Y.; Zhang, X.J.; Li, Q.; Lu, Z.M.; Shi, J.S.; Xu, Z.H.; Ma, Y.H. Monitoring the microbial community during solid-state acetic acid fermentation of Zhenjiang aromatic vinegar. *Food Microbiol.* **2011**, *28*, 1175–1181. [CrossRef] [PubMed]
35. Caporaso, J.G.; Kuczynski, J.; Stombaugh, J.; Bittinger, K.; Bushman, F.D.; Costello, E.K.; Fierer, N.; Pena, A.G.; Goodrich, J.K.; Gordon, J.I.; et al. QIIME allows analysis of high-throughput community sequencing data. *Nat. Methods* **2010**, *7*, 335–336. [CrossRef] [PubMed]
36. Bolger, A.M.; Lohse, M.; Usadel, B. Trimmomatic: A flexible trimmer for Illumina sequence data. *Bioinformatics* **2014**, *30*, 2114–2120. [CrossRef]
37. Edgar, R.C.; Haas, B.J.; Clemente, J.C.; Quince, C.; Knight, R. UCHIME improves sensitivity and speed of chimera detection. *Bioinformatics* **2011**, *27*, 2194–2200. [CrossRef]
38. Koljalg, U.; Larsson, K.H.; Abarenkov, K.; Nilsson, R.H.; Alexander, I.J.; Eberhardt, U.; Erland, S.; Hoiland, K.; Kjoller, R.; Larsson, E.; et al. UNITE: A database providing web-based methods for the molecular identification of ectomycorrhizal fungi. *New Phytol.* **2005**, *166*, 1063–1068. [CrossRef]
39. French, K.E.; Tkacz, A.; Turnbull, L.A. Conversion of grassland to arable decreases microbial diversity and alters community composition. *Appl. Soil Ecol.* **2017**, *110*, 43–52. [CrossRef]
40. Bustin, S.A.; Benes, V.; Garson, J.A.; Hellemans, J.; Huggett, J.; Kubista, M.; Mueller, R.; Nolan, T.; Pfaffl, M.W.; Shipley, G.L.; et al. The MIQE Guidelines: Minimum Information for Publication of Quantitative Real-Time PCR Experiments. *Clin. Chem.* **2009**, *55*, 611–622. [CrossRef]
41. Rodriguez, A.; Luque, M.I.; Andrade, M.J.; Rodriguez, M.; Asensio, M.A.; Cordoba, J.J. Development of real-time PCR methods to quantify patulin-producing molds in food products. *Food Microbiol.* **2011**, *28*, 1190–1199. [CrossRef] [PubMed]
42. Martinez-Blanch, J.F.; Sanchez, G.; Garay, E.; Aznar, R. Development of a real-time PCR assay for detection and quantification of enterotoxigenic members of Bacillus cereus group in food samples. *Int. J. Food Microbiol.* **2009**, *135*, 15–21. [CrossRef] [PubMed]
43. Chen, B.; Wu, Q.; Xu, Y. Filamentous fungal diversity and community structure associated with the solid state fermentation of Chinese Maotai-flavor liquor. *Int. J. Food Microbiol.* **2014**, *179*, 80–84. [CrossRef]
44. Li, H.; Huang, J.; Liu, X.P.; Zhou, R.Q.; Ding, X.F.; Xiang, Q.Y.; Zhang, L.Q.; Wu, C.D. Characterization of Interphase Microbial Community in Luzhou-Flavored Liquor Manufacturing Pits of Various Ages by Polyphasic Detection Methods. *J. Microbiol. Biotechnol.* **2017**, *27*, 130–140. [CrossRef] [PubMed]
45. Li, X.R.; Ma, E.B.; Yan, L.Z.; Meng, H.; Du, X.W.; Zhang, S.W.; Quan, Z.X. Bacterial and fungal diversity in the traditional Chinese liquor fermentation process. *Int. J. Food Microbiol.* **2011**, *146*, 31–37. [CrossRef]
46. Benito, A.; Jeffares, D.; Palomero, F.; Calderon, F.; Bai, F.Y.; Bahler, J.; Benito, S. Selected Schizosaccharomyces pombe Strains Have Characteristics That Are Beneficial for Winemaking. *PLoS ONE* **2016**, *11*, e0151102. [CrossRef]
47. Yuangsaard, N.; Yongmanitchai, W.; Yamada, M.; Limtong, S. Selection and characterization of a newly isolated thermotolerant Pichia kudriavzevii strain for ethanol production at high temperature from cassava starch hydrolysate. *Antonie Van Leeuwenhoek Int. J. Gen. Mol. Microbiol.* **2013**, *103*, 577–588. [CrossRef]
48. Su, C.; Zhang, K.Z.; Cao, X.Z.; Yang, J.G. Effects of Saccharomycopsis fibuligera and Saccharomyces cerevisiae inoculation on small fermentation starters in Sichuan-style Xiaoqu liquor. *Food Res. Int.* **2020**, *137*, 109425. [CrossRef]
49. Yang, Y.R.; Zhong, H.Y.; Yang, T.; Lan, C.H.; Zhu, H. Characterization of the key aroma compounds of a sweet rice alcoholic beverage fermented withSaccharomycopsis fibuligera. *J. Food Sci. Technol.-Mysore* **2021**, *58*, 3752–3764. [CrossRef]

fermentation

Article

Rational Metabolic Engineering Combined with Biosensor-Mediated Adaptive Laboratory Evolution for L-Cysteine Overproduction from Glycerol in *Escherichia coli*

Xiaomei Zhang [1], Zhenhang Sun [1], Jinyu Bian [1], Yujie Gao [2], Dong Zhang [1], Guoqiang Xu [2], Xiaojuan Zhang [2], Hui Li [1], Jinsong Shi [1] and Zhenghong Xu [2,*]

1   Laboratory of Pharmaceutical Engineering, School of Life Science and Health Engineering, Jiangnan University, Wuxi 214122, China; zhangxiaomei@jiangnan.edu.cn (X.Z.); 6211507033@stu.jiangnan.edu.cn (Z.S.); 6191502001@stu.jiangnan.edu.cn (J.B.); d.zhang1993@outlook.com (D.Z.); lihui@jiangnan.edu.cn (H.L.); shijs@jiangnan.edu.cn (J.S.)
2   National Engineering Research Center for Cereal Fermentation and Food Biomanufacturing, Jiangnan University, Wuxi 214122, China; 7200201006@stu.jiangnan.edu.cn (Y.G.); xuguoqiang@jiangnan.edu.cn (G.X.); zhangxj@jiangnan.edu.cn (X.Z.)
*   Correspondence: zhenghxu@jiangnan.edu.cn; Tel./Fax: +86-510-85918206

Abstract: L-Cysteine is an important sulfur-containing amino acid with numerous applications in the pharmaceutical and cosmetic industries. The microbial production of L-cysteine has received substantial attention, and the supply of the precursor L-serine is important in L-cysteine biosynthesis. In this study, to achieve L-cysteine overproduction, we first increased L-serine production by deleting genes involved in the pathway of L-serine degradation to glycine (serine hydroxymethyl transferase, SHMT, encoded by *glyA* genes) in strain 4W (with L-serine titer of 1.1 g/L), thus resulting in strain 4WG with L-serine titer of 2.01 g/L. Second, the serine-biosensor based on the transcriptional regulator NCgl0581 of *C. glutamicum* was constructed in *E. coli*, and the validity and sensitivity of the biosensor were demonstrated in *E. coli*. Then 4WG was further evolved through adaptive laboratory evolution (ALE) combined with serine-biosensor, thus yielding the strain 4WGX with 4.13 g/L L-serine production. Moreover, the whole genome of the evolved strain 4WGX was sequenced, and ten non-synonymous mutations were found in the genome of strain 4WGX compared with strain 4W. Finally, 4WGX was used as the starting strain, and deletion of the L-cysteine desulfhydrases (encoded by *tnaA*), overexpression of serine acetyltransferase (encoded by *cysE*) and the key enzyme of transport pathway (encoded by *ydeD*) were performed in strain 4WGX. The recombinant strain 4WGX-Δ*tnaA-cysE-ydeD* can produce 313.4 mg/L of L-cysteine using glycerol as the carbon source. This work provides an efficient method for the biosynthesis of value-added commodity products associated with glycerol conversion.

Keywords: *Escherichia coli*; biosensor; glycerol; adaptive laboratory evolution; L-cysteine

Citation: Zhang, X.; Sun, Z.; Bian, J.; Gao, Y.; Zhang, D.; Xu, G.; Zhang, X.; Li, H.; Shi, J.; Xu, Z. Rational Metabolic Engineering Combined with Biosensor-Mediated Adaptive Laboratory Evolution for L-Cysteine Overproduction from Glycerol in *Escherichia coli*. *Fermentation* **2022**, *8*, 299. https://doi.org/10.3390/fermentation8070299

Academic Editor: Donatella Cimini

Received: 17 May 2022
Accepted: 23 June 2022
Published: 25 June 2022

1. Introduction

L-cysteine has been widely used in the food, agricultural and pharmaceutical industries. Because of the toxicity of L-cysteine and the complex regulation of its synthesis pathway, efficient microbial production of L-cysteine at the industrial scale has not been achieved [1–4]. Most studies have focused on producing L-cysteine from glucose by recombinant *Escherichia coli* or *Corynebacterium glutamicum* [5–8]. However, *C. glutamicum* grows slowly, thus resulting in a long manufacturing cycle. Compared with *C. glutamicum*, *E. coli* has a higher growth rate, and the its genetic engineering method is well developed, thus suggesting that the production of L-cysteine by *E. coli* has great potential [2,3,7,8]. The precursor of L-cysteine is L-serine in *E. coli*, and the biosynthesis of L-cysteine from L-serine in *E. coli* occurs via a two-step pathway, the catalysis of L-serine acetyltransferase (encoded by *cysE*) and L-cysteine synthase (encoded by *cysK*) (Figure 1). The first reaction catalysed by CysE is the rate limiting step of L-cysteine biosynthesis in *E. coli*. Moreover, multiple L-cysteine desulfhydrases (CD encoded by *tnaA*) catalyse the degradation of L-cysteine [2,3,7–9]. Previous studies have shown that L-serine is an important precursor for the biosynthesis of L-cysteine, and enhancing the L-serine synthesis is a necessary metabolic engineering strategy for L-cysteine accumulation [1,2].

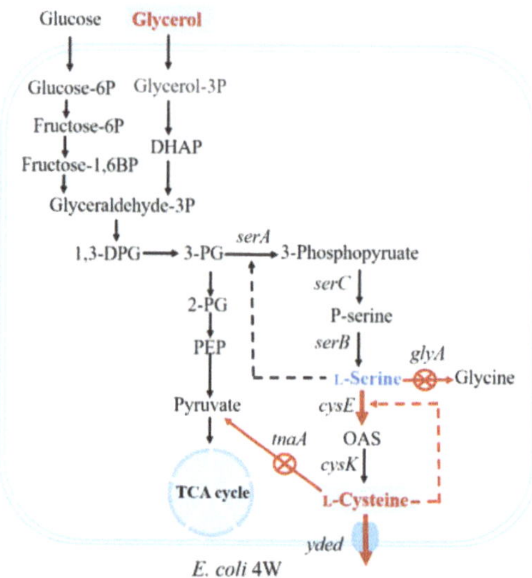

Figure 1. The protocol of constructing strain 4WGX over-producing L-cysterine from glycerol. *glyA* encoded serine hydroxymethyl transferase, *tnaA* encoded L-cysterine transporter, Red crosses on solid lines (⊗) indicated genes that were deleted. *cysE* encoded serine acetyltransferase, the red line indicated gene that were overexpressed. The starting strain *E. coli* 4W had been constructed in our previous study with deletion of *sdaA*, *sdaB*, and *tdcG* (The three genes encoded L-serine deaminases), and the removal of feedback inhibition of *serA* (*serA* encoded 3-phosphoglycerate dehydrogenase).

Simultaneously, in addition to traditional fermentation substrates such as glucose, sucrose and other sugar raw materials, glycerol has become a very competitive new choice [10,11]. As non-renewable fossil energy is increasingly depleted, searching for new alternative energy sources, such as biodiesel, has become a top priority. However, the large-scale development and utilization of biodiesel has brought about another serious problem: the treatment and reuse of crude glycerol, which was the by-product of biodiesel manufacturing. According to statistics, every 10 kg of biodiesel produces approximately 1 kg of crude glycerol by-product [12,13]. If these by-products could be converted into high value-added chemicals, the crude glycerol could be reused, and the chemical cost could also be decreased. Therefore, using glycerol as a carbon source to produce high value-added chemicals has become a major topic in the biodiesel industry [14,15]. The chemicals produced by using glycerol as a substrate mainly include shikimic acid, lactic acid, succinate, lysine, L-phenylalanine and 1,3-propanediol [11,13,16–19]. However, nearly all studies on the production of L-cysteine by *E. coli* have focused on the pathway begin from glucose. Compared with that from glucose, the metabolic pathway from glycerol to L-cysteine is shorter, and the carbon atom economy is also better [20,21].

ALE, also called adaptive evolutionary engineering, has become a valuable tool in metabolic engineering for strain development [22]. However, the traditional screening procedure is cumbersome and time-consuming. A pressing need to develop an efficient approach to screen high-yield mutant strains [23–25]. High-throughput screening methods with biosensors have been used to screen mutant strains, such as those overproduce amino acids and organic acids, and the biosensors, including the ones based on riboswitches, enzymes, and transcription factors, can transform information about a specific metabolite into a graded fluorescence output [26–31]. In our previous study [32], we developed a genetic metabolite biosensor capable of detecting L-serine in single *C. glutamicum* cells, this biosensor is based on the transcriptional regulator *NCgl0581* of *C. glutamicum*, which activates expression of the *NCgl0580* promoter to drive transcription of enhanced yellow fluorescent protein [32]. However, no study has reported screening of L-serine overproducing strains a biosensor in *E. coli*.

In this study, to achieve L-cysteine overproduction, we first enhanced production of the precursor L-serine by deleting the L-serine degradation pathway *glyA* with CRISPR/Cas9 in strain 4W, thus resulting in strain 4WG. Second, in *E. coli*, we constructed a serine-biosensor based on the transcriptional regulator NCgl0581 of *C. glutamicum*. The validity and sensitivity of the biosensor were studied. Subsequently, 4WG was further evolved by using ALE combined with serine-biosensor, thus yielding the evolved strain 4WGX. The whole genome of the evolved strain 4WGX was sequenced, and comparative genomics analysis and reverse mutation were performed. Finally, 4WGX was used as the starting strain, and the deletion of the L-cysteine desulfhydrases (*tnaA*), overexpression of serine acetyltransferase (*cysE*) and the key enzyme in the transport pathway (*ydeD*) were performed. The recombinant strain was successfully constructed and found to produce L-cysteine using glycerol as substrate.

2. Materials and Methods

2.1. Strains and Plasmids

Strains and plasmids used in this study are summarized in Table 1. Primers for gene cloning and deleting are listed in Table 2. Strain 4W, carrying deletions of *sdaA*, *sdaB* and *tdcG*, was constructed in our previous study [20]. The serine-biosensor pDser from *C. glutamicum* was also constructed in our previous study [32]. The plasmids pTarget and pCas were used for knocking out the *glyA* gene.

Table 1. Strains and plasmids used in this study.

Strains or Plasmids	Description	Sources
Strains		
E. coli JM109	recA1, endA1, gyrA96, thi-1, hsd R17(rk- mk+) supE44	Invitrogen
4W	W3110△*tdcG*△*sdaA*△*sdaB serA*^{dr}	Invitrogen
4WG	4W with *glyA* deletion	This study
4W-pDer	4W harboring serine-biosensor pDser	This study
4WGX	A mutant derived from 4W	This study
4WG-pDer	4WG harboring serine- biosensor pDser	This study
4WG-*cysE*	4WG harboring pEtac-*cysE*	This study
4WG-*cysE-ydeD*	4WG harboring pEtac-*cysE-ydeD*	This study
4WG-Δ*tnaA*	4WG with *tnaA* deletion	This study
4WG-Δ*tnaA-cysE*	4WG-Δ*tnaA* harboring pEtac-*cysE*	This study
4WG-Δ*tnaA-cysE-ydeD*	4WG-Δ*tnaA* harboring pEtac-*cyE-ydeD*	This study
Plasmids		
pCas	Carrying Cas9 and λRed System, kan	Invitrogen
pTargetF	Carrying N20 sequence, spc or smr	Invitrogen
pDser	Biosensor, kan	Invitrogen
pEtac	Inducible expression plasmid, *tac*, kan	This study
pEtac-*cysE*	Carrying *cysE* gene from *E. coli*	This study
pEtac-*cysE-ydeD*	Carrying *cysE* and *ydeD* gene from *E. coli*	This study

Table 2. Primers used in this study.

Primers	Sequence
pTargetF-$\triangle$*glyA*1-F	ACTGTGGCAGGCTATGGAGCGTTTTAGAGCTAGAAATAGCAAGTT
pTargetF-$\triangle$*glyA*1-R	GCTCCATAGCCTGCCACAGTACTAGTATTATACCTAGGACTGAGC
pTargetF-$\triangle$*glyA*2-F	AGAAGCCGAAGCGAAAGAACGTTTTAGAGCTAGAAA-TAGCAAGTT
pTargetF-$\triangle$*glyA*2-R	GTTCTTTCGCTTCGGCTTCTACTAGTATTATACCTAGGACTGAGC
glyA-U-F	AGCCCTGCAATGTAAATGGTT
glyA-U-R	ACAGCAAATCACCGTTTCGCCCGCATCTCCTGACTCAGCTA
glyA-D-F	AGCTGAGTCAGGAGATGCGGGCGAAACGGTGATTTGCTGTC
glyA-D-R	TCGCCAGACAGGATTTAACCC
pTargetF:1756F23	CCCTGATTCTGTGGATAACCGTA
pTargetF:78R23	ACATCAGTCGATCATAGCACGAT
cysE-F	TTCACACAGGAAACAGAATTCATGTCGTGTGAAGAACTGGAAATTG
cysE-R	TGCGGCCGCAAGCTTGTCGACTTAGATCCCATCCCCATACTCAA
ydeD-F	GGGATCTAAGTCGACAAGCTTCGCTGAGCAATAACTAGCATAACC
ydeD-R	GTGGTGGTGGTGGTGCTCGAGTTAACTTCCCACCTTTACCGCT
tnaA-U-F	TTGCATATATATCTGGCGAATTAATCGG
tnaA-U-R	GCCACTCTGTAGTATTAAGTATCAAAGAAATAGTTAGAGAACGCCA
tnaA-D-F	ACTTAATACTACAGAGTGGCTATAAGGATGTT
tnaA-D-R	ACGAAAATGGCTGTGCAGAT
pTargetF-Δ*tnaA*-F	CGTTCTCTTTCACATGTTTAACTAGTATTATACCTAGGACTG
pTargetF-Δ*tnaA*-R	TAAACATGTGAAAGAGAACGTTTTAGAGCTAGAAATAGCAA

2.2. Growth Medium and Culture Conditions

Luria-Bertani (LB) medium was used for plasmid construction. When appropriate, streptomycin (50 µg/mL) or kanamycin (50 µg/mL) was added. For L-serine fermentation, mineral AM1 medium [33] supplemented with 10.0 g/L glycerol, 8.6 g/L $(NH_4)_2 \cdot HPO_4$, 3.9 g/L $NH_4H_2PO_4$ and 1 g/L yeast extract was used. *E. coli* was cultured according to our previous study [20].

2.3. Gene Deletion with CRISPR/Cas9

For deletion of *glyA* gene, as an example, the upstream homologous arms of the *glyA* gene were obtained by using primers *glyA*-U-F and *glyA*-U-R, and the downstream homologous arms of *glyA* gene by using primers *glyA*-D-F and *glyA*-D-R. Then both the upstream and downstream arms were used as templates, and the homologous recombination repair templates were obtained by overlap extension PCR using primers *glyA*-U-F and *glyA*-D-R.

Plasmid pTarget was extracted from *E. coli* JM109 and amplified with primers pTargetF-$\triangle$*glyA*1-F, pTargetF-$\triangle$*glyA*1-R and pTargetF-$\triangle$*glyA*2-R and pTargetF-$\triangle$*glyA*2-F. The template was degraded with the *DpnI* enzyme, and then PCR products were transferred to *E. coli* JM109 to repair the cyclization gap. After the clones were grown, pTargetF:1756F23 and pTargetF:78R23 were used for PCR amplification. If the sequencing results were correct, then plasmids were amplified and extracted. Finally, two pTarget plasmids containing N20 sequences specifically targeting *glyA* were obtained. The N20 sequences were predicted in CHOPCHOP (chopchop.cbu.uib.no).

Finally, homologous arm fragments and two pTarget plasmids were electroporated into competent cells containing the pCas plasmid. Transformants were selected on kanamycin and streptomycin plates and verified by PCR using the corresponding primers *glyA*-U-F and *glyA*-D-R. The gene *tnaA* was deleted by this method.

2.4. Gene Overexpression

The plasmid pEtac was used for the expression of foreign genes. This plasmid carries a tac promoter and Kan resistance marker. For the expression of *cysE*, as an example, we amplified the *cysE* gene by using the primers *cysE*-F/*cys*-R. The target gene was digested with *EcoR* I and *Sal* I, then ligated with the linearized plasmid pEtac to construct the inducible expression plasmid pEtac-*cysE*. The gene *ydeD* was overexpressed through this method, thus yielding the plasmid pEtac-*cysE-ydeD*.

2.5. Construction and Verification of the Serine-Biosensor

The pDser plasmid was constructed in our previous study [32]. The pDser plasmid was introduce into *E. coli* 4WG, transformants were selected and verified, and strains containing the pDser plasmid were achieved. The verification of the serine-biosensor was performed according to our previous study [32].

2.6. Biosensor-Driven Evolution Experiment

For strain evolution, *E. coli* 4WG harboring pDser was cultured in LB medium, and 10% (v/v) inoculum was transferred into fresh AM1 medium with 6 g/L L-serine for 24 h. This step was repeated ten times (recorded as ten generations). The evolved strain was approximately 600 generations. Then L-serine concentration were increased to 12, 25 and 50 g/L, and this process was repeated. Successive rounds of ALE were carried out with the L-serine increased stepwise (6, 12, 25 and 50 g/L). According to the generation time (GT) of *E. coli*, the evolved strain was approximately 600, 1200, 1800, 2400 generations, respectively.

At the end of the experiment, an appropriate amount of the evolved strains after 600, 1200, 1800, 2400 generations were diluted to 10^{-5}-10^{-6} fold. Then 100 µL of diluted bacterial solution was spread on kanamycin plates and cultured overnight at 37 °C and single colonies were transferred to 96-well plates. After 24 h of fermentation, the fluorescence intensity of the strain in each well was measured, the most efficient strain was selected by using FACS according to our previous study [32].

2.7. Genome Sequencing

The whole genome of *E. coli* 4WGX was sequenced, and comparative genomics analysis was performed with the parent strain 4WG. Genomic DNAs of the strains were extracted using Molpure Bacterial DNA Kit (Yeasen. Shanghai, China). Library construction and genome sequencing were performed by Genewiz (Suzhou, China) by using Illumina Hiseq2500 sequencing platform. Quality assurance of the output was analyzed by using FastQC software (v.0.10.1) and NGSQC Toolkit software (v.2.3.3). BWA alignment software (v.0.7.17) and SAM tools software (v.1.9) were used for alignment and variant calling, respectively. Variations were annotated by using the SnpEff software (v.4.3i).

2.8. Analytical Methods

Cell growth was measured as the OD_{600} (AOE UV-1200S, China). A triglyceride assay kit for measuring glycerol concentration was purchased from Nanjing Jiancheng Bioengineering Institute. First, a standard curve based on different concentrations of glycerol standard and the corresponding OD_{550} values was constructed, and then the actual glycerol concentration of each sample was calculated according to the OD_{550} value. The fluorescence intensity of bacteria was detected with a microplate reader with an excitation wavelength of 488 nm and emission wavelength of 530 nm. The concentrations of L-serine and L-cysteine were determined with high-performance liquid chromatography (HPLC; Agilent 1100, USA) according to a previously reported method [32].

3. Results

3.1. Improved the Precursor L-Serine Accumulation by Decreasing L-Serine Degradation in E. coli

To achieve L-cysteine overproduction, we first enhanced the L-serine production of strain 4W. In the L-serine degradation pathway, SHMT (*glyA*) were deleted in strain 4W by using CRISPR/Cas9, thus resulting in strain 4WG. As shown in Figure 2, strain 4WG showed cell growth inhibition, with a maximum OD_{600} of 2.37 (Figure 2b). In contrast, the maximum OD_{600} of parent strain 4W was 5.73 (Figure 2a), *glyA* deletion significantly decreased the cell growth. Correspondingly, the L-serine accumulation of strain 4WG was 0.75 g/L, a level significantly lower than that of the parental strain 4W (1.1 g/L).

We inferred that intracellular glycine deficiency caused by knocking out the pathway of L-serine degradation resulted in poorer growth status of the strain, and leaded to lower L-serine accumulation.

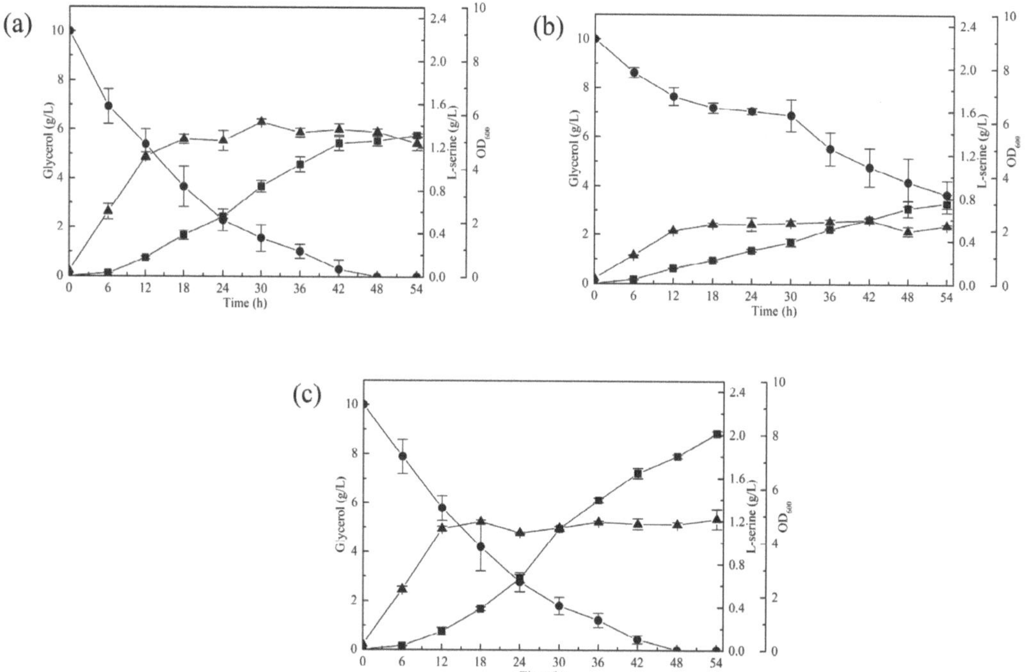

Figure 2. Fermentation profiles of strain 4W, strain 4WG and strain 4WG with 0.15 g/L glycine. (**a**) Profiles of glycerol consumption, cell growth and L-serine production in strain 4W; (**b**) Profiles of glycerol consumption, cell growth and L-serine production in 4WG; (**c**) Profiles of glycerol consumption, cell growth and L-serine production in strain 4WG with 0.15 g/L glycine added. Squares represent cell growth, circles represent residual glycerol, and triangles represent L-serine. Values denote the average of three independent experiments, and error bars indicate standard deviation.

According to a previous study [5], inactivation of SHMT in *E. coli* could effectively reduce the intracellular degradation of L-serine, and exogenous glycine could be added to maintain the cell growth. After 0.15 g/L (2 mM) glycine was added to the medium, the strain 4WG returned to normal growth with a maximum OD_{600} value of 4.87 (Figure 2c). Meanwhile, L-serine accumulation also increased significantly, reaching 2.01 g/L after 54 h of fermentation, which was 53.4% higher than that of the control strain 4W. Simultaneously, the substrate glycerol was completely consumed during fermentation for 48 h, which was similar to that for the parental strain 4W.

Although L-serine titer of strain 4WG increased with the addition of glycine, L-serine was found to be highly toxic to this strain even at low concentrations. As shown in Figure 3, the cell growth of strain 4WG was significantly decreased with the addition of L-serine in the medium. When 6 g/L L-serine was added, the maximum OD_{600} was 2.62. When L-serine addition reached 12, 25 and 50 g/L, strain 4WG showed negligible growth. The strain's tolerance to L-serine was the key to over-producing L-serine.

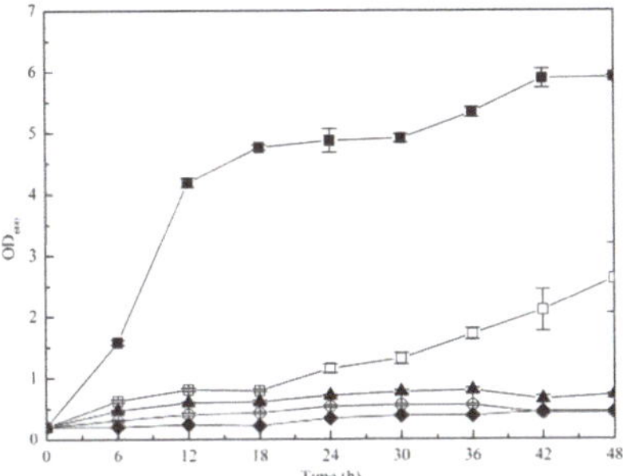

Figure 3. Growth profiles of strains 4WG in AM1 containing different concentrations of L-serine. Squares represent 0 g/L L-serine, open squares represent 6 g/L L-serine, triangles represent 12 g/L L-serine, open circles represent 25 g/L L-serine, and diamonds represent 50 g/L L-serine. Values denote the average of three independent experiments, and error bars indicate standard deviation.

3.2. Increased L-Serine Production through ALE Combined with a Serine-Biosensor

3.2.1. Construction and Verification of a Serine-Biosensor in *E. coli*

ALE was selected to improve L-serine tolerance and L-serine production. We constructed a serine-biosensor to increase the screening efficiency, and verified its efficacy in *E. coli*. The serine-biosensor of *C. glutamicum* was constructed and used to screen L-serine overproducing strains in our previous study [32]. However, there is no research adapting this biosensor in *E. coli*, and whether the heterologous expression of the biosensor was also effective in high-performance screening in *E. coli* needed to be tested. The serine-biosensor pDser was transformed into *E. coli* 4WG, resulting in 4WG-pDser. Afterward, 4WG and 4WG-pDser were photographed under a laser scanning confocal microscope under visible light and UV light. As shown in Figure 4, the parental strain 4WG showed no fluorescence signal, and 4WG-pDser showed substantial fluorescence intensity. This result confirmed that the serine-biosensor was successfully expressed in *E. coli*.

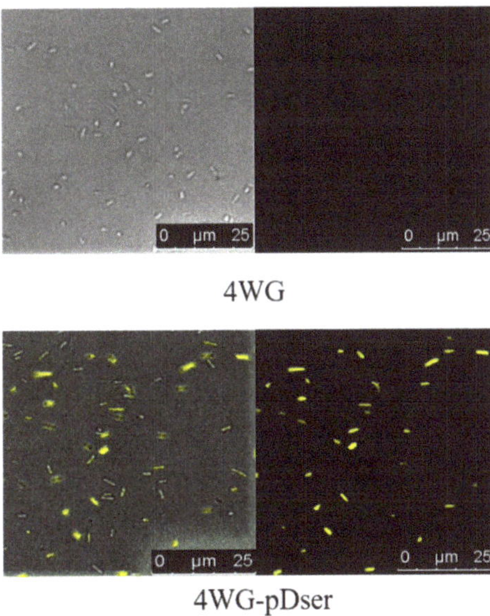

Figure 4. Identification of serine-biosensor pDser in *E. coli*. 4WG-pDser clearly emitted yellow fluorescence, whereas the control *E. coli* 4WG did not emit yellow fluorescence.

The relationship between fluorescence intensity and L-serine titer was then studied. The fluorescence signal from the serine-biosensor correlated with the L-serine titer (Figure 5a). Moreover, in the ALE experiment, L-serine was added to the medium, and the effect of the L-serine addition to the biosensor was studied. No fluorescence significant change was observed with varying amounts of L-serine added (data not shown), indicating that only the cellular L-serine biosynthesized was monitored by serine-biosensor. These results demonstrated the functionality of the serine-biosensor in *E. coli*, which was adapted from *C. glutamicum*. We used this method to screen serine over-producing strain in the rest of this study.

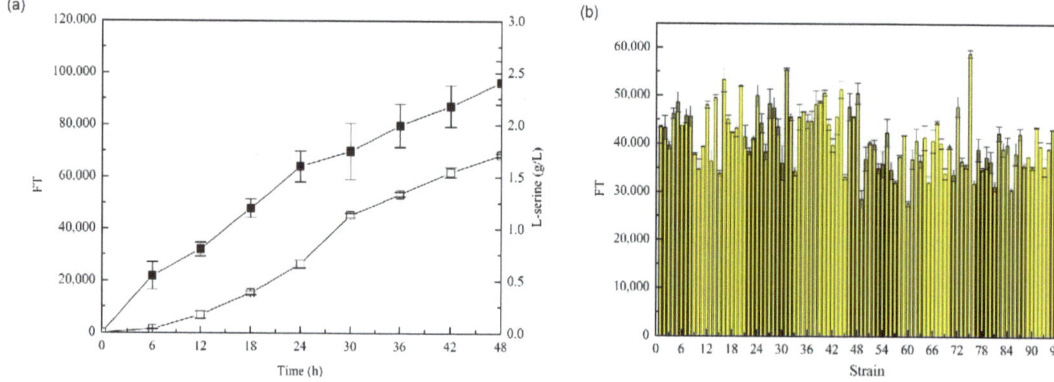

Figure 5. The relationship of L-serine accumulation and the fluorescence intensity. (**a**) Comparison of L-serine accumulation and the fluorescence intensity of *E. coli* 4WG-pDser. Squares represent fluorescence intensity, and open squares represent L-serine titer by HPLC; (**b**) Comparison of the adaptive strains' fluorescence intensity. Cylinders represent fluorescence intensity. Values denote the average of three independent experiments, and error bars indicate standard deviation.

3.2.2. Increased L-Serine Yield Achieved by Biosensor-Driven Evolution

ALE with biosensors was used to improve L-serine tolerance and L-serine production. The strain harbouring serine-biosensor (4WG-pDser) was evolved in the medium with 6, 12, 25 and 50 g/L L-serine, and the evolve strain was approximately 600, 1200, 1800 and 2400 generations respectively, according to the generation time of *E. coli*. In this process, we observed that the cell growth of strain 4WG was significantly inhibited in the medium with 25 g/L and 50 g/L L-serine. With increasing generation number, the cell growth rate clearly increased, as did the maximum OD_{600} increased (data not shown). The final evolved strain was achieved 2400 generations. As shown in Figure 5b, the first strain was the control strain 4WGX-pDser, and the remaining 95 strains were single colonies selected on the plates. Five strains with the highest intensity values were selected for flask fermentation. The resultant ALE strain was named 4WGX. As shown in Figure 6a, after 48 h of fermentation, the maximum OD_{600} of strain 4WGX was 6.87, glycerol was completely consumed at 24 h, the L-serine titer was 4.13 g/L at 48 h, which was 105% higher than that of 4WG (2.01 g/L) and 275% higher than that of 4W (1.1 g/L), and the substrate conversion rate was 41.3%. Strain 4WGX was cultured in medium with the addition of 50 g/L L-serine, the cell growth was shown in Figure 6b, the maximum OD_{600} of strain 4WGX reached 3.65, and the parental strain 4WG showed almost no growth in the same medium (Figure 3).

To clarify the reasons for the greatly improved serine-tolerance of strain 4WGX, we sequenced the whole genome of strain 4WGX. The sequencing results revealed a total of eleven single base mutations in the genome of strain 4WGX compared with strain 4W, including ten non-synonymous mutations (*bamA*, *brnQ*, *ybcJ*, *fepB*, *agp*, *dgcT*, *oppB*, *fliK*, *ygbN and eno*) and one synonymous mutation (*fdrA*). We chose two genes (*agp* encoded glucose-1-phosphatase, and *eno* encoded enolase) not involved in the membrane for further study. First, reverse mutation of *agp* and *eno* was performed in the genome of 4WG, thus yielding mutant strains. However, we did not observe any significant change in L-serine production, cell growth and glycerol consumption in both strains (data not shown).

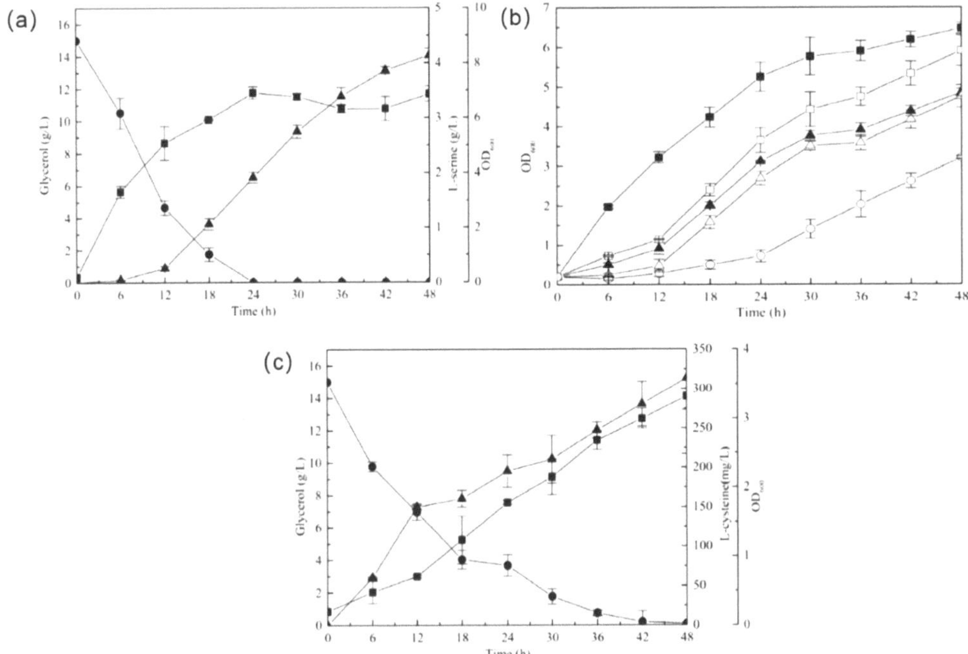

Figure 6. Fermentation profiles of strain 4WGX, strain 4WGX in AM1 containing different concentrations of L-serine, and stain 4WGX-Δ*tnaA-cysE-ydeD*. (**a**) Profiles of glycerol consumption, cell growth and L-serine production in strain 4WGX. Squares represent cell growth, circles represent residual glycerol, and triangles represent L-serine; (**b**) Growth profiles of the evolved strains 4WGX in AM1 containing different concentrations of L-serine. Squares represent 0 g/L L-serine, open squares represent 6 g/L L-serine, triangles represent 12 g/L L-serine, open triangles represent 25 g/L L-serine, and open circles represent 50 g/L L-serine; (**c**) Profiles of glycerol consumption, cell growth and L-serine production in strain 4WGX-Δ*tnaA-cysE-ydeD*. Squares represent cell growth, circles represent residual glycerol, and triangles represent L-cysteine. Values denote the average of three independent experiments, and error bars indicate standard deviation.

3.3. Construction of L-Cysteine-Producing Recombinant Strain

Strain 4WGX was used as the staring strain, in which the L-cysteine desulfhydrases (*tnaA* gene) gene was deleted by the CRISPR/Cas9 method. The 4WGX genome was used as a template, and primers *tnaA-1/tnaA-2* and *tnaA-3/tnaA-4* were used to amplify the upstream and downstream fragments. The pTarget plasmid with the specific N20 sequence and the homology arm fragment were electroporated into competent cells containing the Cas9 plasmid. Clones were selected on LB agar plates containing kanamycin and streptomycin (50 µg/mL). The clones were verified by PCR, the plasmid was finally eliminated, and strain 4WGX-Δ*tnaA* was achieved. Then the plasmid pEtac was used for the overexpression of *cysE* and *ydeD*. The plasmid pEtac-*cysE-ydeD* was constructed and transformed into 4WGX-Δ*tnaA*, thus yielding the L-cysteine-producing strain 4WGX-Δ*tnaA*-pEtac-*cysE-ydeD*. Fermentation by strain 4WGX-Δ*tnaA*-pEtac-*cysE-ydeD* was performed (Figure 6c). The cell growth (OD$_{600}$) reached a maximum value of 3.4 at 48 h, the glycerol was completely consumed at 42 h, and the accumulation of L-cysteine gradually increased after 6 h, exhibiting a cell growth-independent production profile. At 48 h, the final L-cysteine titer was 313.4 mg/L, and the original strain 4WGX did not produce L-cysteine. Compared with those of the original strain 4WGX (Figure 6a), the cell growth and the glycerol consumption rate of 4WGX-$\Delta tnaA$-pEtac-*cysE-ydeD* were significantly lower, thus suggesting that L-cysteine might be toxic to the cell, and improving the strain's tolerance to L-cysteine might be key in the future.

4. Discussion

In this study, we successfully established a biosensor-driven laboratory evolution approach using serine-biosensor from *C. glutamicum* for improving L-cysteine production in *E. coli*. Within several iterative rounds, *E. coli* 4WGX was isolated from a large evolved strain library and found to produce 4.13 g/L L-serine. Furthermore, the L-cysteine producing strain was obtained through deletion of *tnaA*, overexpression of *cysE* and *ydeD* in strain 4WGX. The recombinant strain 4WGX-$\Delta tnaA$-*cysE-ydeD* with 313.4 mg/L of L-cysteine was constructed using glycerol as the carbon source. This is the first report of producing L-cysteine from glycerol. Compared with glucose as carbon source for microbial L-cysteine production, using glycerol has several advantages, including its better carbon atomic economy and higher degree of reduction; glycerol is a highly promising substrate for amino acid production [20]. Although these results showed that L-cysteine titer was lower than that with glucose as the carbon source, glycerol is an alternative substrate providing a variety of economic and metabolic advantages. With further engineering and optimization, fermentation directly using glycerol as carbon source could become competitive. Moreover, this work provides an efficient method for value-added products bioconversion using glycerol as substrate.

L-serine is the precursor of L-cysteine in *E. coli*, and L-serine accumulation is important to efficient produce L-cysteine. However, degradation is a crucial issue in microbial L-serine production. L-serine has two main degradation pathways to either glycine or pyruvate. The conversion of L-serine to pyruvate in *E. coli* is catalyzed by three L-serine deaminases, *sdaA*, *sdaB* and *tdcG*. The conversion of serine to glycine is catalyzed by serine hydroxymethyl transferase (SHMT). Strain 4W was obtained by deleting *sdaA*, *sdaB* and *tdcG* in *E. coli* W3110. In this study, to remove the L-serine degradation pathway in *E. coli* 4W, the gene *glyA* was deleted with CRISPR/Cas9, thus resulting in strain 4WG, which produced 2.01 g/L L-serine with the addition of glycine. Decreasing SHMT activity strongly affects L-serine accumulation was observed in other studies [5–7]. However, in the present experiments, *glyA* deletion resulted in cell growth inhibition and a lower glycerol consumption rate. These results were not completely consistent with Mundhada's study, in which the T1 strain (*E. coli* MG1655 with *tdcG*, *sdaA* and *sdaB* deletion) had a higher glucose consumption rat e and a lower cell growth than the Q1 (strain T1 with *glyA* deletion) [5]. With *glyA* deletion, the cell growth did not significantly change, possibly because of the different carbon source. The low growth rate of strain 4WG limited the its application. To overcome this problem, we used ALE to enhance the strain's tolerance.

ALE or random mutagenesis followed by screening for a non-selectable phenotype is often labor-intensive [32,34]. However, this process can be circumvented by combining ALE with biosensors [23,24]. By using ALE combined with serine-biosensor, we obtained the evolved strain 4WGX, and L-serine production was increased. The maximum OD_{600} of strain 4WGX was 6.87, glycerol was consumed completely at 24 h, faster than the parental strain 4W, with the addition of 2 mM glycine. The L-serine titer was 4.13 g/L at 48 h (Figure 6a), a value 105% higher than that of 4WG and 275% higher than that of 4W. The substrate conversion rate was 41.3%, and the value reported for the strain developed in this study is close to highest yield reported from sugar [5]. Moreover, sequencing results of strain 4WGX revealed a total of eleven single base mutations in the genome of strain 4WGX, including ten non-synonymous mutations and one synonymous mutation (Table S1). Interestingly, most of the mutated genes encoded the membrane protein, such as *brnQ*, encoding the branched chain amino acid transporter BrnQ, *fepB*, *oppB* encoding the ABC transporter, and *fepB*, encoding FepB with a key role in transporting the catecholate siderophore ferric enterobactin from the outer to the inner membrane in Gram-negative bacteria [35]. *ygbN* encodes a putative transporter. *bamA* encodes the outer membrane protein BamA in *E. coli*, and a recent study has reported that the outer membrane fluidity linked to BamA activity [36,37]. The gene *agp* encodes glucose-1-phosphatase, and *eno* encodes enolase. The function of the other mutated genes (*ybcJ*, *fepB*, *dgcT*, *fliK*) were unclear. We chose to study the genes *agp* and *eno*, which were not involved in the membrane. Reverse mutation of *agp* and *eno* were performed in the genome of 4WG, thus yielding mutant strains. However, we did not observe any significant changes in L-serine production and cell growth in both strains (data not shown). The previous study has showed that a site-specific variant of enolase results in the functional and structural changes [38]. Most of the mutated genes (*brnQ,BrnQ*, *fepB*, *oppB*, *ygbN* and *bamA*) encoded transporters in *E. coli*, and these mutations were likely to alter cellular metabolite to help bacteria cope with the toxic metabolite [37,39], thus potentially explained why the final evolved strain grew better than the parent strain in 50 g/L L-serine. Further studies are needed to explore relationship between the membrane protein and phenotypic.

Biosensors have been widely used to develop high throughput screening methods and optimize pathway expression [40–42]. In our previous study, the serine-biosensor pDser, which was based on NCgl0581 (a transcription factor specifically responsive to L-serine in *C. glutamicum*), had been constructed in *C. glutamicum* for high-throughput screening of L-serine high-yield strains. However, we did not know whether this biosensor was suitable for screening L-serine over-producing *E. coli*, because the heterologous expression of the transcriptional regulator might have significantly interfered with the host gene regulatory networks. Moreover, the sensitivity of the biosensor was determined by the rate of promoter occupation by transcription factors through protein-protein interaction [43]. Therefore, the validity and sensitivity of serine-biosensor pDser were verified, and the results showed that biosensor from *C. glutamicum* was effective in selecting L-serine over-producing *E. coli*. On this basis, we developed a high-throughput screening method. Moreover, in evolution experiments, L-serine was added to the medium, and the effect of the L-serine addition to the biosensor was studied. No significant change in fluorescence intensity was observed with varying amount of L-serine added (data not shown), thus indicating that only the cellular L-serine biosynthesized was monitored by serine-biosensor. This work indicates that the serine-biosensor from *C. glutamicum* is useful in selecting serine over-producing *E. coli*, thus expanding the application of biosensor and enabling expanded strategies for screening high performance strains. Moreover, glycerol is a promising carbon source for the production of L-cysteine. In the future, to further increase L-cysteine production, we will focus on the genes involved in biosynthesis and transport of L-cysteine, and in the uptake of sulfur sources.

Supplementary Materials: The following supporting information can be downloaded at: https://www.mdpi.com/article/10.3390/fermentation8070299/s1, Table S1. The eleven single base mutations in the genome of strain 4WGX. References [35,37–39,44–50] are cited in the supplementary materials.

Author Contributions: Conceptualization, X.Z. (Xiaomei Zhang); methodology, D.Z.; validation, Z.S. and J.B.; data curation, Y.G.; writing—original draft preparation, X.Z. (Xiaomei Zhang) and D.Z.; writing—review and editing, G.X.; visualization, X.Z. (Xiaojuan Zhang) and H.L.; supervision, J.S.; project administration, Z.X.; funding acquisition, X.Z. (Xiaomei Zhang). All authors have read and agreed to the published version of the manuscript.

Funding: This work was financially supported by the National Key Research and Development Program of China (2018YFA0901400). The National Natural Science Foundation of China (32171470).

Institutional Review Board Statement: Not applicable.

Informed Consent Statement: Not applicable.

Data Availability Statement: Not applicable.

Conflicts of Interest: The authors declare no conflict of interest.

References

1. Takagi, H.; Ohtsu, I. L-cysteine metabolism and fermentation in microorganisms. *Adv. Biochem.* **2017**, *159*, 129–151.
2. Wendisch, V.F. Metabolic engineering advances and prospects for amino acid production. *Metab. Eng.* **2020**, *58*, 17–34. [CrossRef] [PubMed]
3. Liu, H.; Wang, Y.; Hou, Y.; Li, Z. Fitness of chassis cells and metabolic pathways for L-cysteine overproduction in *Escherichia coli*. *J. Agric. Food. Chem.* **2020**, *68*, 14928–14937. [CrossRef] [PubMed]
4. Yin, J.; Ren, W.; Yang, G.; Duan, J.; Huang, X.; Fang, R.; Li, C.; Li, T.; Yin, Y.; Hou, Y.; et al. L-Cysteine metabolism and its nutritional implications. *Mol. Nutr. Food. Res.* **2016**, *60*, 134–146. [CrossRef] [PubMed]
5. Kishino, K.; Kondoh, M.; Takagi, H. Enhanced L-cysteine production by overexpressing potential L-cysteine exporter genes in an L-cysteine-producing recombinant strain of Corynebacterium glutamicum. *Biosci. Biotechnol. Biochem.* **2019**, *83*, 2390–2393. [CrossRef]
6. Kawano, Y.; Ohtsu, I.; Takumi, K.; Tamakoshi, A.; Nonaka, G.; Funahashi, E.; Ihara, M.; Takagi, H. Enhancement of L-cysteine production by disruption of *yciW* in *Escherichia coli*. *J. Biosci. Bioeng.* **2015**, *119*, 176–179. [CrossRef] [PubMed]
7. Wei, L.; Wang, H.; Xu, N.; Zhou, W.; Ju, J.; Liu, J.; Ma, Y. Metabolic engineering of *Corynebacterium glutamicum* for L-cysteine production. *Appl. Microbiol. Biotechnol.* **2019**, *103*, 1325–1338. [CrossRef] [PubMed]
8. Liu, H.; Fang, G.; Wu, H.; Li, Z.; Ye, Q. L-cysteine production in *Escherichia coli* based on rational metabolic engineering and modular strategy. *Biotechnol. J.* **2018**, *13*, 1–6. [CrossRef]
9. Kawano, Y.; Onishi, F. Improved fermentative L-cysteine overproduction by enhancing a newly identified thiosulfate assimilation pathway in *Escherichia coli*. *Appl. Microbiol. Biotechnol.* **2017**, *101*, 6879–6889. [CrossRef] [PubMed]
10. Gottlieb, K.; Albermann, C.; Sprenger, G.A. Improvement of L-phenylalanine production from glycerol by recombinant *Escherichia coli* strains: The role of extra copies of *glpK*, *glpX*, and *tktA* genes. *Microb. Cell. Fact.* **2014**, *13*, 16. [CrossRef]
11. Nguyen-Vo, T.P.; Liang, Y.; Sankaranarayanan, M.; Seol, E.; Chun, A.Y.; Ashok, S.; Chauhan, A.S.; Kim, J.R.; Park, S. Development of 3-hydroxypropionic-acid-tolerant strain of *Escherichia coli* W and role of minor global regulator yieP. *Metab. Eng.* **2019**, *53*, 48–58. [CrossRef] [PubMed]
12. Sprenger, G.A. Engineering of microorganisms for the production of chemicals and fuels from renewable resources. *Springer Nat. Verl.* **2017**, *4*, 93–123.
13. Luo, X.; Ge, X.; Cui, S.; Li, Y. Value-added processing of crude glycerol into chemicals and polymers. *Bioresour. Technol.* **2016**, *215*, 144–154. [CrossRef] [PubMed]
14. Trondle, J.; Trachtmann, N.; Sprenger, G.A.; Weuster-Botz, D. Fed-batch production of L-tryptophan from glycerol using recombinant *Escherichia coli*. *Biotechnol. Bioeng.* **2018**, *115*, 2881–2892. [CrossRef]
15. Yang, F.X.; Hanna, M.A.; Sun, R.C. Value-added uses for crude glycerol-a byproduct of biodiesel production. *Biotechnol. Biofuels.* **2012**, *5*, 10. [CrossRef]
16. Li, Q.; Wu, H.; Li, Z.; Ye, Q. Enhanced succinate production from glycerol by engineered *Escherichia coli* strains. *Bioresour. Technol.* **2016**, *218*, 217–223. [CrossRef]
17. Litsanov, B.; Brocker, M.; Bott, M. Glycerol as a substrate for aerobic succinate production in minimal medium with *Corynebacterium glutamicum*. *Microb. Biotechnol.* **2013**, *6*, 189–195. [CrossRef]
18. Mazumdar, S.; Blankschien, M.D.; Clomburg, J.M.; Gonzalez, R. Efficient synthesis of L-lactic acid from glycerol by metabolically engineered *Escherichia coli*. *Microb. Cell. Fact.* **2013**, *12*, 11. [CrossRef]
19. Weiner, M.; Albermann, C.; Gottlieb, K.; Sprenger, G.A.; Weuster-Botz, D. Fed-batch production of L-phenylalanine from glycerol and ammonia with recombinant *Escherichia coli*. *Biochem. Eng. J.* **2014**, *83*, 62–69. [CrossRef]
20. Zhang, X.; Zhang, D.; Zhu, J.F.; Liu, W.; Xu, G.Q.; Zhang, X.M.; Shi, J.S.; Xu, Z.H. High-yield production of L-serine from glycerol by engineered *Escherichia coli*. *J. Ind. Microbiol. Biotechnol.* **2019**, *46*, 883–885. [CrossRef]
21. Rennig, M.; Mundhada, H.; Wordofa, G.G.; Gerngross, D.; Wulff, T.; Worberg, A.; Nielsen, A.T.; Nørholm, M.H.H. Industrializing a bacterial strain for L-serine production through translation initiation optimization. *ACS. Synth. Biol.* **2019**, *8*, 2347–2358. [CrossRef] [PubMed]
22. Zhu, Z.M.; Zhang, J.; Ji, X.M.; Fang, Z.; Wu, Z.M.; Chen, J.; Du, G.C. Evolutionary engineering of industrial microorganisms-strategies and applications. *Appl. Microbiol. Biotechnol.* **2018**, *102*, 4615–4627. [CrossRef] [PubMed]
23. Leavitt, J.M.; Wagner, J.M.; Tu, C.C.; Tong, A.; Liu, Y.Y.; Alper, H.S. Biosensor-enabled directed evolution to improve muconic acid production in *Saccharomyces cerevisiae*. *Biotechnol. J.* **2017**, *12*, 9. [CrossRef]
24. Sandberg, T.E.; Salazar, M.J.; Weng, L.L.; Palsson, B.O.; Feist, A.M. The emergence of adaptive laboratory evolution as an efficient tool for biological discovery and industrial biotechnology. *Metab. Eng.* **2019**, *56*, 1–16. [CrossRef]
25. Della, C.D.; Van, B.H.L.; Syberg, F.; Schallmey, M.; Tobola, F.; Cormann, K.U.; Schlicker, C.; Baumann, P.T.; Krumbach, K.; Sokolowsky, S.; et al. Engineering and application of a biosensor with focused ligand specificity. *Nat. Commun.* **2020**, *11*, 4851. [CrossRef] [PubMed]
26. Binder, S.; Schendzielorz, G.; Stabler, N.; Krumbach, K.; Hoffmann, K.; Bott, M.; Eggeling, L. A high-throughput approach to identify genomic variants of bacterial metabolite producers at the single-cell level. *Genome Biol.* **2012**, *13*, 12. [CrossRef]
27. Chen, W.; Zhang, S.; Jiang, P.; Yao, J.; He, Y.; Chen, L.; Gui, X.; Dong, Z.; Tang, S.Y. Design of an ectoine-responsive *AraC* mutant and its application in metabolic engineering of ectoine biosynthesis. *Metab. Eng.* **2015**, *30*, 149–155. [CrossRef]

28. Cress, B.F.; Trantas, E.A.; Ververidis, F.; Linhardt, R.J.; Koffas, M.A.G. Sensitive cells: Enabling tools for static and dynamic control of microbial metabolic pathways. *Curr. Opin. Biotechnol.* **2015**, *36*, 205–214. [CrossRef]

29. Liu, Y.; Li, Q.; Zheng, P.; Zhang, Z.; Liu, Y.; Sun, C.; Cao, G.; Zhou, W.; Wang, X.; Zhang, D.; et al. Developing a high-throughput screening method for threonine overproduction based on an artificial promoter. *Microb. Cell. Fact.* **2015**, *14*, 11. [CrossRef]

30. Schulte, J.; Baumgart, M.; Bott, M. Development of a single-cell GlxR-based cAMP biosensor for *Corynebacterium glutamicum*. *J. Biotechnol.* **2017**, *258*, 33–40. [CrossRef]

31. Wang, Q.Z.; Tang, S.Y.; Yang, S. Genetic biosensors for small-molecule products: Design and applications in high-throughput screening. *Front. Chem. Sci. Eng.* **2017**, *11*, 15–26. [CrossRef]

32. Zhang, X.; Zhang, X.M.; Xu, G.Q.; Shi, J.S.; Xu, Z.H. Integration of ARTP mutagenesis with biosensor-mediated high-throughput screening to improve L-serine yield in *Corynebacterium glutamicum*. *Appl. Microbiol. Biotechnol.* **2018**, *102*, 5939–5951. [CrossRef]

33. Martinez, A.; Grabar, T.B.; Shanmugam, K.T.; Yomano, L.P.; York, S.W.; Ingram, L.O. Low salt medium for lactate and ethanol production by recombinant *Escherichia coli*. B. *Biotechnol. Lett.* **2007**, *29*, 397–404. [CrossRef] [PubMed]

34. Feist, A.M.; Zielinski, D.C.; Orth, J.D.; Schellenberger, J.; Herrgard, M.J.; Palsson, B.O. Model-driven evaluation of the production potential for growth-coupled products of *Escherichia coli*. *Metab. Eng.* **2010**, *12*, 173–186. [CrossRef] [PubMed]

35. Chu, B.C.; Otten, R.; Krewulak, K.D.; Mulder, F.A.; Vogel, H.J. The solution structure, binding properties, and dynamics of the bacterial siderophore-binding protein FepB. *J. Biol. Chem.* **2014**, *289*, 29219–29234. [CrossRef]

36. Storek, K.M.; Vij, R.; Sun, D.; Smith, P.A.; Koerber, J.T.; Rutherford, S.T. The *Escherichia coli* β-barrel assembly machinery is sensitized to perturbations under high membrane fluidity. *J. Bacteriol.* **2018**, *201*, e00517-18. [CrossRef] [PubMed]

37. Doyle, M.T.; Bernstein, H.D. BamA forms a translocation channel for polypeptide export across the bacterial outer membrane. *Mol. Cell.* **2021**, *81*, 2000–2012. [CrossRef] [PubMed]

38. Lee, D.C.; Cottrill, M.A.; Forsberg, C.W.; Jia, Z. Functional insights revealed by the crystal structures of *Escherichia coli* glucose-1-phosphatase. *J. Biol. Chem.* **2003**, *278*, 31412–31418. [CrossRef] [PubMed]

39. Dutta, S.; Corsi, I.D.; Bier, N.; Koehler, T.M. BrnQ-type branched-chain amino acid transporters Influence *Bacillus anthracis* growth and virulence. *mBio* **2022**, *25*, e03640-21. [CrossRef]

40. Eggeling, L.; Bott, M. A giant market and a powerful metabolism: L-lysine provided by *Corynebacterium glutamicum*. *Appl. Microbiol. Biotechnol.* **2015**, *99*, 3387–3394. [CrossRef]

41. Johnson, A.O.; Gonzalez-Villanueva, M.; Wong, L.; Steinbuchel, A.; Tee, K.L.; Xu, P.; Wong, T.S. Design and application of genetically-encoded malonyl-CoA biosensors for metabolic engineering of microbial cell factories. *Metab. Eng.* **2017**, *44*, 253–264. [CrossRef] [PubMed]

42. Xu, P. Production of chemicals using dynamic control of metabolic fluxes. *Curr. Opin. Biotechnol.* **2018**, *53*, 12–19. [CrossRef] [PubMed]

43. Mahr, R.; Von Boeselager, R.F.; Wiechert, J.; Frunzke, J. Screening of an Escherichia coli promoter library for a phenylalanine biosensor. *Appl. Microbiol. Biotechnol.* **2016**, *100*, 6739–6753. [CrossRef] [PubMed]

44. Kim, N.Y.; Lee, Y.J.; Park, J.W.; Kim, S.N.; Kim, E.Y.; Kim, Y.; Kim, O.B. An *Escherichia coli* FdrA Variant Derived from Syntrophic Coculture with a Methanogen Increases Succinate Production Due to Changes in Allantoin Degradation. *mSphere* **2021**, *6*, e0065421. [CrossRef] [PubMed]

45. Volpon, L.; Lievre, C.; Osborne, M.J.; Gandhi, S.; Iannuzzi, P.; Larocque, R.; Cygler, M.; Gehring, K.; Ekiel, I. The solution structure of YbcJ from *Escherichia coli* reveals a recently discovered *alpha1* motif involved in RNA binding. *J. Bacteriol.* **2003**, *185*, 4204–4210. [CrossRef] [PubMed]

46. Tagliabue, L.; Antoniani, D.; Maciag, A.; Bocci, P.; Raffaelli, N.; Landini, P. The diguanylate cyclase YddV controls production of the exopolysaccharide poly-N-acetylglucosamine (PNAG) through regulation of the PNAG biosynthetic *pgaABCD* operon. *Microbiology* **2010**, *156*, 2901–2911. [CrossRef]

47. Masulis, I.S.; Sukharycheva, N.A.; Kiselev, S.S.; Andreeva, Z.S.; Ozoline, O.N. Between computational predictions and high-throughput transcriptional profiling: In depth expression analysis of the OppB trans-membrane subunit of *Escherichia coli* OppABCDF oligopeptide transporter. *Res. Microbiol.* **2020**, *171*, 55–63. [CrossRef]

48. Minamino, T.; Inoue, Y.; Kinoshita, M.; Namba, K. FliK-Driven Conformational Rearrangements of FlhA and FlhB Are Required for Export Switching of the Flagellar Protein Export Apparatus. *J. Bacteriol.* **2020**, *202*, e00637-19. [CrossRef]

49. Lolkema, J.S. Domain structure and pore loops in the 2-hydroxycarboxylate transporter family. *J. Mol. Microbiol. Biotechnol.* **2006**, *11*, 318–325. [CrossRef]

50. Poyner, R.R.; Larsen, T.M.; Wong, S.W.; Reed, G.H. Functional and structural changes due to a serine to alanine mutation in the active-site flap of enolase. *Arch. Biochem. Biophys.* **2002**, *401*, 155–163. [CrossRef]

Article

Application of Ultrafiltration and Ion Exchange Separation Technology for Lysozyme Separation and Extraction

Shanshan Chen, Yaqing Tan, Yaqing Zhu, Liqin Sun, Jian Lin * and Hailing Zhang *

College of Life Sciences, Yantai University, Yantai 264005, China; 18865557198@163.com (S.C.); tanyaqing371122@163.com (Y.T.); zhuyaqing0524@163.com (Y.Z.); sliqin2005@163.com (L.S.)
* Correspondence: linjian3384@163.com (J.L.); hailing1203@hotmail.com (H.Z.); Tel.: +86-0535-6902638 (H.Z.)

Abstract: In this study, the fermentation broth of the recombinant *Pichia pastoris* strain ncy-2 was studied. After pretreatment, separation, and purification, lysozyme was optimized using biofilm and ion exchange separation. Finally, lysozyme dry enzyme powder was prepared by concentrating and vacuum drying. The removal rate of bacterial cells was 99.99% when the fermentation broth was centrifuged at low temperature. The optimum conditions were: transmembrane pressure of 0.20 MPa, pH 6.5, 96.6% yield of lysozyme, enzyme activity of 2612.1 u/mg, which was 1.78 times higher than that of the original enzyme; D152 resin was used for adsorption and elution. Process conditions were optimized: the volume ratio of resin to liquid was 15%; the adsorption time was 4 h; the concentration of NaCl was 1.0 mol/L; the recovery rate of lysozyme activity was 95.67%; the enzyme activity was 3879.6 u/mL; and the purification multiple was 0.5, 3.1 times of the original enzyme activity. The enzyme activity of lysozyme dry enzyme powder was 12,573.6 u/mg, which had an inhibitory effect on microsphere lysozyme. Its enzymatic properties were almost the same as those of natural lysozyme, which demonstrated good application prospects and production potential.

Keywords: lysozyme; biofilm; ion exchange resin; separation; purification

Citation: Chen, S.; Tan, Y.; Zhu, Y.; Sun, L.; Lin, J.; Zhang, H. Application of Ultrafiltration and Ion Exchange Separation Technology for Lysozyme Separation and Extraction. *Fermentation* **2022**, *8*, 297. https://doi.org/10.3390/fermentation8070297

Academic Editors: Xian Zhang and Zhiming Rao

Received: 30 May 2022
Accepted: 17 June 2022
Published: 24 June 2022

Publisher's Note: MDPI stays neutral with regard to jurisdictional claims in published maps and institutional affiliations.

1. Introduction

Lysozyme is a kind of natural lyase, which can specifically hydrolyze the peptidoglycan structure of the cell wall of many microorganisms, especially Gram-positive bacteria. Its antibacterial protection mechanism is significant. It has been a long development process since Nicolle first isolated the dissolving factor from *Bacillus subtilis*, and the World Health Organization (WHO) and many countries and regions determined that lysozyme can be used as a non-toxic and safe additive. Nowadays, lysozyme is widely used in medicine [1], food, scientific research, and other fields [2–4], especially in a variety of industries. Therefore, the production and purification of lysozyme has become very important. Lysozyme is an alkaline protein with stable chemical properties. It can maintain its original structure and activity under a wide temperature and pH range. It has a high isoelectric point and is mostly positively charged. It is a high molecular weight compound with a variety of dissociable multivalent amphoteric electrolytes. It has different amounts of positive or negative charges at different pH values. In acidic and near neutral environments, there are many positive charges. Most other proteins are acidic proteins. When they coexist with other proteins, the interaction between molecules is easy to combine into a certain macromolecule, so it is difficult to separate lysozyme directly.

Lysozyme, because of its non-specificity, can be used as a cellular immune protein in various organisms, such as birds, mammals, and bacteria. Among them, the content inegg white is particularly rich. The traditional process usually adopts the combination of ultrafiltration and chromatography to remove the force of enzyme molecules and other proteins, so as to achieve the purpose of separating lysozyme. The common way to obtain lysozyme is to use egg white as raw material. However, the process of extracting lysozyme from egg white is limited by the source of egg white. Because egg white has only about 3.5% lysozyme, extraction has high production cost, it is a complicated operation, and it has very limited output and low profit, which is not conducive to amplification, and it is difficult to realize large-scale industrial production. Therefore, it is necessary to produce substitutes and reduce production costs to solve the problem of reducing enzyme production. Microbial fermentation with genetically engineered bacteria is an effective way to solve the above problems. However, it is an important challenge to separate lysozyme from the fermentation broth, because the metabolism of microorganisms involves multiple integrated processes, so that in addition to lysozyme, there are many complex components in the fermentation broth: water, residual substrates, by-products, and macromolecules (such as protein and polysaccharide). The first step in the treatment of fermentation broth is to remove various insoluble impurities such as microbial cells and residual substrates, which can be solved by centrifugal filtration. Then, it is necessary to remove the impurity protein and other macromolecular substances in the clear liquid as much as possible. Ultrafiltration can be used for further treatment. Ultrafiltration membrane technology can push water and small molecular substances through the membrane and release them into the permeate according to the molecular weight of the target product and the pressure difference between the two sides of the membrane; lysozyme is intercepted during this process. It is an excellent fermentation liquid purification technology. Important aspects include obtaining clearer permeate; less energy demand; recyclable material; simple and efficient operation; improved purity; and assuredlysozyme activity [5,6]. However, the viscous substances in the feed liquid are easy to adsorb, block the membrane pores, and form a filter cake layer, resulting in concentration polarization on the surface of the ultrafiltration membrane, increasing the resistance and affecting the transmittance, rejection, and membrane flux. Therefore, adjusting the ionic strength in the feed liquid, weakening the force between molecules, selecting the appropriate ultrafiltration membrane pressure, and improving the membrane flux are of great significance to improving the purity of the feed liquid. Lysozyme cannot be completely separated from other impurities in the feed solution by using biofilm, and it needs further refining and purification. Ion exchange technology can complete the refining and purification process, select the appropriate resin as the filler, and use the difference of binding force between lysozyme and exchange groups in the resin to complete the purification of lysozyme in the process of adsorption, binding, and elution. There are many factors affecting the adsorption–desorption process, such as resin type, time, eluent concentration and dosage, etc.,and optimizing the operating conditions of ion exchange is particularly important for the commercial production of lysozyme [7]. The whole process is easy to operate, has low cost, has no need to add any chemical reagent, and is safe and efficient. In particular, ultrafiltration technology has mild conditions; does not cause changes in temperature and pH; can prevent denaturation, inactivation, and autolysis of lysozyme molecules; ensures the activity of lysozyme to the greatest extent; and provides good contact conditions for ion exchange. The macroporous resin has many and large pores, large surface area, many active centers, fast diffusion speed, short distance, high efficiency, short processing time, easy adsorption and exchange, strong pollution resistance, and a stable structure, and is renewable and recyclable [8]. The purpose of this study is to combine ultrafiltration membrane and ion exchange technology, improve the extraction process, seek the best process parameters, ensure the enzyme activity to the greatest extent, extract lysozyme step by step from the fermentation broth, purify and refine lysozyme, and lay a foundation for microbial fermentation to produce lysozyme and realize industrial production.

2. Materials and Methods

2.1. Materials and Reagents

Feed liquid: fermentation broth liquid of recombinant *Pichia pastoris* strain ncy-2. Resin: D152 ion exchange resin, Beijing Solabao Technology Co., Ltd., Beijing, China.

Micrococcus dissolving wall: Institute of Microbiology, Chinese Academy of Sciences. Lysozyme Detection Kit: Nanjing Jiancheng Bioengineering Institute, Nanjing, China.

2.2. Instruments and Equipment

Biofilm: 30 kDapolyethersulfone(PES)polysulfone ultrafiltration membrane, Shanghai Mosu Scientific Equipment Co., Ltd., Shanghai, China. Membrane separation equipment: ro-nf-vf-40, Shanghai Mosu Scientific Equipment Co., Ltd., Shanghai, China. Ultraviolet visible spectrophotometer: uv-5100, Shanghai Yuanxie Instrument Co., Ltd., Shanghai, China. Mechanical stirring bioreactor: 10jsa, Shanghai Baoxing Biological Equipment Engineering Co., Ltd., Shanghai, China. High-speed refrigerated centrifuge: 20 pr-52 d, Hitachi Co., Ltd., Tokyo, Japan. Gradient mixer: th-500, Shanghai Huxi Analytical Instrument Factory Co., Ltd., Shanghai, China. CNC drip automatic part collector: sbs-100, Shanghai Huxi Analytical Instrument Factory Co., Ltd., Shanghai, China. Timing digital display constant flow pump: hl-2d, Shanghai Huxi Analytical Instrument Factory Co., Ltd., Shanghai, China.

2.3. Methods

2.3.1. Preparation of Feed Solution

A recombinant Pichia pastoris strain ncy-2 producing lysozyme was used for continuous fermentation in a 10 L bioreactor for 120 h to obtain yeast fermentation broth. The solid–liquid separation was carried out at 5000 r/min and 4 °C to remove yeast cells and various insoluble impurities in the fermentation broth. The removal rate of the cells was calculated using the blood cell counting plate method.

2.3.2. Preliminary Separation

In addition to secreting the target product, strain ncy-2 could also secrete other extracellular proteases. The purpose of this study was to isolate and extract lysozyme from the fermentation broth. Considering that the feed solution contains a large number of highly viscous substances such as proteins and sugars, in order to ensure the membrane flux of the biofilm, the PESmembrane with a molecular weight of 30 kDa was selected, and macromolecular substances such as protease greater than 30 kDa were preliminarily removed. The enzyme activities of the intercepted solution and permeate were measured to evaluate the separation effect of the biofilm. The supernatant was ultrafilteredusing the operation equipment shown in Figure 1 [9–11].

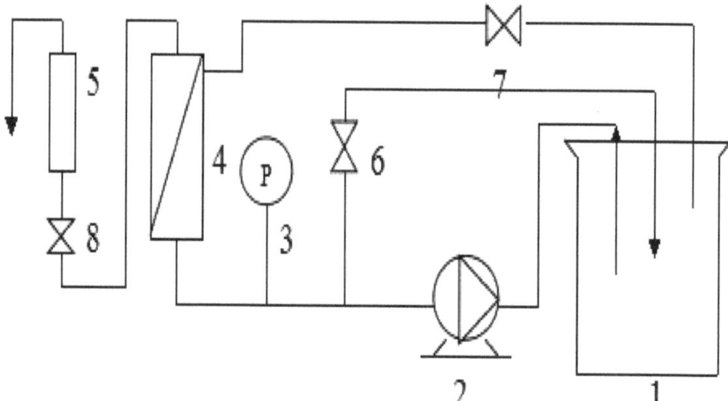

Figure 1. Schematic representation of the ultrafiltration equipment: 1—feed tank; 2—pressure pump; 3—pressure gauge; 4—membrane module; 5—flowmeter; 6—circulating valve; 7—concentration valve; 8—outflow valve; P—Pressure gauge.

Effect of Transmembrane Pressure on Membrane Separation Effect

The pH was fixed at 6.5, and pressure P was changed to 0.10, 0.20, and 0.30 MPa to explore the change of biofilm flux with time and its relationship with lysozyme content. Enzyme activity, membrane flux, rejection rate, transmittance, and yield were also calculated.

Effect of pH on Membrane Separation

Pressure P was set as 0.20 MPa, and pH was varied to 4.5, 6.5, and 11.0 to explore the change in biofilm flux with time and its relationship with lysozyme content. Enzyme activity, membrane flux, rejection rate, transmittance, and yield were also calculated.

2.3.3. Ion Exchange Chromatography

The filtrate obtained from the biofilm was purified using ion exchange chromatography. The purpose of separating and extracting lysozyme was realized using a weak acid group combined with lysozyme and an ion exchange medium. D152 resin was selected to optimize the single factor, design the response orthogonal experiment, construct the model, analyze and verify, and optimize the process parameters.

Single-Factor Experiment

Single-factor experiments were carried out with the ratio of resin dosage to liquid volume, stirring speed, pH, temperature, treatment time, and NaCl concentration as the investigation factors [12,13].

1. Ratio of resin dosage to liquid volume

The amounts of resin, fixed speed, temperature, and other conditions for ion exchange chromatography were varied, and the content of lysozyme in the eluent was determined. The recovery rate of enzyme activity was calculated, and the influence of the ratio of resin amount to solid–liquid volume on the extraction was explored.

2. Stirring speed

The stirring speed, fixed resin dosage, temperature, and other conditions for ion exchange chromatography were used to determine the lysozyme content in the eluent, calculate the enzyme activity recovery, and explore the influence of stirring speed on the extraction.

3. pH

The pH value was changed; the stirring speed, temperature, and other conditions were fixed for ion exchange chromatography; the lysozyme content in the eluent was determined; the recovery of enzyme activity was calculated; and the effect of pH on the extraction was explored.

4. Temperature

The temperature, fixed resin dosage, stirring speed, and other conditions for ion exchange chromatography were varied, and the content of lysozyme in the eluent was determined. The enzyme activity recovery was calculated, and the influence of temperature on the extraction was explored.

5. Processing time

Ion exchange chromatography was carried out under the conditions of fixed resin dosage, stirring speed, and temperature. Samples were procured every 20 min to determine the change in lysozyme content, calculate the recovery rate of enzyme activity, and explore the influence of treatment time on the extraction.

6. NaCl concentration

Ion exchange chromatography was carried out under the conditions of fixed resin dosage, stirring speed, and temperature. The concentration of the NaCl solution was varied for elution. The change in lysozyme content in the eluent was measured, the recovery of enzyme activity was calculated, and the effect of eluent concentration on the lysozyme extraction was explored.

Orthogonal Experiment

Based on the single-factor experiment and referring to the response surface optimization method [14], a three-factor and three-level response surface analysis experiment [15–17] was designed with the enzyme activity recovery of lysozyme as the response value, and a (resin dosage in volume ratio of feed solution) (%), B (adsorption time) (h), and C (NaCl concentration) (mol/L) as variables, as shown in Table 1.

Table 1. Level of response surface experimental factors.

Level	Factor		
	A (Ratio of Resin Dosage to Liquid Volume) in %	B (Adsorption Time) in h	C (NaCl Concentration) in mol/L
−1	15	4	0.5
0	20	6	1.0
1	25	8	1.5

2.3.4. Preparation of Dry Enzyme Powder

Using the eluent obtained via ion exchange chromatography as a raw material, lysozyme concentrate was obtained by treatment with a polysulfone membrane, with a molecular weight of 5 kDa, and vacuum drying at -30 °C to prepare lysozyme dry enzyme powder.

Enzymatic Properties

Using dry enzyme powder as the research object, the changes in lysozyme activity under different temperatures, pH, metal ions, and surfactants were measured [18–20].

1.   Effect of temperature on enzyme activity and thermal stability

Lysozyme solution (pH 6.2) was prepared with 0.01 mol/L phosphate buffer, the temperature was changed, samples were procured every 30 min, the lysozyme activity was measured, and the time change curve of lysozyme at different temperatures was drawn.

2.   Effect of pH on enzyme activity and pH stability

The pH value of the phosphate buffer was adjusted, the lysozyme activity was determined, and the relationship curve between the pH value and lysozyme activity was drawn.

3.   Effect of metal ions on enzyme activity

$FeSO_4$, NaCl, KCl, $CaCl_2$, $MgSO_4$, $CuSO_4$, $ZnSO_4$, $MnSO_4$, and $FeCl_2$ at a concentration of 0.01 mol/L each were used to prepare a lysozyme solution such that the final concentration of each metal ion was 5 mmol/L. Lysozyme activity was determined, and the effects of different metal ions on lysozyme activity were explored.

4.   Effect of surfactants on enzyme activity

Glycerol, Tween 20, Tween 80, and Span 80 were added to the lysozyme solution such that the final concentration of each surfactant was 0.5 mg/mL. Lysozyme activity was measured to explore the effects of different surfactants on lysozyme activity.

Bacteriostatic Test

Lysozyme microspheres were activated to ensure that the strain recovered its original activity. Using lysozyme dry enzyme powder as a sample, the Oxford cup bacteriostatic experiment was carried out, the size of the bacteriostatic circle was measured, and the bacteriostatic effect of lysozyme and natural lysozyme on lysozyme microspheres was explored.

2.4. Analysis Method

2.4.1. Calculation of Enzyme Activity

The lysozyme content was determined according to the kit operation manual [21], that is, the enzyme activity.

$$U = \frac{UT_{15} - OT_{15}}{ST_{15} - OT_{15}} \times 200U \cdot mL^{-1} \times N,$$

where U is in u/mL, $UT_{15}/ST_{15}/OT_{15}$ is the transmittance of the liquid to be tested/standard/blank; 200 u/mL is the standard concentration (2.5 µg/mL); and N is the dilution multiple of the solution to be tested.

$$U = \frac{U_1}{U_0} \times 100\%,$$

where U is in %; U_1 is enzyme activity after treatment in u/mg; and U_0 is primary enzyme activity in u/mg.

$$U = \frac{m_1 - m_2}{V},$$

where U is the enzyme content (mg/mL), m_1 is the mass of the sample before vacuum drying (g), m_2 is the mass of dry mass and constant weight of the sample (g), and V is the total sample volume (mL).

2.4.2. Evaluation Parameters of Membrane Performance

Membrane flux [22] is the volume of permeate per unit membrane area per unit time. The experimental operation parameters were changed, samples were taken after stable operation, and the volume of the permeate was recorded within a certain time. The formula used for calculation is:

$$J_w = \frac{V}{S_m t},$$

where J_w is the membrane flux, $L/(m^2 \cdot h)$; V is the total volume of permeate, L; S_m is the effective area of the membrane, m^2; and t is the filtering time, h.

Retention rate (the retention capacity of membrane to solute) is expressed as decimal or percentage. Concentration polarization exists in the actual membrane separation process, and the real rejection rate is:

$$R_0 = 1 - \frac{C_p}{C_m},$$

Because it is difficult to determine the polarization concentration C_m, the volume concentration of the feed solution was used to replace the polarization concentration, and the apparent rejection rate R was used to replace the rejection rate R_0.

$$R = 1 - \frac{C_p}{C_b},$$

where R_0 is the rejection rate, R is the apparent rejection rate, and C_p, C_b, and cm are the permeate, feed solution, and membrane surface concentrations, respectively (mol/L).

Concentration multiple CF and yield REC [23] are determined as follows:

$$CF = \left(\frac{C_0}{C}\right)^R,$$

$$REC = \left(\frac{V_0}{V}\right)^{R-1},$$

where C_0 and C are the concentration of the feed solution and concentrated solution, respectively (mol/L), and V_0 and V are the volume of the feed solution and concentrated solution, respectively.

3. Results and Discussion

3.1. Preparation of Feed Solution

The cell concentration in the fermentation broth was 4.675×10^9 cells/mL, and after solid–liquid separation, it decreased to 8.25×10^5 cells/mL, and the removal rate of bacterial cells reached 99.98%. All kinds of insoluble impurities such as bacterial cells and fermentation residues in the fermentation broth were preliminarily removed.

3.2. Preliminary Separation

3.2.1. Effect of Transmembrane Pressure on Membrane Separation Effect

Effect of Different Pressures on the Extraction of Microbial Enzymes

As shown in Figure 2, the transmembrane pressure had a significant impact on the membrane flux and lysozyme activity. When the transmembrane pressure was 0.20 MPa, the maximum membrane flux was 38.2 $L \cdot (M^2 \cdot h)^{-1}$, and the enzyme activity was the highest. In the early stage, owing to the increase in transmembrane pressure, the shear force on the membrane surface increased; the feed liquid velocity accelerated, showing a turbulent state;the concentration polarization phenomenon on the membrane surface decreased, resulting in an increase in membrane flux; and agel layer was formed on the surface of the membrane. The higher the pressure, the higher the density and thickness of the gel layer, and the worse the polarization phenomenon; the larger the resistance, the lower the flow rate of the liquid material and the longer the time, and the lower the membrane flux, resulting in aloss of and decrease in enzyme activity. However, excessive pressure leads to serious membrane pollution, difficult membrane cleaning, and increased energy consumption, and it affects the activity of lysozyme. Therefore, it is very important to select the appropriate pressure.

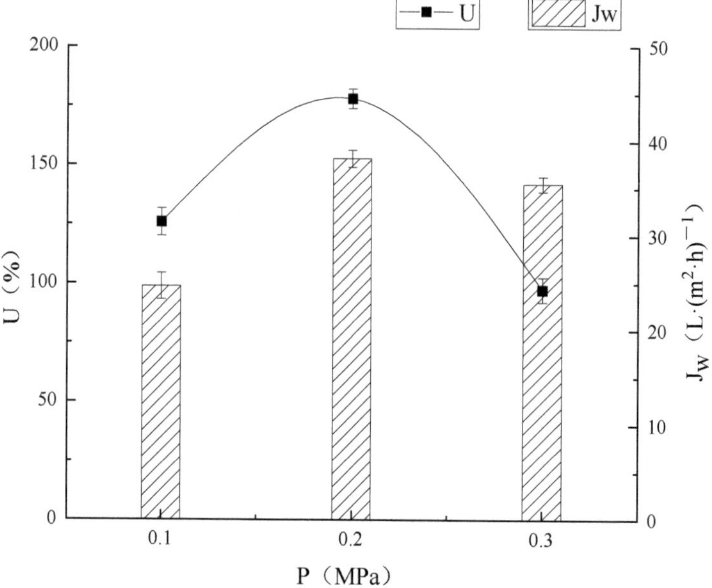

Figure 2. Effect of transmembrane pressure on ultrafiltration separation effect.

Effect of Different Pressures on Membrane Flux

As shown in Figure 3, the membrane flux increased with an increase in pressure. When the pressure rose to 0.30 MPa, the membrane flux was lower than 0.20 MPa, which was due to the concentration polarization on the membrane surface. Under the same pressure, the flux trend of the membrane increased rapidly and then decreased slowly. The higher the initial pressure, the higher the membrane flux. The gel layer formed at a later stage of the membrane. As the resistance of the ultrafiltration membrane increased, the concentration polarization phenomenon on the surface of the membrane increased, and the flux of the membrane decreased.

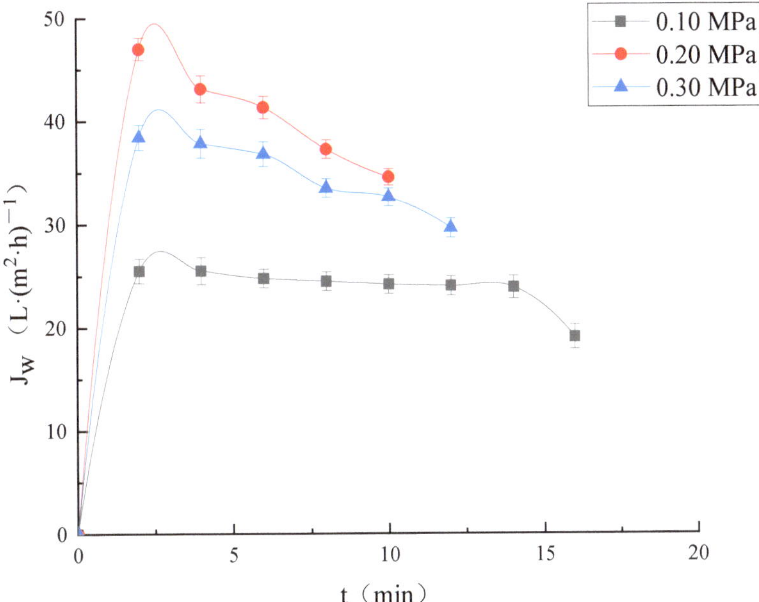

Figure 3. Variation curve of membrane flux with time under different pressures.

Effect of Different Pressures on Ultrafiltration

As shown in Table 2, different operating pressures had different effects on the yield of lysozyme, especially the rejection and transmittance. Owing to the difference in trans-membrane pressure, the thickness and density of the gel layer were different, and the resistance was also different, thus affecting the rejection rate and transmittance. At the pressure of 0.20 MPa, the minimum interception rate of lysozyme was 7.4%. The maximum transmittance was 45.7% and the yield was 96.6%.

Table 2. Effect of transmembrane pressure on ultrafiltration effect.

P (MPa)	R (%)	R' (%)	REC (%)
0.10	11.2	37.4	93.6
0.20	7.4	45.7	96.6
0.30	15.85	37.7	91.9

P—pressure R—rejection rate; R'—transmittance; REC—yield.

3.2.2. Effect of pH on Membrane Separation Effect

Effect of Different pH Values on the Extraction of Lysozyme

As shown in Figure 4, the pH of the feed solution had a significant influence on the membrane separation. When the membrane flux and lysozyme activity in the permeate at pH 4.5 and 11.0 were lower than that at pH 6.5, the maximum membrane flux was 38.2 L·(M²·h)⁻¹, and the lysozyme activity was the highest. Under the condition of pH 4.5, almost no charge, aggregation occurred easily, forming a gel layer on the surface of the membrane and increasing resistance, resulting in a decrease in membrane flux. When the pH value was 6.5, it was negatively charged, and there was an electrostatic repulsion between the molecules. Concurrently, other hetero-proteins were also be negatively charged, which weakened the concentration polarization phenomenon to a certain extent and ensured high membrane flux. In addition, the isoelectric point of lysozyme was 11, without charge, and the repulsive force between molecules was the smallest. It was easy to form a gel layer by molecular flocculation, and the concentration polarization phenomenon was aggravated, and the membrane flux decreased.

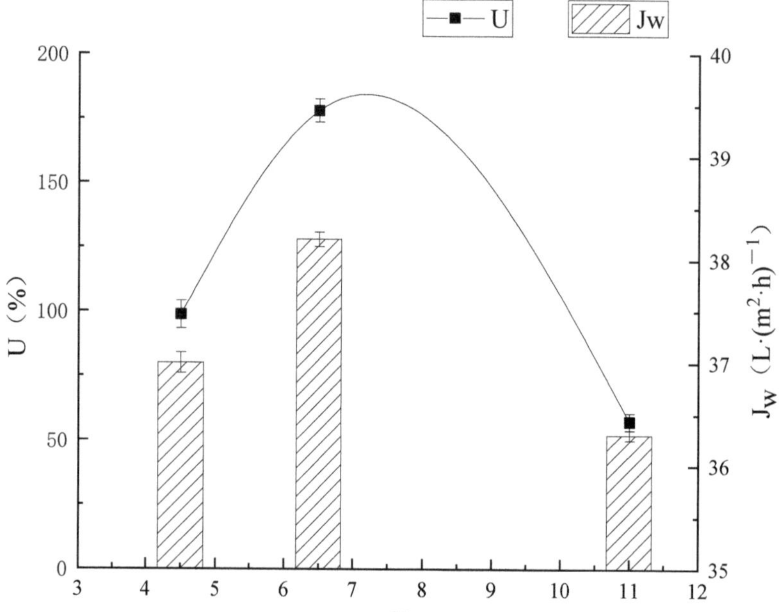

Figure 4. Effect of pH on ultrafiltration separation.

Effect of Different pH on Ultrafiltration Membrane Flux

As shown in Figure 5, the membrane flux decreased with time. At pH 6.5, lysozyme was negatively charged and was relatively stable between molecules, ensuring a high membrane flux. At pH 4.5, molecular aggregation easily occurred and was attached to the membrane surface, resulting in a decrease in the membrane flux. The charge of lysozyme at the isoelectric point was 0, the repulsion between molecules was the smallest, intermolecular flocculation occurred easily, and the membrane flux was reduced.

Effect of Different pH Values on Ultrafiltration

As shown in Table 3, pH had a significant influence on the interception rate of lysozyme. When the pH value was 6.5, the rejection and transmittances were 7.4% and 45.7%, respectively, and the yield of lysozyme was 96.6%. This was due to the negative charge of lysozyme and the electrostatic interaction between enzyme molecules, resulting in concentration polarization on the membrane surface, which affected the membrane separation effect.

Table 3. Effect of pH on ultrafiltration.

pH	R (%)	R′ (%)	REC (%)
4.5	16.2	48.5	92.1
6.5	7.4	45.7	96.6
11	15.3	47.7	93

R—rejection rate; R′—transmittance; REC—yield.

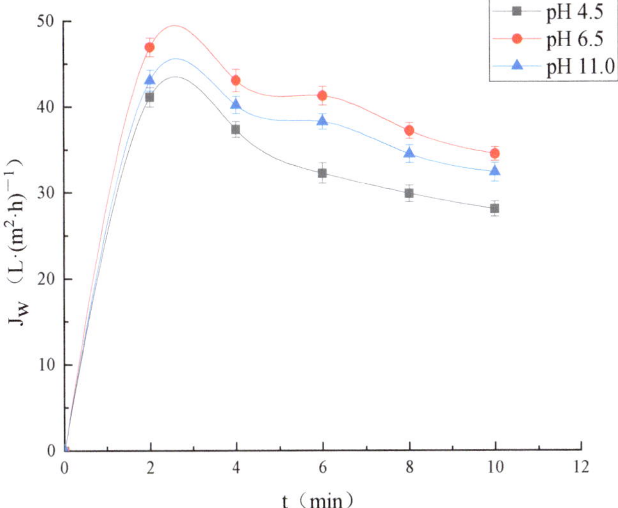

Figure 5. Variation curve of membrane flux with time at different pH values.

3.3. Ion Exchange Chromatography

3.3.1. Single Factor Experiment

Ratio of Resin Dosage to Liquid Volume

As shown in Figure 6, the ability of the resin to adsorb and desorb enzyme molecules first increased and then decreased. In the adsorption stage, when the amount of resin reached 20%, the growth of resin adsorption capacity slowed down; when it exceeded 25%, it did not increase, but decreased, indicating that the resin had been adsorbed and was saturated. In the elution stage, when the amount of resin reached 20%, the elution was complete.

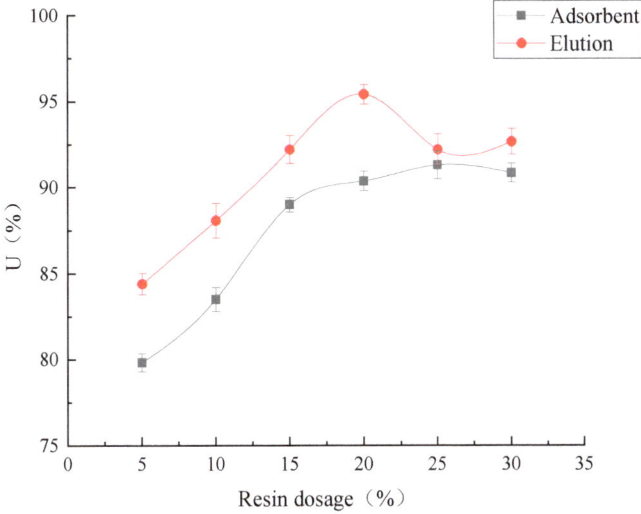

Figure 6. Effect of the ratio of resin dosage to liquid volume on the extraction of lysozyme.

Stirring Speed

As can be seen from Figure 7, the external diffusion of lysozyme increased with an increase in the stirring speed of the feed solution, the resin exchange capacity increased, and the internal diffusion was not affected.

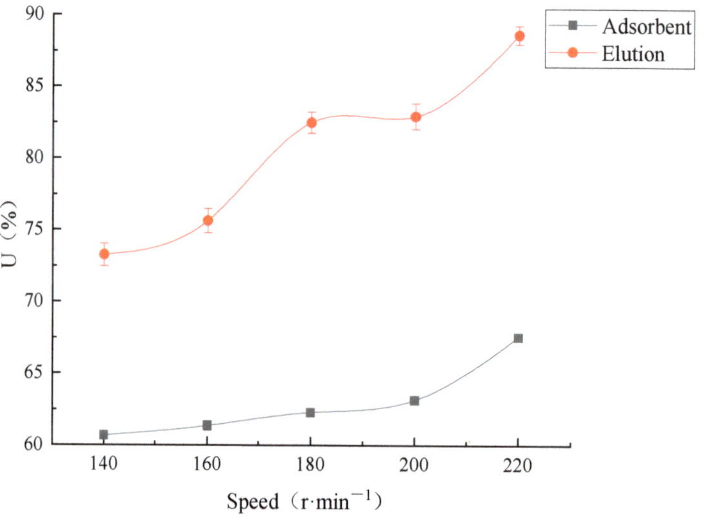

Figure 7. Effect of stirring speed on extraction of lysozyme.

pH

The pH of the feed solution determined the dissociation strength between the resin active group and the exchange ion and affected the exchange capacity of the resin. When the pH changed, the intermolecular forces changed, and the charges differed. As shown in Figure 8, with an increase in the pH of the feed solution, the swelling degree with a high degree of hydration was large, and the resin exchange capacity increased first and then decreased.

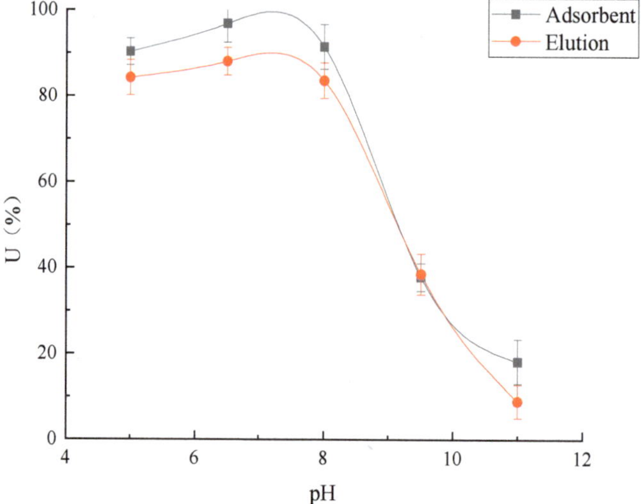

Figure 8. Effect of pH on extraction of lysozyme.

Temperature

As shown in Figure 9, the adsorption saturation of the resin was sensitive to temperature. With an increase in the feed liquid temperature, the diffusion and exchange speed also accelerated, and the enzyme activity decreased when the time was too long at too high a temperature.

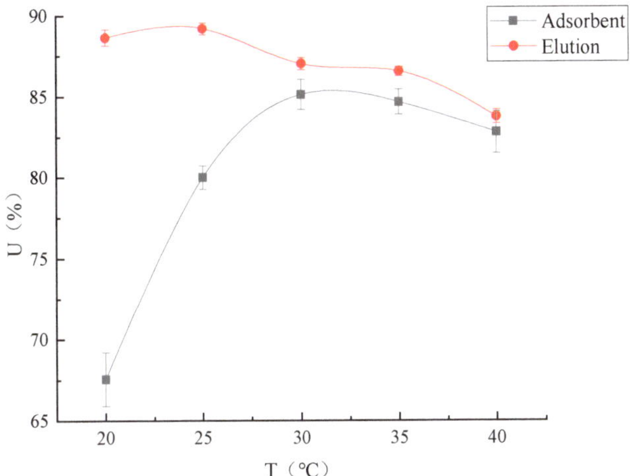

Figure 9. Effect of temperature on extraction of lysozyme.

Processing Time

As shown in Figure 10, after the resin reached saturation, the enzyme activity remained unchanged or even decreased, which was caused by long-time adsorption, partial inactivation of enzymes, or human factors. At the initial stage of analysis, owing to the large concentration difference between the resin and the lysozyme in the eluent, the desorption power was large, and the speed was fast, but the enzyme activity decreased after 80 min of elution and increased after 100 min of elution, which was due to the uneven distribution caused by salting out, resulting in an increase in enzyme activity.

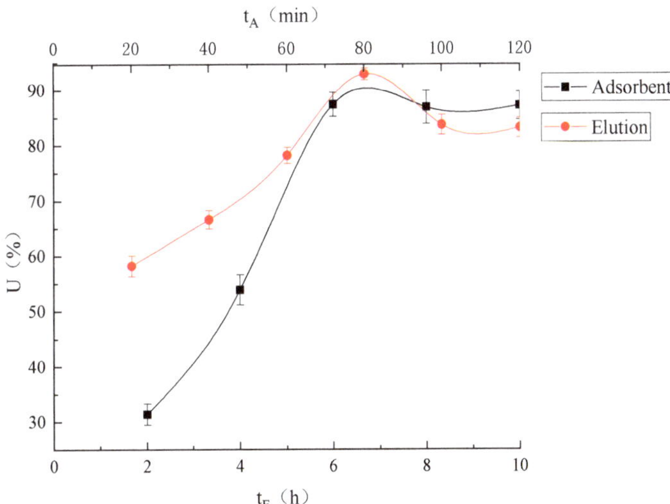

Figure 10. Effect of treatment time on extraction of lysozyme.

NaCl Concentration

As shown in Figure 11, the degree of enzyme desorption was significantly affected by the NaCl concentration. When the NaCl concentration reached 1.0 mol/L, the enzyme activity recovery, reaching the maximum value, and the concentration continued to increase but then decreased. This is because the salt concentration in the feed solution was too high, resulting in the aggregation of lysozyme molecules and affecting desorption.

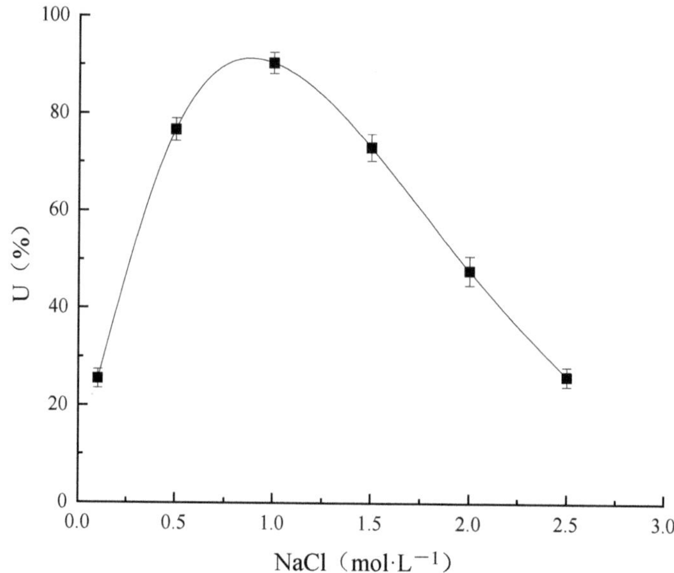

Figure 11. Effect of NaCl concentration on extraction of lysozyme.

3.3.2. Orthogonal Experiment

Establishment of Regression Model

A response surface optimization method was used to design the experimental scheme. The specific results are presented in Tables 4 and 5.

Table 4. Response surface experimental design scheme and results.

Number	A	B	C	U/%
1	0	1	−1	43.432
2	−1	0	−1	66.8512
3	1	1	0	71.8863
4	−1	−1	0	94.5497
5	0	−1	1	77.8369
6	0	0	0	92.4846
7	0	0	0	81.9566
8	0	1	1	67.7667
9	1	0	1	78.9759
10	−1	0	1	71.6521
11	0	0	0	86.7681
12	−1	1	0	76.006
13	0	−1	−1	62.9551
14	1	0	−1	49.2229
15	1	−1	0	89.2804
16	0	0	0	85.6185
17	0	0	0	85.1607

Table 5. Variance analysis of quadratic response surface regression model.

Source	SS	Df	MS	F	p
Model	3195.16	9	355.02	37.15	<0.0001
A	48.48	1	48.48	5.07	0.059
B	536.79	1	536.79	56.17	0.0001
C	680.26	1	680.26	71.18	<0.0001
AB	0.33	1	0.33	0.035	0.8578
AC	155.65	1	155.65	16.29	0.005
BC	22.34	1	22.34	2.34	0.1702
A^2	0.047	1	0.047	4.893×10^{-3}	0.9462
B^2	53.74	1	53.74	5.62	0.0495
C^2	1655.29	1	1655.29	173.19	<0.0001
Residua	66.9	7	9.56		
Lack of Fit	7.85	3	2.62	0.18	0.9065
Pure Error	59.05	4	14.76		
Cor Total	3262.06	16			
R^2	0.9795				
R^2 (adj)	0.9531				
C.V.%	0.041				

A quadratic response surface regression analysis was carried out on the data in Table 4 using Design-Expert software, and amultivariate quadratic regression model of lysozyme activity Y on resin dosage in feed liquid volume ratio a, adsorption time B, and NaCl concentration C was obtained: $Y = 60.132 - 3.32861A + 3.68373B + 112.97956C + 0.028742AB + 2.49521AC + 2.36321BC + 4.21546 \times 10^{-3}A^2 - 0.89312B^2 - 79.31020C^2$ ($p < 0.0001$, $R^2 = 0.9795$, R^2 (adj) = 0.9531 (>0.80), CV = 4.10%.

As shown in Table 5, according to the equation model results, the model significance test value f was 37.15, the significance level $p < 0.0001$, and the model term $p \leq 0.05$, indicating that the established model regression equation was significant and statistically significant; the mismatch term $p > 0.05$, the correlation coefficient $R^2 = 0.9795$, the adjusted correlation coefficient R^2 (adj) = 0.9531 (>0.80), and coefficient of variation CV = 4.10%, indicating that the model had good fit, the proportion of abnormal error between the model and the actual fitting was small, and the mismatch term was not significant. The model equation can be used to preliminarily analyze and predict the process of lysozyme extraction.

Regression Model Analysis

We fixed any factor and analyzed the deformation of the regression equation. The results are as follows.

It can be seen from the response surface and contour map in Table 5 and Figures 12–14 that there was no significant difference in the interaction between the resin dosage volume ratio a and the adsorption time B, the interaction between the resin dosage volume ratio a and the NaCl concentration C was extremely significant, and the interaction between the adsorption time B and the NaCl concentration C was not significant. It can be seen from the contour line that the extraction rate y was less sensitive to the change of adsorption time B than to the change of the ratio of resin dosage to feed liquid volume a, and it was more sensitive to the change of NaCl concentration C than to the change of the ratio of resin dosage to feed liquid volume a. The interaction between adsorption time B and NaCl concentration C was more sensitive than that of the other two groups. Therefore, the sequence of three factors of resin dosage in volume ratio of feed solution a, adsorption time B, and NaCl concentration C affecting the enzyme activity recovery y of lysozyme was C > B> A.

Validation Experiment

The data were analyzed using Design-Expert (StatEase company, Minneapolis, MN, USA) and SPSS (International Business Machines Corporation, Armonk, New York, NY, USA) software, and the comparison results are shown in the table below.

It can be seen from Table 6 that the analysis results of the Design-Expert and SPSS were consistent. The NaCl concentration in the regression analysis of variance table was $p < 0.0001$, the main effect test of SPSS analysis was $p = 0.022$, the multiple comparison results of NaCl concentration were $p = 0.011$ (all < 0.05) when the concentration was 1.0 mol/L, and the most influential factor was NaCl concentration.

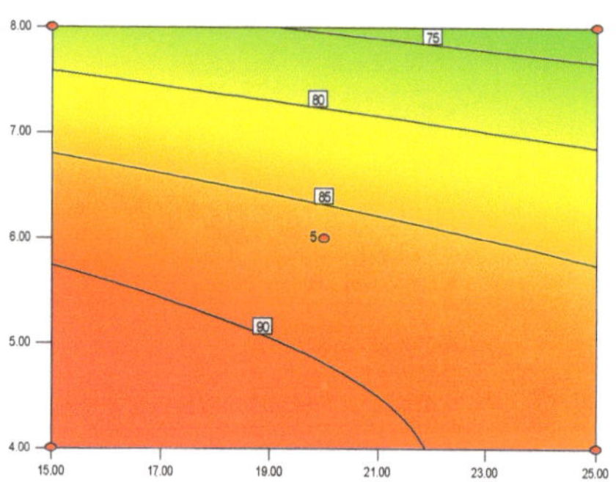
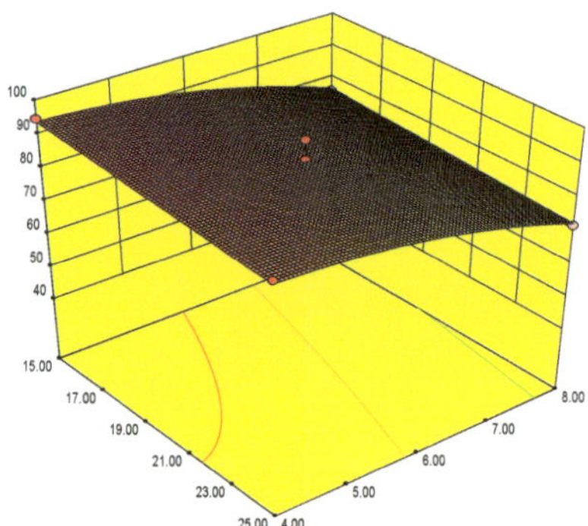

Figure 12. Contour and response surface of resin dosage in volume ratio of material to liquid and adsorption time. The red dots in the figure are the salient points of the contour map and response surface map.

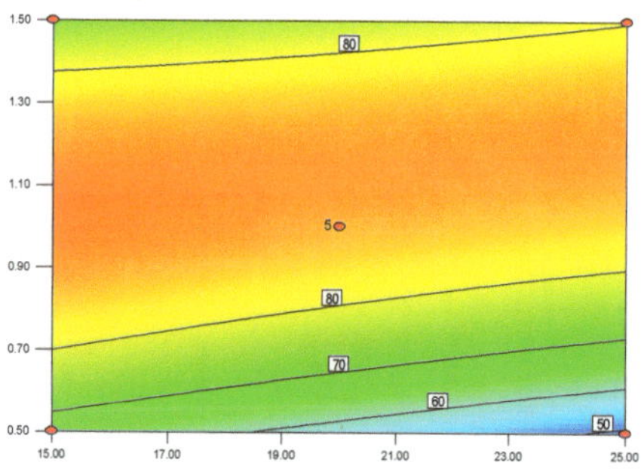
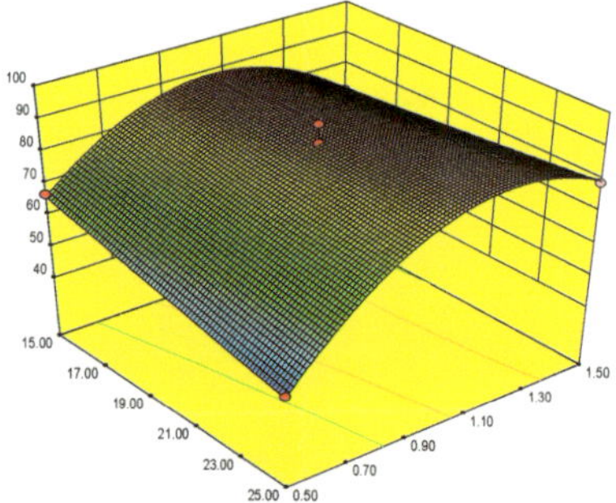

Figure 13. Contour and response surface of resin dosage in volume ratio of material to liquid and NaCl concentration. The red dots in the figure are the salient points of the contour map and response surface map.

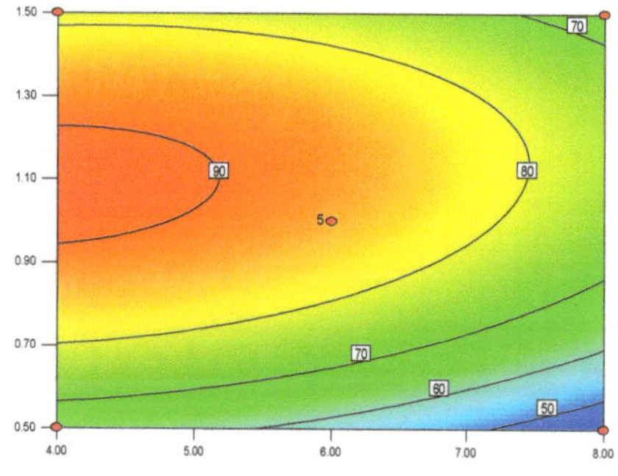
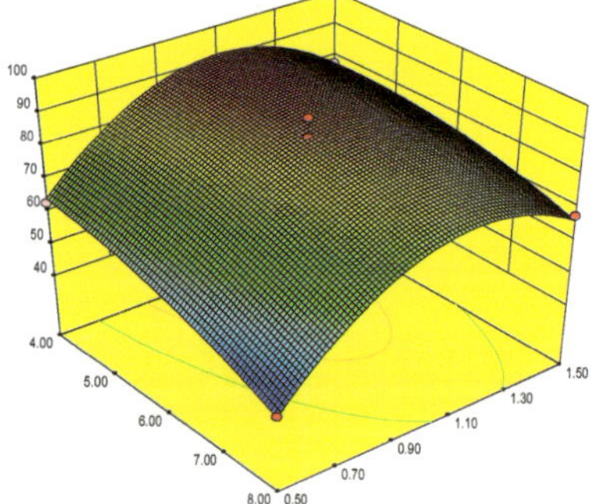

Figure 14. Contour and response surface of adsorption time and NaCl concentration. The red dots in the figure are the salient points of the contour map and response surface map.

Table 6. Comparison of Design-Expert and SPSS software results.

Software	R^2	p	Significant Factor
Design-Expert	0.9795	<0.05	NaCl
SPSS	0.9820	<0.05	NaCl

By solving the regression equation, the optimum theoretical conditions were as follows: the amount of resin accounted for 15% of the volume of the feed solution, the adsorption time was 4.0 h, the concentration of NaCl was 1.01 mol/L, and the recovery of enzyme activity was 93.88%. Considering the actual operation, the amount of resin accounted for 15% of the volume of the feed solution, the adsorption time was 4 h, and the NaCl concentration was 1 mol/L. Under these conditions, the recovery of enzyme activity was 95.67%, and the error was 0.22% (<1%). The process conditions established in this experiment are reliable, and the model fits well with the actual situation.

3.4. Preparation of Dry Enzyme Powder

The prepared lysozyme dry enzyme powder was milky white, and the lysozyme content was 33.8 mg/mL. The enzyme solution was prepared with 0.005 g/mL. The enzyme activity was 12,573.6 u/mg.

3.4.1. Enzymatic Property Test
Effect of Temperature on Enzyme Activity and Thermal Stability

As shown in Figure 15, in the range of 20–50 °C, the activity of lysozyme increased with an increase in temperature, which was relatively stable, and the enzyme activity remained above 50%. The temperature continued to rise, and lysozyme activity decreased. This was due to the denaturation and inactivation caused by the increase in temperature.

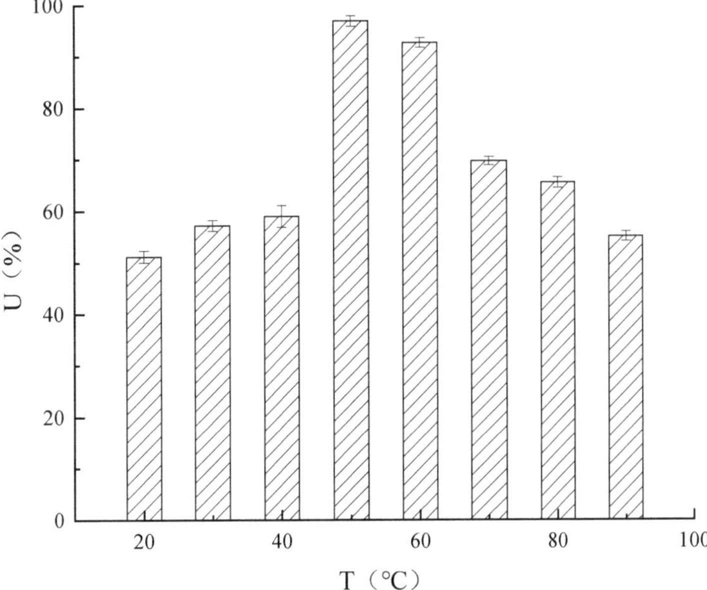

Figure 15. Effect of temperature on enzyme activity.

As shown in Figure 16, temperature had little effect on lysozyme activity in the range of 20–60 °C. Lysozyme had good thermal stability. After holding at 70 °C for 90 min, the enzyme activity decreased to approximately 75%, and the enzyme activity decreased continuously. At 90 °C, lysozyme was extremely unstable, denatured, and inactivated in a short time, and the enzyme activity decreased sharply. When the temperature was lower than 60 °C, lysozyme couldstably maintain its activity.

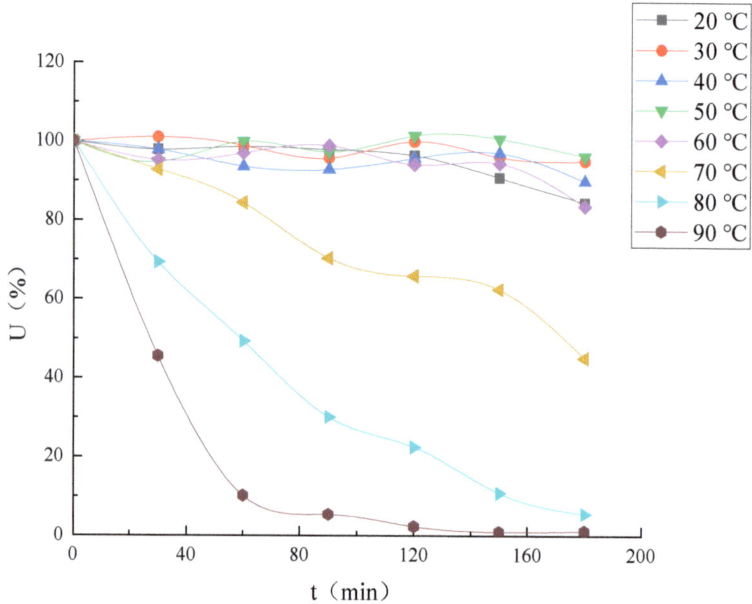

Figure 16. Changes in enzyme activity with time at different temperatures.

Effect of pH on Enzyme Activity and pH Stability

As shown in Figure 17, the activity of lysozyme showed a "peak" shape, which first increased and then decreased with an increase in pH. Under neutral conditions, the activity of lysozyme was the largest and most stable.

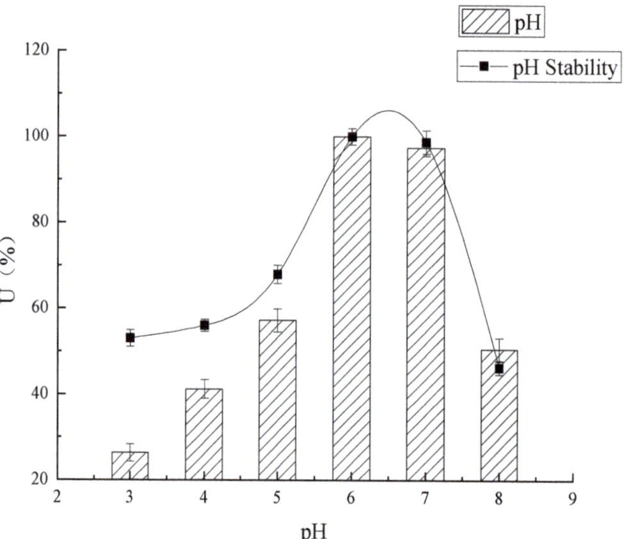

Figure 17. Effect of pH on enzyme activity.

Effect of Metal Ions on Enzyme Activity

As shown in Figure 18, the enzyme activity was inhibited by Fe^{2+}, Fe^{3+}, Zn^{2+}, and Cu^{2+} and activated by Na^+ and Mg^{2+}, whereas Mn^{2+}, K^+, and Ca^{2+} had no significant effect on lysozyme activity. Except for a few metal ions with strong inhibitory effects, most metal ions could pair well with lysozyme and had little effect on enzyme activity.

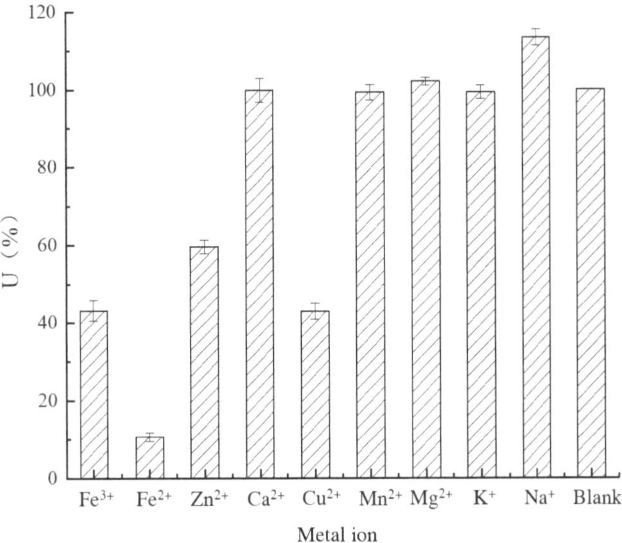

Figure 18. Effect of metal ions on enzyme activity.

Effect of Surfactants on Enzyme Activity

As shown in Figure 19, the overall effects of the four surfactants were similar. Span80 inhibited lysozyme activity, whereas Tween20 and 80 and glycerol activated it.

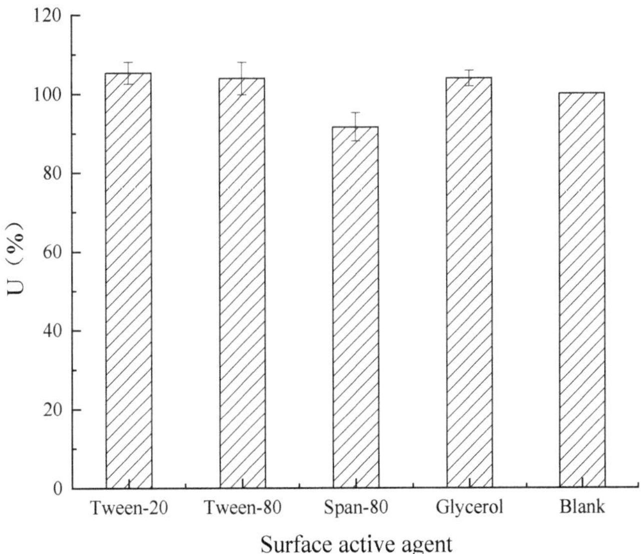

Figure 19. Effect of surfactant on enzyme activity.

3.4.2. Bacteriostatic Test

As shown in Figure 20, lysozyme dry enzyme powder prepared from the fermentation broth of the recombinant yeast strain had the same inhibitory effect on microsphere lysozyme as natural lysozyme.

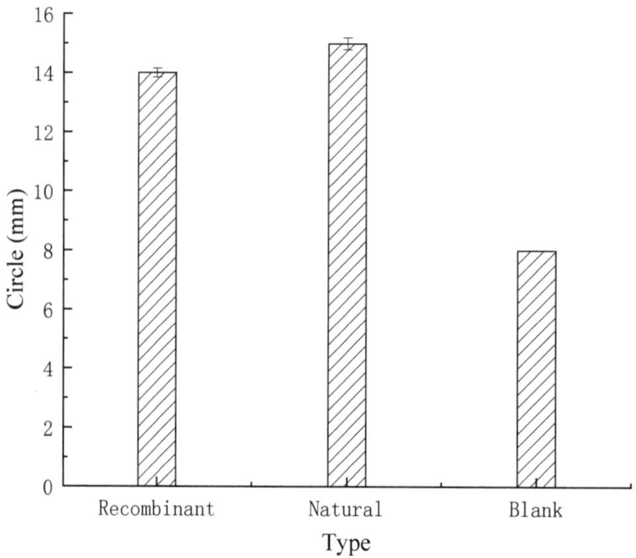

Figure 20. Comparison of bacteriostatic effect.

4. Conclusions

After solid–liquid separation, the removal rate of bacterial cells reached 99.98%. The optimal conditions of biofilm are as follows: transmembrane pressure of 0.20 MPa and feed solution pH of 6.5. The yield of lysozyme was 96.6%, and the enzyme activity was 2612.1 u/mg, which was 1.78 times that of the original enzyme. A regression equation model was established. The optimal process conditions obtained using the model were as follows: The amount of resin accounted for 15% of the volume of material liquid, the adsorption time was 4 h, and the NaCl concentration was 1.0 mol/L. Under these conditions, the recovery of lysozyme was 95.67%, the enzyme activity was 3879.6 u/mL, and the purification multiple was 0.5, which was 3.1 times that of the original enzyme. Lysozyme dry enzyme powder had an inhibitory effect on *Micrococcus* lysozyme, and the enzyme activity was 12,573.6 u/mg. The optimum temperature was 50 °C, and the thermal stability was good, in the range of 20–60 °C. The optimum pH was 6.5. The enzyme activity of lysozyme was inhibited by Fe^{2+}, Fe^{3+}, Zn^{2+}, and Cu^{2+}, while Na^+ and Mg^{2+} activated the enzyme activity, which can also be inhibited by Span 80 and activated by Tween 20, Tween 80, and glycerol. The product obtained by the above process not only has the advantages of natural lysozyme, but also overcomes the disadvantages of low activity, low yield, inconvenient preparation, and inability to be stably preserved. It has a wide application prospect by using microbial fermentation to expand production.

Author Contributions: Conceptualization, H.Z. and S.C.; methodology, Y.T. and Y.Z.; software, S.C.; validation, H.Z. and J.L.; formal analysis, S.C.; investigation, Y.T.; resources, J.L.; data curation, S.C.; writing—original draft preparation, H.Z. and S.C.; writing—review and editing, H.Z.; visualization, J.L.; supervision, H.Z.; project administration, L.S.; funding acquisition, H.Z. and L.S. All authors have read and agreed to the published version of the manuscript.

Funding: This work was supported by the National Natural Science Foundation of China (No. 32001836), the Natural Science Foundation of Shandong Province (ZR201911180224), and the Yantai Science and Technology Development Plan (SK21H266).

Institutional Review Board Statement: Not applicable.

Conflicts of Interest: The authors declare no conflict of interest.

References

1. Li, H.; Wei, X.; Yang, J.; Zhang, R.; Yang, J. The bacteriolytic mechanism of an invertebrate-type lysozyme from mollusk Octopus ocellatus. *Fish Shellfish. Immunol.* **2019**, *93*, 232–239. [CrossRef] [PubMed]
2. Zhe, W.; Hu, S.; Gao, Y.; Chen, Y.; Wang, H. Effect of collagen-lysozyme coating on fresh-salmon fillets preservation. *LWT-Food Sci. Technol.* **2017**, *75*, 59–64. [CrossRef]
3. Satish, L.; Rana, S.; Arakha, M.; Rout, L.; Ekka, B.; Jha, S.; Dash, P.; Sahoo, H. Impact of imidazolium-based ionic liquids on the structure and stability of lysozyme. *Spectrosc. Lett.* **2016**, *49*, 383–390. [CrossRef]
4. Jha, I.; Bisht, M.; Mogha, N.K.; Venkatesu, P. Effect of Imidazolium-Based Ionic Liquids on the Structure and Stability of Stem Bromelain: Concentration and Alkyl Chain Length Effect. *J. Phys. Chem. B Condens. Matter Mater. Surf. Interfaces Biophys.* **2018**, *122*, 7522–7529. [CrossRef]
5. Loske, L.; Nakagawa, K.; Yoshioka, T.; Matsuyama, H. 2D Nanocomposite Membranes: Water Purification and Fouling Mitigation. *Membranes* **2020**, *10*, 295. [CrossRef]
6. Xing, Y.; Ama, C.; Aa, A.; Lfdd, E.; Xz, F.; Sht, G.; Hp, H.; Lk, A. Towards next generation high throughput ion exchange membranes for downstream bioprocessing: A review. *J. Membr. Sci.* **2022**, *647*, 120325. [CrossRef]
7. Tomczak, W.; Gryta, M. Comparison of Polypropylene and Ceramic Microfiltration Membranes Applied for Separation of 1,3-PD Fermentation Broths and Saccharomyces cerevisiae Yeast Suspensions. *Membranes* **2021**, *11*, 44. [CrossRef]
8. Tomczak, W.; Gryta, M. Clarification of 1,3-Propanediol Fermentation Broths by Using a Ceramic Fine UF Membrane. *Membranes* **2020**, *10*, 319. [CrossRef]
9. Ruby-Figueroa, R.; Saavedra, J.; Bahamonde, N.; Cassano, A. Permeate flux prediction in the ultrafiltration of fruit juices by ARIMA models. *J. Membr. Sci.* **2016**, *524*, 108–116. [CrossRef]
10. Chen, C.; Li, X.; Yue, L.; Jing, X.; Yang, Y.; Xu, Y.; Wu, S.; Liang, Y.; Liu, X.; Zhang, X. Purification and characterization of lysozyme from Chinese Lueyang black-bone Silky fowl egg white. *Prep. Biochem. Biotechnol.* **2019**, *49*, 215–221. [CrossRef]
11. Smma, B.; Pda, B.; Kta, B.; Apa, B. Filtration of protein-based solutions with ceramic ultrafiltration membrane. Study of selectivity, adsorption, and protein denaturation. *Comptes Rendus Chim.* **2019**, *22*, 198–205. [CrossRef]
12. Lu, J.; Wan, Y.; Cui, Z. Strategy to separate lysozyme and ovalbumin from CEW using UF. *Desalination* **2006**, *200*, 477–479. [CrossRef]
13. Al-Hakeim, H.K.; Al-Shams, J.K.; Kadhem, M.A. Immobilization of Urease enzyme on ion-exchange resin. *J. Univ. Babylon* **2012**, *20*, 1231–1236.
14. Easton, L.E.; Shibata, Y.; Lukavsky, P.J. Rapid, nondenaturing RNA purification using weak anion-exchange fast performance liquid chromatography. *RNA* **2010**, *16*, 647–653. [CrossRef] [PubMed]
15. Parhi, R.; Suresh, P. Formulation optimization and characterization of transdermal film of simvastatin by response surface methodology. *Mater. Sci. Eng. C* **2016**, *58*, 331–341. [CrossRef] [PubMed]
16. Rabinarayan, P.; Suresh, P.; Patnaik, S. Application of Response Surface Methodology for Design and Optimization of Reservoir-type Transdermal Patch of Simvastatin. *Curr. Drug Deliv.* **2016**, *13*, 742–753. [CrossRef]
17. Mondal, S.C.; Kumar, J. Application of Box-Behnken design for the optimisation of process parameters in dry drilling operation. *Int. J. Product. Qual. Manag.* **2016**, *18*, 456–473. [CrossRef]
18. Yang, D.; Wang, Q.; Cao, R.; Chen, L.; Liu, Y.; Cong, M.; Wu, H.; Li, F.; Ji, C.; Zhao, J. Molecular characterization, expression and antimicrobial activities of two c-type lysozymes from manila clam Venerupis philippinarum. *Dev. Comp. Immunol.* **2017**, *73*, 109–118. [CrossRef]
19. Cugnata, N.M.; Guaspari, E.; Pellegrini, M.C.; Fuselli, S.R.; Alonso-Salces, R.M. Optimal Concentration of Organic Solvents to be Used in the Broth Microdilution Method to Determine the Antimicrobial Activity of Natural Products Against Paenibacillus Larvae. *J. Apic. Sci.* **2017**, *61*, 37–53. [CrossRef]
20. Evli, S.; Akta, D.; Enzymatic, U. Activity of Urokinase Immobilized onto Cu 2+-Chelated Cibacron Blue F3GA-Derived Poly (HEMA) Magnetic Nanoparticles. *Appl. Biochem. Biotechnol.* **2019**, *188*, 194–207. [CrossRef]
21. Bezemer, J.M.; Radersma, R.; Grijpma, D.W.; Dijkstra, P.J.; Feijen, J.; Van Blitterswijk, C.A. Zero-order release of lysozyme from poly(ethylene glycol)/poly(butylene terephthalate) matrices. *J. Control. Release* **2000**, *64*, 179–192. [CrossRef]
22. Czermak, P.; Grzenia, D.L.; Wolf, A.; Carlson, J.O.; Specht, R.; Han, B.; Wickramasinghe, S.R. Purification of the densonucleosis virus by tangential flow ultrafiltration and by ion exchange membranes. *Desalination* **2008**, *224*, 23–27. [CrossRef]
23. Lu, W.; Khan, T.; Mohanty, K.; Ghosh, R. Cascade ultrafiltration bioreactor-separator system for continuous production of F(ab')2 fragment from immunoglobulin G. *J. Membr. Sci.* **2010**, *351*, 96–103. [CrossRef]

fermentation

Article

Breeding of a High-Nisin-Yielding Bacterial Strain and Multiomics Analysis

Leshan Han [1,2], Xiaomeng Liu [1,2], Chongchuan Wang [1,2], Jianhang Liu [2,3], Qinglong Wang [1,2], Shuo Peng [1,2], Xidong Ren [1,2], Deqiang Zhu [1,2,3,*] and Xinli Liu [1,2,*]

[1] State Key Laboratory of Biobased Material and Green Papermaking, Qilu University of Technology, Shandong Academy of Sciences, Jinan 250353, China; lshan9869@126.com (L.H.); mkt1201@163.com (X.L.); wcc17861405023@163.com (C.W.); wangqinglong975@163.com (Q.W.); 19546151248@163.com (S.P.); renxidong1986@126.com (X.R.)

[2] Shandong Provincial Key Laboratory of Microbial Engineering, School of Bioengineering, Qilu University of Technology, Shandong Academy of Sciences, Jinan 250353, China; liujx@tib.cas.cn

[3] Key Laboratory of Systems Microbial Biotechnology, Tianjin Institute of Industrial Biotechnology, Chinese Academy of Sciences, Tianjin 300308, China

* Correspondence: zdq0819@yeah.net (D.Z.); vip.lxl@163.com (X.L.); Tel.: +86-187-6580-9552 (D.Z.); +86-186-6077-3985 (X.L.)

Abstract: Nisin is a green, safe and natural food preservative. With the expansion of nisin application, the demand for nisin has gradually increased, which equates to increased requirements for nisin production. In this study, *Lactococcus lactis* subsp. *lactis* lxl was used as the original strain, and the compound mutation method was applied to induce mutations. A high-yielding and genetically stable strain (*Lactobacillus lactis* A32) was identified, with the nisin titre raised by 332.2% up to 5089.29 IU/mL. Genome and transcriptome sequencing was used to analyse A32 and compare it with the original lxl strain. The comparative genomics results show that 107 genes in the A32 genome had mutations and most base mutations were not located in the four well-researched nisin-related operons, *nisABTCIPRK*, *nisI*, *nisRK* and *nisFEG*: 39 single-nucleotide polymorphisms (SNPs), 34 insertion mutations and 34 deletion mutations. The transcription results show that the expression of 92 genes changed significantly, with 27 of these differentially expressed genes upregulated, while 65 were downregulated. Our findings suggest that the output of nisin increased in *L. lactis* strain A32, which was accompanied by changes in the DNA replication-related gene *dnaG*, the ABC-ATPase transport-related genes *patM* and *tcyC*, the cysteine thiometabolism-related gene *cysS*, and the purine metabolism-related gene *purL*. Our study provides new insights into the traditional genetic mechanisms involved nisin production in *L. lactis*, which could provide clues for a more efficient metabolic engineering process.

Keywords: nisin; *Lactococcus lactis* subsp. *lactis*; mutation breeding; genomics; transcriptomics

Citation: Han, L.; Liu, X.; Wang, C.; Liu, J.; Wang, Q.; Peng, S.; Ren, X.; Zhu, D.; Liu, X. Breeding of a High-Nisin-Yielding Bacterial Strain and Multiomics Analysis. *Fermentation* **2022**, *8*, 255. https://doi.org/10.3390/fermentation8060255

Academic Editors: Flavia Marinelli and Francesca Berini

Received: 26 April 2022
Accepted: 24 May 2022
Published: 27 May 2022

Publisher's Note: MDPI stays neutral with regard to jurisdictional claims in published maps and institutional affiliations.

1. Introduction

Nisin is a small-molecule polypeptide with broad-spectrum antibacterial effects, and is produced by *Lactococcus* ssp. Nisin is internationally recognized as a safe and nontoxic natural food additive. It has no side effects, and has been used as a food preservative for more than 60 years [1,2]. Due to the high biosafety of nisin, research on its application value has long expanded past the field of food preservation. In recent years, nisin applications, including its use as a packaging material [3–5], as a medical antibacterial agent [6–9], as a biosensor [10], in immune regulation [11,12], as an anticancer drug [13,14] and in other aspects, have been explored.

In the process of fermentation, lactic acid-producing bacteria (LAB) produce a variety of metabolites, and nisin is a byproduct of the growth process of *Lactococcus lactis* subspecies. The nisin synthesis process is very complex, and its synthesis efficiency is relatively low [1]. (1) Nisin is generated to inhibit other microorganisms that compete with *L. lactis* ssp. for living space [15,16]. To avoid self-inhibition, *L. lactis* also exerts a series of immune mechanisms. (2) Nisin is produced by ribosomal synthesis and is a post-translationally modified peptide (RiPP); its production process involves precursor peptide-related genes and modified protein synthesis genes. Nisin biosynthesis genes are transcribed in four operons, *nisABTCIPRK*, *nisI*, *nisRK* and *nisFEG* [16–18]. (3) From a production environment standpoint, during their growth processes, LAB produce a variety of acids, such as carboxylic acid, lactic acid and other acidic substances, decreasing the pH of the media [19]. A low pH inhibits the growth of LAB and the production of nisin, requiring production strains that tolerate acidic environments.

The complexity of nisin synthesis makes it difficult to reconstruct the production pathway in other heterogenic hosts; thus, optimizing nisin production by *L. lactis* strains is important and useful. At present, two main methods are used to modify bacterial strains, traditional mutagenesis-type breeding and genetic engineering. Genetic engineering of bacteria can alter the expression of several genes; however, the choice of operation sites is based on the clear understanding of the intracellular pathway and the corresponding regulatory mechanism. Furthermore, it is difficult to achieve the coexpression of a series of genes, and accordingly, the corresponding increase in production is limited. Although a greater effort is needed to induce traditional mutations, a series of genes can be changed, which is more favourable than changes in the whole strain. To screen the strains with high nisin yields, in this study, compound mutation combined with high-concentration-nisin plate screening was used to modify and quickly screen high-nisin-yielding strains. Compound mutagenesis involves the successive or simultaneous use of one or more mutagens on microorganisms, or the same mutagen acting on microorganisms repeatedly. It is a commonly used method to improve the production of microorganisms. Atmospheric room temperature plasma (ARTP) can change the structure and permeability of the cell wall membrane, causing gene damage and mutation when plasma jets with highly active particles are discharged toward microbial cells. The method is simple to use, is highly pure and pollution-free, is harmless to the human body and causes various mutations [11]. Ultraviolet (UV) mutation is the most traditional and useful mutation method, and is often used as the first choice for microbial breeding. It can directly affect double-stranded DNA and improve mutation efficiency. Therefore, we chose the use of ARTP and UV light, which are two traditional and modern methods for compound mutagenesis. A high-nisin-concentration screening method can be used to identify strains with increased nisin yields. Together, these methods can effectively mutate and screen mutant strains. In this study, a high-yielding mutant strain (A32) was screened, and then, by comparing the differences at the transcriptome level between the original strain and the mutant strain, we hoped to find a series of genes related to increasing nisin production, which is very important for future genetic engineering of the producing strain.

A high-yielding and genetically stable mutant strain was obtained, which was named *L. lactis* A32. The yield of the mutant strain was three times that of the original strain *L. lactis* lxl, and its resistance to nisin was twofold greater. By comparing the whole-genome sequence of *L. lactis* A32 with that of the original strain lxl, 107 mutated genes were found, including 39 single-nucleotide polymorphisms, 34 insertion mutations and 34 deletion mutations. Through comparative analysis of the transcriptome sequencing results, it was found that the transcription of 92 genes changed significantly, with 27 of these genes significantly upregulated, while 65 genes were significantly downregulated. Analysis of the metabolic pathways of these genes can provide clues for the further study of the molecular mechanism underlying the nisin synthesis pathway.

2. Materials and Methods

2.1. Strains, Media and Growth Conditions

Lactococcus lactis subsp. *lactis* lxl is a store strain in our laboratory. It was screened from raw milk, with 40 mg/L bromocresol purple and 1000 IU/mL nisin standards in modified M17 medium (10 g/L sucrose as the only carbon source). After morphological and physiological analysis and 16S rDNA sequencing, it was identified as *Lactococcus lactis* subsp. *lactis*. The titre test strain was *Micrococcus luteus* NCIB 8166, which was purchased from China General Microbiological Culture Collection Center, and preserved at $-80\ °C$ in 20% (v/v) glycerin in our laboratory.

The composition of the media used is as follows.

GM17 medium (a seed medium): M17 broth medium (soy peptone 5 g/L, peptone 2.5 g/L, casein peptone 2.5 g/L, yeast extract powder 2.5 g/L, beef extract powder 5 g/L, lactose 5 g/L, sodium ascorbate 0.5 g/L, β-sodium glycerophosphate 19 g/L, magnesium sulfate 0.25 g/L) containing glucose (5 g/L). CM1 medium (a fermentation medium): sucrose (20 g/L), peptone (20 g/L), yeast extract (20 g/L), KH_2PO_4 (10 g/L), NaCl (2 g/L) and $MgCl_2$ (0.2 g/L). S1 medium (a detection medium): peptone (8 g/L), yeast extract (5 g/L), glucose (5 g/L), NaCl (5 g/L) and $NaH_2PO_4 \cdot 12H_2O$ (2 g/L).

2.2. Fermentation in Flasks and Nisin Titre Assays

A nisin standard (Maclin, China) was prepared in a series of standard solutions of 50, 100, 200, 400, 800, 1000, 2000, 3000 and 4000 IU/mL. For this procedure, a sterilized Oxford cup was placed on the test plate, the culture media containing the tested bacteria was added to the plate, and the Oxford cup was removed after cooling to form a uniform small hole. Then, 200 μL of standard solutions containing a series of nisin concentrations were added to the small holes, after which the plates were cultured at 37 °C until there was an obvious bacteriostatic circle around each small hole, and the diameter of the bacteriostatic circle was measured. A nisin titre standard curve was constructed, with the diameter of the bacteriostatic circle representing the abscissa and the logarithm of the titre representing the ordinate.

As a seed solution, a single colony of the original strain lxl was inoculated into GM17 media and cultured overnight at 30 °C and 200 rpm. Then, the seed liquid was inoculated into CM1 media at a concentration of 5%, fermented at 30 °C and centrifuged 200 rpm for 24 h. Samples were collected every 2 h during the fermentation period, and a bacteriostatic circle experiment was carried out. The nisin titre was calculated according to the nisin standard curve (Figure S1) [20].

2.3. Compound Mutation and Strain Breeding

The starting strain was cultured by scribing, and a single colony was picked and cultured in GM17 liquid medium for 6–8 h. The culture was appropriately diluted with normal saline instead of the mutagenic strain solution, and the optical density (OD_{600}) of the mutagenic strain solution was maintained between 0.8 and 1.0 to ensure the proper colony number and growth state of each mutagenic strain. The lethality of the ARTP and UV mutations was determined, and the time when the lethality was 80–90% was selected as the mutation time to ensure mutation efficiency. The nisin tolerance range of the original strain was determined, and a screening plate containing high concentrations of nisin was constructed. The solution of bacteria with different mutations was poured onto the screening plate, which was subsequently cultured at 30 °C. The grown colonies were picked out and cultured overnight in an Erlenmeyer flask containing 50 mL of GM17 medium. The cultured bacterial solution was then inoculated into a triangular flask containing 50 mL of CM1 medium for fermentation at 30 °C and centrifugation at 200 rpm for 24 h. The fermentation broth was subsequently centrifuged at 8000 rpm for 5 min, and the supernatant was collected for a bacteriostatic circle experiment to determine the titre of the fermentation broth. The strain with the highest titre from this batch of colonies was selected as the starting strain for the next mutation, as shown in Figure 1. The abovementioned process was repeated until the high-yielding strains whose parameters were within the target range were screened.

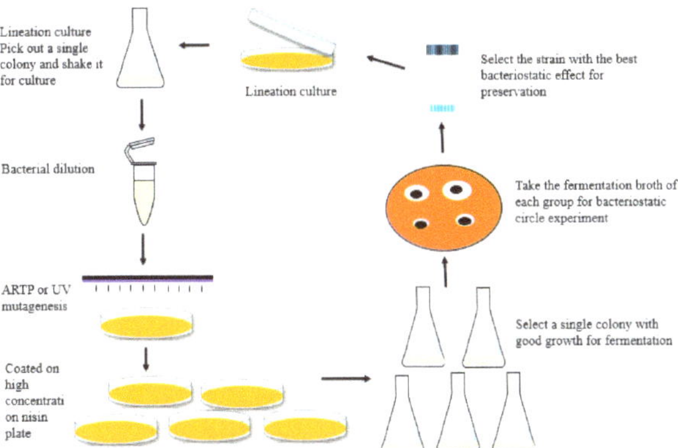

Figure 1. Compound mutation and breeding process.

2.4. Determination of the Genetic Stability of Mutants

To investigate whether the titre decreased and whether the mutant phenotype was reversible, we determined the genetic stability of the mutants. The strains with high nisin titre were screened out. They were transferred onto plates and cultured at 30 °C for 24 h to form the F1 generation. The colonies were then removed from the F1 plate and cultured on another plate to form the F2 generation. This process was repeated to obtain 20 generations in total. Single colonies were picked every three generations from seed liquid culture, then inoculated into liquid fermentation medium CM1 and incubated at 30 °C and 200 rpm for 24 h, and the supernatant of the fermentation broth was collected to determine the nisin titre.

2.5. Comparison of the Fermentation Process between the Mutant Strain and Original Strain

The mutant with the highest titre that was genetically stable and the original strain were inoculated into CM1 fermentation media at the same time and fermented at 30 °C and 200 rpm for 48 h, and samples were collected every 3 h. The biomass of the sample was measured at 600 nm by a spectrometer (Metash, Shanghai, China), and the nisin titre was tested via bacteriostatic circles.

2.6. DNA Library Construction, DNA Sequencing, Assembly and Annotation

Single colonies of *L. lactis* A32 and lxl were selected and cultured in CM1 media at a suitable temperature overnight, and the bacteria were collected during the logarithmic growth period. Genomic DNA of *L. lactis* lxl and A32 was extracted using TIAMP Bacterial DNA Kits (Sparkjade, Jinan, China) according to the manufacturer's instructions. Then, the quality of the DNA was checked to ensure that the quality was sufficient for subsequent sequencing. The concentration of the standard DNA was measured of by a Qubit instrument (Thermo Fisher Scientific, Waltham, MA, USA) (including the following process of establishing the whole library). A Nanodrop spectrophotometer (Thermo Fisher Scientific, Waltham, MA, USA) was used to measure the DNA concentration (between $OD_{260/280}$ and between 1.8 and 2.0), and the DNA integrity was verified by 1% agarose gel electrophoresis. Library construction, Oxford Nanopore Technologies (ONT) sequencing, Illumina sequencing and base determination were carried out by Gene Denovo (Guangzhou, China). The bacterial genome was sequenced by a combination of third-generation ONT sequencing and second-generation Illumina sequencing. This method relies on the long reading characteristics of third-generation sequencers to ensure a more complete genome assembly, and uses second-generation sequencing data for correction to ensure more accurate and reliable assembly results. Fastp software [21] was used to control the ONT and Illumina sequencing data and obtain effective data (clean data). The third-generation sequencing reads were spliced and assembled with Flye (version 2.8.1-b1676) [22]. Pilon (version 1.23) [23] was then used to compare the second-generation sequencing reads to the assembled genome sequence, correct the genome results according to the default parameters of the software, and output the corrected genome sequence and corrected site information. The National Center for Biotechnology Information (NCBI) database was used to predict gene functions. Through BLAST, the predicted gene sequences were compared to the information within the nonredundant (Nr), Swiss-Prot, Gene Ontology (GO), Kyoto Encyclopedia of Genes and Genomes (KEGG) and Clusters of Orthologous Genes (COG) databases to determine the proteins with the highest sequence similarity to those encoded by a given gene, to obtain the protein functional annotation information of the genes. We used Blast2GO to obtain the GO annotation information of the genes, and performed GO functional classification analysis for all of the genes. Diamond was used to compare all the gene sequences with the sequences in the COG database to obtain the annotation results corresponding to the genes and to classify the functions of proteins according to the annotation results. In addition, the gene function databases Antimicrobial Resistance Genes Database (ARDB) and the Virulence Factors of Pathogenic Bacteria Database (VFDB), within which information is subdivided into an increased number of fields, were used to predict secondary metabolic gene clusters and to annotate virulence factor-related genes and drug resistance-related genes. Moreover, the Protein Family (Pfam) database, the Pathogen Host Interactions (PHI)-base database, the Carbohydrate-Active Enzymes (CAZy) database, the Comprehensive Antibiotic Resistance Database (CARD) and other databases were used for secretory protein prediction, type-N secretion system (TNSS) effector protein prediction and two-component system prediction.

2.7. Whole-Genome Comparisons

L. lactis lxl and A32 bacteria were subjected to comparative genomic analysis. MUMmer [24] software was used to detect the SNPs and insertion deletions (indels) in the assembled genomes of the individuals, as well as to perform statistics based on the positional relationships between the mutant genes and the mutation results, and compare the results with information in databases to assess the functions of the mutated genes.

2.8. RNA Extraction, Transcriptome Sequencing and Analysis

Single colonies of *L. lactis* A32 and *L. lactis* lxl were selected and cultured in CM1 media at a suitable temperature overnight, and the bacteria were collected during their logarithmic growth period. An appropriate amount of the cultured bacterial solution was transferred into a suitable centrifuge tube, after which the bacteria were collected by high-speed centrifugation (12,000 rpm). The culture medium in the upper part of the tube was discarded. Total RNA was extracted by the TRIzol (Sevencyd, Jinan, China) method.

The RNA degradation degree and potential contamination were monitored on 1% agarose gels. RNA purity (OD_{260}/OD_{280}, OD_{260}/OD_{230}) was determined using a NanoPhotometer® spectrophotometer (IMPLEN, CA, USA). RNA integrity was measured using a Bioanalyzer 2100 (Agilent, Santa Clara, CA, USA). An Illumina MRZB12424 Ribo Zero rRNA Removal Kit (Bacteria) (Illumina, San Diego, CA, USA) was used to remove rRNA from 1 µg of total RNA. The quality of the RNA samples was detected using an Agilent 2100 Biological Analyser (Agilent Technologies, Santa Clara, CA, USA). The library construction, transcriptome sequencing and RNA sequencing (RNA-seq) analysis of *L. lactis* A32 were carried out in Gene Denovo (Guangzhou, China). FASTP (version 0.18.0) [21] was used to filter the original data produced via the Illumina platform. The filtering criteria were as follows: (1) reads with ≥10% unidentified nucleotides (N); (2) reads in which >50% of the bases had Phred quality scores of ≤20 and (3) reads aligned to the barcode adapter. Bowtie2 (version 2.2.8) [25] was used to compare the retained reads to the reference genome, and the reads associated with rRNA in the comparison were deleted. RSEM [26] was used to identify known genes and calculate their expression. Library construction and transcriptome sequencing of *L. lactis* lxl were carried out by Gene Denovo (Guangzhou, China). The gene expression level was further normalized by using the fragments per kilobase of transcript per million (FPKM) mapped reads method to eliminate the influences of different gene lengths and amounts of sequencing data on the calculation of gene expression. The edgeR package [27] (http://www.r-project.org/ Accessed on 1 January 2022) was then used to identify differentially expressed genes (DEGs) across the samples. Genes were considered differentially expressed if their expression fold-change was ≥ 2 and their false discovery rate-adjusted P (q value) was <0.05. The DEGs were then subjected to GO functional enrichment and KEGG pathway analyses, and q values <0.05 were used as thresholds. The GO functional terms of the DEGs were analysed, the GO functional classification annotations of the DEGs were obtained, and the pathways in which the DEGs were enriched were identified. Using Rockhopper [28] software, the sequencing results were compared to the reference genome sequence and to the annotated genes. Unknown transcript-coding regions were considered new transcript-coding regions. The newly encoded transcripts were compared to the information in the Nr database for annotation, and the annotated transcripts were regarded as new transcripts with coding potential.

2.9. Statistical Analysis

Student's *t*-test was conducted to determine the significant differences in nisin titres between the original *L. lactis* lxl strain and the mutant A32 strain. The data were statistically analysed using Origin 9.1 (OriginLab, Northampton, MA, USA).

3. Results and Discussion
3.1. Titres of Mutant Strains

As the duration of ARTP and UV treatment increased, the mortality of bacteria also increased. When the bacteria were treated with ARTP and UV light for 120 s and 60 s, respectively, the lethality rates were 90% and 92%, respectively, as shown in Figure 2A,B. In the recent literature on mutagenesis, a treatment time with a lethal rate of approximately 90% is usually used to ensure mutation efficiency [17]. In the present study, ARTP and UV light were applied for 120 s and 60 s, respectively.

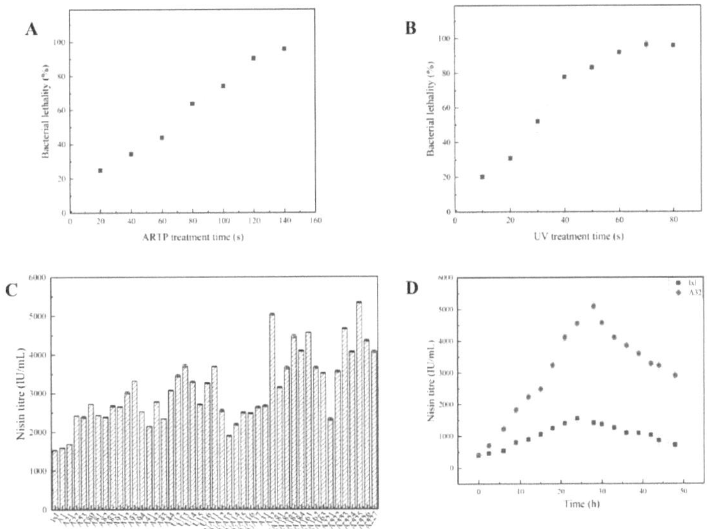

Figure 2. Mutagenesis preparation and results. (**A**) Lethality to *L. lactis* lxl of ARTP mutagenesis; (**B**) lethality to *L. lactis* lxl of UV mutagenesis; (**C**) mutagenic strain titre; (**D**) titre curves of *L. lactis* A32 and lxl with increasing fermentation times.

Through a series of compound mutagenesis and step-by-step screening, some mutants that could grow on plates supplemented with high nisin concentration were selected, fermented, and their nisin production titres were determined. The results are shown in Figure 2C. Among all examined mutants, mutants A32 and A225 had higher titres, and their titre peaks were 5089.29 IU/mL and 5361.16 IU/mL, respectively.

3.2. Evaluation of Mutant Stability

The experimental results regarding the genetic stability of the mutants were shown in Table S2. We observe that the ability of the mutant *L. lactis* A32 to produce nisin was stable, while the ability of A225 to produce nisin was not stable. In the third generation, the nisin titre of A225 was only half that of A32. A32 was also twice as resistant to nisin as the original strain. These results show that the A32 mutant generated by ARTP and UV radiation yielded high amounts of nisin and was genetically stable. Therefore, the mutant strain *L. lactis* A32 was selected for use in the following experiments.

3.3. Comparison of the Fermentation of L. lactis A32 and Lxl

There was a significant difference in growth curves and nisin titres between *L. lactis* A32 and lxl. Mutant A32 took significantly longer to reach the stationary phase than the original lxl strain. The A32 biomass peaked at 12 h, and the lxl biomass peaked in the stationary phase at 9 h, as shown in Figure S2. Since nisin is a growth-coupled secondary metabolite, hypothetically, A32 would accumulate more nisin than lxl during this process. This hypothesis is verified in Figure 2D. As shown in Figure 2D, the nisin titre continued to increase within 0–28 h of A32 fermentation, the peak nisin production intensity of A32 reached up to 181.76 IU/(mL·h), which is 5089.29 IU/mL, and lxl yielded a maximum of 1529.79 IU/mL at 24 h. These results show that the correlation between product synthesis and thallus growth was not shifted, but the regulation around the synthetic pathways flux was improved.

3.4. Genome Assembly and Annotation

The PromethION (Oxford Nanopore Technologies, Oxford, UK) and Agilent 2100 sequencing platforms were used to sequence the whole genome of *L. lactis* A32. A total of 7,244,702 bp of clean reads and 1,086,215,124 bp from A32 were generated. The genome sequencing data of *L. lactis* A32 have been submitted to the Sequence Read Archive database of the NCBI under accession number PRJNA799214. After data filtering, the genome of *L. lactis* A32 was assembled, with a total genome length of 2.40 Mb (Table S2). The GC contents (lxl, 34.44%; A32, 34.57%) were similar (Table 1). The complete genome sequences of *L. lactis* A32 and lxl have been submitted to the GenBank database of the NCBI (accession number for A32, SAMN25149786; accession number for lxl, SAMN25148929). According to the comparative results of five gene function databases (Nr, Swiss-Prot, GO, KEGG and COG), 2278 DNA sequences (CDSs) in lxl and 2285 CDSs in A32 were predicted in the genome (Table 1), and five genes encoded proteins with no predicted function in lxl and A32.

Table 1. Genome characteristics of *L. lactis* A32 and lxl.

Item	A32	lxl
Genome size (bp)	2,399,752	2,399,772
Correction (bp)	3081	3147
GC content (%)	35.24	36.08
No. CDSs	2285	2278
CDS length (bp)	75-4917	75-4917
No. tRNAs	66	66
No. rRNAs	18	18
No. sRNAs	2	2
No. repeats	42	43

3.5. Functional Classification and Comparison

All annotated genes of *L. lactis* A32 and lxl were classified into GO functional categories (Table S3). The GO functional classification system is based on three ontologies, including biological processes, cell components and molecular processes. It includes 58 gene functional classifications, such as cellular processes, metabolic processes, single-organism processes, responses to stimuli, cellular component organization or biogenesis, biological regulation and biological processes. There were some differences in GO cluster analysis between A32 and lxl. Further analysis showed that, compared with lxl, A32 had six more genes related to molecular processes, four more genes related to metabolic processes, four more genes related to catalytic activity, two more genes related to single biological processes, two fewer genes related to responses to stimuli and one less gene related to development processes.

The COG functional categories of all annotated proteins of *L. lactis* A32 and lxl are shown in Table 2, and were divided into 25 main functional categories. The main functional categories were related to translation; ribosomal structure and biogenesis; replication, recombination and repair; cell wall/membrane/envelope biogenesis; carbohydrate transport; and amino acid transport and metabolism. The distribution patterns of the protein clusters of homologous genes in the two strains were very similar. There were four fewer CDSs in the genome of A32 compared with the original lxl strain. In addition, compared with lxl, A32 had one more CDS related to transcription- and signal transduction-related proteins; two fewer CDSs related to replication, recombination and repair; two fewer CDSs related to cell cycle control, cell division and chromosome distribution; two fewer CDSs related to carbohydrate transport and metabolism; and one less CDS related to inorganic ion transport and metabolism.

Table 2. GO functional categories.

Functional Category	No. A32 Genes	No. Lxl Genes
Translation, ribosomal structure and biogenesis	148	148
RNA processing and modification	0	0
Transcription	160	159
Replication, recombination and repair	119	121
Chromatin structure and dynamics	1	1
Cell cycle control, cell division, chromosome partitioning	19	20
Nuclear structure	0	0
Defence mechanisms	50	50
Signal transduction mechanisms	55	53
Cell wall/membrane/envelope biogenesis	104	104
Cell motility	14	14
Cytoskeleton	0	0
Extracellular structures	0	0
Intracellular trafficking, secretion and vesicular transport	25	25
Post-translational modification, protein turnover, chaperones	55	55
Energy production and conversion	74	75
Carbohydrate transport and metabolism	172	174
Amino acid transport and metabolism	195	195
Nucleotide transport and metabolism	76	76
Coenzyme transport and metabolism	75	75
Lipid transport and metabolism	61	61
Inorganic ion transport and metabolism	112	113
Secondary metabolites biosynthesis, transport and catabolism	38	38
General function prediction only	263	263
Function unknown	202	202

3.6. Comparison of Genomes between *L. lactis* lxl and A32

To reveal the relationships between phenotypes and genetic variations, and provide clues for the further study of the molecular mechanism of the nisin synthesis pathway, a genome-wide comparison was performed between the high-nisin-yielding mutant A32 and the original lxl strain. The cyclic images from the genome comparison show the microstructural mutations between *L. lactis* A32 and lxl (Figure 3). A total of 107 genes in the A32 genome were mutated compared with the lxl genome, including 39 single-base mutations, 34 insertion mutations and 34 deletion mutations. The sequences of the mutated genes were compared with the reference genome sequence in the NCBI database of COG functional annotations. While the functions of some genes were not clear, some genes with base mutations had clear functions (Table 3), and some mutated genes, such as *rexB*, *ftsH*, *gntP* and *yfmR*, were related to energy metabolism. Additionally, some mutated genes were related to ion transport, such as *yfmR*, *rbcR*, *zitR*, *adcA* and *copB*; to DNA replication, transcription and translation, such as *dnaG*, *rpsI*, *rex* and *arlR*; and to amino acid transport and metabolism, such as *patM*, *tcyC*, *tcyJ*, *cysS* and *brnQ*. However, a vast improvement in the nisin titre was achieved, except for one mutation in the four nisin-related operons, *nisK*; most base mutations were not located in the well-researched nisin-related four operons, *nisABTCIPRK*, *nisI*, *nisRK* and *nisFEG*. Sequence analysis of the nisin genes *nisA*, *nisB*, *nisC* and *nisP* before and after mutation showed that these genes were not mutated, thus the amino acid sequence of nisin and the mechanism of action of the modified protein did not change. Based on these analyses, it is reasonable to infer that the nisin produced by the mutant is structurally identical to that of the original bacterium, and thus its stability and other properties remain unchanged.

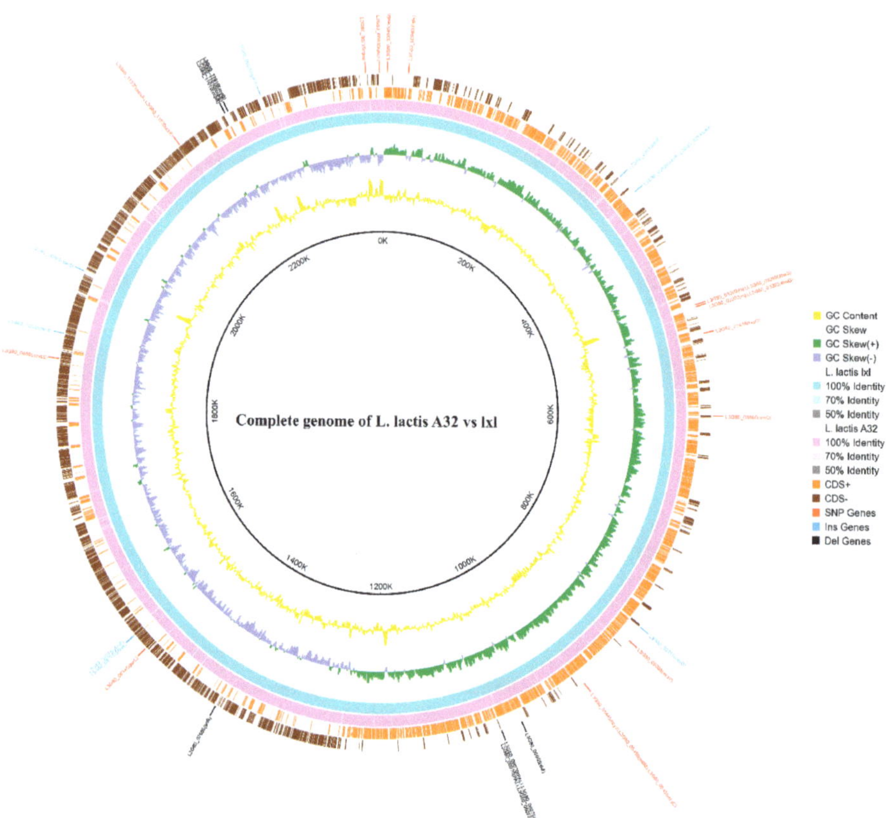

Figure 3. The cyclic images from the genome comparison.

Table 3. COG analysis and type of gene mutation.

Gene ID	Name	Mutation Type	COG Functional Description
L3G80_00845	*rexB*	SNP	ATP-dependent nuclease, subunit B
L3G80_00940	*ftsH*	SNP	ATP-dependent Zn proteases
L3G80_09890	*cysS*	SNP	Cysteinyl-tRNA synthetase
L3G80_00020	*gntP*	Ins	H+/gluconate symporter and related permeases
L3G80_08275	*yfmR*	Ins	ATPase components of ABC transporters with duplicated ATPase domains
L3G80_05905	*dut*	DEL	dUTPase
L3G80_11910	*ccpB*	DEL	Beta-glucosidase/6-phospho-beta-glucosidase/beta-galactosidase
L3G80_00735	*rpsI*	SNP/Del	Ribosomal protein S9
L3G80_08145	*purL*	SNP	Phosphoribosylformylglycinamidine (FGAM) synthase, synthetase domain
L3G80_05450	*patM*	SNP	ABC-type amino acid transport system, permease component
L3G80_05455	*tcyC*	SNP	ABC-type polar amino acid transport system, ATPase component
L3G80_05445	*tcyJ*	SNP	ABC-type amino acid transport/signal transduction systems, periplasmic component/domain
L3G80_11520	*adcA*	SNP	ABC-type metal ion transport system, periplasmic component/surface adhesin
L3G80_03200	*tig*	SNP	FKBP-type peptidyl-prolyl cis-trans isomerase (trigger factor)
L3G80_06070	*PAL*	Del	Cell wall-associated hydrolases (invasion-associated proteins)
L3G80_06075	*rex*	Del	AT-rich DNA-binding protein
L3G80_05015	*copB*	Ins	Cation transport ATPase
L3G80_10025	*rbcR*	Ins	Transcriptional regulator
L3G80_11525	*zitR*	SNP	Transcriptional regulators
L3G80_05095	*mraY*	SNP	UDP-N-acetylmuramyl pentapeptide phosphotransferase/UDP-N-acetylglucosamine-1-phosphate transferase
L3G80_03295	*dnaG*	SNP	DNA primase (bacterial type)
L3G80_03435	*tspO*	SNP	Tryptophan-rich sensory protein (mitochondrial benzodiazepine receptor homologue)
L3G80_03890	*brnQ*	SNP	Branched-chain amino acid permeases
L3G80_00800	*ytqA*	SNP	Predicted Fe-S oxidoreductase
L3G80_02410	*thiT*	Ins	Predicted membrane protein
L3G80_10565	*yadS*	Ins	Predicted membrane protein
L3G80_07665	*arlR*	Del	Response regulators consisting of a CheY-like receiver domain and a winged-helix DNA-binding domain
L3G80_02525	*penA*	Ins	Cell division protein FtsI/penicillin-binding protein 2
L3G80_02535	*ddl*	Ins	D-alanine-D-alanine ligase and related ATP-grasp enzymes

3.7. Transcriptome Sequencing and Analysis

Transcriptome sequencing was used to detect the differences in gene expression between the original *L. lactis* lxl strain and A32, the inserted genes were examined to determine the metabolic pathways that they influenced. After quality screening, 19.58 to 24.42 million clean reads and 1.65 to 2.14 Gb of clean data submitted to the NCBI gene expression comprehensive database (BioProject: PRJNA825060) were obtained from the control strain and the mutant strain, respectively (Table S4). The sequencing reads of the original strain and mutant strain were mapped to the A32 genome assembly sequence, and 2231 known mRNAs were identified in A32. The statistics of gene coverage are shown in Figure 4A. Compared with that in the original *L. lactis* lxl strain, the expression of 92 genes in *L. lactis* A32 changed significantly ($p < 0.005$), of which 27 were significantly upregulated and 65 were significantly downregulated, as shown in Figure 4B. GO enrichment analysis was performed on the significant DEGs (Figure 4C) to understand their biological functions. DEG pathway enrichment analysis is helpful for further understanding the gene-related metabolic networks. The main metabolic pathways of significantly differentially expressed genes were metabolic pathways, ABC transporters, two-component systems, biosynthesis of secondary metabolites, quorum sensing, arginine biosynthesis and butanoate metabolism (Figure 4D).

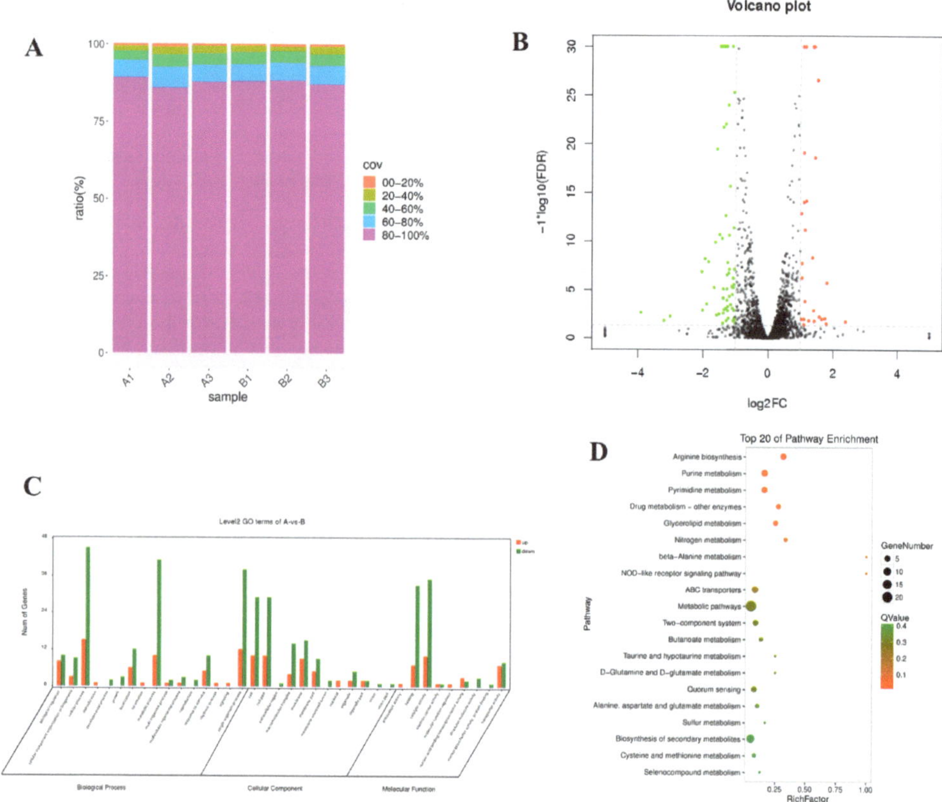

Figure 4. Transcriptome sequencing results and analysis. (**A**) *L. lactis* and A32 gene coverage; (**B**) map of significant DEGs; (**C**) GO classification of DEGs; (**D**) statistical characteristics of DEG enrichment in KEGG pathways. Genes with no significant difference in expression are shown in black, upregulated genes are shown in red and downregulated genes are shown in green.

3.8. Comprehensive Analysis

The fermentation conditions and growth states of *L. lactis* strains A32 and lxl were observed. Comparative genomic analysis and transcriptomic analysis were then performed in combination to identify associated genes, and try to infer the regulatory pathway of nisin biosynthesis.

3.8.1. DEGs Associated with Nisin Immunity Genes

Nisin mainly targets Gram-positive bacteria, and *Lactococcus* itself is also a gram-positive bacterium. The nisin immunity genes *nisEFG* and *nisI* were found to be present in the nisin synthesis-related gene cluster [18,29]. Due to the existence of immunity genes, *Lactococcus* has immune activity against nisin. As also mentioned in the introduction of this paper, as early as 1998, Kim tried to increase the production of nisin by overexpressing the nisin immunity gene [30]. Therefore, the improvement of nisin immunological activity must depend on the upregulation of immunity-related gene expression. The results of transcriptomic analysis verified this conclusion.

3.8.2. DEGs Related to DNA Replication, Transcription and Translation

During the fermentation and growth of *L. lactis* strains A32 and lxl, the logarithmic growth period of A32 obviously lagged behind that of lxl, which made the cumulative nisin production time longer, and the nisin yield higher, in the former. When performing our comparative omics analysis, we focused on the DNA enzyme synthesis-related gene *dnaG*, which had a base mutation. Although the change in its expression was not obvious in transcriptomics analysis, the expression of a single-stranded DNA-binding (SSB) protein bound to *dnaG* decreased significantly. This SSB protein is very important for DNA metabolism. First, it can stabilize the intermediate products of single-stranded DNA (ssDNA) produced during DNA processing. In addition, it can interact with 14 proteins and deliver reserve enzymes to DNA when needed to repair DNA damage [31,32]. It is speculated that the oligonucleotide/oligosaccharide-binding fold (OB-fold) of the *dnaG* protein was changed due to the gene mutation in *dnaG* [33], resulting in a decrease in its ability to bind to the SSB protein.

3.8.3. DEGs Related to ABC ATPase

ABC proteins refer only to proteins containing an ATP-binding box or an ABC-ATPase domain, and these proteins are involved in coupling ATP hydrolysis energy with many physiological processes that are not always, but usually, related to transport. On the other hand, when cytoplasmic ABC-ATPase binds to a hydrophobic membrane domain (MD), it forms an ABC transporter, which is synonymous with a trafficking ATPase (or permeability of the input system). Most (but not all) ABC transporters form six hypothetical helical transmembrane segments [34]. Romano showed that the ABC transporter mcjd has high specificity for its substrate, and the experiment also proved that the nisin transporter *nisT* cannot recognize MccJ25. Similarly, *nisT* is also an ABC transporter, and these transporters depend on the energy generated by ATP hydrolysis as an energy source [35,36]. In this study, many ABC-ATPase-related genes, including *patM, tcyc, tcyj, adcA, yfmr* and *rexb*, were mutated. According to the transcriptomic analysis, the expression of some phosphoamino acid transporter-related genes, such as *phnE, phnC* and *rbsA*, was significantly downregulated. It is speculated that changes occur in the distribution of ATP energy, and a portion of ATP energy flows into the pathway related to ABC transport, which promotes the excretion of nisin outside the cell.

3.8.4. DEGs Related to Cysteine Thiometabolism Translation

Nisin is a small-molecular polypeptide composed of various amino acid molecules. Amino acid metabolism in all cells is closely related to nisin synthesis. Nisin, as a lanthanide compound, has an intramolecular sulfide bridge similar to those in other lanthanide compounds. During the modification process from the nisin precursor to mature nisin, serine and threonine in the core peptide must be hydrated by the cysteine protease NisB, followed by the coupling of these dehydrated residues with cysteine by cyclase (NisC), which is homologous to asparagine synthase, to form a sulfide ring [37]. Studies have shown that the automatic regulation of signal transduction in nisin biosynthesis [38] and the antibacterial mechanism of nisin, including binding with lipid II and the inhibition of bacterial cell wall synthesis [39–41], are related to the presence of sulfide bonds.

In this study, the mutated genes also included some genes involved in cysteine sulfur metabolism, such as cysteine tRNA synthase (CysS) and iron sulfur reductase (YtqA). According to our transcriptomic analysis, the expression of cysteine synthase (CysK), which is a gene related to *cysS*, was significantly upregulated, and the expression of the whole cysteine metabolic pathway was upregulated and enriched. With respect to the biosynthesis of nisin, the cysteine thiometabolism pathway is related to the formation of sulfide bonds. CysS catalyses the CO conversion of cysteine polysulfide [42], which improves the efficiency of nisin modification.

3.8.5. DEGs Related to Purine Metabolism

The de novo biosynthesis of purine plays an important role in various metabolic pathways. It is involved in not only nucleotide synthesis, but also other biosynthetic processes. *Pur*-type purine synthesis-related genes are related to different physiological activities of microorganisms. Yanfei Xia proved that purL, whose product catalyses the transformation of 5′-phosphoribosyl-N-formylglycinamide (FGAR) into 5′-phosphoribosyl-N-formyl-glycinamidine (FGAM) in the purine synthesis pathway, is a gene that affects *Bacillus subtilis* nematocidal activity [43]. Buendia-Claver analysed *Sinorhizobium fredii* with a mutation in the *purL* gene, and found that its specific lipopolysaccharide (LPS) changed [44]. Studies have shown that mutation of the *pur* gene usually leads to significant weakening of the virulence of human and animal pathogens. Mutation of the *purE* gene in *Brucella melitensis* reduces its ability to infect mice [45], and mutations of *purL* and *purF* in *Francisella tularensis* were also shown to significantly reduce the toxicity of the strain [46]. Maegawa analysed the relationship between the production of *Clostridium difficile* toxin and purine synthesis, and found that the production of *C. difficile* toxin was negatively correlated with the expression of the *purL* gene [47].

In this study, comparative genomic analysis revealed base mutations in the *purL* gene. According to the transcriptomic analysis, the expression levels of *purL, purC, purD* and *purE*, which are involved in the purine metabolism pathway, were significantly downregulated. Quanli Liu tested the characteristics of a *purL* mutant (*Bacillus amyloliquefaciens* L4) and proved, for the first time, that purl inactivation can significantly enhance the production of subtilosin A by *B. amyloliquefaciens* L4, resulting in a production level three times higher than that in the wild-type strain [36]. It can be inferred that mutations in the *purL* gene reduce the activity of its protein and improve the ability of *L. lactis* A32 to produce nisin.

4. Conclusions

In this study, the mutant *L. lactis* A32 was screened from among lxl strain bacteria subjected to ARTP and UV irradiation. Mutant A32 presented improved nisin production ability and nisin resistance. The results of comprehensive analysis show that strain A32 obtained in this study had a delayed growth rate compared with that of the original lxl strain. The comparative genomics results show that 107 genes in the A32 genome had mutations, and most base mutations were not located in the four well-researched nisin-related operons. The overall transcriptome sequencing and analysis showed that the output of nisin increased in *L. lactis* strain A32, which was accompanied by changes in the DNA replication-related gene *dnaG*, the ABC-ATPase transport-related genes *patM* and *tcyC*, the cysteine thiometabolism-related gene *cysS*, and the purine metabolism-related gene *purL*. This research provides novel insights into the traditional genetic mechanisms involved nisin production in *L. lactis*, which was far beyond the four traditional well-known nisin-related operons. This study revealed a series of DEGs related to high nisin yields, which provides important genomic information and highlights potential influencing factors for further improving nisin yields.

Supplementary Materials: The following supporting information can be downloaded at: https://www.mdpi.com/article/10.3390/fermentation8060255/s1. Table S1: Results of the nisin production stability of strains *L. lactis* A32 and A225; Table S2: Statistics of the genome sequencing data of *L. lactis* A32 and lxl; Table S3: GO functional categories of *L. lactis* A32 and lxl; Table S4: Numbers and lengths of reads and numbers of expressed genes detected by transcriptome sequencing in control and acid-treated samples; Figure S1: Nisin titre standard curve; Figure S2: Biomass curve of *L. lactis* A32 and lxl with increasing fermentation times.

Author Contributions: The research was conceived and designed by L.H., X.L. (Xiaomeng Liu), D.Z. and X.L. (Xinli Liu); compost experiments and microbial communities were analysed by L.H., X.L. (Xiaomeng Liu), J.L., C.W., Q.W., S.P. and D.Z.; and the manuscript was written and revised by L.H., X.R., D.Z. and X.L. (Xinli Liu). All authors have read and agreed to the published version of the manuscript.

Funding: This research was funded by the National Natural Science Foundation of China (No. 31901821), as well as a grant from the State Key Laboratory of Biobased Material and Green Papermaking, Qilu University of Technology, Shandong Academy of Sciences (No. ZZ20200137).

Institutional Review Board Statement: Not applicable.

Informed Consent Statement: Not applicable.

Data Availability Statement: Not applicable.

Conflicts of Interest: The authors declare that they have no competing financial interests.

References

1. Yi, Y.; Zhang, Q.; Wilfred, A.V. Insights into the evolution of lanthipeptide biosynthesis. *Protein Sci.* **2013**, *22*, 1478–1489. [CrossRef]
2. Joint FAO/WHO Expert Committee on Food Additives. *Specifications for the Identity and Purity of Food Additives and Their Toxicological Evaluation: Some Antibiotics, Twelfth Report of the Joint FAO/WHO Expert Committee on Food Additives, Geneva, 1–8 July 1968*; World Health Organization: Geneva, Switzerland, 1969. Available online: https://extranet.who.int/iris/restricted/handle/10665/40752 (accessed on 16 June 2012).
3. Wu, H.; Teng, C.; Liu, B.; Tian, H.; Wang, J.G. Characterization and long-term antimicrobial activity of the nisin anchored cellulose films. *Int. J. Biol. Macromol.* **2018**, *113*, 487–493. [CrossRef] [PubMed]
4. Wenting, L.; Rong, Z.; Tengteng, J.; Dur, E.; Sameena; Saeed, A.; Wen, Q. Improving nisin production by encapsulated *Lactococcus lactis* with starch/carboxymethyl cellulose edible films. *Carbohydr. Polym.* **2021**, *251*, 117062. [CrossRef]
5. Jung, Y.; Alayande, A.B.; Chae, S.; Kim, I.S. Applications of nisin for biofouling mitigation of reverse osmosis membranes. *Desalination* **2018**, *429*, 52–59. [CrossRef]
6. Andre, C.; de Jesus Pimentel-Filho, N.; de Almeida Costa, P.M.; Vanetti, M.C.D. Changes in the composition and architecture of *Staphylococcal* biofilm by nisin. *Braz. J. Microbiol.* **2019**, *50*, 1083–1090. [CrossRef] [PubMed]
7. Eming, S.A.; Martin, P. Wound repair and regeneration: Mechanisms, signaling, and translation. *Sci. Transl. Med.* **2014**, *6*, 265sr6. [CrossRef]
8. Mouritzen, M.V.; Andrea, A.; Qvist, K.; Poulsen, S.S.; Jenssen, H. Immunomodulatory potential of Nisin A with application in wound healing. *Wound Repair Regen.* **2019**, *27*, 650–660. [CrossRef]
9. Wei, Q.; Kun, Y.; Jun, L.; Ke, L.; Fuqiang, L.; Junhui, J.; Wei, Z. Precise management of chronic wound by nisin with antibacterial selectivity. *Biomed. Mater.* **2019**, *14*, 045008. [CrossRef]
10. Malvano, F.; Pilloton, R.; Albanese, D. A novel impedimetric biosensor based on the antimicrobial activity of the peptide nisin for the detection of *Salmonella* spp. *Food Chem.* **2020**, *325*, 126868. [CrossRef]
11. Malaczewska, J.; Kaczorek-Łukowska, E.; Wójcik, R.; Rękawek, W.; Andrzej, K.S. In vitro immunomodulatory effect of nisin on porcine leucocytes. *J. Anim. Physiol. Anim. Nutr.* **2019**, *103*, 882–893. [CrossRef]
12. Bagde, P.; Nadanathangam, V. Improving the stability of bacteriocin extracted from *Enterococcus faecium* by immobilization onto cellulose nanocrystals. *Carbohydr. Polym.* **2019**, *209*, 172–180. [CrossRef] [PubMed]
13. Rana, K.; Sharma, R.; Preet, S. Augmented therapeutic efficacy of 5-fluorouracil in conjunction with lantibiotic nisin against skin cancer. *Biochem. Biophys. Res. Commun.* **2019**, *520*, 551–559. [CrossRef] [PubMed]
14. Joo, N.E.; Ritchie, K.; Kamarajan, P.; Miao, D.; Kapila, Y.L. Nisin, an apoptogenic bacteriocin and food preservative, attenuates HNSCC tumorigenesis via CHAC1. *Cancer Med.* **2012**, *1*, 295–305. [CrossRef] [PubMed]
15. Siegers, K.; Entian, K.D. Genes involved in immunity to the lantibiotic nisin produced by *Lactococcus lactis* 6F3. *Appl. Environ. Microbiol.* **1995**, *61*, 1082–1089. [CrossRef]
16. Ra, S.R.; Qiao, M.; Immonen, T.; Pujana, I. Genes responsible for nisin synthesis, regulation and immunity form a regulon of two operons and are induced by nisin in *Lactoccocus lactis* N8. *Microbiology* **1996**, *142*, 1281–1288. [CrossRef]
17. Qiao, M.; Ye, S.; Koponen, O.; Ra, R.; Usabiaga, M.; Immonen, T.; Saris, P.E.J. Regulation of the nisin operons in *Lactococcus lactis* N8. *J. Appl. Bacteriol.* **1996**, *80*, 626–634. [CrossRef]
18. Li, H.; O'Sullivan, D.J. Identification of a nisI promoter within the nisABCTIP operon that may enable establishment of nisin immunity prior to induction of the operon via signal transduction. *J. Bacteriol.* **2006**, *188*, 8496–8503. [CrossRef]
19. Patnaik, R. Engineering complex phenotypes in industrial strains. *Biotechnol. Prog.* **2008**, *24*, 38–47. [CrossRef]
20. Jian, Z.; Qinggele, C.; Wenjing, F.; Xiuli, Z.; Bin, Q.; Guangrong, Z.; Jianjun, Q. Enhance nisin yield via improving acid-tolerant capability of *Lactococcus lactis* F44. *Sci. Rep.* **2016**, *6*, 27973. [CrossRef]
21. Shifu, C.; Yanqing, Z.; Yaru, C.; Jia, G. Fastp: An ultra-fast all-in-one FASTQ preprocessor. *Bioinformatics* **2018**, *34*, 884–890. [CrossRef]
22. Kolmogorov, M.; Yuan, J.; Yu, L.; Pevzner, P.A. Assembly of long error-prone reads using de Bruijn graphs. *Proc. Natl. Acad. Sci. USA* **2016**, *113*, 8396–8405. [CrossRef]
23. Walker, B.J.; Abeel, T.; Shea, T.; Priest, M.; Abouelliel, A. Pilon: An Integrated Tool for Comprehensive Microbial Variant Detection and Genome Assembly Improvement. *PLoS ONE* **2014**, *9*, e112963. [CrossRef] [PubMed]
24. Kurtz, S.; Phillippy, A.; Delcher, A.L.; Smoot, M. Versatile and open software for comparing large genomes. *Genome Biol.* **2004**, *5*, R12. [CrossRef] [PubMed]
25. Langmead, B.; Salzberg, S.L. Fast gapped-read alignment with Bowtie 2. *Nat. Methods* **2012**, *9*, 357–359. [CrossRef]
26. Li, B.; Dewey, C.N. RSEM: Accurate transcript quantification from RNA-Seq data with or without a reference genome. *BMC Bioinform.* **2011**, *12*, 323. [CrossRef]
27. Robinson, M.D.; McCarthy, D.J.; Smyth, G.K. EdgeR: A Bioconductor package for differential expression analysis of digital gene expression data. *Bioinformatics* **2010**, *26*, 139–140. [CrossRef]

28. Tjaden, B. De novo assembly of bacterial transcriptomes from RNA-seq data. *Genome Biol.* **2015**, *16*, 1. [CrossRef]
29. Hacker, C.; Christ, N.A.; Duchardt-Ferner, E.; Korn, S.; Berninger, L. The Solution Structure of the Lantibiotic Immunity Protein NisI and Its Interactions with Nisin. *J. Biol. Chem.* **2015**, *290*, 28869–28886. [CrossRef]
30. Kim, W.S.; Hall, R.J.; Dunn, N.W. Improving nisin production by increasing nisin immunity/resistance genes in the producer organism *Lactococcus lactis*. *Appl. Microbiol. Biotechnol.* **1998**, *50*, 429–433. [CrossRef]
31. Huang, Y.H.; Huang, C.Y. Comparing SSB-PriA Functional and Physical Interactions in Gram-Positive and -Negative Bacteria. *Methods Mol. Biol.* **2021**, *2281*, 67–80. [CrossRef]
32. Bianco, P.R. The mechanism of action of the SSB interactome reveals it is the first OB-fold family of genome guardians in prokaryotes. *Protein Sci.* **2021**, *30*, 1757–1775. [CrossRef] [PubMed]
33. Bianco, P.R.; Pottinger, S.; Tan, H.Y.; Nguyenduc, T.; Rex, K.; Varshney, U. The IDL of *E. coli* SSB links ssDNA and protein binding by mediating protein-protein interactions. *Protein Sci.* **2017**, *26*, 227–241. [CrossRef] [PubMed]
34. Young, J.; Holland, I.B. ABC transporters: Bacterial exporters-revisited five years on. *Biochim. Biophys. Acta (BBA) Biomembr.* **1999**, *1461*, 177–200. [CrossRef]
35. Romano, M.; Fusco, G.; Choudhury, H.G.; Mehmood, S.; Robinson, C.V.; Zirah, S.; Hegemann, J.D. Structural basis for natural product selection and export by bacterial ABC Transporters. *ACS Chem. Biol.* **2018**, *13*, 1598–1609. [CrossRef]
36. Beis, K.; Rebuffat, S. Multifaceted ABC transporters associated to microcin and bacteriocin export. *Res. Microbiol.* **2019**, *170*, 399–406. [CrossRef]
37. Hegemann, J.D.; Fouque, K.J.D.; Santos-Fernandez, M.; Fernandez-Lima, F. A bifunctional leader peptidase/ABC transporter protein is involved in the maturation of the lasso peptide cochonodin I from *Streptococcus suis*. *J. Nat. Prod.* **2021**, *84*, 2683–2691. [CrossRef]
38. Kuipers, O.P.; Beerthuyzen, M.M.; Pascalle, G.; Luesink, E.J.; Willem, M.V. Autoregulation of Nisin Biosynthesis in *Lactococcus lactis* by Signal Transduction. *J. Biol. Chem.* **1995**, *270*, 27299–27304. [CrossRef]
39. Sahl, H.G.; Jack, R.W.; Bierbaum, G. Biosynthesis and biological activities of lantibiotics with unique post-translational modifications. *Eur. J. Biochem.* **1995**, *230*, 827–853. [CrossRef]
40. Driessen, A.J.M.; Hooven, H.W.; Kuiper, W.; Kamp, M.; Sahl, H.G.; Konings, R.N.H.; Konings, W.N. Mechanistic studies of lantibiotic-induced permeabilization of phospholipid vesicles. *Biochemistry* **1995**, *34*, 1606–1614. [CrossRef]
41. Moll, G.N.; Konings, W.N.; Driessen, A.J. The lantibiotic nisin induces transmembrane movement of a fluorescent phospholipid. *J. Bacteriol.* **1998**, *180*, 6565–6570. [CrossRef]
42. Akaike, T.; Ida, T.; Wei, F.Y.; Nishida, M.; Kumagai, Y.; Alam, M.M. Cysteinyl-tRNA synthetase governs cysteine polysulfidation and mitochondrial bioenergetics. *Nat. Commun.* **2017**, *8*, 177. [CrossRef] [PubMed]
43. Xia, Y.F.; Xie, S.S.; Ma, X.; Wu, H.J.; Wang, X.; Gao, X.W. The purL gene of *Bacillus subtilis* is associated with nematicidal activity. *FEMS Microbiol. Lett.* **2011**, *322*, 99–107. [CrossRef] [PubMed]
44. Buendía-Clavería, A.M.; Moussaid, A.; Ollero, F.J.; Vinardell, J.M.; Torres, A. A purL mutant of Sinorhizobium fredii HH103 is symbiotically defective and altered in its lipopolysaccharide. *Microbiology* **2003**, *149*, 1807–1818. [CrossRef]
45. Crawford, R.M.; Verg, L.V.D.; Yuan, L.; Hadfield, T.L.; Warren, R.L.; Drazek, E.S.; Houng, H.H.; Hammack, C. Deletion of purE attenuates Brucella melitensis infection in mice. *Infect. Immun.* **1996**, *64*, 2188–2192. [CrossRef]
46. Kadzhaev, K.; Zingmark, C.; Golovliov, I.; Bolanowski, M.; Shen, H. Identification of genes contributing to the virulence of Francisella tularensis SCHU S4 in a mouse intradermal infection model. *PLoS ONE* **2009**, *4*, e5463. [CrossRef] [PubMed]
47. Maegawa, T.; Karasawa, T.; Ohta, T.; Wang, X.; Kato, H.; Hayashi, H.; Nakamura, S. Linkage between toxin production and purine biosynthesis in *Clostridium difficile*. *J. Med. Microbiol.* **2002**, *51*, 34–41. [CrossRef] [PubMed]

 *fermentation*

Article

Dynamic Regulation of Transporter Expression to Increase L-Threonine Production Using L-Threonine Biosensors

Sumeng Wang, Ruxin Hao , Xin Jin, Xiaomeng Li, Qingsheng Qi * and Quanfeng Liang *

State Key Laboratory of Microbial Technology, Shandong University, Jinan 250100, China;
smwang001@126.com (S.W.); hrx970625@163.com (R.H.); jinx.in@outlook.com (X.J.); lleky0713@163.com (X.L.)
* Correspondence: addresses: qiqingsheng@sdu.edu.cn (Q.Q.); liangquanfeng@sdu.edu.cn (Q.L.)

Abstract: The cytotoxicity of overexpressed transporters limits their application in biochemical production. To overcome this problem, we developed a feedback circuit for L-threonine production that uses a biosensor to regulate transporter expression. First, we used IPTG-induced *rhtA* regulation, L-threonine exporter, to simulate dynamic regulation for improving L-threonine production, and the results show that it had significant advantages compared with the constitutive overexpression of *rhtA*. To further construct a feedback circuit for *rhtA* auto-regulation, three L-threonine sensing promoters, P_{cysJ}, P_{cysD}, and P_{cysJH}, were characterized with gradually decreasing strength. The dynamic expression of *rhtA* with a threonine-activated promoter considerably increased L-threonine production (21.19 g/L) beyond that attainable by the constitutive expression of *rhtA* (8.55 g/L). Finally, the autoregulation method was used in regulating *rhtB* and *rhtC* to improve L-threonine production and achieve a high titer of 26.78 g/L (a 161.01% increase), a yield of 0.627 g/g glucose, and a productivity of 0.743 g/L/h in shake-flask fermentation. This study analyzed in detail the influence of dynamic regulation and the constitutive expression of transporters on L-threonine production. For the first time, we confirmed that dynamically regulating transporter levels can efficiently promote L-threonine production by using the end-product biosensor.

Keywords: biosensor; dynamic regulation; exporter; L-threonine production

Citation: Wang, S.; Hao, R.; Jin, X.; Li, X.; Qi, Q.; Liang, Q. Dynamic Regulation of Transporter Expression to Increase L-Threonine Production Using L-Threonine Biosensors. *Fermentation* **2022**, *8*, 250. https://doi.org/10.3390/fermentation8060250

Academic Editors: Mohamed Koubaa and Xian Zhang

Received: 19 April 2022
Accepted: 24 May 2022
Published: 26 May 2022

Publisher's Note: MDPI stays neutral with regard to jurisdictional claims in published maps and institutional affiliations.

1. Introduction

L-threonine (Thr, CAS 72-19-5), an essential amino acid for mammals, is widely used in food, cosmetics, animal feed, and pharmaceuticals [1]. L-threonine is one of four major commercial amino acids, the others being L-glutamic acid, L-lysine, and DL-methionine. The global production of L-threonine totals 700,000 metric tons annually [2]. Currently, microbial fermentation with catalyzation by six enzymes from oxalacetate—including *aspC*-encoded aspartate aminotransferase, *thrA* (*metL* or *lysC*)-encoded aspartate kinase and homoserine dehydrogenase, *asd*-encoded aspartyl semialdehyde dehydrogenase, *thrB*-encoded homoserine kinase, and *thrC*-encoded threonine synthase—is the main method of L-threonine production (Figure 1) [3]. L-threonine belongs to the aspartate family of amino acids. First, L-aspartate is generated from oxaloacetate by L-aspartate aminotransferase. L-aspartate is then catalyzed by aspartate kinase to aspartate phosphate. This is followed by a two-step reduction reaction catalyzed by aspartyl semialdehyde dehydrogenase and homoserine dehydrogenase to generate L-homoserine. The homoserine kinase catalyzes the phosphorylation of L-homoserine to L-homoserine phosphate, which is finally catalyzed by threonine synthase to L-threonine.

Many strategies have been explored to synthesize L-threonine, including cofactor engineering, energy adjustment, the overexpression of rate-limiting steps, the removal of feedback inhibition, and the removal of competitive pathways [1,3–5]. Transporter engineering has also emerged as an attractive approach to improve L-threonine production [6–10]. The overexpression of transporters is a commonly used tactic to increase strain tolerance by pumping out products [3,7,8,11]. However, the overexpression of membrane proteins seriously hinders the homeostasis of proteins in the cytoplasm [12,13], which is not conducive to efficient catalytic anabolites. For example, overexpressing membrane proteins can reduce the levels of the respiratory chain complexes succinate dehydrogenase and cytochrome *bd* and *bo₃* oxidases in *Escherichia coli* [13]. Therefore, the expression of transporters should be carefully modified for chemical production without disturbing the physiological metabolism. In addition, some transporter engineering strategies were also performed in microbial manufacturing, such as omics-guided transporter mining, transporter evolution to enhance transport efficiency and substrate specificity, and the heterologous expression of transporter [9]. However, transporters obtained from the above strategies still need to be overexpressed to improve product outflow. One promising method of performing this is to dynamically regulate transporter expression through a general biosensor. Dynamic regulation can automatically control the level of transporters according to the needs of cells, which can not only alleviate the adverse effects of overexpressed transporters on cells, but also timely discharge intracellular metabolites. To produce n-butanol and 1-alkene, previous studies have induced transporter expression by responding to membrane stress and exhausted glucose [14,15]. However, such a general biosensor needs to be coupled with the cell growth state to perform its regulatory functions. The cell growth state can be affected by many factors, such as hyperosmotic stress condition [16], which lead to the non-precise regulation of transporter expression by using the cell-growth-state-dependent biosensor. In contrast to the general biosensor, a product-specific biosensor finely regulates transporter expression without being affected by culture conditions and changes in cell physiology. Many end-product-specific biosensors have been designed and utilized to dynamically regulate metabolic pathways to improve chemical production [17,18]. Although Lange et al. found that the exporter BrnFE can be activated by the transcription factor *lrp* and that its expression increased with the addition of branched-chain amino acids [19], the effects of dynamically regulated transporter expression by using end-product biosensors have not been studied in detail. The comparison of the effects of dynamic regulation and the constitutive expression of transporters on product synthesis has not been reported.

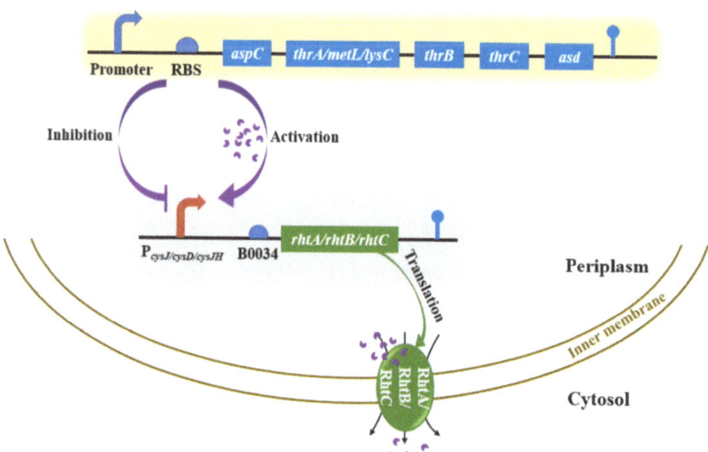

Figure 1. Scheme of sensing L-threonine to regulate transporter expression. The L-threonine production pathway consists of five genes: aspartate aminotransferase encoded by *aspC*; aspartate kinases encoded by *thrA*, *metL*, or *lysC*; semialdehyde dehydrogenase encoded by *asd*; homoserine dehydrogenase encoded by *thrA*; homoserine kinase encoded by *thrB*; and threonine synthase encoded by *thrC*. Three transporters were encoded by *rhtA*, *rhtB*, *and rhtC*. L-threonine transporters belong to the inner membrane. P_cysJ, P_cysD, and P_cysJH are L-threonine-sensed promoters for activating gene translation.

In this study, we use the native L-threonine-sensing promoters P_{cysJ}, P_{cysD}, and P_{cysJH} to study in detail the effect of dynamically regulated transporter expression for L-threonine production (Figure 1). Compared with the constitutive expression of transporters, dynamic regulation significantly improved L-threonine production with an increase of approximately 147%. This regulation can also be extended to two other native transporters, *rhtB* and *rhtC*.

2. Materials and Methods

2.1. Chemicals and Reagents

The L-threonine standard was purchased from Sigma-Aldrich (St. Louis, MO, USA). Phanta HS Super-Fidelity DNA Polymerase was purchased from Vazyme Biotech (Nanjing, China). High-performance liquid chromatography grade acetonitrile (75-05-8) was obtained from Tedia (ACN, Tedia Company, Inc., Fairfield, OH, USA). Isopropyl-β-d-thiogalactoside (IPTG; CAS, 367-93-1) and spectinomycin dihydrochloride pentahydrate (CAS, 22189-32-8) were purchased from Sangon Biotech (Shanghai, China) Co., Ltd. Triethylamine (CAS, 121-44-8) and phenyl isothiocyanate (CAS, 103-72-0) were purchased from Aladdine. Other reagents were purchased from Sinopharm Chemical Reagent Beijing Co. Ltd. (Shanghai, China).

2.2. Construction of Strains and Plasmids

All strains and plasmids used in this study are listed in Tables S1 and S2. *E. coli* DH5α was used to reconstruct plasmids, and *E. coli* K-12 MG1655 was used for L-threonine sensor characterization. The L-threonine-producing Tm strain was provided by Fufeng Group. The strains Tm-100AR, Tm-101AR, and Tm-109AR containing plasmids cl-100AR, cl-101AR, and cl-109AR, respectively, were constructed to study the effect of *rhtA* levels on L-threonine production. We introduced pCL1920 and cl-*rhtA* into Tm to construct strains Tm-cl and Tm-AR to study the effects of *rhtA* expression times on cell growth and L-threonine production. MG-cl, MG-P_{cysJ}, MG-P_{cysD}, and MG-P_{cysJH} were constructed by transforming the plasmids pCL1920, cl-JAR, cl-DAR, and cl-JHAR into *E. coli* K-12 MG1655. Tm-P_{cysJ}, Tm-P_{cysD}, and Tm-P_{cysJH} were constructed by transforming the plasmids cl-JAR, cl-DAR, and cl-JHAR into Tm. Tm-JAR, Tm-DAR, Tm-JHAR, Tm-105AR, Tm-110AR, and Tm-116AR were engineered by introducing cl-JAR, cl-DAR, cl-JHAR, cl-105AR, cl-110AR, and cl-116AR into Tm to analyze the influence of the dynamic regulation of *rhtA* on L-threonine production. Tm-JB and Tm-JC were constructed by introducing cl-JB or cl-JC into Tm and used to further dynamically regulate the native transporters to improve L-threonine production. All engineered strains were constructed by transforming the corresponding plasmids into *E. coli* K-12 MG1655 or Tm with the calcium chloride (CaCl$_2$) or electroporation transformation methods.

The primers used for plasmid reconstruction are listed in Table S3. The low copy number plasmid pCL1920 with the pSC101 replicon was used as the cloning vector. All plasmids in this study were constructed using the Gibson assembly method [20]. To construct cl-*rhtA*, the backbone was amplified from pCL1920 with the primers cl-F1 and cl-R1, and *rhtA* was amplified from the *E. coli* K-12 MG1655 genome with the primers cl-rhtA-F and cl-rhtA-R by using Phanta HS Super-Fidelity DNA Polymerase (Vazyme Biotech, Nanjing, China). cl-*rhtA* was assembled by fusing the amplified backbone and *rhtA* with the 2X MultiF Seamless Assembly Mix (ABclonal Technology Co., Ltd. Wuhan, China). P*cysJ*-*rfp*, P*cysD*-*rfp*, and P*cysJH*-*rfp* were constructed through fusing the promoters P_{cysJ}, P_{cysD}, and P_{cysJH} amplified from the *E. coli* K-12 MG1655 genome with the reporter *rfp* and then was inserted into amplified pCL1920 backbone. The dynamically regulated *rhtA* plasmids containing cl-JAR, cl-DAR, and cl-JHAR were constructed with the frame of the $P_{cysJ/cysD/cysJH}$-B0034-*rhtA*-terminator BBa_B1006-P_{cysJ} B0034-*rfp*. To constitutively express *rhtA*, we selected three different transcription levels of promoters (J23105, J23110, and J23116) from iGEM (http://parts.igem.org/Main_Page /(accessed on December 2021)). cl-105AR, cl-110AR, and cl-116AR were constructed according to the form of the $P_{J23105/J23110/J23116}$-B0034-*rhtA*-terminator BBa_B1006-P_{cysJ} B0034-*rfp*. The native L-threonine exporters *rhtB* and *rhtC* were amplified from the *E. coli* K-12 MG1655 genome and cloned into pCL1920 to generate the recombinant plasmids cl-JB and cl-JC for dynamic regulation.

2.3. Characterization of the L-Threonine-Sensing Promoters

To characterize the promoters P_{cysJ}, P_{cysD}, and P_{cysJH}, the fluorescence intensity of the reporter gene *rfp* was monitored using a multidetection microplate reader (Synergy HT, BioTek, Winooski, VT, USA). Single colonies were grown in 12-well microassay plates with 2 mL of LB medium supplemented with spectinomycin dihydrochloride pentahydrate at 37 °C for approximately 12 h. The feeds were then transferred to a 96-well microassay plate containing 200 µL Luria–Bertani (LB) medium with 2% (*v/v*) inoculation at 37 °C for 48 h. Spectinomycin dihydrochloride pentahydrate was added during the cultivation process to maintain the stabilization of the plasmids. The fluorescence intensity of *rfp* was measured through excitation at 590 nm and emission at 645 nm. The characterization was calculated with the ratio of the fluorescence intensity of *rfp* to the optical density at 600 nm (OD_{600}). All measurements were performed in triplicate.

2.4. Culture Medium and L-Threonine Fermentation

A LB medium containing 10 g/L tryptone, 5 g/L yeast extract, and 10 g/L NaCl was used for plasmid construction and L-threonine sensor characterization. The L-threonine fermentation medium consisted of 15 g/L of $(NH_4)_2SO_4$, 2 g/L of KH_2PO_4, 1 g/L of $MgSO_4 \cdot 7H_2O$, 2 g/L of yeast extract, and 0.02 g/L of $FeSO_4$ [8]. We also added 40 g/L of glucose as the initial carbon source and 20 g/L $CaCO_3$ to adjust the pH during fermentation. For shake-flask fermentation, single colonies were cultured in fresh LB medium at 37 °C for approximately 12 h. The precultured seeds were subsequently transformed into a 300 mL shake flask attached with flaps containing a 20 mL fermentation medium with 1% (*v/v*) inoculation. Fermentation was performed at 220 rpm, 37 °C for 36 h. When glucose was lower than 10 g/L, it was supplemented up to 40 g/L. *rhtA* expression was induced with 0.1/0.2/0.5 mM of IPTG when necessary.

Analytical Methods

A 1 mL culture was vigorously vortexed, and 0.1 mL of it was transferred to a 1.5 mL centrifuge tube. We added 0.9 mL of 1 mM HCl and mixed the solution to remove the residual $CaCO_3$. Subsequently, OD_{600} was measured using a spectrophotometer (Shimazu, Kyoto, Japan). For glucose and L-threonine detection, the culture was centrifuged at 12,000 rpm for 2 min to collect the supernatant. The collected supernatant was filtered with a 0.22 µm aqueous membrane for glucose analysis. Glucose was quantified with a high-performance liquid chromatography (HPLC) system (Shimadzu, Kyoto, Japan) equipped with a refractive index detector (RID-10A; Shimadzu, Kyoto, Japan) and an Aminex HPX-87H ion exclusion column (Bio-Rad Laboratories, Hercules, CA, USA); 5 mM H_2SO_4 was used as the mobile phase at a flow rate of 0.6 mL/min [21].

For the detection of L-threonine, the collected supernatant was initially deproteinized with 5% trichloroacetic acid. Subsequently, the pretreated supernatant was derived with triethylamine and phenyl isothiocyanate and then extracted with n-hexane. Briefly, 0.2 mL of samples and standard L-threonine were pretreated with the mixture of triethylamine–acetonitrile (1.4 mL of triethylamine mixed with 8.6 mL of acetonitrile). Next, we added phenyl isothiocyanate–acetonitrile (25 µL of phenyl isothiocyanate mixed with 2 mL of acetonitrile) to pretreat the samples and standard L-threonine for 1 h at room temperature. We added 0.4 mL of n-hexane and vigorously shocked. The bottomed liquid (0.2 mL) was collected and diluted with 0.8 mL of deionized water. The solution was filtered with a 0.22 µm organic membrane, and samples were detected using an HPLC equipped with a diode array detector (SPD-M20A; Shimadzu, Kyoto, Japan), with the VenusilAA (4.6 × 250 mm, 5 µm, Agela Techanologies) column applied at 40 °C. The analysis was performed at a flow rate of 1 mL/min with mobile phases consisting of (A) 15.2 g of sodium acetate dissolved in 1850 mL of ultrapure water and mixed with 140 mL of acetonitrile and

(B) 80% (v/v) acetonitrile and 20% (v/v) ultrapure water. The concentration of L-threonine was quantified with the peak area according to the corresponding standard curve.

2.5. Statistical Analysis

Results are presented as the mean $\pm$ SEM. Differences between the means were evaluated using a one-way ANOVA. $p < 0.05$ was considered statistically significant.

3. Results and Discussion

3.1. Simulating Dynamic rhtA Regulation with IPTG Induction

The membrane protein RhtA consists of 10 predicted transmembrane segments, which were identified and characterized as L-threonine and L-homoserine transporters [22]. Constitutive overexpressing *rhtA* is the commonly method used to improve L-threonine production and removing the toxic intermediate metabolite L-homoserine [22–26]. In this study, we used the industrial L-threonine production host Tm, engineered from *E. coli* K-12 MG1655, from the Fufeng Group (http://www.fufeng-group.com/ (accessed on January 2021)) as the parent strain.

To verify whether the dynamic regulation of transporter expression can improve L-threonine production, we firstly explored the effect of constitutive overexpressed *rhtA* on L-threonine production. Three promoters at strong (J23100, 1348), medium (J23101, 399), and weak (J23109, 18) from the Registry of Standard Biological Parts (http://parts.igem.org/Main_Page(accessed on January 2021)) were characterized in strains Tm-100AR, Tm-101AR, Tm-109AR, respectively (Figure 2A). As illustrated in Figure 2A, compared with the control (strain Tm-cl without *rhtA* overexpression), overexpressing *rhtA* enhanced L-threonine production due to relief from the toxicity of the accumulated products through the timely export of L-threonine [9,27]. Tm-100AR and Tm-109AR significantly increased the titer of L-threonine ($p < 0.0001$) than Tm-101AR. However, there was an irregular correlation between the increase in production and the intensity of *rhtA* levels. Then, we used IPTG-induced *rhtA* regulation to simulate the availability of dynamic regulation. *rhtA* was induced with 0.5 mM of IPTG in different growth phases (OD 0, 3, 6, or 9) (Figure 2B). When enhancing *rhtA* expression at different growth stages, host cells with delayed *rhtA* induction achieved a higher L-threonine titer than constitutive expression (Figure 2B). *rhtA* expressed at OD 3 achieved higher titers at 15.4 g/L than early at OD 0 ($p < 0.05$) with 13.4 g/L or delayed induction at OD 6 ($p < 0.01$) or 9 ($p < 0.001$) with 13.4 g/L and 10.7 g/L, respectively. The premature expression of high-level *rhtA* causes membrane burden. *rhtA* expresses too late to excrete the accumulated L-threonine with low-level *rhtA* in the cell timely, resulting in metabolic burden. Compared to the strength of the constitutive promoter shown in Figure 2A, the promoter strength of Plac-lacO (RFP/OD$_{600}$ values from 6.7 to 55.4) can be evaluated to be roughly between J23101 (RFP/OD$_{600}$ value, 399) and J23109 (RFP/OD$_{600}$ value, 18), suggesting that increased L-threonine productivity at OD 0–9 may be not due to higher intensity of Plac-lacO. Therefore, the dynamic regulation of *rhtA* in different fermentation phases can improve L-threonine production, and it had significant advantages compared with the constitutive overexpression of *rhtA*. Although the transporter *rhtA* was successfully regulated to increase L-threonine production, the addition of the exogenous inducer IPTG is impracticable in industrial production due to the high production cost [28].

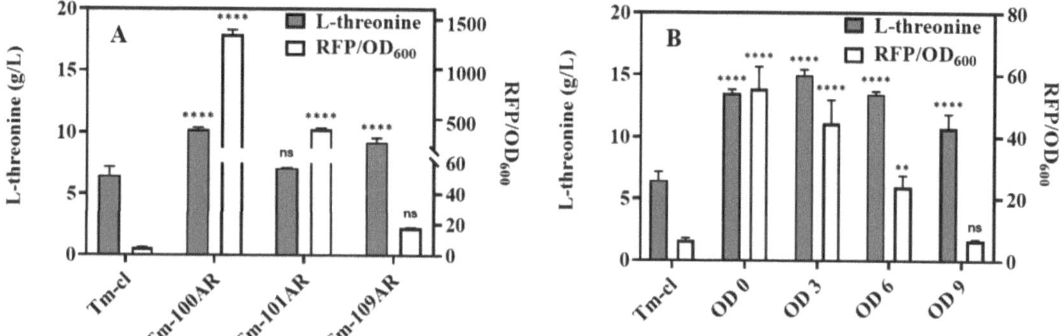

Figure 2. Adding IPTG at different growth stages to dynamically control *rhtA* expression. (**A**) *rhtA* was constitutively expressed with strong (promoter J23100), medium (promoter J23101), and weak (promoter J23109) intensities in strains Tm-100AR, Tm-101AR, and Tm-109AR, respectively. Strain Tm-cl without overexpressed *rhtA* as the control. (**B**) The Tm-AR strain containing overexpressed *rhtA* was induced with 0.5 mM of IPTG at OD 0, 3, 6, and 9. L-threonine titers were measured at 36 h of fermentation. All results were calculated with three (n = 3) independent replicates. Error bars represent mean ± SEM. **** $p < 0.0001$, ** $p < 0.01$. ns: non-significance.

3.2. Characterizing the L-Threonine-Sensing Promoters

Biosensors have emerged as an outstanding tool for dynamically adjusting gene expression in metabolic engineering [18,29]. Several *E. coli*-derived L-threonine-sensing promoters have been excavated through omics analysis [30]. In this paper, the L-threonine sensor was characterized by placing the report gene *rfp* under the promoter P_{cysJ}, P_{cysD}, or P_{cysJH}. We initially tested the response in *E. coli* K-12 MG1655 with the strains MG-P_{cysJ}, MG-P_{cysD}, and MG-P_{cysJH}. After removing the background value of RFP/OD$_{600}$ from strain MG-cl, MG1655 with empty plasmid pCL1920, all three promoters can be activated by adding L-threonine (Figure 3A). The intensity of expression continuously decreased for P_{cysJ}, P_{cysD}, and P_{cysJH}. To demonstrate that the three promoters were practicable in the L-threonine-producing strain and determine the lowest activation threshold, the strains Tm-P_{cysJ}, Tm-P_{cysD}, and Tm-P_{cysJH} were constructed by introducing the plasmid P_{cysJ}-*rfp*, P_{cysD}-*rfp*, or P_{cysJH}-*rfp* into the industrial strain Tm. The results demonstrate that the three promoters can be used to activate gene expression during the L-threonine production process (Figure 3B). Furthermore, expression levels increased with increased L-threonine concentration (Figure 3B,C). As indicated in Table S4, the three promoters had different minimum sensing thresholds during fermentation: P_{cysJ} and P_{cysJH} at 3 h with 0.05 g/L and P_{cysD} at 5 h with 0.25 g/L of L-threonine. The highest concentrations of L-threonine sensing were not measured because the fluorescence intensity of *rfp* continued to increase at the final concentration of L-threonine (Figure 3C). Nonetheless, we quantified the highest sensing threshold (~60 g/L) with MG-P_{cysJ}, MG-P_{cysD}, and MG-P_{cysJH} by exogenously adding different concentrations of L-threonine (Figure 3A).

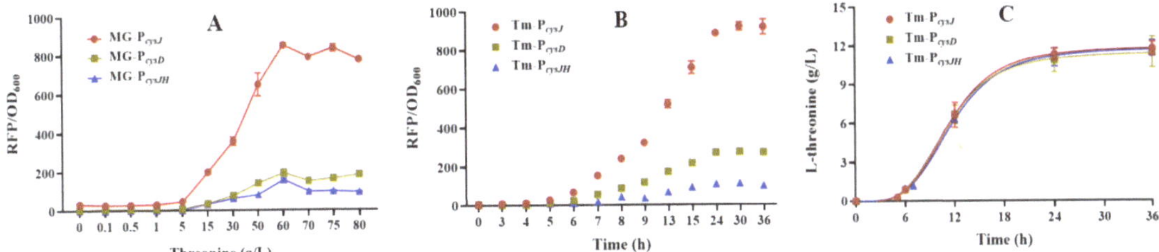

Figure 3. Characterization of the L-threonine-sensing promoters P*cysJ*, *cysD*, and *cysJH*. (**A**) Plasmids P*cysJ*-*rfp*, P*cysD*-*rfp*, and P*cysJH*-*rfp* were introduced into *E. coli* K-12 MG1655 to construct MG-P*cysJ*, MG-P*cysD*, and MG-P*cysJH* for characterization by adding 0–80 g/L L-threonine. Strain MG-cl containing the empty plasmid pCL1920 was the control for removing the background value of RFP/OD$_{600}$. (**B**) The sensing curves of the three promoters were monitored in the strains Tm-J, Tm-D, and Tm-JH containing the plasmids P*cysJ*-*rfp*, P*cysD*-*rfp*, and P*cysJH*-*rfp*. (**C**) Monitoring of the L-threonine concentrations of Tm-J, Tm-D, and Tm-JH during the fermentation process. All results were calculated with three (*n* = 3) independent replicates.

3.3. Improving L-Threonine Production through the Dynamic Regulation of rhtA Expression

The characterized sensing promoters P$_{cysJ}$, P$_{cysD}$, and P$_{cysJH}$ were used for gradually increasing *rhtA* expression in the L-threonine-producing strain Tm, corresponding to the strains Tm-JAR, Tm-DAR, and Tm-JHAR, respectively. To estimate the expression of *rhtA*, the linked *rfp* gene was activated by L-threonine-responsive promoters. Varying strengths of promoters J23105, J23110, and J23116 were selected to constitutively express *rhtA* in the strains Tm-105AR, Tm-110AR, and Tm-116AR, respectively. The strain Tm-cl contained the empty vector as the control. The expression levels of *rhtA* in strains with L-threonine-responsive promoters were gradually increased during the fermentation process with the analysis of RFP/OD$_{600}$ (Figure 4A). As illustrated in Figure 4A, the promoter J23110 had the same effect as P$_{cysJ}$ on *rhtA* expression at 36 h. Comparing the constitutive expression of *rhtA* with the promoter J23100 in the strain Tm-110AR, dynamically regulating the strain Tm-JAR with P$_{cysJ}$ resulted in a significantly improved L-threonine production of up to 21.19 g/L of titers, an increase of approximately 147% than the unexpressed strain (Figure 4B). The other L-threonine-sensing promoters P$_{cysD}$ and P$_{cysJH}$ had lower L-threonine titers than P$_{cysJ}$, 19.11 and 9.49 g/L, respectively, which are a result of the weaker expression of *rhtA*. These results suggest that stronger dynamic expression of *rhtA* is more conducive to L-threonine production. The difference in expression levels of *rhtA* may be due to the different transcription and translation strengths of P$_{cysJ}$, P$_{cysD}$ and P$_{cysJH}$. It may also be caused by the different affinity for L-threonine of the currently unknown transcription regulators. Although a stronger constitutive expression of *rhtA* with the promoter J23116 can increase the L-threonine titer, it is still lower than the dynamic expression with the promoter P$_{cysJ}$ and with a slower cell growth (Figure 4C). These results indicate that the dynamic regulation of *rhtA* expression is more conducive to promoting L-threonine production. Dynamic regulation is a successful strategy to improve biochemical production [4,31–34]. Recently, a study has developed a thermal switch system to increase L-threonine yield through regulating carbon distribution and cofactor supplementation by sensing temperature [4]. Apart from the generally utilized metabolic flux engineering, in this study, we confirmed in the first time that dynamically engineering the transporter *rhtA* is more effective than constitutive expression for improving L-threonine production.

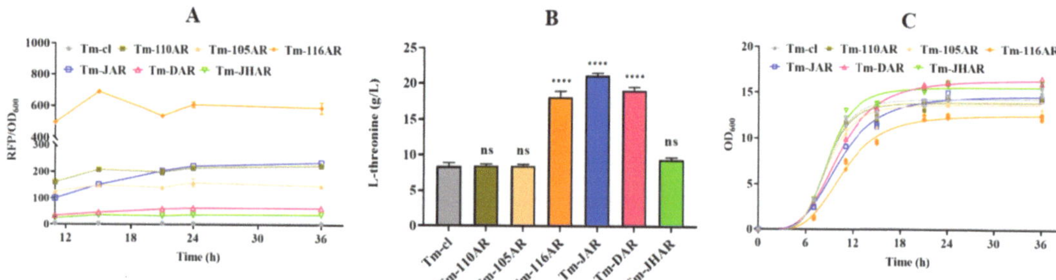

Figure 4. Analysis of the effects of dynamically and constitutively expressed *rhtA* on L-threonine production. (**A**) Analysis of the expression levels of *rhtA* during the whole fermentation process. Characterization of *rhtA* expression levels of the constitutive promoters (J23110, J23105, and J23116) and L-threonine-activated promoters (P_{cysJ}, P_{cysD}, and P_{cysJH}) with fluorescence intensity to OD_{600}. (**B**) L-threonine fermentation with dynamically or constitutively expressed *rhtA*. The constitutive promoters J23105, J23110, and J23116 were used to express *rhtA* in the strains Tm-105AR, Tm-110AR, and Tm-116AR, respectively. The L-threonine sensing promoters P_{cysJ}, P_{cysD}, and P_{cysJH} were used to control *rhtA* expression in the strains Tm-JAR, Tm-DAR, and Tm-JHAR, respectively. (**C**) Comparison of cell growth on varying strains with L-threonine-sensing promoters or constitutive promoters. All results were calculated with three ($n = 3$) independent replicates. Error bars represent mean ± SEM. **** $p < 0.0001$. ns: non-significance.

3.4. Regulating rhtB and rhtC to Further Improve L-Threonine Production

In *E. coli*, three native permeases—RhtA, RhtB, and RhtC encoded with *rhtA*, *rhtB*, and *rhtC*, respectively—were confirmed as the membrane proteins to pump out L-threonine [3]. RhtA belong to the category of nonspecific transporters for exporting L-threonine and L-homoserine. RhtB also transports homoserine lactone (HSL) [7,35]. By contrast, RhtC is the specific transporter for L-threonine [7]. Studies have reported that the expression of all these exporters results in the efficient pumping of L-threonine and increased production [1,3,36]. In this study, we demonstrated that L-threonine titers were considerably increased through the dynamic regulation of *rhtA* expression (Figure 4B). The sensing promoter P_{cysJ} had the greatest impact on increasing L-threonine production. To explore the influence of the autoregulated expression of *rhtB* and *rhtC* on L-threonine production, *rhtB* and *rhtC* were controlled by P_{cysJ} by responding to L-threonine (Figure 5). The plasmids P_{cysJ}-*rhtB* and P_{cysJ}-*rhtC* were transformed into the parent strain Tm and formed Tm-JB or Tm-JC. As illustrated in Figure 5A and Table 1, compared with the unexpressed transporter and overexpressed *rhtC*, the dynamically expressed *rhtB* resulted in the highest L-threonine titers of 26.78 g/L, an approximately 161% increase, and a productivity up to 0.734 g/L/h. Because Tm-JB consumed more glucose during the 36 h fermentation (Figure 5C), the yield (0.627 g/g) was slightly lower than that of Tm-JC (0.665 g/g). Although Tm-JB grew less than Tm-JC during the 0 to 18 h period, the final biomass at 36 h was higher (Figure 5B). These results demonstrate that dynamically regulating *rhtB* expression is relatively beneficial for L-threonine production.

Table 1. Titers, titer increase, yield, and productivity of L-threonine under dynamically regulated RhtB and RhtC.

	Titer (g/L)	Titer Increase (%)	Yield (g/g)	Productivity (g/L/h)
Tm-cl	10.26 (±0.68)	——	0.402 (±0.006)	0.285 (±0.019)
Tm-JB	26.78 (±0.67)	161.01 (±6.61)	0.627 (±0.023)	0.743 (±0.018)
Tm-JC	25.44 (±0.31)	147.95 (±3.02)	0.665 (±0.010)	0.706 (±0.008)

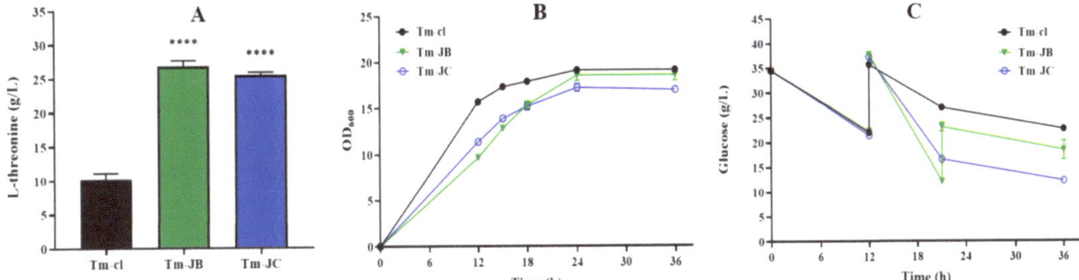

Figure 5. Dynamic regulation of *rhtB* and *rhtC* expression to improve L-threonine production. Strain Tm-cl containing the nonexpressed transporter as the control, *rhtB*, and *rhtC* was overexpressed in the strains Tm-JB and Tm-JC, respectively. (**A**) L-threonine titers were detected at 36 h of fermentation. (**B,C**) Glucose consumption and cell growth were monitored for 36 h of fermentation. All results were calculated with three ($n = 3$) independent replicates. Error bars represent mean $\pm$ SEM. **** $p < 0.0001$.

4. Conclusions

The toxicity of transporter expression to cells limits the application of transporter engineering in improving metabolite production. For the first time, we compared the effects of dynamic and constitutive expression transporters on product synthesis. We analyzed the effect of the dynamic regulation of transporters on L-threonine production based on end-product biosensors. These results demonstrate that the dynamic regulation of transporter expression was more conducive than constituent expression for L-threonine production, which increased by 147%. Therefore, the dynamic regulation of transporters is a promising strategy for improving biochemical production.

Supplementary Materials: The following supporting information can be downloaded at: https://www.mdpi.com/article/10.3390/fermentation8060250/s1, Table S1. Plasmids used in this study; Table S2. Strains used in this study; Table S3. Primers used in this study; Table S4. Analysis of the initial sensing times and L-threonine concentrations of the three promoters in the L-threonine-producing strains [37].

Author Contributions: S.W.: Investigation, conceptualization, and writing original draft. R.H., X.J. and X.L.: Data curation and formal analysis. Q.Q. and Q.L.: Conceptualization, funding acquisition, resource management, project administration, supervision, and writing—review and editing. All authors have read and agreed to the published version of the manuscript.

Funding: This work was supported by the National Key Research and Development Program of China (2019YFA0706900) and National Natural Science Foundation of China (31961133014, 31770095, 31971336).

Institutional Review Board Statement: Not applicable.

Informed Consent Statement: Not applicable.

Data Availability Statement: Not applicable.

Acknowledgments: Strain Tm was kindly provided by the Fufeng Group. We thank Chengjia Zhang and Nannan Dong from the Analysis and Testing Center of the State Key Laboratory for Microbial Technology (Shandong University) for assistance in the fermentation experiment.

Conflicts of Interest: The authors declare no conflict of interest.

References

1. Du, H.; Zhao, Y.; Wu, F.; Ouyang, P.; Chen, J.; Jiang, X.; Ye, J.; Chen, G.Q. Engineering Halomonas bluephagenesis for L-Threonine production. *Metab. Eng.* **2020**, *60*, 119–127. [CrossRef] [PubMed]
2. Wendisch, V.F. Metabolic engineering advances and prospects for amino acid production. *Metab. Eng.* **2020**, *58*, 17–34. [CrossRef] [PubMed]

3. Dong, X.; Quinn, P.J.; Wang, X. Metabolic engineering of Escherichia coli and Corynebacterium glutamicum for the production of L-threonine. *Biotechnol. Adv.* **2011**, *29*, 11–23. [CrossRef] [PubMed]

4. Fang, Y.; Wang, J.; Ma, W.; Yang, J.; Zhang, H.; Zhao, L.; Chen, S.; Zhang, S.; Hu, X.; Li, Y.; et al. Rebalancing microbial carbon distribution for L-threonine maximization using a thermal switch system. *Metab. Eng.* **2020**, *61*, 33–46. [CrossRef]

5. Lee, K.H.; Park, J.H.; Kim, T.Y.; Kim, H.U.; Lee, S.Y. Systems metabolic engineering of Escherichia coli for L-threonine production. *Mol. Syst. Biol.* **2007**, *3*, 149. [CrossRef]

6. Zhao, L.; Lu, Y.; Yang, J.; Fang, Y.; Zhu, L.; Ding, Z.; Wang, C.; Ma, W.; Hu, X.; Wang, X. Expression regulation of multiple key genes to improve L-threonine in Escherichia coli. *Microb. Cell Factories* **2020**, *19*, 46. [CrossRef]

7. Diesveld, R.; Tietze, N.; Fürst, O.; Reth, A.; Bathe, B.; Sahm, H.; Eggeling, L. Activity of exporters of Escherichia coli in Corynebacterium glutamicum, and their use to increase L-threonine production. *J. Mol. Microbiol. Biotechnol.* **2009**, *16*, 198–207. [CrossRef]

8. Kruse, D.; Krämer, R.; Eggeling, L.; Rieping, M.; Pfefferle, W.; Tchieu, J.H.; Chung, Y.J.; Saier, M.H., Jr.; Burkovski, A. Influence of threonine exporters on threonine production in Escherichia coli. *Appl. Microbiol. Biotechnol.* **2002**, *59*, 205–210.

9. Zhu, Y.; Zhou, C.; Wang, Y.; Li, C. Transporter Engineering for Microbial Manufacturing. *Biotechnol. J.* **2020**, *15*, e1900494. [CrossRef]

10. van der Hoek, S.A.; Borodina, I. Transporter engineering in microbial cell factories: The ins, the outs, and the in-betweens. *Curr. Opin. Biotechnol.* **2020**, *66*, 186–194. [CrossRef]

11. Pereira, R.; Wei, Y.; Mohamed, E.; Radi, M.; Malina, C.; Herrgård, M.J.; Feist, A.M.; Nielsen, J.; Chen, Y. Adaptive laboratory evolution of tolerance to dicarboxylic acids in Saccharomyces cerevisiae. *Metab. Eng.* **2019**, *56*, 130–141. [CrossRef]

12. Gubellini, F.; Verdon, G.; Karpowich, N.K.; Luff, J.D.; Boël, G.; Gauthier, N.; Handelman, S.K.; Ades, S.E.; Hunt, J.F. Physiological response to membrane protein overexpression in E. coli. *Mol. Cell. Proteom. MCP* **2011**, *10*, M111.007930. [CrossRef]

13. Wagner, S.; Baars, L.; Ytterberg, A.J.; Klussmeier, A.; Wagner, C.S.; Nord, O.; Nygren, P.A.; van Wijk, K.J.; de Gier, J.W. Consequences of membrane protein overexpression in Escherichia coli. *Mol. Cell. Proteom. MCP* **2007**, *6*, 1527–1550. [CrossRef]

14. Boyarskiy, S.; Davis López, S.; Kong, N.; Tullman-Ercek, D. Transcriptional feedback regulation of efflux protein expression for increased tolerance to and production of n-butanol. *Metab. Eng.* **2016**, *33*, 130–137. [CrossRef]

15. Zhou, Y.J.; Hu, Y.; Zhu, Z.; Siewers, V.; Nielsen, J. Engineering 1-Alkene Biosynthesis and Secretion by Dynamic Regulation in Yeast. *ACS Synth. Biol.* **2018**, *7*, 584–590. [CrossRef]

16. Su, Y.; Guo, Q.Q.; Wang, S.; Zhang, X.; Wang, J. Effects of betaine supplementation on L-threonine fed-batch fermentation by Escherichia coli. *Bioprocess Biosyst. Eng.* **2018**, *41*, 1509–1518. [CrossRef]

17. Xu, P.; Li, L.; Zhang, F.; Stephanopoulos, G.; Koffas, M. Improving fatty acids production by engineering dynamic pathway regulation and metabolic control. *Proc. Natl. Acad. Sci. USA* **2014**, *111*, 11299–11304. [CrossRef]

18. Hossain, G.S.; Saini, M.; Miyake, R.; Ling, H.; Chang, M.W. Genetic Biosensor Design for Natural Product Biosynthesis in Microorganisms. *Trends Biotechnol.* **2020**, *38*, 797–810. [CrossRef]

19. Lange, C.; Mustafi, N.; Frunzke, J.; Kennerknecht, N.; Wessel, M.; Bott, M.; Wendisch, V.F. Lrp of Corynebacterium glutamicum controls expression of the brnFE operon encoding the export system for L-methionine and branched-chain amino acids. *J. Biotechnol.* **2012**, *158*, 231–241. [CrossRef]

20. Gibson, D.G.; Young, L.; Chuang, R.Y.; Venter, J.C.; Hutchison, C.A., 3rd; Smith, H.O. Enzymatic assembly of DNA molecules up to several hundred kilobases. *Nat. Methods* **2009**, *6*, 343–345. [CrossRef]

21. Li, Y.; Li, M.; Zhang, X.; Yang, P.; Liang, Q.; Qi, Q. A novel whole-phase succinate fermentation strategy with high volumetric productivity in engineered Escherichia coli. *Bioresour. Technol.* **2013**, *149*, 333–340. [CrossRef]

22. Livshits, V.A.; Zakataeva, N.P.; Aleshin, V.V.; Vitushkina, M.V. Identification and characterization of the new gene rhtA involved in threonine and homoserine efflux in Escherichia coli. *Res. Microbiol.* **2003**, *154*, 123–135. [CrossRef]

23. Lee, J.H.; Sung, B.H.; Kim, M.S.; Blattner, F.R.; Yoon, B.H.; Kim, J.H.; Kim, S.C. Metabolic engineering of a reduced-genome strain of Escherichia coli for L-threonine production. *Microb. Cell Factories* **2009**, *8*, 2. [CrossRef] [PubMed]

24. Lee, J.H.; Jung, S.C.; Bui, L.M.; Kang, K.H.; Song, J.J.; Kim, S.C. Improved production of L-threonine in Escherichia coli by use of a DNA scaffold system. *Appl. Environ. Microbiol.* **2013**, *79*, 774–782. [CrossRef] [PubMed]

25. Zhao, L.; Zhang, H.; Wang, X.; Han, G.; Ma, W.; Hu, X.; Li, Y. Transcriptomic analysis of an l-threonine-producing Escherichia coli TWF001. *Biotechnol. Appl. Biochem.* **2020**, *67*, 414–429. [PubMed]

26. Reinscheid, D.J.; Kronemeyer, W.; Eggeling, L.; Eikmanns, B.J.; Sahm, H. Stable Expression of hom-1-thrB in Corynebacterium glutamicum and Its Effect on the Carbon Flux to Threonine and Related Amino Acids. *Appl. Environ. Microbiol.* **1994**, *60*, 126–132. [CrossRef] [PubMed]

27. Van Dyk, T.K. Bacterial efflux transport in biotechnology. *Adv. Appl. Microbiol.* **2008**, *63*, 231–247.

28. Xu, S.; Zhang, L.; Zhou, S.; Deng, Y. Biosensor-Based Multigene Pathway Optimization for Enhancing the Production of Glycolate. *Appl. Environ. Microbiol.* **2021**, *87*, e0011321. [CrossRef]

29. Mazumder, M.; McMillen, D.R. Design and characterization of a dual-mode promoter with activation and repression capability for tuning gene expression in yeast. *Nucleic Acids Res.* **2014**, *42*, 9514–9522. [CrossRef]

30. Liu, Y.; Li, Q.; Zheng, P.; Zhang, Z.; Liu, Y.; Sun, C.; Cao, G.; Zhou, W.; Wang, X.; Zhang, D.; et al. Developing a high-throughput screening method for threonine overproduction based on an artificial promoter. *Microb. Cell Factories* **2015**, *14*, 121. [CrossRef]

31. Tan, S.Z.; Prather, K.L. Dynamic pathway regulation: Recent advances and methods of construction. *Curr. Opin. Chem. Biol.* **2017**, *41*, 28–35. [CrossRef]
32. Zhou, S.; Yuan, S.F.; Nair, P.H.; Alper, H.S.; Deng, Y.; Zhou, J. Development of a growth coupled and multi-layered dynamic regulation network balancing malonyl-CoA node to enhance (2S)-naringenin biosynthesis in Escherichia coli. *Metab. Eng.* **2021**, *67*, 41–52. [CrossRef]
33. Zhang, Y.; Wei, M.; Zhao, G.; Zhang, W.; Li, Y.; Lin, B.; Li, Y.; Xu, Q.; Chen, N.; Zhang, C. High-level production of l-homoserine using a non-induced, non-auxotrophic Escherichia coli chassis through metabolic engineering. *Bioresour. Technol.* **2021**, *327*, 124814. [CrossRef]
34. Liang, C.; Zhang, X.; Wu, J.; Mu, S.; Wu, Z.; Jin, J.M.; Tang, S.Y. Dynamic control of toxic natural product biosynthesis by an artificial regulatory circuit. *Metab. Eng.* **2020**, *57*, 239–246. [CrossRef]
35. Zakataeva, N.P.; Aleshin, V.V.; Tokmakova, I.L.; Troshin, P.V.; Livshits, V.A. The novel transmembrane Escherichia coli proteins involved in the amino acid efflux. *FEBS Lett.* **1999**, *452*, 228–232. [CrossRef]
36. Li, Y.; Wei, H.; Wang, T.; Xu, Q.; Zhang, C.; Fan, X.; Ma, Q.; Chen, N.; Xie, X. Current status on metabolic engineering for the production of l-aspartate family amino acids and derivatives. *Bioresour. Technol.* **2017**, *245*, 1588–1602. [CrossRef]
37. Lerner, C.G.; Inouye, M. Low copy number plasmids for regulated low-level expression of cloned genes in Escherichia coli with blue/white insert screening capability. *Nucleic Acids Res.* **1990**, *18*, 4631. [CrossRef]

 *fermentation*

Article

Medium Optimization for GA4 Production by *Gibberella fujikuroi* Using Response Surface Methodology

Bingxuan Wang [1], Kainan Yin [1], Choufei Wu [2], Liang Wang [3], Lianghong Yin [1,*] and Haiping Lin [1,*]

[1] Local and National Joint Engineering Laboratory of Biopesticide High-Efficient Preparation, Zhejiang Agriculture and Forestry University, Hangzhou 311300, China; wbx000731@163.com (B.W.); inyang@163.com (K.Y.)

[2] College of Life Science, Huzhou Normal University, Huzhou 311300, China; wcf@zjhu.edu.cn

[3] Linyi Inspection and Testing Center, Linyi 276002, China; wwwyyyly@163.com

* Correspondence: ylh4@163.com (L.Y.); zjlxylhp@163.com (H.L.); Tel.: +86-15058103796 (L.Y.); +86-13396557805 (H.L.)

Abstract: Gibberellin is an important plant growth regulator that has been widely used in agricultural production with great market prospects. However, the low yield from *Gibberella fujikuroi* restricts its application. To improve the production of gibberellin A4 (GA4), the response surface methodology was used in this study to explore the effect of different types and concentrations of vegetable oil and precursors on the production of GA4. Based on a single factor experiment, the Behnken box and central composite designs were used to establish the fermentation condition model, and the response surface method was used for analysis. The results indicated that the optimum formula was 0.55% palm oil, 0.60% cottonseed oil, 0.64% sesame oil, 0.19 g/L pyruvic acid, 0.21 g/L oxaloacetic acid, and 0.21 g/L citric acid for 48 h, which produced a yield 4.32 times higher than that without optimization. This suggests that the mathematical model is valid for predicting GA4 production in *Gibberella fujikuroi* QJGA4-1.

Keywords: medium optimization; GA4; *Gibberella fujikuroi*; response surface methodology; fermentation

Citation: Wang, B.; Yin, K.; Wu, C.; Wang, L.; Yin, L.; Lin, H. Medium Optimization for GA4 Production by *Gibberella fujikuroi* Using Response Surface Methodology. *Fermentation* **2022**, *8*, 230. https://doi.org/ 10.3390/fermentation8050230

Academic Editors: Francesca Raganati and Alessandra Procentese

Received: 14 April 2022
Accepted: 11 May 2022
Published: 17 May 2022

Publisher's Note: MDPI stays neutral with regard to jurisdictional claims in published maps and institutional affiliations.

1. Introduction

Gibberellin is a natural plant hormone that regulates the growth and development of higher plants. Currently, 136 species have been discovered, which are sequentially termed GA1–GA136 in the order of discovery [1]. The activity of gibberellin is GA3 > GA7 > GA4 in promoting stem growth [2–5]. However, due to its high activity, GA3 often causes excessive growth of the plant hypocotyl when it breaks plant dormancy and reduces the lodging resistance of the plant. At the same time, it promotes the rapid growth of epidermal cells, resulting in a thinner cuticle layer of the epidermis and rupture of spots of the fruit. GA7 can strongly inhibit the formation of flower buds, whereas GA4 promotes the formation of flower buds [2]. In addition, GA4 exhibits high biological activity in terms of promoting plant growth, appearance, and shelf life in apples, sweet northern potatoes, and grapes. Although the biological activities of GA4 have been documented, commercial use in agriculture has remained limited compared to GA3, presumably due to low yield and cost competitiveness. Therefore, there is an urgent need to improve GA4 productivity.

GA4 production by fermentation with *Gibberella fujikuroi* (*G. fujikuroi*) has been reported [6–9]. *G. fujikuroi* belongs to *Fusarium*, which is an industrial strain producing gibberellin. At present, there are two engineering strategies for enhancing GA4 production. One way to improve GA4 yield is by regulating the carbon metabolic flux to GA4 in *G. fujikuroi*. For example, Tudzynski et al. [10] deleted two genes at the end of the GA biosynthesis pathway, DES and P450-3, in order to construct an engineered strain that produces only GA4. It was shown that the DES knockout mutant strain did not produce GA3 and GA7, while GA1 and GA4 production increased significantly, with a yield ratio of approximately 5:1. However, the P450-3 knockout mutant strain did not produce GA3 and GA1, simultaneously improving the yield of GA4 and GA7 significantly compared with the wild-type strain. When both DES and P450-3 were simultaneously deleted, the mutant strain produced only GA4 and the yield was 7–8 times higher than that of the starting strain. However, the first method is difficult to perform, owing to a lack of genetic tools for this organism.

The optimization of fermentation conditions is a powerful alternative to GA4 production. The fermentation process is a complex system that is regulated by many factors. Vegetable oil, as a delayed effect carbon source, can provide a carbon source with strain reproduction, which significantly improves the yield of certain secondary metabolites, such as cephalosporin C [11], spiramycin [12], and erythromycin [13]. Zhuang et al. found that adding sesame oil, olive oil, and soybean oil to the fermentation medium effectively increased GA3 production [14]. This indicated that vegetable oil not only reduced the inhibition of carbon metabolism, but also provided the acetyl-coenzyme A and precursor of gibberellin GA3 [15]. However, the effect of vegetable oil on GA4 production has not yet been reported. Moreover, there are few kinds of microorganisms that can effectively use vegetable oil as a carbon source. Therefore, it is very important to explore the ability of GA4 production by *G. fujikuroi* using vegetable oil as carbon source and to determine the appropriate type and concentration of vegetable oil for GA4 fermentation. In addition, some studies have shown that the concentration of a product is closely related to the amount of supplied primary metabolite precursors (fatty acids, citric acids, sugars, and C_1 compounds) [16], which can be directly used for cell growth and target product formation. However, the amount of precursor added should not be too high, as this exerts a toxic effect on microorganisms and affects target metabolite formation. If the added precursors were too low, the effect was not significant. Therefore, further studies regarding the precursor addition strategy are necessary to achieve a high yield of GA4.

Response surface methodology (RSM) is a common and effective tool for medium optimization and can be used to design various fermentation parameters simultaneously [17]. In view of this, this study determined the main factors affecting the production of GA4 through a single-factor test and Plackett–Burman screening test to optimize the key factors based on the Box–Behnken design principle and central composite design optimization method, providing a reference for the optimization of GA4 and other terpenoid fermentation conditions.

2. Materials and Methods

2.1. Raw Materials

Soybean meal, corn steep liquor, corn gluten meal, and GA4 standard were provided by Zhejiang Qianjiang Biochemistry Company(Zhejiang China). Vegetable oil was purchased from a local supermarket. Methanol used was chromatographically pure, and the other reagents were analytically pure.

2.2. Strains and Media

The *G. fujikuroi* QJGA4-1 used in this study was provided by Zhejiang Qianjiang Biochemical Limited Corporation (Zhejiang China) and was kept in the China General Microbiological Culture Collection Center, and the preservation number was CGMCC17793.

In this study, potato dextrose agar (PDA) medium, containing 200 g/L potato, 20 g/L dextrose, and 20 g/L agar, was used. *G. fujikuroi* QJGA4-1 was incubated at 28 °C for five to six days in PDA medium. The hyphae of *G. fujikuroi* were placed into the inclined side of the test tube and underwent vibration with glass beads. The liquid (4%) was then inoculated into a triangular bottle. The samples were cultivated at 220 rpm and 33 °C for approximately 51 h in the seed media. The seed medium (per liter: 30 g sucrose, 12 g soybean meal, 3 g NH_4Cl, 3 g $NaNO_3$, 1.5 g KH_2PO_4, 0.2 g K_2SO_4, 0.2 g $MgSO_4 \cdot 7H_2O$) was adjusted to pH 6.8 and the seed liquid (4%) was inoculated into a shaking flask. The production of GA4 was measured at 220 rpm and 33 °C for approximately 9 days. The fermentation medium (per liter: 140 g lactose, 10 g soybean meal, 10 g corn steep liquor, 5 g corn gluten meal, 2 g K_2SO_4, 1 g $MgSO_4 \cdot 7H_2O$, 0.0005 g $ZnSO_4 \cdot 7H_2O$, 0.0005 g $CoCl_2 \cdot 6H_2O$, 1.25 g carbamide) was adjusted to pH 7.0.

2.3. Experimental Design

In a previous study, various nutritional conditions were evaluated for their effects on GA production [18,19]. The results revealed that the major variables affecting the fermentation performance in terms of GA3 production were vegetable oil and precursors. Therefore, ten types of vegetable oils (canola, soybean, rice, olive, peanut, sunflower, corn, palm, cottonseed, and sesame oils) and eight precursors (pyruvate, oxaloacetic acid, calcium gluconate, L-isoleucine, citric acid, glycerin, L-glutamic acid, and riboflavin) were selected for further optimization studies.

2.4. Box-Behnken Design (BBD)

The BBD is an efficient method for evaluating the nonlinear relationship between indexes and factors [20]. In this study, three vegetable oils that significantly increased gibberellin yield were chosen based on a single factor experiment, and then further designed and analyzed by Expert Design 8.0. The experimental factors and their levels are listed in Table 1.

Table 1. Experimental design of three independent variables with their coded and actual values using BBD.

Variables	Levels		
	−1 Level	0 Level	1 Level
A (%)	0.45	0.65	0.85
B (%)	0.45	0.65	0.85
C (%)	0.45	0.65	0.85

A: Palm oil concentration; B: cottonseed oil concentration; C: sesame oil concentration.

2.5. Plackett–Burman Design (PBD)

The PBD was employed to screen the most suitable precursors for GA4 production by *G. fujikuroi* QJGA4-1. Each independent variable was tested at two levels, high and low, which are denoted by (+) and (−), respectively. The experimental design with the factor name, symbol code, and actual level of the variables is shown in Table 2.

Table 2. Experimental design of eight independent variables and their coded values using the PBD.

Variables	Code	Low Level (g/L)	High Level (g/L)
Pyruvate	A	0.02	0.05
Oxaloacetic acid	B	0.02	0.05
Calcium gluconate	C	0.02	0.05
L-isoleucine	D	0.02	0.05
Citric acid	E	0.02	0.05
Glycerin	F	0.02	0.05
L-glutamic acid	G	0.02	0.05
Riboflavin	H	1×10^{-5}	4×10^{-5}

Two dummy variables were studied in the 12 experiments to calculate the standard error. GA4 production was performed in duplicate, and the average value was taken as the response. Variables with confidence levels above 90% were considered to have a significant effect on GA4 production, and thus were used for further optimization.

2.6. Path of the Steepest Ascent Experiment

To move rapidly toward the optimal response, we used the steepest ascent method. The experiments were adopted to determine a suitable direction by increasing or decreasing the concentrations of variables according to the results of the PBD [21].

2.7. Central Composite Designs (CCD) and Response Surface Methodology

A central composite design and response surface methodology were used to describe the nature of the response surface in the optimum region. The levels of each factor and the design matrix are listed in Table 3.

Table 3. Experimental design using CCD of three independent variables with their coded and actual value.

No.	A (g/L)	B (g/L)	E (g/L)
1	0.18	0.22	0.22
2	0.18	0.18	0.18
3	0.22	0.22	0.22
4	0.20	0.20	0.23
5	0.20	0.17	0.20
6	0.20	0.20	0.17
7	0.20	0.20	0.20
8	0.22	0.18	0.22
9	0.17	0.20	0.20
10	0.20	0.20	0.20
11	0.20	0.20	0.20
12	0.18	0.22	0.18
13	0.22	0.18	0.18
14	0.18	0.18	0.22
15	0.23	0.20	0.20
16	0.22	0.22	0.18
17	0.20	0.23	0.20

2.8. Gibberellin Analysis

GA4 was analyzed using high performance liquid chromatography (HPLC; Thermo Fisher Scientific, Waltham, MA, USA). A reversed-phase C18 column with 150 mm Lichrospher and 5 μm particle size was used for analysis (Waters, Waltham, MA, USA). Detection was performed at 201 nm. Quantification was achieved using the external standard method with the peak area. Samples were filtered through 0.22-μm membrane filters and directly injected into the HPLC using the 20 μL loop of a Rheodyne injector. If required, the samples were diluted to <40 mg/L.

3. Results and Discussion

3.1. Screening of Vegetable Oils Affecting GA4 Production

As a slow-acting carbon source, vegetable oil can increase the yield of secondary metabolites. As shown in Figure 1, ten different vegetable oils were selected. The experimental treatments numbered 1–11 were without vegetable oil, canola, soybean, rice, olive, peanut, sunflower, corn, palm, cottonseed, and sesame oils to investigate the effect of 0.25% oil on GA4 production by *G. fujikuroi*.

The results showed that adding vegetable oil to the medium increased the GA4 production, among which the production of palm oil, cottonseed oil, and sesame oil were significantly higher than those of other oils. It can be found from Figure 1 that there is no significant difference in improving the production of GA4 by comparing the three oils. Therefore, palm oil, cottonseed oil, and sesame oil were selected for further research.

3.2. Effect of Palm, Cottonseed, and Sesame Oils on the Production of GA4

Based on the above results, the impact of different concentrations of palm, cottonseed, and sesame oils on GA4 production was investigated. The results are presented in Figure 2. Palm oil, cottonseed oil, and sesame oil (0.05%, 0.25%, 0.45%, 0.65%, 0.85%, and 1.05%) were added to the medium. As the oil concentration increased, the GA4 value first increased and then decreased gradually. When 0.65% oil was added, the highest yield of GA4 was achieved. Therefore, 0.65% palm, cottonseed, and sesame oils were determined as optimal concentrations.

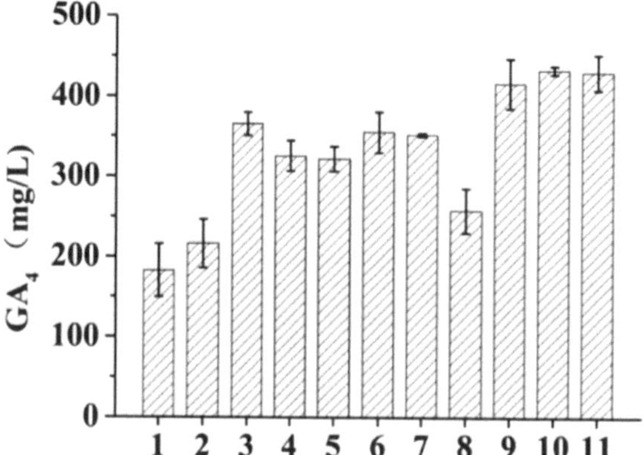

Figure 1. Effect of different vegetable oils on GA4 production by *G. fujikuroi* QJGA4-1.

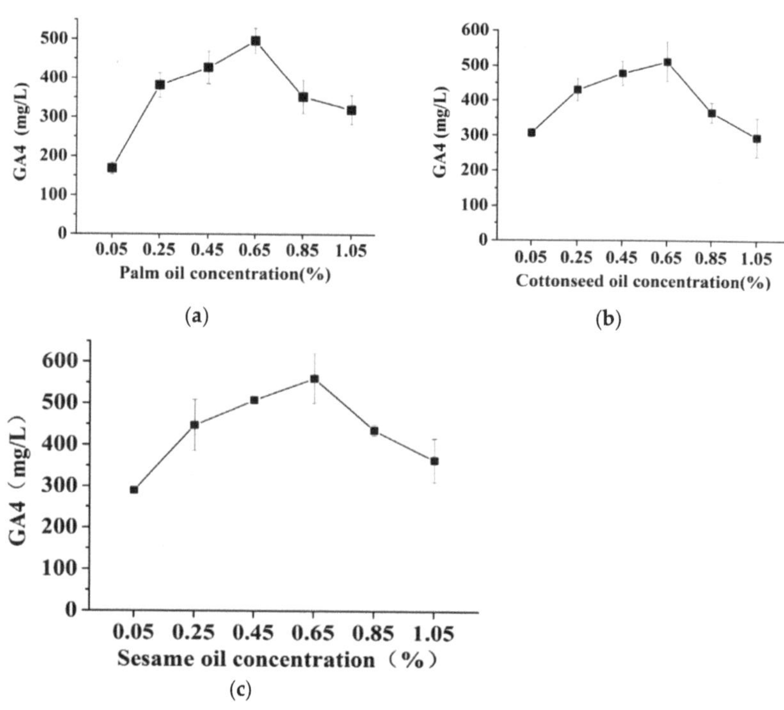

Figure 2. Effect of oil concentrations on GA4 production (**a**): palm oil, (**b**): cottonseed oil, (**c**): sesame oil.

3.3. Interaction of Palm Oil, Cottonseed Oil, and Sesame Oil on GA4 Production Using BBD

Palm oil (A), cottonseed oil (B), and sesame oil (C) were independent variables, and GA4 (Y) was the response value in BBD. The Box–Behnken experimental design and results are shown in Table 4.

Table 4. Experimental design of three independent variables and actual values of GA4 production using BBD.

Test Number	A (%)	B (%)	C (%)	Y (mg/L)
1	0.65	0.85	0.45	438.30
2	0.65	0.65	0.65	800.93
3	0.45	0.65	0.85	645.54
4	0.85	0.85	0.65	548.54
5	0.45	0.45	0.65	758.06
6	0.65	0.65	0.65	976.88
7	0.65	0.65	0.65	661.92
8	0.65	0.45	0.85	495.44
9	0.85	0.45	0.65	395.96
10	0.85	0.65	0.45	480.55
11	0.85	0.65	0.85	325.60
12	0.45	0.85	0.65	518.87
13	0.65	0.65	0.65	837.61
14	0.65	0.85	0.85	267.78
15	0.65	0.65	0.65	1136.46
16	0.65	0.45	0.45	542.49
17	0.45	0.65	0.45	631.38

The Box–Behnken experimental design and results are presented in Table 4. The analysis of variance of the test results is presented in Table 5, and the p value of the model was less than 0.05, indicating that the ANOVA test was significant. The p value of lack of fit was 0.9507, indicating a high degree of fit between the predicted and actual values. Therefore, the model could be used to analyze and predict the production of GA4.

Table 5. Analysis of variance of response surface quadratic model for GA4 production.

Source	Sum of Squares	df	Mean Square	F Value	p Value
Model	7.043×10^5	9	78,256.94	3.87	0.0439 *
A	80,645.53	1	80,645.53	3.99	0.0858
B	21,886.57	1	21,886.57	1.08	0.3325
C	16,052.62	1	16,052.62	0.79	0.4022
AB	38,376.48	1	38,376.48	1.90	0.2105
AC	7149.36	1	7149.36	0.35	0.5706
BC	3811.91	1	3811.91	0.19	0.6770
A2	61,978.17	1	61,978.17	3.07	0.1233
B2	1.788×10^5	1	1.788×10^5	8.85	0.0206
C2	2.439×10^5	1	2.439×10^5	12.08	0.0103
Residual	1.414×10^5	7	20,196.33		
Lack of Fit	10,647.26	3	3549.09	0.11	0.9507
Pure Error	1.307×10^5	4	32,681.76		
Cor Total	8.457×10^5	16			

Note: * means $p < 0.05$.

The data were analyzed using Design Expert 8.0 software, and a quadratic response surface regression model was established. The fitting quadratic regression equation was obtained as follows:

$$Y = 882.76 - 100.40 \times A - 52.31 \times B - 44.79 \times C + 97.95 \times A \times B - 42.28 \times A \times C - 30.87 \times B \times C - 121.33 \times A^2 - 206.09 \times B^2 - 240.67 \times C^2. \tag{1}$$

To better estimate the influence of the interactions between any two independent parameters on GA4 production, three-dimensional response surfaces and two-dimensional contour plots are plotted in Figure 3. The effect of palm oil concentration and cottonseed oil concentration on GA4 production is shown in the three-dimensional response surface plot in Figure 3a. Figure 3c depicts the influence of the interactions between sesame oil concentration and palm oil concentration on GA4 production. A response surface plot was used to depict the effect of cottonseed oil concentration and sesame oil concentration on GA4 production (Figure 3e). The downward opening of the response surface (Figure 3a,c,e) revealed that GA4 production was improved by promoting palm, cottonseed, and sesame oil levels until the optimum value was obtained. However, continuing to increase the oil levels resulted in a linear decrease in GA4 production. The highest GA4 production was obtained using palm, cottonseed, and sesame oil values of 0.55%, 0.60%, and 0.64%, respectively. Figure 3b,d,f shows the two-dimensional contour plot of the interaction between palm, cottonseed, and sesame oils, respectively, and the effect of variables on GA4 production. From the contour plot, we observe that the interactive contour map between palm, cottonseed, and sesame oil is nearly circular, indicating that the interaction has no significant effect on GA4 production.

Overall, analysis of the response surface showed that the optimal conditions for GA4 production were 0.55% palm oil, 0.60% cottonseed oil, and 0.64% sesame oil, the GA4 production was 915.35 mg/L under the above conditions (predicted with RSM).

Confirmation of the optimized medium component was performed in shake-flask cultures using the optimum predicted values of the various variables. To verify the accuracy of the model, GA4 fermentation experiments were performed on the optimum medium. The average production of GA4 obtained from three parallel experiments was 875.51 mg/L, which is close to the predicted value of the regression equation. This finding demonstrated that the constructed model was valid for predicting the generation of GA4 using QJGA 4-1. Similarly, Wang et al. reported that adding soybean oil to the fermentation medium effectively improved GA3 production [22], because oil as a carbon source does not inhibit carbon metabolism but can provide the necessary acetyl-coenzyme for gibberellin synthesis [18].

3.4. Screening of Precursors for GA4 Production Using the PBD

Pyruvate (A), oxaloacetic acid (B), calcium gluconate (C), L-isoleucine (D), citric acid (E), glycerin (F), L-glutamic acid (G), and riboflavin (H) were used as screening factors in the 12 fermentation experiments. The concentration of each factor and the experimental results are listed in Table 6.

Table 6. Experimental design using PB of eight independent variables with their concentration and actual values of GA4 production.

No.	A (g/L)	B (g/L)	C (g/L)	D (g/L)	E (g/L)	F (g/L)	G (g/L)	H (g/L)	GA4 (mg/L)
1	0.05	0.2	0.2	0.2	0.05	0.05	0.05	4×10^{-5}	594.17
2	0.2	0.05	0.2	0.2	0.2	0.05	0.05	1×10^{-5}	754.32
3	0.05	0.05	0.05	0.05	0.05	0.05	0.05	1×10^{-5}	593.47
4	0.05	0.2	0.05	0.2	0.2	0.05	0.2	4×10^{-5}	690.35
5	0.2	0.05	0.05	0.05	0.2	0.05	0.2	4×10^{-5}	801.70
6	0.05	0.05	0.05	0.2	0.05	0.2	0.2	1×10^{-5}	545.66
7	0.05	0.05	0.2	0.05	0.2	0.2	0.05	4×10^{-5}	679.13
8	0.05	0.2	0.2	0.05	0.2	0.2	0.2	1×10^{-5}	710.36
9	0.2	0.2	0.05	0.05	0.05	0.2	0.05	1×10^{-5}	864.39
10	0.2	0.2	0.05	0.2	0.2	0.2	0.05	1×10^{-5}	982.36
11	0.2	0.2	0.2	0.05	0.05	0.05	0.2	1×10^{-5}	747.22
12	0.2	0.05	0.2	0.2	0.05	0.2	0.2	4×10^{-5}	479.35

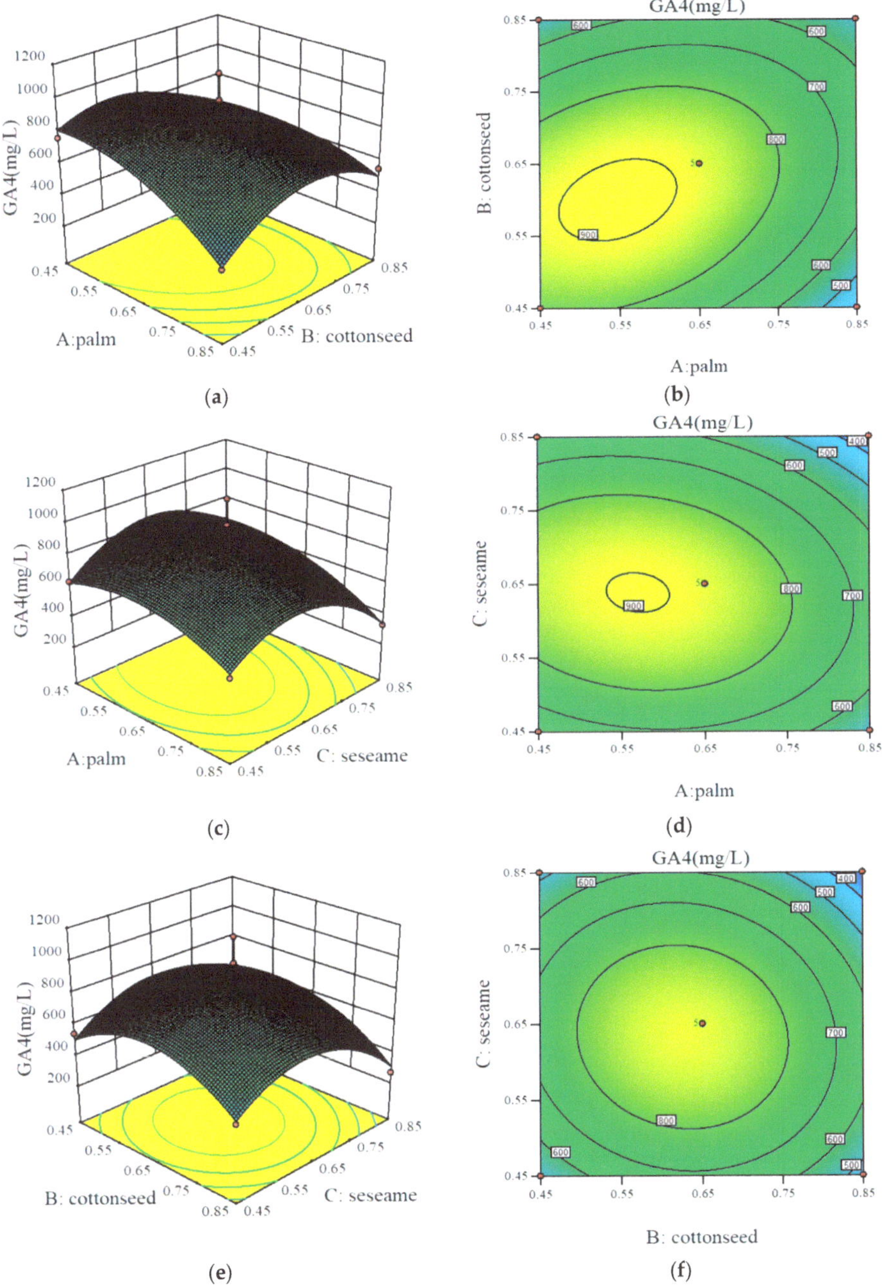

Figure 3. Three-dimensional response surface plots and corresponding two-dimensional contour plot showing the effects of variables and interaction with GA4 production. Three-dimensional response surface interaction plots of palm and cottonseed oil (**a**), palm and sesame oil (**c**), and cottonseed and sesame oil (**e**). Corresponding two-dimensional contour plot interaction of palm and cottonseed oil (**b**), palm and sesame oil (**d**), and cottonseed and sesame oil (**f**).

To investigate the influence of different precursors on GA4 production, p value analysis was conducted using Design Expert 8.0. As shown in Table 7, the p value was $0.0453 < 0.05$, and the R^2 value was 0.962, which indicates that the fitted model is significant and reliable. Among the eight factors, pyruvic acid, citric acid, and oxaloacetic acid were the most important factors for GA4. From the positive and negative effects of the statistical analysis coefficients of various factors in Table 7, we observed that the concentrations of pyruvate, oxaloacetic acid, and citric acid should be increased to enhance the production of GA4.

Table 7. Analysis of variance of the response surface quadratic model for GA4 production.

Variable	Sum of Squares	Prob > F	Significance
Model	2.096×10^5	0.0453	Significance
A	48,646.20	0.0246	*
B	39,041.67	0.0328	*
C	18,133.57	0.0828	
D	7803.59	0.1909	
E	55,775.46	0.0205	*
F	137.77	0.8374	
G	16,624.40	0.0911	
H	3483.7	0.3426	

Note: R^2 represents the fitting degree of the regression equation, * means $p < 0.05$.

3.5. Steepest Ascent Design

To find the proper direction of changing variables, the steepest ascent design experiment and the corresponding results are shown in Table 8. Pyruvic acid (A), oxaloacetic acid (B), and citric acid (E) were important factors affecting the production of GA4.

Table 8. The steepest ascent experiment design and results; the values represent the mean $\pm$ SD of three replicates.

No.	A (g/L)	B (g/L)	E (g/L)	GA4 (mg/L)
1	0.10	0.10	0.10	854.40 $\pm$ 40.21
2	0.15	0.15	0.15	879.98 $\pm$ 18.18
3	0.2	0.2	0.2	938.09 $\pm$ 63.17
4	0.25	0.25	0.25	903.65 $\pm$ 45.55
5	0.30	0.30	0.30	845.29 $\pm$ 24.92

The steepest ascent test showed that the maximum GA4 concentration near to the central point was A, 0.2 (g/L), B, 0.2 (g/L), and E, 0.2 (g/L).

3.6. CCD and Statistical Optimization of GA4 Production

Based on the central point, the central composite design, with three factors and five levels, was determined (Table 9).

Design-Expert software (version 8.0) was used to fit the experimental data with quadratic polynomial regression. The following regression fitting equation is obtained:

$$Y(GA4) = 73899.20 + 22421.45\ A + 21009.31\ B - 1684.45\ E - 17.84\ AB - 3412.57\ AE + 5235.86\ BE - 24566.76\ A^2 - 51962.34\ B^2 - 3691.15\ E^2. \tag{2}$$

The results of the variance analysis are presented in Table 10 ($p = 0.0169$), indicating that the regression model was significant. The fitting degree was 0.0707, which was not significant, indicating that the fitting degree was good. This model was found to be effective. The data can be analyzed using the model, and the GA4 content can be predicted using the regression equation. Table 10 shows that factors A, B, AE, BE, and B^2 had significant effects on GA4 production.

Table 9. Experimental design using CCD of three independent variables with their coded and actual values of GA4 production.

No.	A (g/L)	B (g/L)	E (g/L)	GA4 (mg/L)
1	0.18	0.22	0.22	992.57
2	0.18	0.18	0.18	814.97
3	0.22	0.22	0.22	861.32
4	0.20	0.20	0.23	869.82
5	0.20	0.17	0.20	796.45
6	0.20	0.20	0.17	912.60
7	0.20	0.20	0.20	953.15
8	0.22	0.18	0.22	787.74
9	0.17	0.20	0.20	845.29
10	0.20	0.20	0.20	943.20
11	0.20	0.20	0.20	965.31
12	0.18	0.22	0.18	792.20
13	0.22	0.18	0.18	872.31
14	0.18	0.18	0.22	812.17
15	0.23	0.20	0.20	895.41
16	0.22	0.22	0.18	844.40
17	0.20	0.23	0.20	824.27

Design-Expert software (version 8.0) was used to obtain three-dimensional response surface plots and corresponding two-dimensional contour plots of the relationship between pyruvic acid, oxaloacetic acid, and citric acid on GA4 production (Figure 4). From Figure 4a,c,e, we observed the downward opening of the response surface, which proves that the model has a maximum value and GA4 has a maximum production. Figure 4d shows that the contour plot of pyruvic acid and oxaloacetic acid on GA4 production is nearly circular, indicating that the interaction between pyruvic acid and oxaloacetic acid has no significant effect on GA4 production. Figure 4b,f show the interactive contours of oxaloacetic acid with citric acid and pyruvic acid with citric acid, respectively. The contours are elliptical, which proves that the interactions between oxaloacetic acid and citric acid, and pyruvic acid and citric acid exerted a significant effect on GA4 production. When the concentration of A was 0.19 g/L, B 0.21 g/L, and E 0.21 g/L, the predicted production of GA4 was 960.34 mg/L.

Table 10. Regression coefficients and their significance for response surface quadratic model.

Factor	Sum of Squares	df	F Value	p Value	Significance
Model	60,790.68	9	5.57	0.0169	Significant
A	16,676.95	1	13.76	0.0076	*
B	11,504.28	1	9.50	0.0178	*
E	3204.59	1	2.64	0.1479	
AB	1566.46	1	1.29	0.2929	
AE	8793.30	1	7.26	0.0309	*
BE	11,602.44	1	9.58	0.0175	*
A^2	7826.45	1	6.46	0.0386	*
B^2	31,502.05	1	26.00	0.0014	*
E^2	6644.86	1	5.48	0.0517	
Residual	8481.11	7			
Lack of Fit	8235.81	5	13.43	0.0707	Not significant
Pure Error	245.30	2	122.65		
Cor Total	69,271.79	16			

Note: * means $p < 0.05$.

3.7. Experimental Validation of the Model

To verify the predicted results of the model, three repeated shaking flask fermentation experiments were conducted under the optimized conditions. The actual GA4 production in the optimum medium was 972.95 mg/L, which is consistent with the predicted value of 960.34 mg/L. This demonstrated that the model could predict GA4 production well. Compared with the case without the precursor, GA4 production increased by 11.1%.

The type and concentration of the precursors affect the synthesis of gibberellin. At present, most studies have focused on the influence of the precursor on the synthesis of GA3. For example, Wang and et al. [23] found that oxaloacetic acid, calcium gluconate, and riboflavin exerted obvious effects on the synthesis of GA3. Through experiments and metabolic flux analysis, some scholars found that oxaloacetate also had a significant effect on the formation of GA3 [24]. This may be as a result of the metabolic flux distribution change. However, relevant studies on the optimization of fermentation conditions for GA4 have not been reported. Based on the relevant studies on GA3, we can further explore the influence of the precursor on the synthesis of GA4. In this study, oxaloacetic acid was also found to promote the synthesis of GA4. This may be a result of the metabolic flux distribution change between the HMP pathway and TCA cycle with the addition of oxaloacetic acid [25–27]. Acetyl-CoA is a key regulator of metabolite synthesis by *Fusarium fujikuroi*. Adding oxaloacetic acid relieved the inhibition of the EMP pathway. At the same time, it was decarboxylated to pyruvic acid and further metabolized to acetyl CoA, which is a substance promoting GA4 production. Some studies have shown that pyruvic acid, as the intermediate product of the tricarboxylic acid (TCA) cycle and the final product of the glycolysis pathway, can be converted into fatty acids, amino acids, and polysaccharides through coenzyme A and TCA cycle metabolites [28]. Oxaloacetic acid, citric acid, and pyruvic acid are all important metabolites of the TCA cycle and the EMP pathway. Adding an appropriate amount of pyruvic acid, oxaloacetic acid, and citric acid may accelerate the metabolism of bacteria, promote the formation of acetyl CoA, and further improve GA4 production. In addition, the strain may have different affinities for different precursors. For example, in this study, calcium gluconate and riboflavin did not significantly affect the production of GA4.

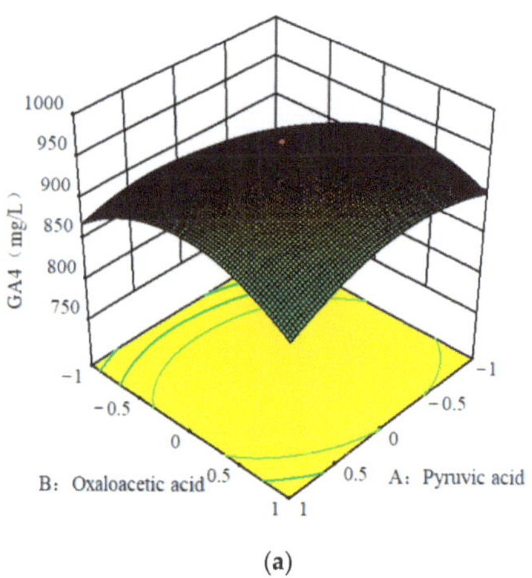

(a)

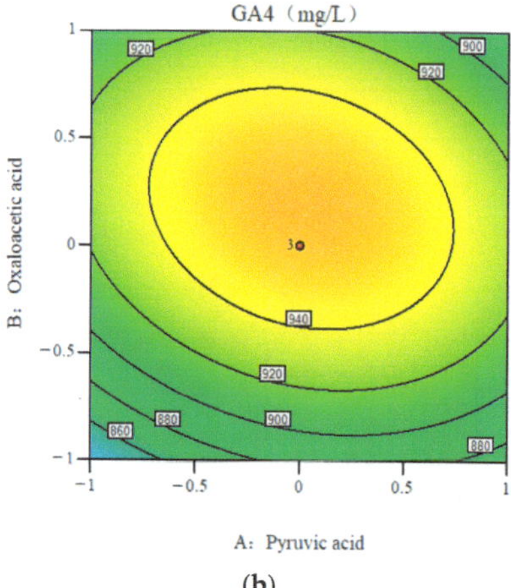

(b)

Figure 4. *Cont.*

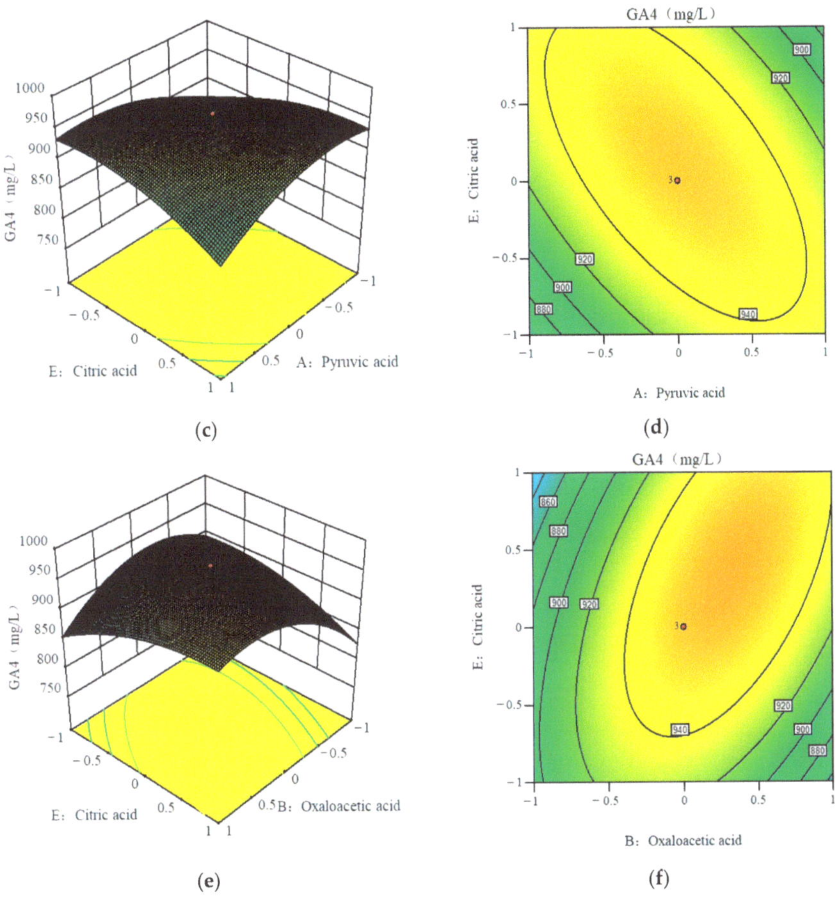

Figure 4. Three-dimensional response surface plots and corresponding two-dimensional contour plot showing the effects of variables and the interaction with GA4 production (mg/L); (**a,b**) interaction between pyruvic acid and oxaloacetic acid; (**c,d**) interaction between oxaloacetic acid and citric acid; (**e,f**) interaction between pyruvic acid and citric acid.

4. Conclusions

G. fujikuroi QJGA4-1 is a GA4 producing strain, but its production level is not competitive. To further improve the production of GA4, the nutritional conditions of the vegetable oil and precursors were optimized using response surface methodology. Based on these results, response surface models for vegetable oils and precursors were determined. To the best of our knowledge, this GA4 production level is the highest by *G. fujikuroi* QJGA4-1 under the optimized conditions until now.

Regarding the industrial GA4 fermentation process, a large amount of vegetable oil is required, which will inevitably lead to an increase in production cost and pollution probability. Therefore, vegetable oil utilization efficiency needs to be improved to obtain a cheap substitute for GA4 production. In the future, we will analyze the type and content of oleic acid in vegetable oil and explore the mechanism of promoting GA4 synthesis via oleic acid metabolism. In addition, the precursor for the synthesis of GA4 was determined based on RSM. Subsequently, transcriptomics, proteomics, and metabolomics will be used to analyze GA4 biosynthesis, through global regulatory analysis, to explore the product synthesis mechanism, in order to identify new targets for increasing GA4 production.

Author Contributions: Conceptualization, H.L. and L.Y.; formal analysis, B.W., methodology, B.W. and K.Y.; validation, B.W., K.Y., C.W. and L.W.; formal analysis, B.W.; data curation, B.W.; writing—original draft preparation, B.W.; writing—review and editing, B.W.; supervision, H.L. and L.Y.; project administration, H.L. All authors have read and agreed to the published version of the manuscript.

Funding: This work was financially supported by the Basic Public Research Project of Zhejiang Province (LGG22C140001); the Key Research Program of Zhejiang Province (2020C02005); Zhejiang Provincial Natural Science Foundation of China (grant no. LZ22C200001), National Natural Science Foundation of China (No. 31600070).

Institutional Review Board Statement: Not applicable.

Informed Consent Statement: Not applicable.

Data Availability Statement: Not applicable.

Conflicts of Interest: The authors declare no conflict of interest.

References

1. Sponsel, V.M.; Hedden, P. Gibberellin Biosynthesis and Inactivation. In *Plant Hormones Biosynthesis Signal Transduction Action*; Springer: Dordrecht, The Netherlands, 2010; pp. 63–94. [CrossRef]
2. Yan, F.G.; Qin, J.; He, Z.G.; Li, J.L. Research progress of gibberellin A4, A7. *J. Microbiol.* **1994**, *21*, 163–167.
3. Hedden, P.; Thomas, S.G. Gibberellin biosynthesis and its regulation. *Biochem. J.* **2012**, *444*, 11–25. [CrossRef]
4. Bömke, C.; Tudzynski, B. Diversity, regulation, and evolution of the gibberellin biosynthetic pathway in fungi compared to plants and bacteria. *Phytochemistry* **2009**, *70*, 1876–1893. [CrossRef] [PubMed]
5. Cen, Y.-K.; Lin, J.-G.; Wang, Y.-L.; Wang, J.-Y.; Liu, Z.-Q.; Zheng, Y.-G. The Gibberellin Producer *Fusarium fujikuroi*: Methods and Technologies in the Current Toolkit. *Front. Bioeng. Biotechnol.* **2020**, *8*, 232. [CrossRef]
6. Tudzynski, B.; Mihlan, M.; Rojas, M.C.; Linnemannstöns, P.; Gaskin, P.; Hedden, P. Characterization of the Final Two Genes of the Gibberellin Biosynthesis Gene Cluster of *Gibberella fujikuroi*. *J. Biol. Chem.* **2003**, *278*, 28635–28643. [CrossRef]
7. Yan, F.; Lin, J. Gibberella fujikuroi Strain Used for Industrial Fermentation Production of Gibberellin A4 and A7. China Patent CN122257514, 14 July 1999.
8. Wu, Y.F.; Li, Z.Y. A New Strain of Fusarium fujikuroi and Its Fermentation Method for Producing Gibberellin A4. China Patent CN105524840 A, 27 April 2016.
9. Hedden, P. The Current Status of Research on Gibberellin Biosynthesis. *Plant Cell Physiol.* **2020**, *61*, 1832–1849. [CrossRef] [PubMed]
10. Gallazzo, J.L.; Lee, M.D. Production of High Titers of Gibberellins, GA4 and GA7 by Strain LTB-1027. USA Patent 6287800, 11 September 2001.
11. Luo, H.; Zhang, J.; Yuan, G.; Zhao, Y.; Liu, H.; He, Z.; Shi, Z. Performance improvement of cephalosporin C fermentation by *Acremonium chrysogenum* with DO-Stat based strategy of co-feeding soybean oil and glucose. *Process Biochem.* **2013**, *48*, 1822–1830. [CrossRef]
12. Sun, X.Q.; Meng, G.S.; Jin, Z.H.; Jin, Y.P.; Wang, P. Effect of soybean oil on spiramycin fermentation. *Chin. J. Antibiot.* **2002**, *27*, 524–528. [CrossRef]
13. Shen, Z.B.; Chen, G.H.; Chen, C.H. Study on effect of soybean oil on fermentation of erythromycin and its mechanism. *Chin. J. Antibiot.* **2006**, *31*, 657–660. [CrossRef]
14. Zhuang, M.K.; Wang, J.G.; Ji, Z.X.; Chen, S.W. Effects of Vegetable Oils on Ferment of Gibberellin Acid. *J. Chin. Cereals Oils Assoc.* **2008**, *23*, 77–80.
15. Peng, H.; Shi, T.Q.; Nie, Z.K.; Guo, D.S.; Huang, H.; Ji, X.J. Fermentative production of gibberellins: A review. *Chem. Ind. Eng. Prog.* **2016**, *35*, 3611–3618. [CrossRef]
16. Chu, J. The role of special precursors in antibiotic biosynthesis. *World Notes Antibiot.* **1999**, *5*, 202–212.
17. Singh, A.K.; Mukhopadhyay, M. Lipase-catalyzed glycerolysis of olive oil in organic solvent medium: Optimization using response surface methodology. *Korean J. Chem. Eng.* **2016**, *33*, 1247–1254. [CrossRef]
18. Gancheva, V.; Dimova, T.; Kamenov, K.; Futekova, M. Biosynthesis of gibberellins: III. Optimization of nutrient medium for biosynthesis of gibberellins upon using mathematical methods for planning the experiment. *Acta. Microbiol. Bulg.* **1984**, *14*, 80–84. [PubMed]
19. Cihangir, N. Stimulation of the gibberellic acid synthesis by Aspergillus niger in submerged culture using a precursor. *World J. Microbiol. Biotechnol.* **2002**, *18*, 727–729. [CrossRef]
20. Chhabra, M.; Dwivedi, G.; Baredar, P.; Shukla, A.K.; Garg, A.; Jain, S. Production & optimization of biodiesel from rubber oil using BBD technique. *Mater. Today Proc.* **2020**, *38*, 69–73. [CrossRef]
21. Gheshlaghi, R.; Scharer, J.; Moo-Young, M.; Douglas, P. Medium optimization for hen egg white lysozyme production by recombinantAspergillus niger using statistical methods. *Biotechnol. Bioeng.* **2005**, *90*, 754–760. [CrossRef]
22. Wang, J.G.; Chen, S.W.; Yu, Z.N. Effects of Soybean Oil on Fermentation of GA3 Produced by *Gibberella fujikuroi* 978. *J. Huazhong Agric. Univ.* **2005**, *24*, 470–473.

23. Karaffa, L.; Szentirmai, A. Cephalosporin-C production, morphology and alternative respiration of *Acremonium chrysogenum* in glucose-limited chemostat. *Biotechnol. Lett.* **1996**, *18*, 701–706. [CrossRef]
24. Wang, J.F.; Wu, H. Effect of Precursor on the Avermectins Biosynthesis. *J. South China Univ. Technol. (Nat. Sci.)* **2002**, *30*, 16–18. [CrossRef]
25. Salazar-Cerezo, S.; Martínez-Montiel, N.; García-Sánchez, J.; Pérez-y-Terrón, R.; Martínez-Contreras, R.D. Gibberellin biosynthesis and metabolism: A convergent route for plants, fungi and bacteria. *Microbiol. Res.* **2018**, *208*, 85–98. [CrossRef] [PubMed]
26. Wang, B.; Si, W.; Wu, Y.; Zhang, X.; Wang, S.; Wu, C.; Lin, H.; Yin, L. Research progress in biosynthesis and metabolism regulation of gibberellins in *Gibberella fujikuroi*. *Chin. J. Biotechnol.* **2020**, *36*, 189–200. [CrossRef]
27. Zhang, X.W.; Zhang, D. Recent advances of secondary metabolites in genus Fusarium. *Plant Physiol. J.* **2013**, *49*, 201–216. [CrossRef]
28. Wang, W. Screening High Gibberellin Acid Producing Strains from *Gibberella fujikuroi* and Optimization on the Fermentation Condition. Ph.D. Thesis, Central South University of Forestry and Technology, Changsha, China, 2015. [CrossRef]

fermentation

Article

Regulation of β-Disaccharide Accumulation by β-Glucosidase Inhibitors to Enhance Cellulase Production in *Trichoderma reesei*

Tingting Long [†], Peng Zhang [†], Jingze Yu [†], Yushan Gao, Xiaoqin Ran and Yonghao Li *

Chongqing Key Laboratory of Industrial Fermentation Microorganism, School of Chemistry and Chemical Engineering, Chongqing University of Science and Technology, Chongqing 401331, China; tt_long@cqust.edu.cn (T.L.); p.zhang@cqust.edu.cn (P.Z.); jzzzz_y@cqust.edu.cn (J.Y.); ys_gao@hotmail.com (Y.G.); xq_ran@hotmail.com (X.R.)

* Correspondence: yh_li@cqust.edu.cn; Tel.: +86-023-6502-2211

† These authors contributed equally to this work.

Abstract: *Trichoderma reesei* is a high-yield producer of cellulase for applications in lignocellulosic biomass conversion, but its cellulase production requires induction. A mixture of glucose and β-disaccharide has been demonstrated to achieve high-level cellulase production. However, as inducers, β-disaccharides are prone to be hydrolyzed by β-glucosidase (BGL) during fermentation, therefore β-disaccharides need to be supplemented through feeding to overcome this problem. Here, miglitol, an α-glucosidase inhibitor, was investigated as a BGL inhibitor, and exhibited an IC_{50} value of 2.93 μg/mL. The cellulase titer was more than two-fold when miglitol was added to the fermentation medium of *T. reesei*. This method was similar to the prokaryotic expression system using unmetabolized isopropyl-β-D-thiogalactopyranoside (IPTG) as the inducer instead of lactose to continuously induce gene expression. However, cellulase activity was not enhanced with BGL inhibition when lactose or cellulose was used as an inducer, which demonstrated that the transglycosidase activity of BGL is important for the inducible activity of lactose and cellulose. This novel method demonstrates potential in stimulating cellulase production and provides a promising system for *T. reesei* protein expression.

Keywords: *Trichoderma reesei*; cellulase; β-glucosidase inhibitors; sophorose; miglitol

Citation: Long, T.; Zhang, P.; Yu, J.; Gao, Y.; Ran, X.; Li, Y. Regulation of β-Disaccharide Accumulation by β-Glucosidase Inhibitors to Enhance Cellulase Production in *Trichoderma reesei*. *Fermentation* **2022**, *8*, 232. https://doi.org/10.3390/fermentation8050232

Academic Editors: Xian Zhang, Seraphim Papanikolaou and Fabrizio Beltrametti

Received: 26 March 2022
Accepted: 16 May 2022
Published: 17 May 2022

Publisher's Note: MDPI stays neutral with regard to jurisdictional claims in published maps and institutional affiliations.

1. Introduction

Excessive exploitation and utilization of petroleum, coals, natural gas and other fossil fuels have resulted in a series of problems, including diminishing reserves, increasingly severe pollution, climate changes and the frequent occurrence of natural disasters. These problems largely restrict socioeconomic sustainability [1]. Lignocellulosic biorefineries, which involve the production of biofuels and biochemicals from forestry and agricultural residues, are important alternatives for addressing the energy crisis and sustainable economic development [2–4]. However, the robust supramolecular structures of lignocellulosic biomasses can only be hydrolyzed into fermentable sugars by efficient cellulase [5,6]. High-yield cellulase strains mainly originate from *Trichoderma reesei*, but the biosynthesis of the cellulases must be induced [7,8].

Sophorose is the most efficient inducer for *T. reesei* cellulase synthesis, and its inducing ability is more than 2500 times higher than that of cellobiose [9]. Unfortunately, sophorose is so expensive that it has never been used as a sole inducer for cellulase production. According to the cellulase production process developed by NREL for evaluating the biochemical conversion of lignocellulosic biomass to ethanol, the mixture glucose-sophorose has been used as an inducer of *T. reesei*. When grown on this substrate, *T. reesei* has been shown to productively secrete cellulase [10]. In preliminary studies, a mixture of glucose and β-disaccharide (MGD) was prepared from glucose through the transglycosylation reaction catalyzed by β-glucosidase (BGL) and used as an inducer for the efficient synthesis of cellulases [11–14]. A maximum cellulase titer of 102.63 IU/mL was achieved by engineered *T. reesei,* using MGD as an inducer [15]. Moreover, transcriptomic and iTRAQ proteomic analyses showed that the secretory volumes of major cellulases are higher than those of other commonly used inducers [16,17], but the BGL was low. To overcome this disadvantage, *aabgl1* encoding BGL in *Aspergillus aculeatus* was heterologously expressed in *T. reesei* Rut C30. The BGL activity of recombinant *T. reesei* increased 71-fold compared to the parent strain [16], however, we found that the cellulase activity was significantly reduced.

It is speculated that the BGL expressed by *T. reesei* could degrade sophorose into glucose; consequently, sophorose does not maintain induction concentrations in cells or is completely degraded, thus losing its inducing ability. Hence, this sophorose containing inducer must be supplemented through continual feeding. Although the transglycosylation activity of BGL in *T. reesei* has been confirmed to be irreplaceable during cellulase fermentation when cellulose or lactose is adopted as the inducer [18,19], it is no longer needed when sophorose is directly used. We assume that if the BGLs are inhibited, an appropriate intracellular concentration of sophorose for induction can be theoretically maintained, which decreases the dosage and consumption of inducers.

Although BGL is pivotal in the degradation of cellulose components into glucose [20], research generally indicates that the BGL secreted by *T. reesei* is insufficient, and the efficient hydrolysis of cellulose components is achieved only when enough BGL is added [21–25]. Therefore, the actions of the background BGL are generally not considered. Hence, the inhibition or knockout of BGL genes does not significantly affect the quality of *T. reesei* cellulase.

In the present study, miglitol, an α-glucosidase inhibitor [26], was used to inhibit BGL in *T. reesei* fermentation, and we investigated whether intracellular β-disaccharide can be maintained at an appropriate induction concentration to stabilize cellulase production. This is an alternative strategy is to induce the synthesis of cellulase, providing a viable method for further efficient cellulase production by fed-batch fermentation.

2. Materials and Methods

2.1. Materials and Microorganisms

Trichoderma reesei Rut C30 was a gift from the USDA ARS Culture Collection. Recombinant *T. reesei* PB-3 and PB-4 were developed by overexpressing *aabgl1* (GenBank: D64088.1) using the *pdc1* promoter (encoding pyruvate decarboxylase) in *T. reesei* RUT C30, which had four and three gene copies, respectively [16]. The spores were preserved in 50% glycerol at $-80\,^{\circ}$C. β-Glucosidase SUNSON® was purchased from Sunson Industry Group Co., Ltd. (Yinchuan, Ningxia, China).

The mixture of glucose and β-disaccharides (MGD) was synthesized from glucose by β-glucosidase (BGL) through a transglycosylation reaction [11]. In detail, BGL was added to a substrate containing 600 g/L glucose at 20 IU/g (glucose), and the transglycosylation reaction was performed at 65 $^{\circ}$C and pH 4.8 for 72 h, after which the BGL was deactivated by incubating the mixture at 100 $^{\circ}$C for 5 min. Finally, the MGD composition was 410.20 g/L glucose, 60.56 g/L gentiobiose, 9.34 g/L cellobiose and 13.66 g/L sophorose.

2.2. Production of Cellulase and β-Glucosidase from T. reesei C30, PB3 and PB4

T. reesei Rut C30, PB-3 and PB-4 were applied separately in the production of cellulase and BGL, using MGD, lactose or cellulose as an inducer. Lignocellulase production in shake flasks was performed, according to a previous work [11]. *T. reesei* was cultured on malt extract agar plates (3% malt extract and 1.5% agar) for 7 d. Spores were then harvested and inoculated into 250-mL Erlenmeyer flasks containing 50 mL of a medium that was composed of 4 g/L glucose and 10 g/L corn steep liquor. After 24 h, mycelium was inoculated at 8% (*v*/*v*) into 250-mL flasks containing 50 mL of a fermentation medium for cellulase and BGL production. The fermentation medium contained 10 g/L carbon source (MGD, lactose or cellulose) supplemented with 1 g/L peptone, 0.3 g/L urea, 0.8 g/L CaCl$_2$, 0.5 mL/L Tween 80, 4 g/L KH$_2$PO$_4$, 0.6 g/L MgSO$_4\cdot$7H$_2$O, 2.8 g/L (NH4)$_2$SO$_4$, 10 mg/L FeSO$_4\cdot$7H$_2$O, 3.4 mg/L MnSO$_4\cdot$H$_2$O, 2.8 mg/L ZnSO$_4\cdot$7H$_2$O, 4 mg/L CoCl$_2$ and 500 mL/L

0.2 M Na_2HPO_4-citric acid (pH 5.0) was formulated, as described previously. Each flask was incubated at 28 °C under shaking at 150 rpm. Triplicate was applied. Time-courses for the production of cellulase and BGL were monitored at 24, 36, 48 and 60 h.

2.3. Development and Validation of β-Glucosidase Inhibitors

Miglitol inhibitor with a concentration gradient (0–5 mg/mL) was added to the reaction mixtures, which contained 200 μL of 15 mM cellobiose as a substrate and 200 μL properly diluted enzyme solution. The reaction mixtures were incubated at 50 °C for 30 min and then boiled for 2 min to stop the reaction. BGL activity was determined by the same method described in Section 2.9. GraphPad Prism software (version 8; San Diego, CA, USA) software was used to draw the IC_{50} curves, and the IC_{50} of miglitol on BGL inhibition was obtained.

Aesculin was used to identify β-glucosidase activity in a fermentation broth (24, 36, 48, 60 and 72 h) from *T. reesei* Rut C30 supplemented with 1 g/L miglitol using 10 g/L MGD as the carbon source [27], and the nitrogen source and inorganic salts were the same as the fermentation medium described in Section 2.2. Then, 0.5 g/L aesculin and 1.0 g/L $FeSO_4 \cdot 7H_2O$ were dissolved in 1 mL of a 0.2 M HAC-NaAC buffer. The fermentation broth (200 μL) was dropped into 1 mL of the test solution and incubated at 50 °C for 1 h. Finally, 200 μL of the mixture was extracted and placed into a 96-well plate for imaging. The time-course of BGL production was monitored at 24, 36, 48, 60 and 72 h, with water added as a negative control.

2.4. Molecular Docking

Two crystal structures of β-glucosidase (BGL) were downloaded from the RCSB PDB. The PDB IDs of BGL were named 3ZYZ and 3AHY. The structures of miglitol, sophorose and cellobiose were exported to PubChem.

All the water molecules, co-crystal ligands and heteroatoms were deleted from the receptor using AutoDock. Hydrogen atoms and Gasteiger charges were added. The output files were saved in the format of pdbqt. The ligands were energy-minimized before they were converted to the format of pdbqt. Then, the active site of the molecular docking was determined by referring to the original ligand coordinates in the target protein, and a $10 \times 10 \times 10$ (Å3) molecular pocket was set, according to the BGL protein active pocket, as an effective structure domain for docking with ligands. AutoDock was used to perform molecular docking and analyze the best binding modes for receptor–ligand interactions. The conformers with the lowest Gibbs free binding energy (estimated as ΔG in kcal/mol) were selected for PostDock analysis. Miglitol, sophorose and cellobiose were used as ligands and BGL was used as a receptor. Finally, the docking results were visualized on PyMOL [28].

2.5. Effects of Miglitol on the Growth and Cellulase Induction of T. reesei

To verify whether miglitol can be used as a carbon source for *T. reesei*, the spores were inoculated into plates (35 mm in diameter) containing 10 g/L miglitol as the sole carbon source and the rest of the ingredients were the same as the fermentation medium described in Section 2.2, but the peptone was removed. Glucose as the sole carbon source was used as a control. The plates were cultured at 28 °C for 48 h.

To test the effect of miglitol on the cellulase activity of *T. reesei* Rut C30, we measured cellulase activity in the fermentation medium with or without miglitol (1 g/L), supplemented using 10 g/L glucose as the carbon source. Enzyme activity was detected after 48 h of shake flask incubation. The fermentation broth was centrifuged at 5000 rpm for 4 min to obtain the supernatant for the cellulase activity assay. The method for cellulase production in shake flasks was described in Section 2.2, and the enzyme activity assay method was described in Section 2.9.

To detect the effect of miglitol on *T. reesei* growth, we inoculated 1 μL of spores in the plates with fermentation medium containing different concentrations of miglitol (0,

0.125, 0.25, 0.5, 1 and 2 g/L) and cultured them at 28 °C for 96 h. The nitrogen source and inorganic salts were the same as the fermentation medium described in Section 2.2, but 10 g/L MGD was used as the carbon source. Meanwhile, three groups were tested in parallel, and mycelium growth was measured at 12 h intervals.

2.6. Enzyme Production Induced by Different Inducers with/without Miglitol Supplemented

For cellulase and β-glucosidase production after the addition of miglitol, the spores of *T. reesei* Rut C30 were inoculated into 250-mL Erlenmeyer flasks containing 50 mL medium composed of 5 g/L glucose and 10 g/L corn steep liquor. After 24 h of cultivation at 28 °C and 150 rpm on a rotary shaker, mycelium was inoculated at 8% (*v/v*) into 250-mL Erlenmeyer flasks containing 50 mL fermentation medium for cellulase production. The fermentation medium was formulated based on the recipe developed in method 2.2, but 10 g/L MGD, 10 g/L lactose or 10 g/L cellulose was used as the sole carbon source. In the groups with different carbon sources, 1 g/L miglitol was added to one group, and the other group without the miglitol addition was used as a control. Cellulase activity and β-glucosidase activity were measured at 12 h intervals until after 96 h of culture. Each experiment was conducted in triplicate.

2.7. qPCR

Total RNA was extracted using a fungal total RNA isolation kit (Sangon Biotech, Shanghai, China), according to the manufacturer's instructions. Reverse transcription was carried out using a PrimeScript RT reagent kit (Takara, Tokyo, Japan), according to the manufacturer's instructions. Quantitative PCR was performed on a Bio-Rad myIQ2 thermocycler (Bio-Rad, Richmond, CA, USA). Amplification reactions were performed using TB Green® Premix Ex Taq™ II (Tli RNaseH Plus) (Takara), according to the manufacturer's instructions. The *sar1* gene was used as a normalized control, and the primers are listed in Table S1. The expression of genes was calibrated by the $2^{-\Delta Ct}$ method, and at least three biological replicates were carried out for each experiment.

2.8. Protein Analysis

The fermentation broth (prepared according to Section 2.6) was centrifuged at 4 °C and 5000 rpm for 4 min to remove hyphal pellets, and the supernatant was transferred to a new tube for further analysis. Then, 8 µL of the supernatant was mixed with 2 µL of 5× native sample loading buffer and boiled for 10 min for denaturation. After that, the samples and a PageRuler™ prestained protein ladder (10 to 180 kDa) were loaded onto 10% SDS-polyacrylamide separating gels. Then, the gels were pre-run at 80 V for 30 min and run at 120 V for 100 min in Tris-glycine buffer. Finally, clear bands were obtained after staining and destaining.

2.9. Analysis Methods

Specifically, 2 mL of a mixture of shake flask fermentation was sampled. Then, the sample was centrifuged at 4 °C and 5000 rpm for 4 min to remove hyphal pellets, and the supernatant was transferred to another tube for further analysis. The activities of the cellulase and β-glucosidase (BGL) were determined using standard protocols [29]. In brief, 50 mg of filter paper (Whatman No. 1, GE healthcare, Sheffield, UK) was added to 1 mL of a 0.1 M HAC-NaAC buffer at pH 4.8. Then, 500 µL of an enzyme solution diluted to the appropriate concentration was added to the mixture and incubated at 50 °C for 60 min. The reaction was stopped by adding 2 mL of alkaline 3,5-dinitrosalicylic (DNS) and boiling for 5 min, followed by immediate ice incubation. Then, the mixture was diluted fourfold, and the absorbance at 540 nm was detected for correction of cellulase activity. One unit of cellulase activity was defined as the amount of the enzyme needed to release 1 µmol of reducing sugar per minute.

BGL activity was measured using 15 mM cellobiose (Sigma, St. Louis, MO, USA) as a substrate [29]. Each assay was carried out in 400 µL of a reaction mixture containing

200 μL of the enzyme solution diluted to the appropriate concentration and 200 μL of the respective substrate and incubated at 50 °C for 30 min. One unit of β-glucosidase activity was defined as 1 mL of the enzyme solution needed to hydrolyze cellobiose to produce 1 μmol of glucose per minute.

Cellobiohydrolase activity was assayed according to the methods described elsewhere, which used *p*-nitrophenol-D-cellbioside (Sigma–Aldrich, St. Louis, MO, USA) as a substrate [30]. The assays were carried out in 150 μL of a reaction mixture containing 100 μL of the culture supernatant and 50 μL of 1 mg/mL pNPC with incubation at 50 °C for 30 min. Then, 150 μL of 10% Na_2CO_3 was added to terminate the reaction. Afterwards, the mixture was diluted fourfold, and the absorbance at 415 nm was detected for correction of cellobiohydrolase activity. One unit of cellobiohydrolase activity was defined as the amount of the enzyme required to release 1 μg of pNP from the substrate per minute under the standard assay conditions.

For xylanase activity determination [31], 180 μL of 1% oat spelt xylan (TCL, Tokyo, Japan) in 50 mM sodium citrate buffer at pH 4.8 was mixed with 20 μL of the diluted enzyme and incubated for 5 min. The following steps were similar to the cellulase activity analysis. One unit of xylanase activity was defined as the amount of the enzyme needed to release 1 μmol of reducing sugar per minute.

Glucose was determined using the biological sensor S-1 (Shenzhen Sieman Technology Co., Ltd., Shenzhen, China).

All results are presented as the mean of triplicates and replications, and are from three independent experiments with standard deviation significance set as $p < 0.05$.

3. Results and Discussion

3.1. Effect of Different Inducers on the Synthesis of Cellulase from T. reesei Rut C30, PB3 and PB4

T. reesei, as an industrial fermentation strain of cellulase, has a low level of β-glucosidase (BGL) expression. Hence, BGL must be supplemented through commercial enzymes or through genetic engineering to overcome the lack of BGL [32]. In our previous research, *aabgl1* encoding β-glucosidase 1 from *A. aculeatus* with high specific activity was used as the target gene to achieve higher BGL activity in *T. reesei* Rut C30, forming a series of genetically engineered *T. reesei* strains. Of these, four gene copies in *T. reesei* PB3 were integrated into the genome, enhancing the BGL activity by 73 times. Recombinant *T. reesei* PB4 integrated three copies, resulting in lower BGL activity than that of *T. reesei* PB3. However, we found that the cellulase activity of *T. reesei* PB3 and PB4 was significantly reduced when MGD was used as an inducer [16]. It was speculated that the sophorose in MGD was degraded into glucose, thus losing its inducing ability, so that the cellulase and BGL were evaluated using *T. reesei* Rut C30 and the two recombinant *T. reesei* with different inducers (MGD, lactose or cellulose) (Figure 1).

The BGL activities of both *T. reesei* PB3 and PB4 were significantly improved, regardless of the inducer (Figure 1A). Compared to *T. reesei* Rut C30, the BGL activity of *T. reesei* PB3 under the induction of MGD, lactose and cellulose was improved by 14.88, 7.45 and 7.73 times, respectively, and the BGL activity of *T. reesei* PB4 was enhanced by 10.96, 5.68 and 6.77 times, respectively.

Improving the BGL level relieves the inhibitory effects of cellobiose on cellobiohydrolase and endoglucanase, thereby significantly improving the cellulase activity [33]. In the present study, cellulase activity was improved under induction by cellulose or lactose, but was weakened when MGD was used as an inducer (Figure 1B). Under induction by cellulose or lactose, the cellulase activity of PB3 relative to that of *T. reesei* Rut C30 increased by 17.86% and 22.86%, respectively, and the cellulase activity of *T. reesei* PB4 relative to that of *T. reesei* Rut C30 increased by 15.71% and 3.7%, respectively. Unexpectedly, the cellulase activity of *T. reesei* PB3 and *T. reesei* PB4 relative to *T. reesei* Rut C30 under induction by MGD decreased by 26.36% and 32.73%, respectively.

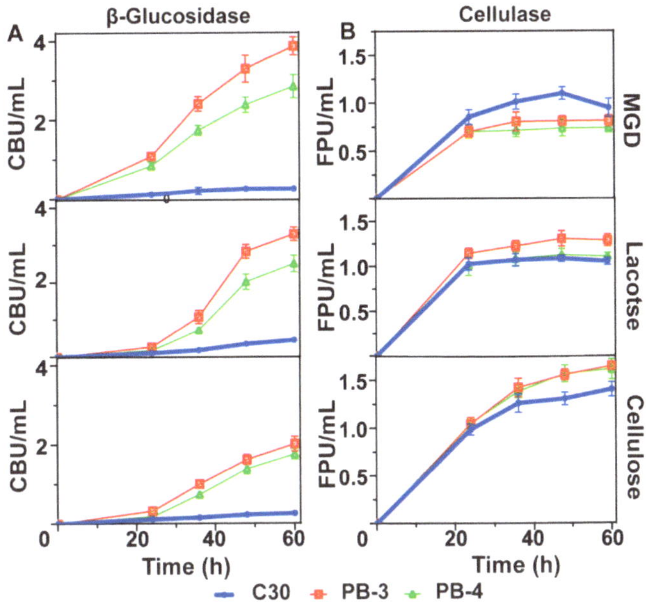

Figure 1. The cellulase and β-glucosidase production of recombinant *Trichoderma reesei* PB-3, PB-4 with *aabgl1* overexpressed and the parent strain Rut C30. The transformants *T. reesei* PB-3 and PB-4 had four and three gene copies, respectively. (**A**) The β-glucosidase activities and (**B**) the cellulase activities of *T. reesei* Rut C30, PB-3 and PB-4 cultured with 10 g/L MGD, lactose or cellulose. For each experiment, three individual replicates were performed. The cellulase and β-glucosidase activity were expressed in filter paper units (FPU) and cellobiase units (CBU), respectively.

Sophorose is the strongest inducer for cellulase production by *T. reesei* known thus far, but BGL degrades sophorose into glucose, which makes it lose its inducing ability and ultimately leads to a decrease in the expression levels of the cellulase gene. With cellulose as the inducer, cellobiohydrolase and endoglucanase synergistically degrade cellulose into cellobiose, which can enter cells. The inducing ability of cellobiose is 2000 times lower than that of sophorose, but BGL has the ability to catalyze cellobiose into sophorose by a transglycosidation reaction to efficiently induce cellulase production [19]. Thus, when cellulose is used as an inducer, the transglycosylation activity of BGL is needed. The induction process of lactose is similar to that of cellulose [18]. Hence, when cellulose or lactose is used as an inducer, the transglycosylation activity of BGL is indispensable for maintaining its inducing ability. When MGD is used, the transglycosylation activity of BGL is not needed, rather, its hydrolytic activity degrades sophorose into glucose and loses its inducibility. Moreover, the sophorose as an inducer was needed at a very low concentration, and extra sophorose tended to be hydrolysed into glucose by β-glucosidase, which is used as a carbon source [14].

Thus, we propose an innovative method of cellulase production. Namely, inhibition of the BGL of *T. reesei* deprives the sophorose metabolizing ability of *T. reesei*, which allows BGL-deficient cellulase to be continually produced with MGD as the inducer. This process would not affect the use of cellulase because the cellulase produced by *T. reesei* sufficiently supplements BGL during practical cellulase applications. Moreover, during BGL supplementation, the background activity of BGL can be ignored, indicating that the innovative method of cellulase production does not affect the quality of cellulase. This idea is similar to the prokaryotic expression system using isopropyl-β-D-thiogalactopyranoside (IPTG), which cannot be metabolized to replace lactose as the inducer, resulting in the persistent induction of the gene expression without the addition of continual flow inducers;

therefore, this new method demonstrates potential in stimulating cellulase production and a promising system of *T. reesei* protein expression.

3.2. Development and Validation of β-Glucosidase Inhibitors

To rapidly validate the feasibility of the above-mentioned method for producing cellulase, we tried BGL inhibitor and added this inhibitor to test whether MGD improves the induction of cellulase production by *T. reesei*. Although there is no commercial β-glucosidase (BGL) inhibitor, α-glucosidase inhibitors are commonly used to decrease oral blood sugar to treat type II diabetes. Commercially available drugs include miglitol, acarbose and voglibose. We preliminarily tested the inhibitory effects of these three compounds on BGL and found that miglitol had the best inhibitory effects on BGL.

A miglitol concentration gradient (0–5 mg/mL) was used with cellobiose as the substrate, and the inhibition rates on BGL were detected under the optimal enzyme reaction conditions. Miglitol inhibited extracellular BGL activity with an IC_{50} value of 2.93 μg/mL (Figure 2A). Furthermore, a quick screening method was employed to identify the BGL activity of fermentation broth from *T. reesei* Rut C30 using MGD as an inducer at 24, 36, 48, 60 or 72 h. Aesculin solutions that turned black indicated the presence of BGL. We found that the enzyme solution detected by the aesculin solution turned black in the miglitol-free medium, however, it did not turn black when miglitol was supplemented (Figure 2B). The color changes were consistent with those in the negative control, suggesting that miglitol efficiently suppresses the β-glucosidase activity of *T. reesei*.

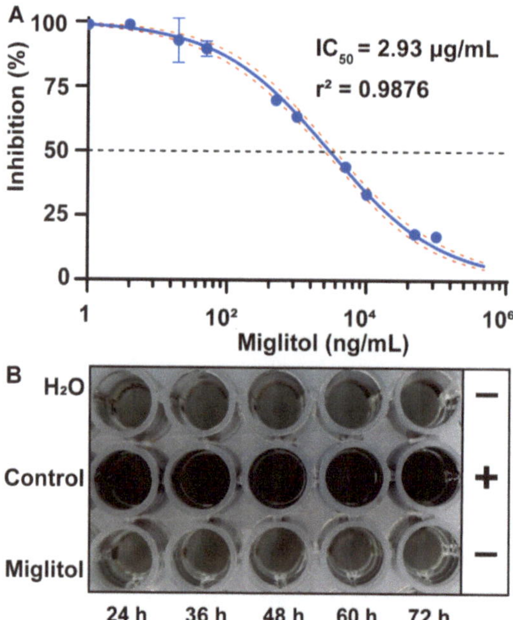

Figure 2. Developing and validating the inhibitory effect of miglitol on β-glucosidase. (**A**) Inhibition of β-glucosidase activity by miglitol at different concentrations. The figures also show the corresponding IC_{50} values and correlation r^2 values; (**B**) Effect of miglitol on the β-glucosidase activity of *T. reesei* Rut C30 using 10 g/L MGD as a carbon source and sampled at 24, 36, 48, 60 or 72 h. A quick detection system containing aesculin was employed in 96-well plates with no miglitol or water added as a negative control.

3.3. Molecular Mechanism by which Miglitol Inhibits β-Glucosidase

To analyze the molecular mechanism by which miglitol inhibits BGL, we molecularly docked miglitol to two BGLs (Bgl1 and Bgl2) previously reported in *T. reesei* and set the substrates (cellobiose and sophorose) as control tests. Bgl1 and Bgl2 are the most important extracellular and intracellular BGLs in *T. reesei*, respectively. Figure 3A shows the associations of miglitol with the two BGLs. Figure 3B,C illustrate the docking between the substrates (sophorose and cellobiose) and the two BGLs. The results showed that the binding sites between miglitol and the two BGLs are located at the active sites and are competitive inhibitors.

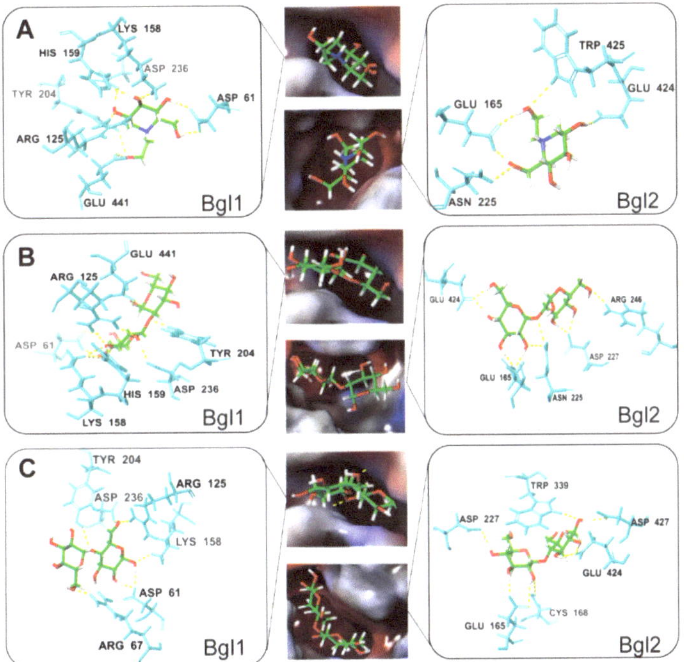

Figure 3. Binding modes of three substrates with Bgl1 and Bgl2. The interactions between miglitol (**A**); sophorose (**B**); and cellobiose (**C**) with the associated residues at the active site of β-glucosidase. 3D surface representation of the ligand-binding pocket (ligands displayed in stick mode) is in the middle of the figure. The black labels and yellow lines show the residue names and hydrogen bonds, respectively.

Molecular docking experiments showed that miglitol binds with several key active residues (aspartic acid, glutamate, arginine and tyrosine) in the active site of BGL to form hydrogen bonds. The binding sites between miglitol and BGL were magnified, showing that miglitol binds with seven key amino acid residues of Bgl1 and four key amino acid residues of Bgl2. Notably, the main interactive affinity is attributed to the presence of hydroxyl groups, and it is the key to anchoring the substrate to the site catalytic pocket [34,35]. The formation of more types and larger quantities of hydrogen bonds indicates that the ligand- and protein-binding ability is stronger. Molecular docking experiments indicated that the quantity of bonds in Bgl1 was greater than that in Bgl2, suggesting that the affinity of miglitol and Bgl1 was stronger with more stable docking and a higher matching degree. Overall, miglitol matched well with the two types of BGLs, indicating that miglitol is an efficient BGL inhibitor and an excellent experimental material with intracellular functionality.

Based on the molecular docking experiments, we predicted that miglitol is a potential BGL competitive inhibitor. The substrates (cellobiose and sophorose) of the two BGLs were docked with Bgl1 and Bgl2, and the active site between the different ligands and proteins were the same (Table 1).

Table 1. Verification of the docking results for miglitol with β-glucosidase using the binding energy as the metric.

Protein	Ligand	Binding Energy/(Kcal/mol)
Bgl1	Miglitol	−6.341
	Sophorose	−5.680
	Cellobiose	−5.868
Bgl2	Miglitol	−7.511
	Sophorose	−7.586
	Cellobiose	−6.396

The docking results showed that the binding energy values of Bgl1 or Bgl2 with miglitol were less than −6 kcal/mol, which indicated that the binding force between the two BGLs and miglitol was stronger. In addition, the binding energy of miglitol was significantly higher than those of the two substrates, suggesting that the miglitol and Bgl1 affinity is the strongest. For Bgl2, the scores of miglitol were similar to those of sophorose, and they were both better than those of cellobiose, indicating that the affinity between cellobiose and Bgl2 is the lowest. Hence, the binding energy of miglitol with both BGLs was higher than that with the two substrates (cellobiose and sophorose). The above analyses indicated that the miglitol and BGL interactive force is strong enough to inhibit BGL. With the coexistence of miglitol, sophorose and cellobiose, the affinity of miglitol for Bgl1 or Bgl2 was stronger and competitively inhibited enzyme and substrate binding, further inhibiting the activity of BGL and the decomposition of sophorose. As a result, the intracellular sophorose concentration, which continually induces cellulase gene expression, was maintained.

3.4. Effects of Miglitol on the Growth and Induction of T. reesei

To validate whether *T. reesei* uses miglitol as the carbon source, we conducted plate growth experiments using MGD or miglitol as the sole carbon source, with the other medium components and culture conditions being completely consistent (Figure 4). The strains grew well in MGD (Figure 4A), however, when miglitol was used as the sole carbon source, the strains did not grow (Figure 4B), indicating that *T. reesei* cannot metabolize miglitol as the carbon source. To further validate whether miglitol acts as an inducer of cellulase synthesis, 1 g/L miglitol was added to the fermentation medium, with 10 g/L glucose as a carbon source. In the absence of the inducer, a small amount of cellulase was detected with an activity of 0.085 IU/mL, and the cellulase activity was then further reduced to 0.024 IU/mL when miglitol was added to the fermentation medium (Figure 4C). The reason was that miglitol could not induce *T. reesei* to generate cellulase, but instead inhibited β-glucosidase and thereby inactivated cellulase. The results suggested that miglitol cannot be used by *T. reesei* as a carbon source or as an inducer of cellulase synthesis.

To further verify whether miglitol affects the growth of strains, we added different concentrations of miglitol to solid plates with MGD as the carbon source. At the same culture time, the concentration of miglitol, even up to 2 g/L, did not significantly affect the growth diameter of the strains (Figure 4D). Although the above results suggested that miglitol does not affect the growth of *T. reesei* or the induced synthesis of cellulase, it significantly inhibits the activity of BGL. Moreover, the activity of BGL produced by *T. reesei* was low. The results of the half maximal inhibitory concentration (IC$_{50}$) showed that trace miglitol inactivated BGL. However, to severely inactivate BGL, we added 1 g/L miglitol in subsequent experiments to validate that the intracellular β-disaccharide concentration of *T. reesei* can be stabilized, which continually induces the synthesis of cellulase.

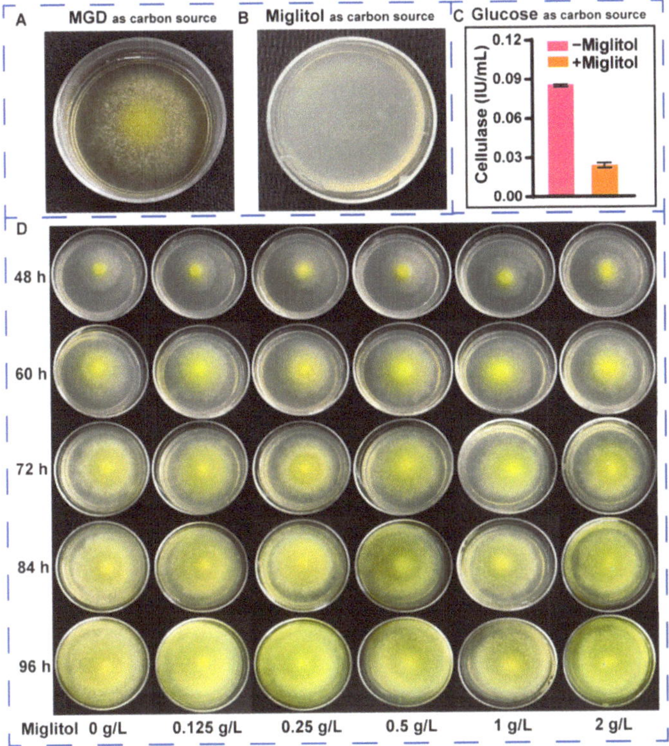

Figure 4. Effects of miglitol on the growth and induction of *T. reesei*. Colony morphology of *T. reesei* on plates containing (**A**) 10 g/L MGD or (**B**) 10 g/L miglitol as the sole carbon source; (**C**) Effect of adding miglitol versus no miglitol on the cellulase activity of *T. reesei* Rut C30 with 10 g/L glucose as the carbon source at 48 h; (**D**) Time-course of the colony morphology of *T. reesei* grown using 10 g/L MGD as a carbon source on different concentrations of miglitol. The mycelium growth was measured at 12 h intervals.

3.5. β-Glucosidase Suppression Promotes MGD to Persistently Induce Cellulase Synthesis in T. reesei

Figure 5 illustrates the production of cellulase and BGL by *T. reesei* cultured in fermentation medium with MGD, lactose or cellulose as the sole inducer. The activity of BGL in the presence of miglitol was effectively inhibited regardless of the inducer, which was consistent with the results above (Figure 5A–C). We then analyzed whether inducer degradation can be prevented so that the cellulase synthesis by *T. reesei* can be continually stimulated after the suppression of BGL.

Figure 5D shows cellulase production induced by MGD with the addition of miglitol. Interestingly, cellulase activity was persistently enhanced. In the absence of miglitol, the cellulase activity was maximized to 1.54 IU/mL at 48 h, then decreased after 60 h and increased sometime thereafter, which was due to the extracellular release of intracellular cellulase after cell death. In contrast, the cellulase activity after miglitol addition reached 1.8 IU/mL at 48 h, but further increased afterwards. The maximum cellulase activity achieved 3.22 IU/mL at 84 h, which was 2.09 times ($p < 0.01$) higher than that without the addition of miglitol. This result suggested that the addition of miglitol inactivated BGL, which prevented the degradation of the inducers in MGD (e.g., sophorose), allowing continual induction of cellulase synthesis in *T. reesei*. This is equivalent to endowing *T. reesei* with a phenotype that cannot metabolize sophorose but does not lose its ability to synthesize sophorose-induced cellulase. The above result provides a new clue for cellulase production in *T. reesei*. In the future, genetically engineered *T. reesei* can be developed to avoid the use of miglitol or other similar inhibitors. Moreover, the BGL activity of *T. reesei* is extremely low, and extra supplementation is needed during the hydrolysis of straw biomass. Hence, this method of cellulase production does not affect the enzyme quality.

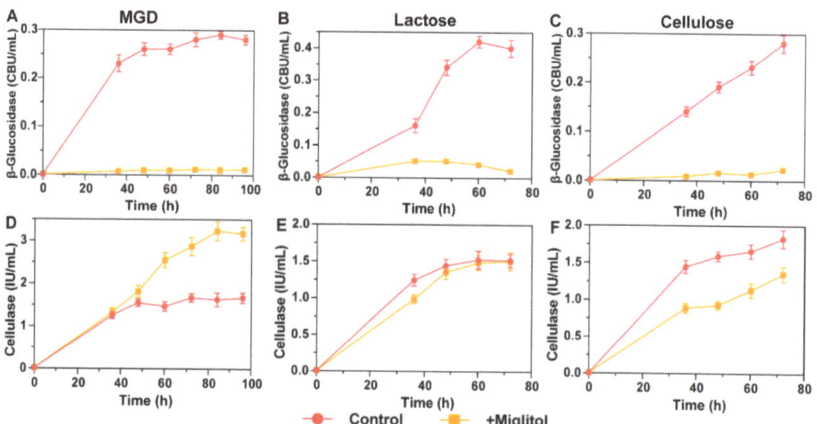

Figure 5. For cellulase and β-glucosidase activity after the addition of miglitol, time-course profiles of β-glucosidase (**A–C**) and cellulase activity (**D–F**) for *T. reesei* Rut C30 taking 10 g/L MGD (**A,D**); 10 g/L lactose (**B,E**); or 10 g/L cellulose (**C,F**) as the carbon source with/without miglitol. Bars denote the standard deviations for three independent experiments.

However, the results of cellulase production with lactose or cellulose as the inducer were opposite to the result with MGD (Figure 5D–F). Although BGL activity was also inhibited after the addition of miglitol (Figure 5B,C), the cellulase activity was not further enhanced (Figure 5E,F), which was consistent with the results obtained without the addition of miglitol, and the enzymatic activity was even reduced. The cellulase activity was not significantly decreased after the addition of miglitol compared to the lactose control (Figure 5E); however, the activity was significantly weakened by 26% in the presence of miglitol when cellulose was used as an inducer (Figure 5F). These results indicated that BGL is indispensable when cellulose or lactose act as an inducer. Reportedly, the transglycosylation activity of BGL is critical for the inducibility of lactose or cellulose [18,19]. The reason is that a strong cellulase inducer, sophorose, can be produced in *T. reesei*, which was directly demonstrated by our previous findings. MGD contains a certain amount of sophorose and does not require the transglycosylation activity of BGL, which can induce *T. reesei* to persistently synthesize cellulase after BGL inhibition.

Overall, the cellulase activity reached its highest level at 48 h and started to decrease after 60 h without miglitol addition, indicating that the carbon source and inducer in the medium were already metabolized by *T. reesei*, leading to the retarded growth of the strain. To further explore whether the activities of cellobiohydrolase and xylanase were also enhanced after the addition of miglitol, we measured the activities of these two enzymes at the two tested time points (Figure 6A,B). The results showed that the changing trends were consistent with that of cellulase activity, as the two enzymatic activities without the addition of miglitol started to decrease after 60 h. After the addition of miglitol, however, the activities of cellobiohydrolase and xylanase at 60 h were 1.36 and 1.80 times higher than the corresponding results without miglitol, respectively ($p < 0.01$), indicating that the enzyme would be more efficient for hydrolyzing lignocellulosic biomass pretreated by alkali, in which the hemicelluloses and xylan remain hydrolyzed together with the cellulose component [36]. This comparison further validates that the inducing activity of MGD can be maintained after BGL inhibition by miglitol. The experimental results to test whether the addition of miglitol affects the biomass of the strain at the two tested time points (Figure 6C) were consistent with those shown in Figure 4D. Namely, the addition of miglitol did not affect the growth or metabolism of the strain and did not significantly change the biomass at either time point. Moreover, SDS-PAGE showed that the extracellular protein concentrations of *T. reesei* were significantly improved after the addition of miglitol (Figure 6D), further verifying the actions of miglitol.

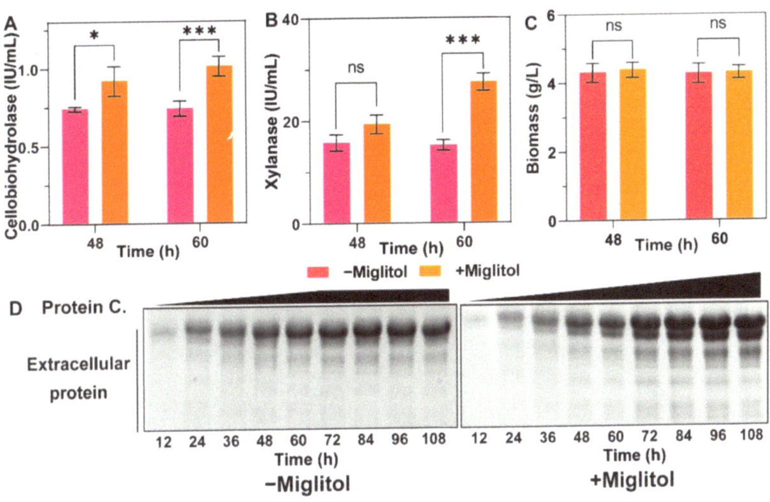

Figure 6. Effect of miglitol on lignocellulase production of *T. reesei* Rut C30 with 10 g/L MGD as the carbon source at 48 h and 60 h ((**A**) Cellobiohydrolase; (**B**) Xylanase; (**C**) Biomass); (**D**) SDS-PAGE analysis of supernatant. Values are the mean ± SD of the results from three independent experiments. Asterisks indicate a significant difference (* $p < 0.05$, *** $p < 0.01$, Student's *t*-test).

Figure 7A–C illustrates the expression of two major cellulase genes (*cbh1* and *bgl1*) and the most important transcription activator (*xyr1*), which were tested to explore the molecular mechanism by which MGD induces cellulase production in *T. reesei* after the addition of miglitol. The results showed that the three genes were all expressed under induction by MGD. When miglitol was not added, the expression levels of the three genes all peaked after 36 h, but significantly decreased after 48 h and nearly approached 0, which indicated that the inducer was fully metabolized after 48 h of culture. These results were consistent with the cellulase activity results shown in Figure 5. The genetic transcription level of cellulase was maximized at 36 h, hence, cellulase activity maximized at 48 h. After the addition of miglitol, the gene expression of *cbh1*, *bgl1* and *xyr1* can still be started after 48 or 60 h, further indicating that the inducer exists in the medium and can induce cellulase synthesis after the suppression of BGL. Notably, miglitol only inhibited the enzymatic activity but did not affect the gene expression of BGL (Figure 7B).

Recently, to improve the efficiency of lignocellulose hydrolysis by cellulase, the overexpression of BGL in *T. reesei* has been extensively used [21–25]. However, the present study showed that BGL is important for the inducing role of cellulose or lactose because its transglycosylation activity must be utilized to produce the strongest inducer of *T. reesei*, sophorose, to maximally induce cellulase production. However, MGD contains a certain amount of sophorose and thus does not require the transglycosylation activity of BGL, conversely, its hydrolyzing ability degrades sophorose into glucose, thereby completely inhibiting the inducing activity of sophorose. Hence, this sophorose containing inducer must be continually supplemented in a fed-batch manner. In the present study, the long-term inducing ability of MGD was maintained after the inhibition of BGL, and the dosage of the inducer was decreased. In addition, although BGL is pivotal in the degradation of cellulose components into glucose, researchers have suggested that the BGL secreted by *T. reesei* is largely insufficient, indicating that the efficient hydrolysis of cellulose components can be achieved only when enough BGL is formulated. Hence, the present new method does not largely reduce the quality of cellulase of *T. reesei* and is helpful for overcoming the problem of the inducer needing to be supplemented through feeding, faced in high titer cellulase production by fed-batch fermentation in the future.

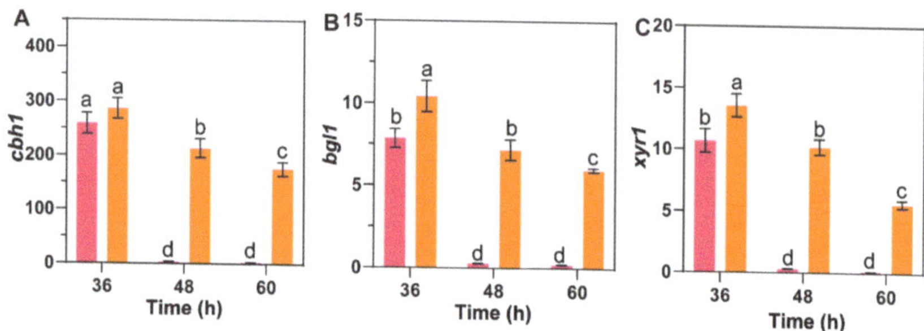

Figure 7. Effect of miglitol on the transcription of genes encoding cellulase ((**A**) *cbh1*; (**B**) *bgl1*); and transcription factor ((**C**) *xyr1*) at 36 h, 48 h and 60 h of *T. reesei* Rut C30 with 10 g/L MGD as the carbon source at 36 h, 48 h and 60 h. Values are the mean $\pm$ SD of the results from three independent experiments. One-way ANOVA was performed to reveal significant ($p < 0.05$) differences between the mean values, which are indicated by different letters.

4. Conclusions

The present study confirmed that miglitol competitively inhibits β-glucosidase, with an IC_{50} of 2.93 μg/mL. Adding miglitol into a sophorose-containing culture medium significantly weakened β-glucosidase activity to levels near 0, which prevented the degradation of sophorose, thereby allowing it to continually induce the production of cellulase by *T. reesei*. However, the addition of miglitol is ineffective when the inducer is lactose or cellulose, indicating that the transglycosylation activity of BGL is crucial for the activities of these two inducers. Our findings indicated that this alternative strategy was developed to induce the synthesis of cellulase and provides a promising system for *T. reesei* protein expression.

Supplementary Materials: The following supporting information can be downloaded at: https://www.mdpi.com/article/10.3390/fermentation8050232/s1, Table S1. Primers for qRT-PCR analysis.

Author Contributions: Conceptualization, Y.L.; methodology, T.L.; software, P.Z. and T.L.; validation, J.Y., Y.G. and X.R.; formal analysis, J.Y.; investigation, T.L.; resources, P.Z.; data curation, T.L.; writing—original draft preparation, Y.L.; writing—review and editing, T.L.; visualization, P.Z.; supervision, J.Y.; project administration, Y.G. and X.R.; funding acquisition, Y.L. All authors have read and agreed to the published version of the manuscript.

Funding: This research was financially funded by the National Natural Science Foundation of China with the grant reference numbers of 21808022, the Natural Science Foundation Project of Chongqing, the Chongqing Science and Technology Commission (CN) (cstc2018jcyjAX0064), the Science and Technology Research Program of Chongqing Municipal Education Commission (KJQN201901523), the Postgraduate Research and Innovation Project of Chongqing University of Science and Technology (YKJCX2120505 and YKJCX2120523), and also supported by the National Undergraduate Training Programs for Innovation and Entrepreneurship of China (No. 202111551022).

Institutional Review Board Statement: Not applicable.

Informed Consent Statement: Not applicable.

Data Availability Statement: Not applicable.

Acknowledgments: The authors appreciate the helpful discussion with Xinqing Zhao at Shanghai Jiao Tong University, and also appreciate the helpful discussion with Ruimeng Gu at the Tianjin Institute of Industrial Biotechnology, Chinese Academy of Sciences.

Conflicts of Interest: The authors declare that they have no competing interests.

References

1. Himmel, M.E.; Edward, A.B. Lignocellulose conversion to biofuels: Current challenges, global perspectives. *Curr. Opin. Biotechnol.* **2009**, *20*, 316–317. [CrossRef] [PubMed]
2. Reaney, M.; Wiens, D.; Tse, T. Production of bioethanol-A review of factors affecting ethanol yield. *Fermentation* **2021**, *7*, 268. [CrossRef]
3. Liu, C.G.; Xiao, Y.; Xia, X.X.; Zhao, X.Q.; Peng, L.; Srinophakun, P.; Bai, F. Cellulosic ethanol production: Progress, challenges and strategies for solutions. *Biotechnol. Adv.* **2019**, *37*, 491–504. [CrossRef] [PubMed]
4. Prasoulas, G.; Gentikis, A.; Konti, A.; Kalantzi, S.; Kekos, D.; Mamma, D. Bioethanol production from food waste applying the multienzyme system produced on-site by *Fusarium oxysporum* F3 and mixed microbial cultures. *Fermentation* **2020**, *6*, 39. [CrossRef]
5. Bhati, N.; Sharma, A.K. Cost-effective cellulase production, improvement strategies, and future challenges. *J. Food Process Eng.* **2021**, *44*, e13623. [CrossRef]
6. Bhardwaj, N.; Bikash, K.; Komal, A.; Pradeep, V. Current perspective on production and applications of microbial cellulases: A review. *Bioresour. Bioprocess.* **2021**, *8*, 95. [CrossRef]
7. Peterson, R.; Nevalainen, H. *Trichoderma reesei* RUT-C30–thirty years of strain improvement. *Microbiology* **2012**, *158*, 58–68. [CrossRef]
8. Fang, H.; Li, C.; Zhao, J.; Zhao, C. Biotechnological advances and trends in engineering *Trichoderma reesei* towards cellulase hyperproducer. *Biotechnol. Bioprocess Eng.* **2021**, *26*, 517–528. [CrossRef]
9. Mandels, M.; Parrish, F.W.; Reese, E.T. Sophprpse as an inducer of cellulase in *Trichoderma viride*. *J. Bacteriol.* **1962**, *83*, 400–408. [CrossRef]
10. Humbird, D.; Davis, R.; Tao, L.; Kinchin, C.; Hsu, D.; Aden, A.; Schoen, P.; Lukas, J.; Olthof, B.; Worley, M.; et al. *Process Design and Economics for Biochemical Conversion of Lignocellulosic Biomass to Ethanol: Dilute-Acid Pretreatment and Enzymatic Hydrolysis of Corn Stover*; Technical Report NREL/TP-5100-47764; National Renewable Energy Laboratory: Golden, CO, USA, 2012. [CrossRef]
11. Li, Y.; Liu, C.; Bai, F.; Zhao, X. Overproduction of cellulase by *Trichoderma reesei* RUT C30 through batch-feeding of synthesized low-cost sugar mixture. *Bioresour. Technol.* **2016**, *216*, 503–510. [CrossRef]
12. Xia, Y.; Yang, L.; Xia, L. High-level production of a fungal β-glucosidase with application potentials in the cost-effective production of *Trichoderma reesei* cellulase. *Process Biochem.* **2018**, *70*, 55–60. [CrossRef]
13. Li, Y.; Zhang, X.; Xiong, L.; Mehmood, M.A.; Zhao, X.; Bai, F. On-site cellulase production and efficient saccharification of corn stover employing *cbh2* overexpressing *Trichoderma reesei* with novel induction system. *Bioresour. Technol.* **2017**, *238*, 643–649. [CrossRef] [PubMed]
14. Li, Y.; Zhang, P.; Zhu, D.; Yao, B.; Hasunuma, T.; Kondo, A.; Zhao, X. Efficient preparation of soluble inducer for cellulase production and saccharification of corn stover using in-house generated crude enzymes. *Biochem. Eng. J.* **2022**, *178*, 108296. [CrossRef]
15. Chen, Y.; Wu, C.; Fan, X.; Zhao, X.; Wang, W. Engineering of *Trichoderma reesei* for enhanced degradation of lignocellulosic biomass by truncation of the cellulase activator ACE3. *Biotechnol. Biofuels* **2020**, *13*, 62. [CrossRef] [PubMed]
16. Li, Y.; Zhang, X.; Zhang, F.; Peng, L.; Zhang, D.; Kondo, A.; Bai, F.; Zhao, X. Optimization of cellulolytic enzyme components through engineering *Trichoderma reesei* and on-site fermentation using the soluble inducer for cellulosic ethanol production from corn stover. *Biotechnol. Biofuels* **2018**, *11*, 49. [CrossRef] [PubMed]
17. Li, Y.; Yu, J.; Zhang, P.; Long, T.; Mo, Y.; Li, J.; Li, Q. Comparative transcriptome analysis of *Trichoderma reesei* reveals different gene regulatory networks induced by synthetic mixtures of glucose and β-disaccharide. *Bioresour. Bioprocess.* **2021**, *8*, 57. [CrossRef]
18. Xu, J.; Zhao, G.; Kou, Y.; Zhang, W.; Zhou, Q.; Chen, G.; Liu, W. Intracellular β-glucosidases CEL1a and CEL1b are essential for cellulase induction on lactose in *Trichoderma reesei*. *Eukaryot. Cell* **2014**, *13*, 1001–1013. [CrossRef]
19. Zhou, Q.; Xu, J.; Kou, Y.; Lv, X.; Zhang, X.; Zhao, G.; Zhang, W.; Chen, G.; Liu, W. Differential involvement of β-glucosidases from *Hypocrea jecorina* in rapid induction of cellulase genes by cellulose and cellobiose. *Eukaryot. Cell* **2012**, *11*, 1371–1381. [CrossRef]
20. Qi, K.; Chen, C.; Yan, F.; Feng, Y.; Bayer, E.A.; Kosugi, A.; Cui, Q.; Liu, Y. Coordinated β-glucosidase activity with the cellulosome is effective for enhanced lignocellulose saccharification. *Bioresour. Technol.* **2021**, *337*, 125441. [CrossRef]
21. Dashtban, M.; Qin, W. Overexpression of an exotic thermotolerant β-glucosidase in *Trichoderma reesei* and its significant increase in cellulolytic activity and saccharification of barley straw. *Microb. Cell Factories* **2012**, *11*, 63. [CrossRef]
22. Treebupachatsakul, T.; Shioya, K.; Nakazawa, H.; Kawaguchi, T.; Morikawa, Y.; Shida, Y.; Ogasawara, W.; Okada, H. Utilization of recombinant *Trichoderma reesei* expressing *Aspergillus aculeatus* β-glucosidase I (JN11) for a more economical production of ethanol from lignocellulosic biomass. *J. Biosci. Bioeng.* **2015**, *120*, 657–665. [CrossRef] [PubMed]
23. Zhang, J.; Zhong, Y.; Zhao, X.; Wang, T. Development of the cellulolytic fungus *Trichoderma reesei* strain with enhanced β-glucosidase and filter paper activity using strong artifical cellobiohydrolase 1 promoter. *Bioresour. Technol.* **2010**, *101*, 9815–9818. [CrossRef] [PubMed]
24. Nakazawa, H.; Kawai, T.; Ida, N.; Shida, Y.; Kobayashi, Y.; Okada, H.; Tani, S.; Sumitani, J.-I.; Kawaguchi, T.; Morikawa, Y.; et al. Construction of a recombinant *Trichoderma reesei* strain expressing *Aspergillus aculeatus* β-glucosidase 1 for efficient biomass conversion. *Biotechnol. Bioeng.* **2012**, *109*, 92–99. [CrossRef] [PubMed]

25. Li, C.; Lin, F.; Li, Y.; Wei, W.; Wang, H.; Qin, L.; Zhou, Z.; Li, B.; Wu, F.-G.; Chen, Z. A β-glucosidase hyper-production *Trichoderma reesei* mutant reveals a potential role of cel3D in cellulase production. *Microb. Cell Factories* **2016**, *15*, 151. [CrossRef] [PubMed]
26. Sels, J.P.J.; Huijberts, M.S.; Wolffenbuttel, B.H. Miglitol, a new α-glucosidase inhibitor. *Expert Opin. Pharmacother.* **1999**, *1*, 149–156. [CrossRef]
27. Pérez, G.; Farina, L.; Barquet, M.; Boido, E.; Gaggero, C.; Dellacassa, E.; Carrau, F. A quick screening method to identify β-glucosidase activity in native wine yeast strains: Application of Esculin Glycerol Agar (EGA) medium. *World J. Microbiol. Biotechnol.* **2011**, *27*, 47–55. [CrossRef]
28. Eberhardt, J.; Santos-Martins, D.; Tillack, A.F.; Forli, S. AutoDock Vina 1.2.0: New Docking Methods, Expanded Force Field, and Python Bindings. *J. Chem. Inf. Model.* **2021**, *61*, 3891–3898. [CrossRef]
29. Ghose, T.K. Measurement of cellulase activities. *Pure Appl. Chem.* **1987**, *59*, 257–268. [CrossRef]
30. Jalak, J.; Väljamäe, P. Mechanism of initial rapid rate retardation in cellobiohydrolase catalyzed cellulose hydrolysis. *Biotechnol. Bioeng.* **2010**, *106*, 871–883. [CrossRef]
31. Törrönen, A.; Mach, R.L.; Messner, R.; Gonzalez, R.; Kalkkinen, N.; Harkki, A.; Kubicek, C.P. The two major xylanases from *Trichoderma reesei*: Characterization of both enzymes and genes. *Nat. Biotechnol.* **1992**, *10*, 1461–1465. [CrossRef]
32. Xia, Y.; Yang, L.; Xia, L. Combined strategy of transcription factor manipulation and β-glucosidase gene overexpression in *Trichoderma reesei* and its application in lignocellulose bioconversion. *J. Ind. Microbiol. Biotechnol.* **2018**, *45*, 803–811. [CrossRef] [PubMed]
33. Hirano, K.; Saito, T.; Shinoda, S.; Haruki, M.; Hirano, N. In vitro assembly and cellulolytic activity of a β-glucosidase-integrated cellulosome complex. *FEMS Microbiol. Lett.* **2019**, *366*, fnz209. [CrossRef]
34. Du, X.; Li, Y.; Xia, Y.; Ai, S.; Liang, J.; Sang, P.; Ji, X.; Liu, S. Insights into protein–ligand interactions: Mechanisms, models, and methods. *Int. J. Mol. Sci.* **2016**, *17*, 20144. [CrossRef] [PubMed]
35. He, M.; Zhai, Y.; Zhang, Y.; Xu, S.; Yu, S.; Wei, Y.; Xiao, H.; Song, Y. Inhibition of α-glucosidase by trilobatin and its mechanism: Kinetics, interaction mechanism and molecular docking. *Food Funct.* **2022**, *13*, 857–866. [CrossRef] [PubMed]
36. Behera, S.; Arora, R.; Nandhagopal, N.; Kumar, S. Importance of chemical pretreatment for bioconversion of lignocellulosic biomass. *Renew. Sustain. Energy Rev.* **2014**, *36*, 91–106. [CrossRef]

Article

Construction of L-Asparaginase Stable Mutation for the Application in Food Acrylamide Mitigation

Bing Yuan [1,†], Pengfei Ma [2,†], Yuxuan Fan [1], Bo Guan [1], Youzhen Hu [1], Yan Zhang [1], Wenli Yan [1], Xu Li [1,*] and Yongqing Ni [1,*]

1   School of Food Science and Technology, Shihezi University, Shihezi 832003, China; ybiice@163.com (B.Y.); sprinklefan@163.com (Y.F.); guanbo@shzu.edu.cn (B.G.); yzhu@shzu.edu.cn (Y.H.); zhangyanzqgh@163.com (Y.Z.); wenliyan62@sina.com (W.Y.)
2   Changji Hui Autonomous Prefecture Institute for Drug Control, Changji 831100, China; mpf2551241@163.com
*   Correspondence: leeluok@163.com (X.L.); niyqlzu@sina.com (Y.N.)
†   These authors contributed equally to this work.

Abstract: Acrylamide, a II A carcinogen, widely exists in fried and baked foods. L-asparaginase can inhibit acrylamide formation in foods, and enzymatic stability is the key to its application. In this study, the *Escherichia coli* L-asparaginase (ECA) stable variant, D60W/L211R/L310R, was obtained with molecular dynamics (MD) simulation, saturation mutation, and combinatorial mutation, the half-life of which increased to 110 min from 60 min at 50 °C. Furthermore, the working temperature (maintaining the activity above 80%) of mutation expanded from 31 °C–43 °C to 35 °C–55 °C, and the relative activity of mutation increased to 82% from 65% at a pH range of 6–10. On treating 60 U/mL and 100 U/g flour L-asparaginase stable mutant (D60W/L211R/L310R) under uncontrolled temperature and pH, the acrylamide content of potato chips and bread was reduced by 66.9% and 51.7%, which was 27% and 49.9% higher than that of the wild type, respectively. These results demonstrated that the mutation could be of great potential to reduce food acrylamide formation in practical applications.

Keywords: L-asparaginase; stability; mutation; acrylamide; food safety

Citation: Yuan, B.; Ma, P.; Fan, Y.; Guan, B.; Hu, Y.; Zhang, Y.; Yan, W.; Li, X.; Ni, Y. Construction of L-Asparaginase Stable Mutation for the Application in Food Acrylamide Mitigation. *Fermentation* **2022**, *8*, 218. https://doi.org/10.3390/fermentation8050218

Academic Editors: Odile Francesca Restaino and Yi-Huang Hsueh

Received: 23 March 2022
Accepted: 28 April 2022
Published: 11 May 2022

Publisher's Note: MDPI stays neutral with regard to jurisdictional claims in published maps and institutional affiliations.

1. Introduction

Potatoes and flour, two of the most important staple foods, are rich in carbohydrates. However, while these high-carbohydrate foods are processed at high temperatures (above 120 °C), a large amount of acrylamide is formed due to the Maillard reaction between reducing sugars and amino acids [1]. The acrylamide content in microwaved snacks and French fries, respectively, reached 20,336 µg/kg and 10,712 µg/kg [2,3], which far exceeded the limit of acrylamide in daily drinking water set by the World Health Organization by 0.5 µg/L, triggering international health alerts.

Some strategies such as raw material selection, processing optimization, addition of plant extracts, and enzymatic treatment were researched to reduce the acrylamide content in food [4–7]. Among these, L-asparaginase (EC 3.5.1.1), which was found to effectively inhibit the acrylamide formation in food by removing the acrylamide precursor (L-asparagine) without changing the food senses [8–11], has attracted extensive attention. On a laboratory scale, different sources of L-asparaginase have been used to inhibit acrylamide formation under restricted reaction conditions [2,12–14]. Wang et al. (2021) pretreated French fries with 10 U/mL *Palaeococcus ferrophilus* L-asparaginase at 85 °C for 10 min to reduce the acrylamide content by 80% [2]. Farahat et al. (2020) reduced acrylamide in French fries by 82% using 20 U/mL *Cobetia amphilecti* L-asparaginase at 40 °C for 30 min [15]. Ran et al. (2017) used *Paenibacillus barengoltzii* L-asparaginase to pretreat French fries and mooncake at 45 °C for 20 and 60 min, and found that the acrylamide content was lowered by 86% and 52%, respectively [13]. Commercial L-asparaginase (10 U/mL; Acrylaway® L-asparaginase) was used to pretreat French fries at 75 °C for 10 min to reduce the acrylamide content by 60% [16].

Many investigations have been conducted to inhibit acrylamide formation in food with L-asparaginase on a laboratory scale, but few have been performed on an industrial scale [9]. Operational temperature, pH, and time are the crucial parameters for the successful application of L-asparaginase on an industrial scale. Hence, L-asparaginase, with better stability, wider action temperature, and pH, has more industrial application potential. In this study, the key residues of *E. coli* L-asparaginase (ECA) were identified by molecular dynamics (MD) simulation. Then, the stability of ECA was improved through saturation and compound mutations, and its application temperature and pH were expanded. Finally, we evaluated its application effect in inhibiting acrylamide formation in potato chips and bread under restricted and non-restricted reaction conditions.

2. Materials and Methods

2.1. Strains, Plasmids, and Chemicals

E. coli BL21 was used as the host strain for gene cloning and expression. The shuttle expression plasmid pET-28a was used for expression and mutagenesis studies. All strains and plasmids were preserved in our laboratory. The restriction enzymes, PrimeSTAR® HS DNA Polymerase and T4 DNA ligase, were purchased from TaKaRa Bio Co. (Dalian, China), and the Mini Plasmid Rapid Isolation Kit and Mini DNA Rapid Purification Kit were obtained from Sangon Biotech Co., Ltd. (Shanghai, China). A HisTrapTM HP column was purchased from GE Healthcare, Inc. (Little Chalfont, Buckinghamshire, UK). All other high-grade chemicals were commercially sourced.

2.2. Construction of Recombinant Strains

The plasmid pET-28a-ansE harboring the ECA gene, obtained from our lab stock [17], was used as the template for cloning mutation genes. With overlap extension PCR, site-saturation mutagenesis was introduced using corresponding primers (Table S1). Using pET-28a-D60W (constructing with site-saturation mutagenesis and harboring the ECA asparaginase mutation D60W gene) as the template, a combinatorial mutant D60W/L211R/L310R gene was constructed by two rounds of PCR using primer pairs F4 and R4 and F5 and R5. All mutations were linked to linearized pET-28a and transferred into *E. coli* BL21 for gene cloning expression. All recombinant plasmids were sequenced by Sangon Biotech Co., Ltd. (Shanghai, China).

2.3. Expression, Purification, and Activity Assay

The expression of ECA and its mutations in *E.coli* BL21 were performed as described by Zhang [18]. Recombinant *E. coli* BL21 were cultured at 37 °C to OD = 1.0 (approximately 3 h) and were induced with 1 mM IPTG, after which time the culture was incubated for 10 h at 16 °C. The cell paste was suspended in Tris-HCl buffer (pH 8.0) and disrupted on ice by sonication to obtain the intracellular proteins (crude enzyme).

The purification and property determination of all proteins were carried out as described in our previous study [17,19]. Ni^{2+}-affinity chromatography and an AKTA purifier system (GE Healthcare, Danderyd, Sweden) were used to purify the crude enzyme. The crude enzyme was loaded onto a 1-mL HisTrapTM HP column with Binding Buffer (0.02 M Tris-HCl buffer and 0.5 M NaCl, pH 7.4) with a 0.5 mL/min loading rate. L-asparaginase was eluted at 1 mL/min with a linear gradient of imidazole concentrations ranging from 0 to 0.5 M. Then, the purified enzyme was dialyzed with Tris-HCl buffer (0.05 M, pH 7.0) to remove imidazole and recycled for SDS-PAGE analysis. The enzyme with only a single SDS-PAGE target band was the purified enzyme at the end of the process.

The activity of ECA II and its mutations were assayed as described by Li [17,20]. The reaction mixture (1 mL) containing L-asparagine (25 mM) and Tris-HCl (50 mM, pH 8.0) was preheated at optimum temperature. Then, 100 µL of enzyme solution was added and reacted with the substrate for 10 min. One hundred µL of 15% trichloroacetic acid (TCA) was used to stop the reaction. The reaction mixture was centrifuged at $20,000 \times g$, mixed gently with 200 µL of the clear supernatant, 4.8 mL of deionized water and 200 µL of Nessler's reagent, and the amount of ammonia released was measured. All measurements were performed spectrophotometrically at 450 nm. TCA and enzyme solution were successively added to the reaction mixture and were used as a blank during the spectrophotometric enzyme activity assays. One unit of L-asparaginase activity was defined as the amount of enzyme required to release 1 µmol of ammonia per minute under assay conditions.

2.4. Determination of Optimum Temperature, Optimum pH, and Thermal Stability

The optimum temperature of L-asparaginase was examined using 50 mM Tris-HCl buffer (pH 7.5) with temperatures ranging from 20 to 60 °C. The optimum pH was measured by assaying the enzyme activity at various pH values (0.05 M acetate buffer, pH 4.0–6.0; 0.05 M phosphate buffer, pH 6.0–7.0; 0.05 M Tris-HCl buffer, pH 7.0–9.0; and 0.05 M glycine-NaOH buffer, pH 9.0–10.0) at optimum temperature. The thermal stability of L-asparaginase was determined by incubating the enzyme in Tris-HCl buffer (50 mM, pH 7.0) for 15–120 min at 50 °C. After incubation, the protein was refolded on ice for 15 min, and the residual enzyme activity was measured at optimum temperature and pH.

2.5. Structure Modeling and MD Simulation

The ECA crystal conformation (PDB ID: 6PAB) as a template [21] and ECA mutation model were acquired by homology modeling using SWISS-MODEL (http://swissmodel. expasy.org/, accessed on 20 March 2022). The molecular structures of all proteins were analyzed using the Program PyMOL [22].

MD simulations were conducted using GROMACS software in a similar manner as in our previous study, to analyze the stability of protein structures [17]. The protein model was immersed in a dodecahedron box, and the distance between any protein atom and the edge of the box was set at >1.2 nm. Following the addition of Na^+ (0.15 M) to balance the negative charges, the system was minimized using the steepest descent method. After MD simulations of ECA conducted at 310 K and 323 K reference temperatures for 30 ns, the root mean square fluctuation (RMSF) values of residues were calculated.

2.6. Application of ECA in French Fries and Bread

The treatment of potatoes was modified based on the study of Farahat et al. [15]. Potatoes (Fovorita, pH 7.3) and bread flour (pH 6.4) were purchased from the local supermarket in Shihezi, Xinjiang Province. The potatoes were peeled and cut into strips ($0.5 \times 0.5 \times 10$ cm^3), and then the strips were immersed in distilled water for 2 min to remove the starch from the surface. To investigate the effect of enzyme on the acrylamide formation in French fries under different conditions, the raw fries were submerged in an enzyme solution (50 mM, pH 7.5 Tris-HCl buffer or tap water, with enzyme concentrations of 10, 20, 40, 60, and 80 U/mL) at 37 °C, 45 °C, and uncontrolled temperature for 20 min each, while the control group was submerged in a similar solution (50 mM, pH 7.5 Tris-HCl buffer or tap water) without the enzyme for 20 min. All samples were fried at 160 °C for 10 min in an electric fryer. After frying, the fries were cooled on a paper at an ambient temperature, and then the acrylamide was extracted for analysis.

The bread dough was prepared using flour (300 g), yeast (3 g), and an enzyme solution (200 g, 50 mM, pH 7.5 Tris-HCl buffer or tap water, with enzyme concentrations of 40, 60, 80, 100, and 120 U/g flour enzyme, and without enzyme as the control group). The dough was kneaded and allowed to rest for 60 min at different temperatures (37 °C, 45 °C, or room temperature). Finally, the bread dough was baked in an oven at 180 °C for 20 min. The bread was cooled at an ambient temperature, and then the acrylamide was extracted for analysis.

2.7. Assay of Concentration Assay

The acrylamide in French fries and bread was extracted by the method described by Wang et al. [2]. One gram of the crushed sample (French fries or bread) was accurately weighed in a 50-mL centrifuge tube, vortexed with 10 mL of hexane for 1 min and centrifuged at 10,000 *g* for 5 min at 4 °C, and the hexane layer was removed. The aforementioned operation was repeated thrice to get rid of the long-chain fatty acids from the sample. Thereafter, 10 mL of methanol, 500 µL of Carrez I (3.6 g potassium ferricyanide/100 mL ultrapure water), and 500 µL of Carrez II (7.2 g zinc sulfate/100 mL ultrapure water) were added to the centrifuge tube and shaken at 30 °C for 30 min. The homogenates were centrifuged at 10,000 *g* for 30 min at 4 °C, and the supernatant was filtered through 0.22-µm Millipore filters. The extracted samples were detected by high-performance liquid chromatography (HPLC) using a C18 chromatographic column (Atlantis TM, 150 × 2.1 mm^2) and a UV detector. The HPLC operating conditions included a mobile phase of 70% (*v/v*) methanol, UV wavelength set at 210 nm, injection volume of 20 µL, and a column temperature of 30 °C. Different concentrations of acrylamide (50–4000 g/L) were used as the reference for HPLC detection.

3. Results and Discussion

3.1. Identification of ECA Stability Key Domains

In our previous study, ECA was expressed by *E. coli* BL21, with a half-life of 6.2 h and 1 h at 37 °C and 50 °C, respectively [17]. The half-life of the enzyme was shortened by 5.2 h when the temperature increased by only 13 °C, which attracted our attention. MD software imparted significant guidance in analyzing enzyme's structural characteristics and rational design [23–26]. The GROMACS software effectively calculated the RMSF of protein amino acids at a simulated temperature and then showed the flexibility of the residue domain at different temperatures [27–30]. To find out why ECA was unstable at higher temperatures at a protein structure level, the RMSF of ECA residues were calculated at 37 °C and 50 °C based on the ECA crystal conformation (PDB ID: 6PAB) [21]. The RMSF of domains G57-T80, P202-K213, and N298-T311 increased by 1.01 nm, 1.00 nm, and 1.08 nm, respectively, which was much higher than the average RMSF increase of 0.60 nm (Supplementary Data and Figure 1A), indicating that these regions fluctuated greatly at higher temperatures.

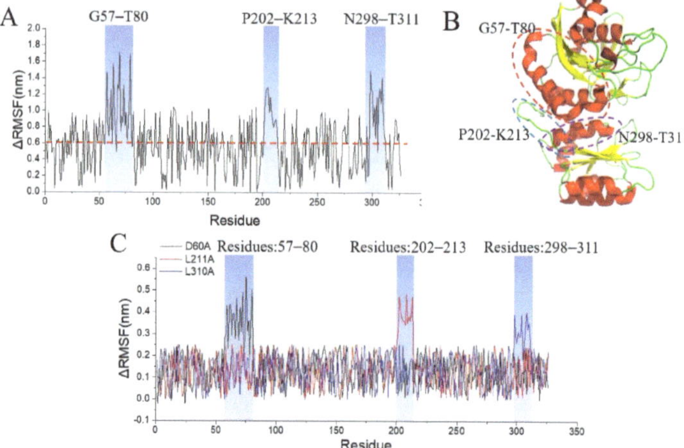

Figure 1. Molecular dynamics simulation and structural analysis of ECA. (**A**) RMSF value of ECA at 37 °C and 50 °C. (**B**) Stability key domains in ECA tertiary structure. G57-T80: the α-helix in red-dashed box; P202-K213: the loop in blue-dashed box; N298-T311: the α-helix in purple-dashed box. (**C**) ECA RMFS with "alanine scanning".

The subunit of L-asparaginase is composed of a large N-terminal and a small C-terminal, and the dimer assembly of the N- and C-terminals "head-to-tail" constitutes the basic functional unit [31–33]. As found in the ECA model (Figure 1B), at the interface of L-asparaginase subunit N- and C-terminals, domains G57-T80 and N298-T311 were, respectively, located as α-helix, and P202-K213 located as the link loop connecting the N- and C-terminals. These interface function domains (G57-T80, P202-K213, and N298-T311) fluctuated at 50 °C, which might be the reason for the sharp shortening of the protein's half-life at 50 °C than that at 37 °C, and were the key domains of thermal stability [29,34].

3.2. Identification of Stability Key Residues and Construction of Stability Mutation

To determine the unstable key residues in the three domains, these domains' residues (except for alanine) were computationally substituted with alanine (alanine scanning; if the original residue was alanine, it was replaced with glycine) at 50 °C [35]. As shown in the results (Supplementary Data and Figure 1C), the RMSF of the mutations D60A, L211A, and L310A showed a maximum rise in the domains G57-T80, P202-K213, and N298-T311, respectively. The results showed that D60, L211, and L310 might be the key residues for the ECA stability.

Saturation mutations were carried out to further verify the effect of these residues (D60, L211, and L310) on the ECA stability and improve the stability (Table 1). The thermal stability of mutations D60W, L211R, and L310R were improved, the half-life of the combinatorial mutant (D60W/L211R/L310R) was extended from 60 to 110 min at 50 °C, while other enzyme characteristics showed no significant changes (Table 1).

Table 1. The enzyme characteristics of ECAII and its mutations.

Enzyme	Optimum Temperature (°C)	$T_{(1/2,\,50\,°C)}$ (Min)	Optimum pH	K_m (μM)	Specific Activity (U/mg)
ECA II	37	60 ± 5	7.5	18 ± 5	235 ± 21
D60W	40	85 ± 5	7.0	15 ± 6	245 ± 31
L211R	42	95 ± 5	7.5	24 ± 8	290 ± 33
L310R	40	80 ± 5	8.0	14 ± 6	217 ± 20
D60W/L211R/L310R	45	110 ± 10	7.5	26 ± 6	281 ± 29
D60A	35	40 ± 5	7.0	28	271 ± 29
D60I	37	50 ± 5	7.0	41	301 ± 41
D60V	37	60 ± 5	7.0	56	223 ± 21
D60F	40	75 ± 5	7.0	32	189 ± 31
D60M	37	60 ± 5	7.5	15	233 ± 33
D60Q	40	75 ± 5	7.5	66	199 ± 24
D60T	37	55 ± 5	7.0	90	211 ± 17
D60N	37	45 ± 5	7.0	55	273 ± 19
D60Y	40	70 ± 5	7.5	24	248 ± 24
D60E	37	60 ± 5	7.0	33	221 ± 44
L211G	35	40 ± 5	7.5	77	281 ± 37
L211A	35	45 ± 5	7.5	63	277 ± 21
L211I	37	55 ± 5	7.5	69	249 ± 11
L211V	37	45 ± 5	7.5	93	211 ± 18
L211P	40	75 ± 5	7.0	45	198 ± 55
L211F	40	65 ± 5	7.5	51	255 ± 23
L211W	37	65 ± 5	7.0	101	294 ± 36
L211S	40	60 ± 5	7.0	67	211 ± 19
L211T	37	45 ± 5	7.5	32	234 ± 12
L211N	37	55 ± 5	8.0	55	189 ± 11
L211D	40	65 ± 5	8.0	41	243 ± 28
L211E	40	70 ± 5	7.5	27	257 ± 31
L211K	40	70 ± 5	7.5	91	231 ± 37
L310A	37	45	7.5	33	221 ± 41
L310I	37	55	7.5	48	232 ± 21
L310P	40	70	7.5	19	203 ± 19

Table 1. *Cont.*

Enzyme	Optimum Temperature (°C)	$T_{(1/2,\,50\,°C)}$ (Min)	Optimum pH	K_m (µM)	Specific Activity (U/mg)
L310F	40	75	8.0	36	199 ± 23
L310M	37	60	8.0	52	198 ± 41
L310W	40	65	8.0	20	188 ± 28
L310Q	37	50	7.5	13	236 ± 23
L310T	37	55	7.5	54	210 ± 31
L310C	37	50	7.5	33	179 ± 51
L310N	37	55	7.5	64	198 ± 31
L310Y	40	65	7.5	88	204 ± 42
L310K	37	60	7.5	91	218 ± 16
L310H	37	65	7.5	31	221 ± 12

To fully understand how the residue mutations D60W, L211R, and L310R affected the thermostability, the combinatorial mutant (D60W/L211R/L310R) was modeled using the crystal structure of ECA (PDB ID: 6PAB). As shown in Figure 2A, the 60th, 211th, and 310th residues were located in the key domains G57-T80, P202-K213, and N298-T311, respectively. After the 60th residue Asp was mutated into Trp, an additional hydrogen bond was formed with the 249th residue Leu on the adjacent subunit. Compared with L211, L211R formed additional hydrogen with Asp63 and Gln307 on G57-T80 and N298-T311, respectively. After the 310th residue, Leu mutated into Arg. Although the connection with the 307th residue was lost, an additional hydrogen bond was formed with the 306th residue Leu and 237th residue Asp. Meanwhile, the RMSF of site-mutations (D60W, L211R, and L310R) and combinatorial mutant (D60W/L211R/L310R) were calculated at 50 °C (Supplementary Data and Figure 2B). The RMSF of site-mutations D60, L211R, and L310R decreased by 0.15 nm, 0.20 nm, and 0.21 nm and the RMSF of their respective regions (G57-T80, P202-K213, and N298-T311) were, respectively, reduced by 0.19 nm, 0.32 nm, and 0.27 nm compared with the wild type. Furthermore, compared with the wild type, the RMSF of the combinatorial mutant (D60W/L211R/L310R) decreased by 0.21 nm, and its RMSF of G57-T80, P202-K213, and N298-T311 decreased by 0.28 nm, 0.26 nm, and 0.25 nm, respectively. These results suggested that all the three residue mutations formed more hydrogen bonds with nearby residues after mutation, which made the connection between the N- and C-terminals of ECA and the connection between the subunits more inseparable, and in turn made the protein structure more difficult to be destroyed. Hence, the thermostability of the combinatorial mutant was improved.

Considering that the extensive application without controlling the treatment temperature and pH was more favorable, we measured the relative activity in the temperature range of 10 °C–70 °C and pH range of 6–10 before and after mutation (Figure 3). Compared with the wild type, the working temperature (the temperature of relative activity > 80%) range of D60W/L211R/L310F was expanded from 31 °C–43 °C to 35 °C–55 °C. Meanwhile, at a pH range of 6–10, the relative activity of D60W/L211R/L310F remained above 82%, while that of the wild type was only 65%. The mutant D60W/L211R/L310F had more hydrogen bonds in the key domains and better stability [36–39], so it could remain stable under adverse conditions (such as high temperature or highly acidic alkaline conditions), which also indirectly widened its working conditions.

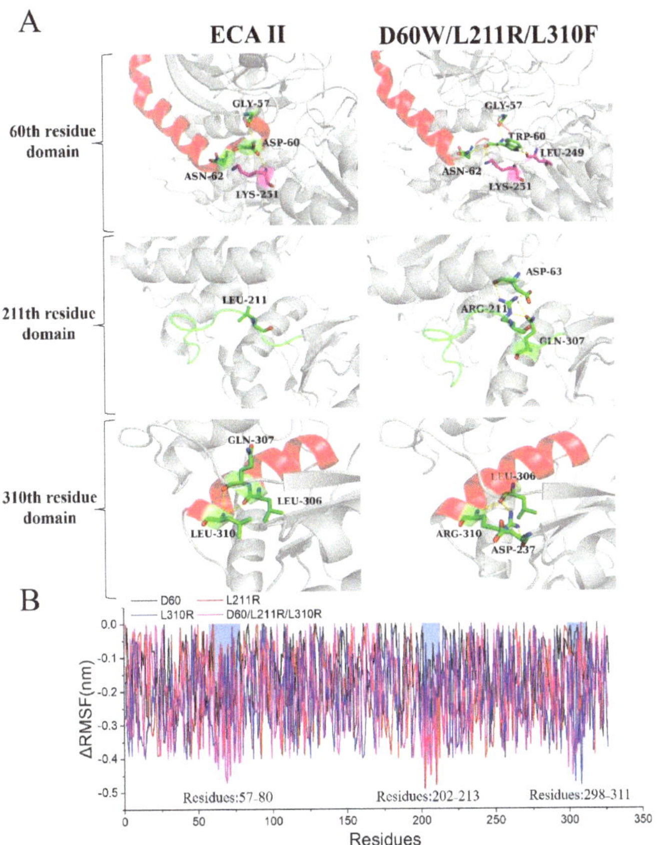

Figure 2. Tertiary structure around ECA stability key residues (**A**) and the changes of RMSF between ECA and its mutants (**B**).

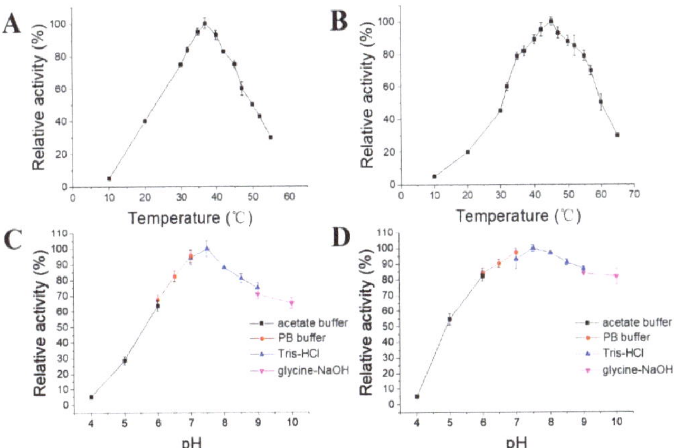

Figure 3. Activity of ECA and its mutation D60W/L211R/L310F at different temperatures and pH. (**A**,**B**) Activity of ECA (**A**) and its mutation D60W/L211R/L310F (**B**) at different temperatures. (**C**,**D**): Activity of ECA (**C**) and its mutation D60W/L211R/L310F (**D**) at different pH.

3.3. Effect of the Enzyme on Acrylamide Formation in French Fries and Bread under Controlled Conditions

Potatoes and flour are two staple foods used to produce fried potatoes and bakery products. In Europe, fried potatoes and bakery products contribute 50% and 20% of humanity's ingestion of acrylamide, respectively [9,11,40,41]. Hence, the degradation of acrylamide content in fried potatoes and bakery products can effectively reduce the daily intake of acrylamide, which is of great significance to a healthy diet. In this study, we investigated the mitigation effect of L-asparaginase on acrylamide formation in common fried potatoes (French fries) and bakery products (bread).

Without enzyme treatment, the acrylamide content in French fries reached 3223 μg/kg. With different concentrations (10, 20, 40, 60, and 80 U/mL) of ECA and its mutant D60W/L211R/L310F, potatoes were, respectively, treated at pH 7.5 and optimum temperatures (ECA 37 °C and D60W/L211R/L310F 45 °C) for 20 min. The mitigation effect on acrylamide formation of French fries is shown in Figure 4A. After treating potatoes with 60 U/mL ECA and its mutant, the acrylamide content in potato chips decreased by 75.5% and 84.1%, respectively; also, the acrylamide content did not further decrease significantly when the enzyme amount was increased to 80 U/mL. For the sake of the production cost, 60 U/mL L-asparaginase was used for the subsequent research.

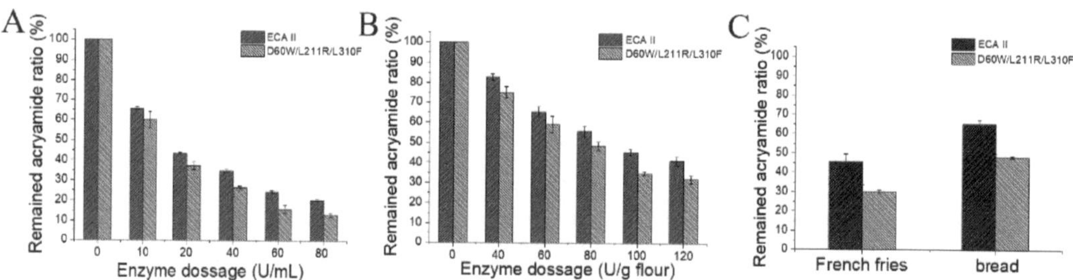

Figure 4. Effect of ECA and its mutation D60W/L211R/L310F on acrylamide formation in French fries and bread. (**A,B**) Effect of different concentrations of the enzyme on acrylamide formation in French fries (**A**) and bread (**B**) under optimum conditions. (**C**) Effect of the enzyme on acrylamide formation in French fries and bread under uncontrolled conditions.

To complete the enzyme treatment of the flour, Tris-HCl (50 mM, pH 7.5) containing different concentrations (40, 60, 80, 100, and 120 U/g) of L-asparaginase were used for kneading the dough, and the optimum temperatures (ECA 37 °C and D60W/L211R/L310F 45 °C) of the dough and fermentation were maintained in an incubator. The mitigation effect on acrylamide formation of bread is shown in Figure 4B. Without enzyme treatment, the acrylamide content in the bread reached 931 μg/kg. After treatment with 100 U/g flour ECA and D60W/L211R/L310F, the acrylamide content in the bread decreased by 54.5% and 65.1%, respectively, while the acrylamide content did not further decrease significantly when the enzyme concentration was increased. Hence, in the subsequent research, 100 U/g flour L-asparaginase was used to inhibit acrylamide formation in bread.

Furthermore, compared with the same dose of ECA, the acrylamide content of potato chips and bread treated with D60W/L211R/L310F was further reduced, showing a better application effect, which might be due to the better stability of the mutant and reduced loss of enzyme activity in the pretreatment time. In addition, when the amount of enzyme reached a certain level (60 U/mL in potatoes and 100 U/g in flour), further increasing the enzyme concentration did not reduce the acrylamide content significantly in food (Figure 4A,B). It might be because L-asparagine, which can be contacted with enzymes, was already degraded. Thus, increasing the enzyme amount hardly increased the reaction between enzyme and L-asparagine further and hence did not reduce the subsequent formation of acrylamide. For reducing the formation of acrylamide in food using the enzyme in the future, the treatment effect may be strengthened by increasing the contact between the enzyme and food raw materials.

3.4. Effect of the Enzyme on Acrylamide Formation in French Fries and Bread under Uncontrolled Conditions

In practical application, it is difficult to strictly control the reaction temperature and pH as in the laboratory; extensive experiments are more consistent with reality. Therefore, we studied the treatment of potatoes and flour with the enzyme in tap water without controlling the reaction temperature and pH to verify its effect on the degradation of acrylamide in potato chips and bread.

In raw materials, the pH of tap water, potatoes, and flour was 6.5, 7.3, and 6.4, respectively. Potatoes and flour were treated with 60 U/mL and 100 U/g flour, respectively, and the residual acrylamide content of French fries and bread was measured (Figure 4C). Compared with the treatment under constant-temperature and constant-pH conditions (Figure 4A,B), the effect of enzyme treatment decreased due to the lack of the optimal conditions. The acrylamide in French fries and bread treated with D60W/L211R/L310F decreased by 69.9% and 51.7%, respectively, which was 27% and 49.9% higher than that of the wild type. Due to the lack of a buffer solution and temperature control, the temperature and pH of the treatment were not constant, while the working temperature and pH of the mutant were wider and more stable, meeting the application requirements, so the mutation had a better application effect than that of the wild type and exhibited a great application potential.

4. Conclusions

Through MD simulation and mutation of *E. coli* L-asparagine, we obtained a mutant with wider application temperature and pH and better stability, verifying the effect of acrylamide control in French fries and bread. Without controlled treatment temperature and pH, the mutant could reduce the acrylamide content in French fries and bread by 69.9% and 51.7%, respectively, with 60 U/mL and 100 U/g flour enzyme, and showed the potential to reduce food acrylamide formation in practical applications.

Supplementary Materials: The following supporting information can be downloaded at: https://www.mdpi.com/article/10.3390/fermentation8050218/s1, Supplementary Data: The original data of RMSF of ECA and its mutants; Table S1: Primers used in this study.

Author Contributions: Methodology, B.Y., P.M., and B.G.; investigation, B.Y., P.M. and Y.H.; data curation, B.Y. and Y.F.; writing—original draft preparation, B.Y. and W.Y.; writing review and editing, X.L.; supervision, X.L. and Y.N.; project administration, Y.Z. and X.L.; funding acquisition, X.L. and Y.N. All authors have read and agreed to the published version of the manuscript.

Funding: This research was funded by the Science and Technology Innovation Team Project of Xinjiang Production and Construction Corps, grant number 2020CB007, the Science and Technology Special Project of Shihezi Municipal Government, grant number 2020PT01, Special program for innovation and development of Shihezi University, grant number CXFZ202109, Launch research project for high-level talents of Shihezi University, grant number RCZK201939, Independently Research Funded by Shihezi University, grant number ZZZC201909A.

Institutional Review Board Statement: Not applicable.

Informed Consent Statement: Not applicable.

Data Availability Statement: Not applicable.

Acknowledgments: The authors would like to thank the members of Academic Committee School of Food Science and Technology, Shihezi University, for their advice and feedback during the manuscript drafting process.

Conflicts of Interest: The authors declare no conflict of interest.

References

1. Cunha, M.C.; Dos Santos Aguilar, J.G.; de Melo, R.R.; Nagamatsu, S.T.; Ali, F.; de Castro, R.J.S.; Sato, H.H. Fungal L-asparaginase: Strategies for production and food applications. *Food Res. Int.* **2019**, *126*, 108658. [CrossRef]
2. Wang, Y.; Wu, H.; Zhang, W.; Xu, W.; Mu, W. Efficient control of acrylamide in French fries by an extraordinarily active and thermo-stable l-asparaginase: A lab-scale study. *Food Chem.* **2021**, *360*, 130046. [CrossRef]
3. Exon, J.H. A review of the toxicology of acrylamide. *J. Toxicol. Environ. Health B Crit. Rev.* **2006**, *9*, 397–412. [CrossRef] [PubMed]
4. Daniali, G.; Jinap, S.; Hajeb, P.; Sanny, M.; Tan, C.P. Acrylamide formation in vegetable oils and animal fats during heat treatment. *Food Chem.* **2016**, *212*, 244–249. [CrossRef]
5. Anese, M.; Nicoli, M.C.; Verardo, G.; Munari, M.; Mirolo, G.; Bortolomeazzi, R. Effect of vacuum roasting on acrylamide formation and reduction in coffee beans. *Food Chem.* **2014**, *145*, 168–172. [CrossRef]
6. Zhu, Y.; Luo, Y.; Sun, G.; Wang, P.; Hu, X.; Chen, F. Role of glutathione on acrylamide inhibition: Transformation products and mechanism. *Food Chem.* **2020**, *326*, 126982. [CrossRef]
7. Khalil, N.M.; Rodríguez-Couto, S.; El-Ghany, M.N.A. Characterization of Penicillium crustosum l-asparaginase and its acrylamide alleviation efficiency in roasted coffee beans at non-cytotoxic levels. *Arch. Microbiol.* **2021**, *203*, 2625–2637. [CrossRef]
8. Kukurová, K.; Morales, F.J.; Bednáriková, A.; Ciesarová, Z. Effect of L-asparaginase on acrylamide mitigation in a fried-dough pastry model. *Mol. Nutr. Food Res.* **2009**, *53*, 1532–1539. [CrossRef]
9. Jia, R.; Wan, X.; Geng, X.; Xue, D.; Xie, Z.; Chen, C. Microbial L-asparaginase for Application in Acrylamide Mitigation from Food: Current Research Status and Future Perspectives. *Microorganisms* **2021**, *9*, 1659. [CrossRef]
10. Bhagat, J.; Kaur, A.; Chadha, B.S. Single step purification of asparaginase from endophytic bacteria Pseudomonas oryzihabitans exhibiting high potential to reduce acrylamide in processed potato chips. *Food Bioprod. Process.* **2016**, *99*, 222–230. [CrossRef]
11. Meghavarnam, A.K.; Janakiraman, S. Evaluation of acrylamide reduction potential of l-asparaginase from Fusarium culmorum (ASP-87) in starchy products. *LWT* **2018**, *89*, 32–37. [CrossRef]
12. Zuo, S.; Zhang, T.; Jiang, B.; Mu, W. Reduction of acrylamide level through blanching with treatment by an extremely thermostable l-asparaginase during French fries processing. *Extremophiles* **2015**, *19*, 841–851. [CrossRef] [PubMed]
13. Shi, R.; Liu, Y.; Mu, Q.; Jiang, Z.; Yang, S. Biochemical characterization of a novel L-asparaginase from Paenibacillus barengoltzii being suitable for acrylamide reduction in potato chips and mooncakes. *Int. J. Biol. Macromol.* **2017**, *96*, 93–99. [CrossRef] [PubMed]
14. Jia, M.M.; Xu, M.J.; He, B.B.; Rao, Z.M. Cloning, Expression, and Characterization of L-Asparaginase from a Newly Isolated Bacillus subtilis B11-06. *J. Agric. Food Chem.* **2013**, *61*, 9428–9434. [CrossRef]
15. Farahat, M.G.; Amr, D.; Galal, A. Molecular cloning, structural modeling and characterization of a novel glutaminase-free L-asparaginase from Cobetia amphilecti AMI6. *Int. J. Biol. Macromol.* **2020**, *143*, 685–695. [CrossRef] [PubMed]
16. Hendriksen, H.V.; Kornbrust, B.A.; Østergaard, P.R.; Stringer, M.A. Evaluating the Potential for Enzymatic Acrylamide Mitigation in a Range of Food Products Using an Asparaginase from Aspergillus oryzae. *J. Agric. Food Chem.* **2009**, *57*, 4168–4176. [CrossRef] [PubMed]
17. Li, X.; Zhang, X.; Xu, S.; Xu, M.; Yang, T.; Wang, L.; Zhang, H.; Fang, H.; Osire, T.; Rao, Z. Insight into the thermostability of thermophilic L-asparaginase and non-thermophilic L-asparaginase II through bioinformatics and structural analysis. *Appl. Microbiol. Biotechnol.* **2019**, *103*, 7055–7070. [CrossRef]
18. Zhang, J.; Xu, M.; Ge, X.; Zhang, X.; Yang, T.; Xu, Z.; Rao, Z. Reengineering of the feedback-inhibition enzyme N-acetyl-l-glutamate kinase to enhance l-arginine production in Corynebacterium crenatum. *J. Ind. Microbiol. Biotechnol.* **2017**, *44*, 271–283. [CrossRef]
19. Li, X.; Zhang, X.; Xu, S.; Zhang, H.; Xu, M.; Yang, T.; Wang, L.; Qian, H.; Zhang, H.; Fang, H.; et al. Simultaneous cell disruption and semi-quantitative activity assays for high-throughput screening of thermostable L-asparaginases. *Sci. Rep.* **2018**, *8*, 7915. [CrossRef]
20. Roberts, J.; Burson, G.; Hill, J.M. New Procedures for Purification of l-Asparaginase with High Yield from *Escherichia coli*. *J. Bacteriol.* **1968**, *95*, 2117–2123. [CrossRef]
21. Lubkowski, J.; Wlodawer, A. Geometric considerations support the double-displacement catalytic mechanism of l-asparaginase. *Protein Sci.* **2019**, *28*, 1850–1864. [CrossRef] [PubMed]
22. Alexander, N.; Woetzel, N.; Meiler, J. Bcl::Cluster: A method for clustering biological molecules coupled with visualization in the Pymol Molecular Graphics System. In Proceedings of the 2011 IEEE 1st International Conference on Computational Advances in Bio and Medical Sciences (ICCABS), Orlando, FL, USA, 3–5 February 2011; Volume 2011, pp. 13–18. [CrossRef]
23. Jiménez-Osés, G.; Osuna, S.; Gao, X.; Sawaya, M.; Gilson, L.; Collier, S.J.; Huisman, G.W.; Yeates, T.O.; Tang, Y.; Houk, K.N. The role of distant mutations and allosteric regulation on LovD active site dynamics. *Nat. Chem. Biol.* **2014**, *10*, 431–436. [CrossRef] [PubMed]
24. Boehr, D.D.; Nussinov, R.; Wright, P.E. The role of dynamic conformational ensembles in biomolecular recognition. *Nat. Chem. Biol.* **2009**, *5*, 789–796. [CrossRef] [PubMed]
25. Boehr, D.D.; Dyson, H.J.; Wright, P.E. An NMR Perspective on Enzyme Dynamics. *Chem. Rev.* **2006**, *106*, 3055–3079. [CrossRef]
26. Alley, E.C.; Khimulya, G.; Biswas, S.; AlQuraishi, M.; Church, G.M. Unified rational protein engineering with sequence-based deep representation learning. *Nat. Methods* **2019**, *16*, 1315–1322. [CrossRef]
27. Lindorff-Larsen, K.; Piana, S.; Palmo, K.; Maragakis, P.; Klepeis, J.L.; Dror, R.O.; Shaw, D.E. Improved side-chain torsion potentials for the Amber ff99SB protein force field. *Proteins Struct. Funct. Bioinform.* **2010**, *78*, 1950–1958. [CrossRef]

28. Hess, B.; Kutzner, C.; van der Spoel, D.; Lindahl, E. GROMACS 4: Algorithms for Highly Efficient, Load-Balanced, and Scalable Molecular Simulation. *J. Chem. Theory Comput.* **2008**, *4*, 435–447. [CrossRef]

29. Xia, Y.; Cui, W.; Cheng, Z.; Peplowski, L.; Liu, Z.; Kobayashi, M.; Zhou, Z. Improving the Thermostability and Catalytic Efficiency of the Subunit-Fused Nitrile Hydratase by Semi-Rational Engineering. *ChemCatChem* **2018**, *10*, 1370–1375. [CrossRef]

30. Phillips, J.C.; Braun, R.; Wang, W.; Gumbart, J.; Tajkhorshid, E.; Villa, E.; Chipot, C.; Skeel, R.D.; Kalé, L.; Schulten, K. Scalable molecular dynamics with NAMD. *J. Comput. Chem.* **2005**, *26*, 1781–1802. [CrossRef]

31. Swain, A.L.; Jaskolski, M.; Housset, D.; Rao, J.K.M.; Wlodawer, A. Crystal-Structure of *Escherichia coli* L-Asparaginase, an Enzyme Used in Cancer-Therapy. *Proc. Natl. Acad. Sci. USA* **1993**, *90*, 1474–1478. [CrossRef]

32. Tomar, R.; Sharma, P.; Srivastava, A.; Bansal, S.; Ashish; Kundu, B. Structural and functional insights into an archaeal L-asparaginase obtained through the linker-less assembly of constituent domains. *Acta Crystallogr. Sect. D Biol. Crystallogr.* **2014**, *70*, 3187–3197. [CrossRef] [PubMed]

33. Miller, M.; Rao, J.K.M.; Wlodawer, A.; Gribskov, M.R. A Left-Handed Crossover Involved in Amidohydrolase Catalysis—Crystal-Structure of Erwinia-Chrysanthemi L-Asparaginase with Bound L-Aspartate. *FEBS Lett.* **1993**, *328*, 275–279. [CrossRef]

34. Zhou, J.; Zhang, R.; Yang, T.; Liu, Q.; Zheng, J.; Wang, F.; Liu, F.; Xu, M.; Zhang, X.; Rao, Z. Relieving Allosteric Inhibition by Designing Active Inclusion Bodies and Coating of the Inclusion Bodies with Fe3O4 Nanomaterials for Sustainable 2-Oxobutyric Acid Production. *ACS Catal.* **2018**, *8*, 8889–8901. [CrossRef]

35. Yang, B.; Wang, H.; Song, W.; Chen, X.; Liu, J.; Luo, Q.; Liu, L. Engineering of the Conformational Dynamics of Lipase to Increase Enantioselectivity. *ACS Catal.* **2017**, *7*, 7593–7599. [CrossRef]

36. Brosnan, M.P.; Kelly, C.T.; Fogarty, W.M. Investigation of the Mechanisms of Irreversible Thermoinactivation of Bacil-lus-Stearothermophilus Alpha-Amylase. *Eur. J. Biochem.* **1992**, *203*, 225–231. [CrossRef]

37. Vidya, J.; Ushasree, M.V.; Pandey, A. Effect of surface charge alteration on stability of l-asparaginase II from *Escherichia* sp. *Enzym. Microb. Technol.* **2014**, *56*, 15–19. [CrossRef]

38. Niu, C.; Zhu, L.; Xu, X.; Li, Q. Rational Design of Disulfide Bonds Increases Thermostability of a Mesophilic 1,3-1,4-beta-Glucanase from Bacillus terquilensis. *PLoS ONE* **2016**, *11*, e0154036. [CrossRef]

39. Long, S.; Zhang, X.; Rao, Z.; Chen, K.; Xu, M.; Yang, T.; Yang, S. Amino acid residues adjacent to the catalytic cavity of tetramer l-asparaginase II contribute significantly to its catalytic efficiency and thermostability. *Enzym. Microb. Technol.* **2016**, *82*, 15–22. [CrossRef]

40. Keramat, J.; LeBail, A.; Prost, C.; Jafari, M. Acrylamide in Baking Products: A Review Article. *Food Bioprocess Technol.* **2011**, *4*, 530–543. [CrossRef]

41. Nguyen, H.T.; Van der Fels-Klerx, H.J.; Peters, R.J.B.; Van Boekel, M.A.J.S. Acrylamide and 5-hydroxymethylfurfural formation during baking of biscuits: Part I: Effects of sugar type. *Food Chem.* **2016**, *192*, 575–585. [CrossRef]

Article

Fermented *Myriophyllum aquaticum* and *Lactobacillus plantarum* Affect the Distribution of Intestinal Microbial Communities and Metabolic Profile in Mice

Yueyang Li [1], Yuxi Ling [1], Jia Liu [2], Michael Zhang [3], Zuming Li [1,*], Zhihui Bai [4], Zhenlong Wu [5], Ran Xia [2], Zhichao Wu [2], Yingxin Wan [1] and Qiyun Zhou [1]

[1] Beijing Key Laboratory of Bioactive Substances and Functional Foods, Department of Food Science, College of Biochemical Engineering, Beijing Union University, Beijing 100023, China; milulikeyou1@163.com (Y.L.); yuxiling2022@163.com (Y.L.); wyx@buu.edu.cn (Y.W.); qiyunzhou@buu.edu.cn (Q.Z.)

[2] Internal Trade Food Science and Technology (Beijing) Co., Ltd., Beijing 102209, China; damao3721@126.com (J.L.); xiaran@cofco.com (R.X.); wuzhichao@cofco.com (Z.W.)

[3] Department of Physics and Astronomy, University of Manitoba, Winnipeg, MB R3T 2N2, Canada; umzhan00@gmail.com

[4] Research Center for Eco-Environmental Sciences, Chinese Academy of Sciences, Beijing 100085, China; zhbai@rcees.ac.cn

[5] College of Animal Science and Technology, China Agricultural University, Beijing 100193, China; wuzhenlong@cau.edu.cn

* Correspondence: zmli20130522@163.com; Tel.: +86-10-5207-2219

Citation: Li, Y.; Ling, Y.; Liu, J.; Zhang, M.; Li, Z.; Bai, Z.; Wu, Z.; Xia, R.; Wu, Z.; Wan, Y.; et al. Fermented *Myriophyllum aquaticum* and *Lactobacillus plantarum* Affect the Distribution of Intestinal Microbial Communities and Metabolic Profile in Mice. *Fermentation* **2022**, *8*, 210. https://doi.org/10.3390/fermentation8050210

Academic Editor: Xian Zhang

Received: 10 March 2022
Accepted: 26 April 2022
Published: 5 May 2022

Publisher's Note: MDPI stays neutral with regard to jurisdictional claims in published maps and institutional affiliations.

Abstract: This research explores the effects of fermented *Myriophyllum aquaticum* (F) and *Lactobacillus plantarum* BW2013 (G) as new feed additives on the gut microbiota composition and metabolic profile of mice. Crude protein ($p = 0.045$), lipid ($p = 0.000$), and ash ($p = 0.006$) contents in *Myriophyllum aquaticum* (N) were improved, whereas raw fiber ($p = 0.031$) content was decreased after solid-state fermentation by G. Mice were fed with no additive control (CK), 10%N (N), 10%N + G (NG), 10%F (F), and 10%F + G (FG). High-throughput sequencing results showed that, compared with the CK group, *Parabacteroides goldsteinii* was increased in treatment groups and that *Lactobacillus delbrueckii*, *Bacteroides vulgatus*, and *Bacteroides coprocola* were increased in the F and FG groups. *Bacteroides vulgatus* and *Bacteroides coprocola* were increased in the F group compared with the N group. Metabolomic results showed that vitamin A, myricetin, gallic acid, and luteolin were increased in the F group compared with the N group. Reduction in LPG 18:1 concentration in the N and F groups could be attenuated or even abolished by supplementation with G. Furthermore, 9-oxo-ODA was upregulated in the FG group compared with the F group. Collectively, N, F, and G have beneficial effects on gut microbiota and metabolic profile in mice, especially intake of FG.

Keywords: *Myriophyllum aquaticum*; *Lactobacillus plantarum*; solid-state fermentation; gut microbiota; metabolomics

1. Introduction

Myriophyllum aquaticum is a heterophyllic amphibious aquatic plant species commonly found in streams, canals, and freshwater lakes [1]. *M. aquaticum* has strong reproductive ability and can be cultivated in most natural water bodies, particularly those enriched with nutrients commonly found in wastewater from pigs [2]. It has been shown that *M. aquaticum* can be used to treat wastewaters high in NH_4^+-N by absorbing nutrients [3]. However, *M. aquaticum* has a prominent population advantage when it is suitable for water habitat conditions, which is manifested in efficient reproduction. It can pose a threat to native aquatic species diversity and fauna composition by passing from non-invaded to invaded habitats [4]. Administration of this species can be quite expensive, and *M. aquaticum* can produce new plants by reproducing in its own waste [5]. Fortunately, *M. aquaticum* contains a high content of crude protein and crude fiber and is rich in essential amino acids and minerals [3]. Therefore, it can be used as an animal feed material or filler, which can not only effectively relieve the shortage of feed resources but also reduce its threat to native aquatic biodiversity.

Probiotics are active microorganisms that provide a benefit to their host by changing the composition of a certain part of gut microbiota [6]. Liu et al. reported that supplementation of *Lactobacillus plantarum* Y44 may have potential for alleviating lipid metabolism disorders and intestinal inflammation in association with modulating gut microbiota [7]. Furthermore, the nutritional quality of feed fermentation can be improved by using probiotics such as *Lactobacillus* [8,9]. *L. plantarum* is frequently used in the food and feed industries as an inoculant, which positively influences different quality parameters, such as pH value, organic acid, dry matter, and protein content in feed [10,11]. In addition, research has shown that food and animal feed fermented by probiotics have beneficial effects on the body, including stabilizing intestinal barrier function [12] as well as maintaining gut microbial balance [13]. Zhong et al. reported that probiotic-fermented blueberry juice may have anti-obesity and anti-hyperglycemia benefits by modulating the gut microbiota [14]. Intakes of kimchi fermented by *L. plantarum* PNU was shown to regulate metabolic parameters and colon health [15]. Therefore, fermentation by *L. plantarum* is not only a sustainable method for preserving food and feed but also a biotechnology that is increasingly used for improving the nutritional content of food and feed. Currently, *M. aquaticum*, a nutrient-rich plant with significant biomass, is used as an animal feed crude material [16]. However, low-cost roughage cannot be entirely utilized by animals [17], and the palatable flavours and potential health-promoting properties of plant-based fermented food and feed are increasing in popularity [18]. In this study, the *L. plantarum* BW2013 strain, extracted from fermented Chinese cabbage, was used as a starter culture for *M. aquaticum* solid-state fermentation in this study.

Analysing metabolites in a biological system is possible using metabolomics, a new method that delivers detailed quantitative profiles of metabolites [19]. Additionally, high-throughput sequencing can be used to determine changes in microbial community composition within the intestines. The application of these two methods can effectively evaluate the impact of feed on animal intestines. A study showed the effects of polysaccharides from fermented *Momordica charantia* L. with *Lactobacillus plantarum* NCU116 on gut microbiota and fecal metabolic profile in obese rats using the above two methods [20]. Being a novel candidate feed, the effects of fermented *M. aquaticum* and *L. plantarum* on the gut microbiota and metabolites of mice have not been fully elucidated. In this study, we determined the effect of *M. aquaticum* and *L. plantarum* BW2013, as a dietary supplement, on the distribution of gut microbiota and metabolites of mice.

2. Materials and Methods

2.1. Bacterial Cultures

The strain *L. plantarum* BW2013 (CGMCC NO.9462) used for solid-state fermentation was isolated from fermented Chinese cabbage. The strains were grown under anaerobic conditions in de Man–Rogosa–Sharpe (MRS, Beijing Land Bridge Technology Co., LTD., China) medium at 37 °C. The bacteria were incubated and grown to the maximum concentration in shaking flasks. For *L. plantarum* BW2013 strain cultures, a centrifuge with $8000 \times g$ was used for 15 min, followed by two phosphate-buffered saline (PBS) washings, and a suspension of 1×10^8 CFU/mL in PBS.

2.2. Solid-State Fermentation and Conditions Optimization

The *M. aquaticum* used in this study was supplied by the Institute of Subtropical Agriculture, Chinese Academy of Sciences. The washed *M. aquaticum* was dried to adjust its moisture content to 65%, and then cut into 1.0 ± 0.5 cm sections. The count of *L. plantarum* BW2013 inoculated in samples was about 1.0×10^8 CFU per g. The samples were then mixed with 6% (*w/w*) sucrose and anaerobically incubated at 30 °C and 35 °C. The pH value, dry matter, raw protein, and organic acid (lactic acid, acetic acid, propionic acid, and butyric acid) contents were measured in samples from 0–10 days (0 d, 1 d, 3 d, 5 d, 7 d, and 10 d). The contents of protein, raw fat, crude fiber, ash, phosphorus, and calcium were determined after the fresh *M. aquaticum* and fermented production were dried.

The pH value was determined using a pH meter (Sanxin, Shanghai, China). The dry matter was determined by oven drying at 105 °C for 16 h. The organic acids were determined in a L-3000 HPLC (RIGOL Co., LTD., Beijing, China) with an Shodex RSpak KC-811 column (8.0 mmI.D. × 300 mm) and a UV detector, using 210 nm as the determining wavelength. Separation was conducted using a gradient elution with two mobile phases at a flow rate of 1.0 mL/min at 50 °C. Samples were injected at a volume of 5.0 µL after filtration. Mobile phase A was 3 mM $HClO_4$ and mobile phase B was methanol. Ammonia nitrogen (NH_3-N) was determined by an indophenol blue method using a continuous flow chemistry analyzer [21].

After digestion with concentrated sulfuric acid, the total protein content of *M. aquaticum* was determined by the Kjeldahl procedure [22]. The Soxhlet extraction was used for raw fat extraction from samples [23]. Crude fiber is the loss on ignition of the dried residue remaining after digestion of the sample with 1.25% H_2SO_4 and 1.25% NaOH solutions under specific conditions [24]. The dried *M. aquaticum* were mineralized at 550 ± 25 °C for about 30 min and then weighed to determine the ash content. The phosphorus content of the ash was determined using molybdenum blue spectrophotometry [25]. To determine the calcium content, the ash was dissolved in HCl (50%) plus HNO_3 (50%), filtered, and filled to volume (25 mL) with distilled water. Extracts were analysed by atomic absorption spectroscopy using an Inductively Coupled Plasma Optical Emission Spectrometer (ICP-OES) [26].

2.3. Animal Experiment

Male ICR mice (4-week-old) were purchased from the Beijing Vital River Laboratory Animal Technology Co., Ltd. (Beijing, China). All mice were kept in a specific pathogen-free (SPF) facility in the National Health Food Function Testing Center of Beijing Union University and were allowed free access to food and water under a 12 h light cycle. After a 7-day adaptation period, sixty ICR mice were randomly assigned to 5 groups (12 for each group): the CK group (normal control group), the N group (10% *M. aquaticum*), the NG group (10% *M. aquaticum* + 2×10^9 CFU/mL/d *L. plantarum* BW2013), the F group (10% *M. aquaticum* fermentation products), and the FG group (10% *M. aquaticum* fermentation products + 2×10^9 CFU/mL/d *L. plantarum* BW2013). *M. aquaticum* and its fermentation products were added to the normal mouse feed at 10% addition. Mice in the FG and NG groups were intragastrically administered the same *L. plantarum* BW2013 during the whole experimental period. The weights of mice were recorded every week. All groups were treated for 5 weeks, and blood was collected from the eyeballs before slaughter. Centrifugation at $3000 \times g$ for 15 min collected a serum sample for measurement of alkaline phosphatase (ALP), aspartate aminotransferase (AST), creatinine (CRE), urea (UREA), cholesterol (CHO), and blood glucose (GLU) using an automatic biochemical analyzer (ACA, Hitachi Co., Ltd., Tokyo, Japan).

2.4. 16S rDNA Sequencing

The fecal DNA was extracted using the CTAB/SDS method. Concentration and integrity of extracted DNA were measured using agarose gel electrophoresis. Analysis of the data and sequences was performed by Beijing Novogene (Beijing, China). The 16S rRNA genes of distinct regions (16S V4) were amplified with specific primers 515F (5'-GTGCCAGCMGCCGCGGTAA-3') and 806R (5'-GGACTACHVGGGTWTCTAAT-3'). The purified amplicons were pooled in equidensity ratios. Using the TruSeq® DNA PCR-Free Sample Preparation Kit (Illumina, San Diego, CA, USA), sequencing libraries were prepared and index codes were added. Quality assessment of the library was conducted using a Qubit@ 2.0 Fluorometer (Thermo Fisher Scientific, Waltham, MA, USA) and an

Agilent Bioanalyzer 2100 system. Paired-end sequencing of the library was performed on the Illumina NovaSeq platform.

Uparse software (Uparse v7.0.1001, http://drive5.com/uparse/ (accessed on 19 January 2020)) was used to analyze sequences [27]. OTUs (Operational Taxonomic Units) cluster with 97% identity. The taxonomy of representative sequences was determined based on bacterial SILVA data sets [28]. QIIME (Version 1.7.0) and R software (Version 2.15.3) were used to calculate the alpha diversities (the indicators ACE and Shannon represent richness and diversity of intestinal bacterial, respectively). Based on weighted UniFrac distance metrics analysis, non-metric multi-dimensional scaling (NMDS) was performed to distinguish the individuals of the five groups. Weighted UniFrac were calculated by QIIME software (Version 1.9.1). Non-metric multi-dimensional scaling (NMDS) analysis was performed by the vegan package in R software (Version 2.15.3). Permutation tests were completed at each classification level (Phylum, Class, Order, Family, Genus, Species) using the R software (Version 1.9.1) to test differences in the gut microbiota of the mice and to obtain the *p*-value.

2.5. Metabolomics

One-hundred milligrams of liquid nitrogen-ground samples were placed in an Eppendorf tube. The homogenate was resuspended with 500 μL of prechilled 80% methanol and 0.1% formic acid by vortexing and shaking. The samples were incubated in an ice bath for 5 min and then were centrifuged at $15,000 \times g$ at a temperature of 4 °C for 5 min. The amount of supernatant was diluted with LC–MS grade water to a methanol concentration of 53%. Afterwards, the samples were transferred to a new Eppendorf tube and centrifuged for 10 min at $15,000 \times g$, 4 °C. The supernatant was collected and injected into LC–MS for analysis. LC–MS/MS analyses were performed using a Vanquish UHPLC system (Thermo Fisher Scientific, Waltham, MA, USA) coupled with an Orbitrap Q Exactive series mass spectrometer (Thermo Fisher Scientific). A 16 min linear gradient flow rate of 0.2 mL/min was used to inject samples into a Hyperil Gold column (100×2.1 mm, 1.9 μm). Eluents A (0.1% FA in water) and B (methanol) were used for the positive polarity mode. Eluent A (5 mM ammonium acetate, pH 9.0) and eluent B (methanol) were used for the negative polarity mode. Solvent gradient: 2% B, 1.5 min; 2–100% B, 14.0 min; 100% B, 14.1 min; 100%–2% B, 14.1 min; 2% B, 17 min. The Q Exactive series mass spectrometer operated with 3.2 kV spray voltage, 320 °C capillary temperature, 35 arb sheath gas flow rate, and 10 arb aux gas flow rate. Analysis of the data followed the same approach as Cao et al. [29]. The metabolites with VIP > 1 and a *p*-value < 0.05 and fold change (FC) ≥ 1.5 or FC ≤ 0.6 were differentially expressed.

2.6. KEGG Pathways

KEGG database was used to study the functions of these metabolites and metabolic pathways. When the *p*-value of a metabolic pathway was less than 0.05, the metabolic pathway was considered to enrich differential metabolites with statistical significance.

2.7. Statistical Analysis

Statistical analyses were performed using SPSS software (Version 26). The results for the analysis of variance (ANOVA) were considered significant with $p < 0.05$.

3. Results and Discussion

3.1. Analysis of Solid-State Fermentation of M. aquaticum

The fermentation and nutritional quality of *M. aquaticum* were analysed. After fermentation, the stem and leaf structure of the fermented *M. aquaticum* in each group was complete; the colour was yellow-green, there was no mildew, no stickiness, a good texture, and an obvious sour fragrance. The three most essential indices of fermentation quality evaluation are pH, lactic acid content, and the ratio of ammonia nitrogen/total nitrogen (NH$_3$-N/N) [10]. The changes in pH value, dry matter, crude protein, organic acid content, and NH_3-N/N over the fermentation period are shown in Table 1. On the fifth day at 30 °C, the pH value (4.11) was lowest, so sustained fermentation by bacteria was achieved. As soon as the pH value exceeded 5.0, fermentation was considered to have failed [10]. The concentration of organic acid increased during fermentation. It can be concluded that the bacterium grew well on *M. aquaticum* and decreased the pH by secreting these types of organic acids in *M. aquaticum*. Protein from the fermentation substrate is converted into NH_3-N by microorganisms, thus the value of NH_3-N reflects the amount of protein decomposition during fermentation. Generally, a high-quality fermented feed should have a ratio of ammonia nitrogen (NH_3-N) to total nitrogen (N) of less than 7 [10]. The contents of dry matter and crude protein were directly related to the nutritional quality of the fermented feed. Fermented feeds that had high dry matter and crude protein contents had better nutritional quality. As a result of the increased dry matter and crude protein, fewer nutrients were lost during fermentation. The higher dry matter contents were observed on the first day at 35 °C and the fifth day at 30 °C. The highest crude protein contents were observed on the fifth and seventh days at 30 °C. Complicated assessments involving multiple indices and different donations were generally processed by a weighted mean. The assessment score was calculated by the weighted mean [30]. Based on the highest score, the optimal fermentation conditions were as follows: fermentation at 30 °C for 5 days.

Table 1. Fermentation quality of *M. aquaticum*.

Fermentation Conditions		pH	Lactic Acid (µg/mL)	Acetic Acid (µg/mL)	Propionic Acid (µg/mL)	Dry Matter (g/30 g)	Crude Protein (%)	NH_3-N/N (%)	Score
0 d	30 °C	5.56 ± 0.04 [a]	0.11 ± 0.01 [b]	43.7 ± 2.78 [e]	1.84 ± 0.12 [b]	8.44 ± 0.15 [ab]	21.6 ± 0.38 [ab]	3.30 ± 0.17 [c]	00.0
	35 °C	5.56 ± 0.04 [a]	0.11 ± 0.01 [b]	43.7 ± 2.78 [e]	1.84 ± 0.12 [b]	8.44 ± 0.15 [ab]	21.6 ± 0.38 [ab]	3.30 ± 0.17 [c]	00.0
1 d	30 °C	5.05 ± 0.02 [b]	0.42 ± 0.05 [b]	61.5 ± 9.70 [cde]	2.06 ± 0.17 [ab]	8.53 ± 0.12 [ab]	21.8 ± 0.51 [ab]	4.56 ± 0.32 [bc]	40.0
	35 °C	5.08 ± 0.03 [b]	0.42 ± 0.01 [b]	51.9 ± 2.06 [de]	2.20 ± 0.31 [ab]	9.71 ± 0.94 [a]	22.2 ± 0.90 [ab]	4.68 ± 0.56 [bc]	53.0
3 d	30 °C	4.34 ± 0.03 [d]	1.09 ± 0.35 [a]	85.4 ± 23.7 [bcd]	2.53 ± 0.87 [ab]	8.26 ± 0.49 [b]	21.0 ± 0.40 [ab]	5.38 ± 0.12 [ab]	56.9
	35 °C	4.26 ± 0.03 [d]	1.13 ± 0.10 [a]	92.0 ± 17.4 [bc]	2.31 ± 0.13 [ab]	8.48 ± 0.34 [ab]	21.8 ± 0.90 [ab]	5.69 ± 0.61 [ab]	59.1
5 d	30 °C	4.11 ± 0.03 [e]	1.11 ± 0.10 [a]	110 ± 16.8 [ab]	2.37 ± 0.16 [ab]	9.18 ± 0.52 [ab]	22.3 ± 0.40 [a]	5.60 ± 0.37 [ab]	76.4
	35 °C	4.22 ± 0.03 [d]	1.16 ± 0.10 [a]	94.8 ± 6.20 [bc]	2.44 ± 0.47 [ab]	8.40 ± 0.25 [ab]	20.0 ± 0.75 [b]	6.70 ± 0.33 [a]	55.3
7 d	30 °C	4.43 ± 0.03 [c]	1.24 ± 0.10 [a]	114 ± 17.1 [ab]	2.57 ± 0.35 [ab]	8.70 ± 0.33 [ab]	22.3 ± 0.78 [a]	5.53 ± 1.41 [ab]	73.0
	35 °C	4.42 ± 0.04 [c]	1.22 ± 0.10 [a]	117 ± 12.2 [ab]	2.98 ± 0.48 [a]	8.44 ± 0.34 [ab]	21.3 ± 0.97 [ab]	6.96 ± 0.52 [a]	60.3
10 d	30 °C	4.45 ± 0.03 [c]	1.08 ± 0.04 [a]	135 ± 14.0 [a]	2.82 ± 0.30 [ab]	8.73 ± 0.78 [ab]	21.5 ± 0.42 [ab]	6.90 ± 0.07 [a]	60.4
	35 °C	4.46 ± 0.03 [c]	1.24 ± 0.04 [a]	134 ± 12.0 [a]	2.52 ± 0.23 [ab]	8.99 ± 0.60 [ab]	21.2 ± 1.22 [ab]	6.91 ± 1.14 [a]	62.3

[a–e] Means with different superscripts within the same row differ based on Tukey's test ($p < 0.05$).

This study detected the major constituents and the mineral element (phosphorus and calcium) contents of *M. aquaticum* (fermented and non-fermented) (Table 2). A significantly higher proportion of crude proteins ($p = 0.045$), lipids ($p = 0.000$), and ash ($p = 0.006$) were found in fermented *M. aquaticum* than in non-fermented *M. aquaticum*. The raw fiber ($p = 0.031$) content dropped after fermentation from 13.20% to 11.20% (w/w). There is a possibility that fiber serves as a nutrient source for microbes. Raw fiber is associated with digestibility and feed intake, and reduction in raw fiber content could promote the feed intake of animals [11]. These results indicated that fermentation appears to have the ability to alter the nutritional composition of *M. aquaticum*.

3.2. Effects of Dietary Intervention on Serum Markers, Body Weight (BW), and Intestinal Length

Alkaline phosphatase (ALP) and aspartate aminotransferase (AST) were analysed to check for possible effects of the diets and/or probiotic supplementation on liver function. Meanwhile, creatinine (CRE) and urea (UREA) are related to kidney function. No significant differences were detected in the levels of AST, ALP, UREA, CRE, and cholesterol (CHO) among all groups. However, blood glucose (GLU) was significantly reduced in the NG and F groups compared with the CK group (Figure 1). BW of all experimental mice can give an overview of their overall health and provide a rough description of their physical condition. During the first 4 weeks of the study, all mice continued to gain weight. In the fifth week, BW was increased markedly in the F group compared with the N group (Figure 2a). The intestinal length of the mice was analysed to check for possible effects of the dietary interventions on the digestion and absorption capacity of the intestines. The results showed that the small intestine length of the mice increased slightly in the NG and FG groups compared with the CK group (Figure 2b). Therefore, it is possible to replace part of the mice feed with *M. aquaticum* and its fermentation products.

Table 2. Nutritional components of *M. aquaticum* and its fermentation products.

Analysis	NFM	FM	Trends	*p*-Value
	Values Shown Per 100 g			
Crude protein	18.1 ± 1.20 g	20.5 ± 0.80 g	↑ 13.3%	0.045 *
Lipids	3.00 ± 0.14 g	4.60 ± 0.22 g	↑ 53.3%	0.000 **
Ash	12.4 ± 0.90 g	15.9 ± 0.72 g	↑ 28.2%	0.006 **
Raw Fiber	13.2 ± 0.80%	11.2 ± 0.70%	↓ 15.2%	0.031 *
Phosphorus	653 ± 6.00 mg	657 ± 4.00 mg	↑ <0.01%	0.391
Calcium	2.07 ± 0.03 g	2.09 ± 0.04 g	↑ <0.01%	0.510

* Indicates statistical significance at $p < 0.05$. ** Indicates statistical significance at $p < 0.01$.

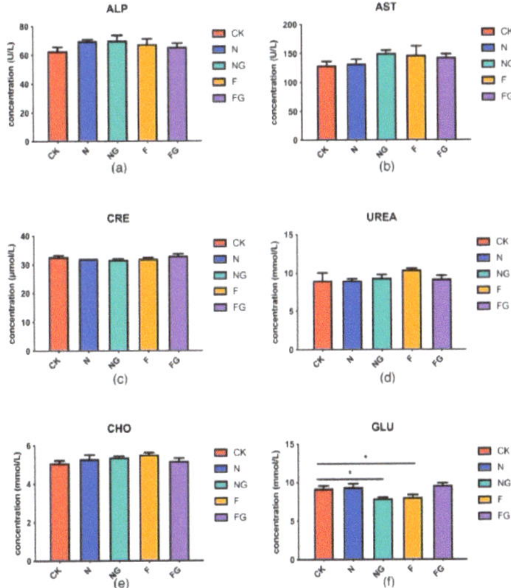

Figure 1. (**a–f**) Effects of dietary intervention on serum markers. * Indicates statistical significance at $p < 0.05$.

3.3. Effects of Dietary Intervention on Gut Microbiota in Mice

High-throughput sequencing analysis was performed to investigate the effect of *M. aquaticum* and *L. plantarum* BW2013 supplementation on gut microbiota composition in mice. α-diversity was measured using two metrics: the ACE index (Figure 2c) and the Shannon index (Figure 2d), which reflect community richness and species diversity, respectively, did not significantly change among the five groups. A beta diversity analysis of the gut microbiota discovered no significant changes among the five groups at OTU levels (Figure 2e). The NMDS plot of beta diversity showed clear separation of the CK and FG groups of mice based on their fecal microbiota (Figure 2f). Furthermore, the potential differences in the groups at week 5 were investigated through a MetaStat analysis. At the phylum level (Figure 3a–c), the relative abundance of *Bacteroidetes* was significantly increased in the NG, F, and FG groups ($p = 0.023$ for the NG group, $p = 0.006$ for the F group, and $p = 0.011$ for the FG group), whereas the relative abundance of *Firmicutes* was significantly decreased in the F and FG groups ($p = 0.008$ for the F group and $p = 0.026$ for the FG group) compared with the CK group. Meanwhile, *Proteobacteria* was significantly reduced in the FG group compared with the CK and F groups ($p = 0.001$ for the CK group and $p = 0.035$ for the F group). At the genus level (Figure 3d,e), the relative abundance of *Faecalibacterium* was significantly increased in the NG and F groups ($p = 0.023$ for the NG group and $p = 0.006$ for the F group) compared with the CK group, while the relative abundance of *Parabacteroides* was significantly increased in the FG group relative to the CK group. The relative abundance of *Faecalibacterium* was significantly increased in the NG and F groups compared with the N group ($p < 0.001$ for the NG group and $p = 0.09$ for the F group). However, the relative abundance of *Faecalibacterium* was reduced in the FG group compared with the F group ($p = 0.023$). At the species level (Figure 3f–i), the relative abundance of *Parabacteroides goldsteinii* was significantly increased in the N, NG, F, and FG groups when compared with the CK group ($p = 0.015$ for the N group, $p = 0.002$ for the NG group, $p = 0.032$ for the F group, and $p < 0.001$ for the FG group). The relative abundance of *Lactobacillus delbrueckii* was significantly increased in the F and FG groups when compared with the CK group ($p = 0.044$ for the F group and $p = 0.006$ for the FG group). The relative abundance of *Bacteroides vulgatus* and *Bacteroides coprocola* was significantly increased in the F ($p = 0.004$ for *B. vulgatus* and $p = 0.020$ for *B. coprocola*) and FG ($p = 0.021$ for *B. vulgatus* and $p = 0.031$ for *B. coprocola*) groups when compared with the CK group. In addition, compared with the N group, the relative abundances of *B. vulgatus* ($p = 0.015$) and *B. coprocola* ($p = 0.020$) were significantly increased in the F group.

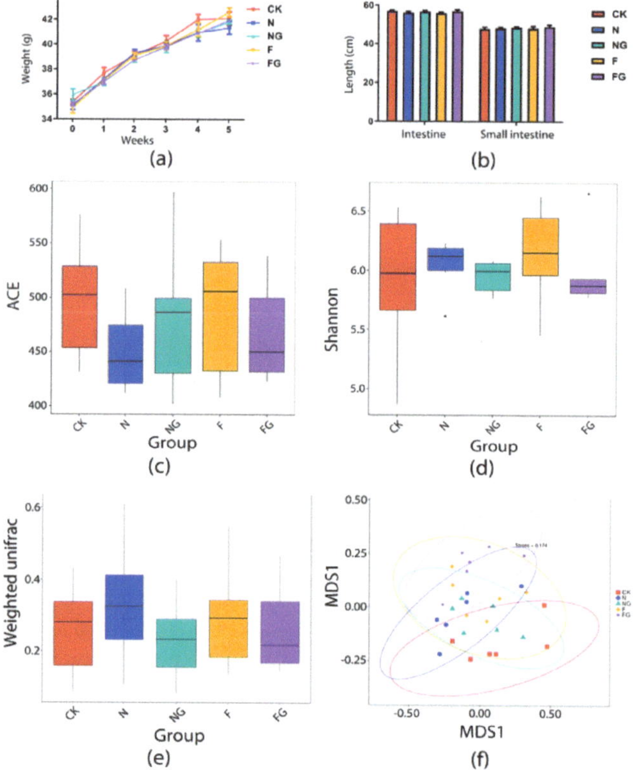

Figure 2. Changes in weight (**a**) and intestinal length (**b**) in the mice. Alpha diversity index of gut microbiota in mice: (**c**) ACE index; (**d**) Shannon index. Changes in global gut microbiota after intervention in each group: (**e**) Beta diversity on weighted UniFrac. (**f**) NMDS score plot based on Bray–Curtis distance at the operational taxonomic unit (OTU) level. Stress < 2 means that NMDS can accurately reflect the degree of difference between groups.

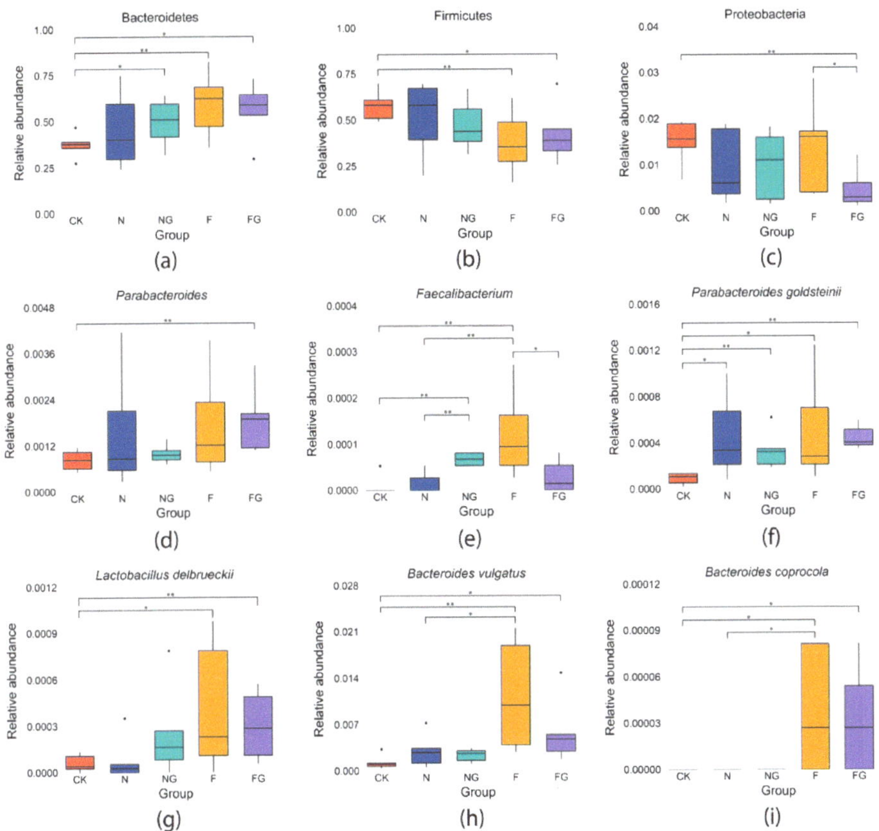

Figure 3. Changes in the relative abundance of individual phyla (**a–c**), genera (**d,e**), and species (**f–i**) among groups. * Indicates statistical significance at $p < 0.05$. ** Indicates statistical significance at $p < 0.01$.

The composition of gut microbiota in mice was improved after the treatment with *M. aquaticum* and *L. plantarum* BW2013 for 5 weeks. *Firmicutes* and *Bacteroidetes* were two dominant bacterial phyla which represented more than 90% of the total gut microbiome. *Proteobacteria* is a group of bacteria causing chronic colitis that are reported to have a low relative abundance in healthy individuals [31]. *Faecalibacterium* is a functionally important genus containing anti-inflammatory bacteria [32]. *Bacteroides*, *Parabacteroides*, and *Faecalibacterium* were the main genera responsible for donor engraftment in studies on fecal microbiota transplantation for *Clostridium difficile* infection [33]. *P. goldsteinii* is a novel probiotic bacterium with the potential to treat obesity as well as metabolic syndrome [34]. Probiotic *L. delbrueckii* could efficiently hydrolyze casein and modulate the intestinal immune system [35,36]. Moreover, treatment with live *B. vulgatus* and *B. dorei* may help prevent coronary artery disease by preventing microbial lipopolysaccharide synthesis [37]. It is believed that diet has an important impact on gut microbiota. Our results showed that different kinds and amounts of components in feed may affect gut microbiota differently. There have been multiple studies exploring the effects of different carbohydrate sources, especially fiber, on gut microbiota [38]. A higher abundance of *Bacteroidetes* and a lower abundance of *Firmicutes* were associated with a positive effect of fiber derived from apple in intestinal microbiota in obese rats [39]. Compared with the CK group, the relative abundances of *Bacteroidetes*, *Faecalibacterium*, *L. delbrueckii*, *B. vulgatus*, and *B.coprocola* in the F group increased, whereas a reduction in the relative abundance of *Firmicutes* was found in this study. In addition, *P. goldsteinii* was significantly increased in the N and F groups compared with the CK group. These results may be due to *M. aquaticum* and its fermented production, which contains a higher fiber content. Similarly, intake of kimchi increased the abundance of *Faecalibacterium* [15]. Bacteroides species could break down food to produce bioactive compounds and energy [40]. *B. coprocola* produces extracellular enzymes to help its host break down some polysaccharides in plants, including cellulose and hemicellulose [41]. We found that the relative abundances of *Faecalibacterium*, *B. vulgatus*, and *B. coprocola* were increased in the F group compared with the N group. Thus, *L. plantarum* BW2013 fermentation may help *M. aquaticum* in regulating the gut environment. A higher abundance of *Faecalibacterium* was associated with a strengthening of epithelial defense functions among piglets supplemented with *L. plantarum* ZLP001 [42]. The NG group had a higher level of *Faecalibacterium* than the N group. In contrast, the relative abundance of *Faecalibacterium* was higher in the F group compared to the FG group. Thus, the synergy effects of *M. aquaticum* and *L. plantarum* BW2013 should be explored. In contrast to the CK group, the relative abundances of *Bacteroidetes*, *Faecalibacterium*, *P. goldsteinii*, *L. delbrueckii*, *B. vulgatus*, and *B. coprocola* were increased, while the relative abundance of *Firmicutes* was decreased in the NG group. Meanwhile, the relative abundances of *Bacteroidetes*, *Parabacteroides*, *P. goldsteinii*, *L. delbrueckii*, *B. vulgatus*, and *B. coprocola* were increased, whereas the relative abundances of *Firmicutes* and *Proteobacteria* were decreased in the FG group. It is worth noting that the relative abundance of *P. goldsteinii* was increased in all treatment groups compared with the CK group. These results indicated that *M. aquaticum*, as a feed additive, has beneficial effects on gut microbiota in mice, especially intakes of fermented *M. aquaticum* and *L. plantarum* BW2013.

3.4. Effects of Dietary Intervention on Fecal Metabolites in Mice

Fecal samples were analyzed with untargeted metabolomics to further explore the effects of *M. aquaticum* and *L. plantarum* on the intestinal metabolic profile of mice. The PCA score plots showed a clear separation among the CK, N, NG, F, and FG groups (Figure 4). The supervised PLS-DA analysis showed differences in fecal metabolic characteristics between each of the two comparison groups (Figure 5). These results suggested that *M. aquaticum* and *L. plantarum* BW2013 intervention significantly affected the fecal metabolic profile in mice. Individual metabolite analysis identified nine significantly changed fecal metabolites: L-aspartate, L-threonine, vitamin A, myricetin, gallic acid, luteolin, lysophosphatidylglycerol 18:1 (LPG (18:1)), and 9-oxo-10,12-octadecadienoic acid (9-oxo-ODA) (Table 3). L-aspartic acid, L-threonine, vitamin A, myricetin, gallic acid, and luteolin were significantly upregulated in all treatment groups compared with the CK group. The F group showed significantly higher levels of vitamin A, myricetin, gallic acid, and luteolin than the N group. LPG (18:1) was downregulated in the N and F groups. However, LPG (18:1) was significantly upregulated in the NG group compared with the N group. Additionally, LPG (18:1) was significantly upregulated in the FG group compared with the F group. Furthermore, 9-oxo-ODA was upregulated in the FG group compared with the F group.

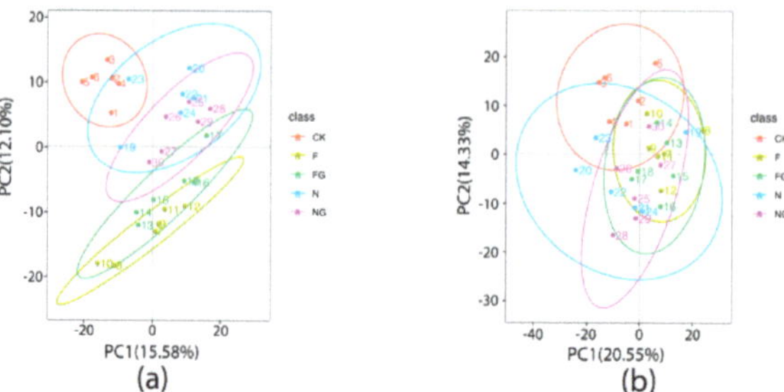

Figure 4. PCA score plots for the CK, F, FG, N, and NG groups in positive (**a**) and negative (**b**) mode.

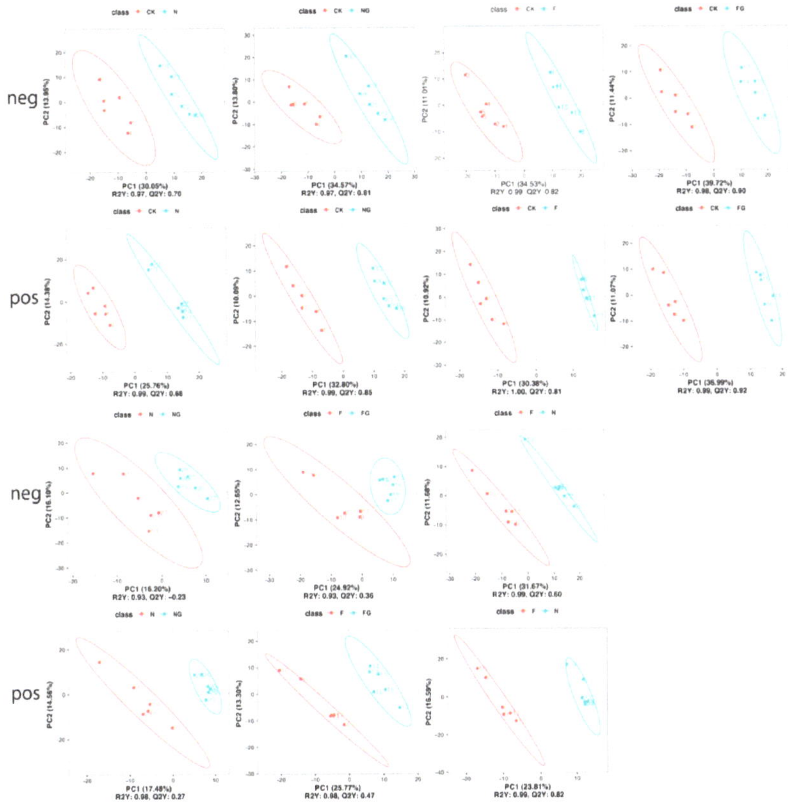

Figure 5. PLS-DA score plots comparing the fecal metabolites between group pairings. R2Y and Q2Y represent the interpretation rate and predictive ability of the PLS-DA model. Higher R2Y than Q2Y values indicate that the PLS–DA model is stable.

Table 3. Fold changes (FCs) in differential metabolites in mice after five weeks of feeding.

No.	Metabolites	Group N-CK		Group NG-CK		Group F-CK		Group FG-CK		Group F-N		Group NG-N		Group FG-F	
		FC	Sig.	FC	Sig.	FC	Sig.	FC	Sig.	FC	Sig.	FC	Sig.	FC	Sig.
1	L-Aspartate	2.01	0.04	2.10	0.04	2.06	0.04	2.09	0.03	1.02	ns	1.04	ns	1.02	ns
2	L-Threonine	2.45	0.01	1.96	0.03	2.28	0.02	2.74	0.01	0.93	ns	0.80	ns	0.83	ns
3	Vitamin A	3.50	<0.01	4.53	<0.01	86.4	<0.01	10.35	<0.01	24.7	0.03	1.30	ns	8.35	ns
4	Myricetin	3.43	0.01	6.39	<0.01	73.2	<0.01	97.5	<0.01	21.3	<0.01	1.86	ns	0.75	ns
5	Gallic acid	3.23	<0.01	5.56	<0.01	16.7	<0.01	9.09	<0.01	5.16	0.03	1.72	ns	0.54	ns
6	Luteolin	4.21	<0.01	5.03	<0.01	9.49	<0.01	8.70	<0.01	2.26	0.03	1.19	ns	1.09	ns
7	LPG 18:1	0.31	0.02	0.52	ns	0.18	<0.01	0.81	ns	0.60	ns	1.70	0.02	4.46	<0.01
8	9-oxo-ODA	1.01	ns	0.70	ns	0.72	ns	1.12	ns	0.71	ns	0.69	ns	1.56	0.049

ns: no significance.

L-aspartic acid and L-threonine metabolites are converted from ingested dietary protein and endogenous protein by intestinal microbes [43]. L-aspartic acid is one major fuel in the intestine that yields ATP for enterocytes, protecting the intestinal barrier from lipopolysaccharide damage [44]. An increased level of threonine could also lead to an increase in mucin synthesis, which strengthens the interaction between microbiota and the metabolome on the surface of the small intestine for more efficient gut function and immune development [45]. Vitamin A is converted from dietary proteinoid carotenoids, which may play a role in regulating gut microbiota composition, relieving inflammation and enhancing the intestinal epithelial barrier in necrotizing enterocolitis [46]. Myricetin, gallic acid, and luteolin belong to the polyphenol famliy, and are bioactive compounds found in fruits and vegetables. Myricetin exhibits therapeutic effects against many diseases, including cancers of different types, inflammatory diseases, atherosclerosis, thrombosis, cerebral ischemia, diabetes, Alzheimer's disease, and pathogenic bacterial infections [47]. The beneficial effects of gallic acid can be observed in cardiovascular protection, immune regulation, and gastrointestinal protection [48]. Luteolin is a flavonoid found in plants and may improve intestinal dysbiosis by inhibiting α-glucosidase. Luteolin has shown anti-cancer activity in cancer cell lines and in vivo models [49,50]. Notably, relative quantities of vitamin A, myricetin, gallic acid, and luteolin were upregulated in the F group compared with the N group. It is possible that the more favorable results found in the F group of our study might be due to the fermented *M. aquaticum*. Furthermore, LPG 18:1 was upregulated in the NG group compared with the N group. Similar results were found in the FG group compared with the F group. LPG, a lysophospholipid, was found to be important in some physiological processes [51]. Ye et al. reported that reduction in LPG in oleate-treated macrophages could be attenuated or even abolished by WY-14643 and/or pioglitazone treatment (two drugs used to treat metabolic diseases) [52]. In this study, LPG (18:1) was significantly downregulated in the N and F groups compared with the CK group. However, LPG (18:1) was upregulated in the NG group compared with the N group and in the FG group compared with the F group. It is suggested that reduction in LPG concentration in the N and F groups could be attenuated or even abolished by supplementation with *L. plantarum* BW2013. In addition, 9-oxo-ODA was upregulated in the FG group compared with the F group. As a PPARα agonist, 9-oxo-ODA could promote fatty acid oxidation to consequently inhibit triglyceride accumulation [53]. Therefore, four dietary interventions may have potential for intestinal protection as well as anti-inflammatory and anti-cancer benefits. Intakes of fermented *M. aquaticum* may be more efficient than *M. aquaticum* with respect to anti-inflammatory, anti-cancer, and intestinal protection by regulating vitamin A, myricetin, gallic acid, and luteolin favorably. Intakes of *M. aquaticum* and *L. plantarum* BW2013 may be more efficient than *M. aquaticum* in metabolic balance by regulating LPG (18:1). Similarly, intakes of fermented *M. aquaticum* and *L. plantarum* BW2013 may be more efficient than fermented *M. aquaticum* in metabolic balance by regulating LPG (18:1). In addition, intakes of fermented *M. aquaticum* and *L. plantarum* BW2013 may be more efficient than fermented *M. aquaticum* in anti-obesity by regulating 9-oxo-ODA.

3.5. The Correlation of Gut Microbiota and Fecal Metabolites

Biochemical metabolic pathways involved in differential metabolites can be identified using KEGG pathway enrichment analysis. The top five enriched pathways were identified by KEGG pathway analysis between each group pairing. The N, NG, F, and FG groups displayed significantly higher enrichment of one, one, three, and four pathways, respectively, compared to the CK group (Figure 6). Pathway enrichment analysis showed that the pathway "Histidine metabolism" was enriched with statistical significance in the N group ($p < 0.05$); "Histidine metabolism" was enriched with statistical significance in the NG group ($p < 0.05$); "Porphyrin and chlorophyll metabolism", "Aminobenzoate degradation", and "Monobactam biosynthesis" were enriched with statistical significance in the F group ($p < 0.05$); "Taurine and hypotaurine metabolism", "Porphyrin and chlorophyll metabolism", "Glycine, serine, and threonine metabolism", and "Histidine metabolism" were enriched with statistical significance in the FG group ($p < 0.05$) compared with the CK group. Moreover, "Biosynthesis of unsaturated fatty acids" was significantly enriched in the F group compared with the N group ($p < 0.05$). "Nitrotoluene degradation", "Sulfur relay system", "Degradation of aromatic compounds", and "Tyrosine metabolism" were significantly enriched in the NG group compared with the N group ($p < 0.05$). "Arginine and proline metabolism" was significantly enriched in the FG group compared with the F group ($p < 0.05$) (Figure 7).

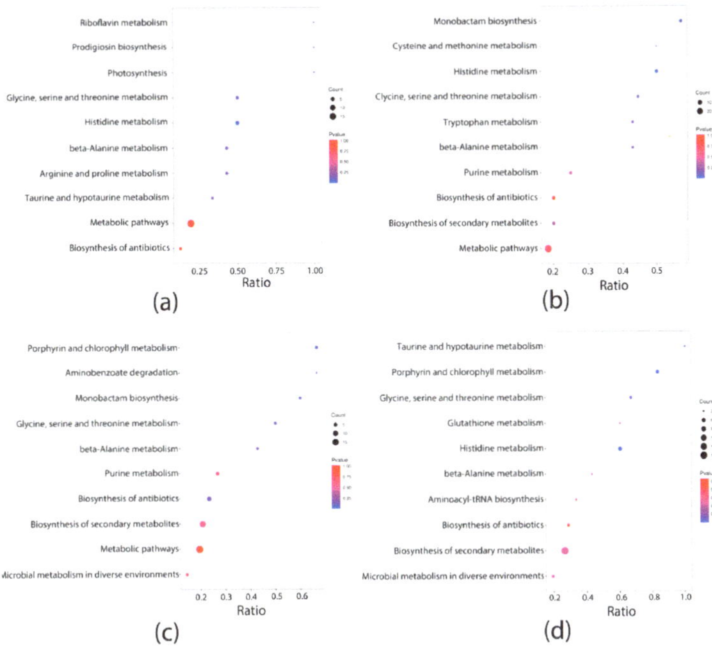

Figure 6. KEGG pathways enrichment integrative analysis between groups: (**a**) N vs. CK; (**b**) NG vs. CK; (**c**) F vs. CK; (**d**) FG vs. CK.

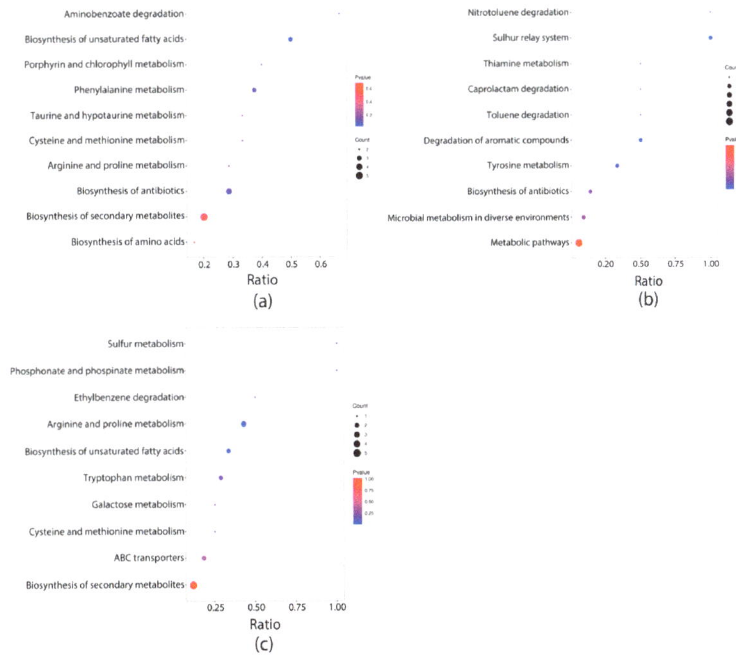

Figure 7. KEGG pathways enrichment integrative analysis between groups: (**a**) F vs. N; (**b**) NG vs. N; (**c**) FG vs. G.

The correlation between gut microbiota and fecal metabolites is helpful in explaining the close relationship between gut microbiota and hosts. Spearman correlation analysis (Figure 8) showed that the change in *P. goldsteinii* was positively associated with changes in vitamin A (r = 0.60, *p* = 0.038), myricetin (r = 0.59, *p* = 0.043), gallic acid (r = 0.58, *p* = 0.046), and luteolin (r = 0.70, *p* = 0.012) contents in the N group compared with the CK group. The change in *P. goldsteinii* was positively associated with change in myricetin (r = 0.83, *p* < 0.001), gallic acid (r = 0.85, *p* < 0.001), and luteolin (r = 0.59, *p* = 0.045) in the NG group compared with the CK group. The change in *P. goldsteinii* was positively associated with change in vitamin A (r = 0.84, *p* < 0.001); the change in *L. delbrueckii* was positively associated with change in vitamin A (r = 0.74, *p* = 0.006), myricetin (r = 0.58, *p* = 0.048), gallic acid (r = 0.83, *p* < 0.001), and luteolin (r = 0.81, *p* = 0.001); and the change in *B. vulgatus* was positively associated with change in vitamin A (r = 0.66, *p* = 0.021) and luteolin (r = 0.83, *p* < 0.001) in the F group compared with the CK group. The change in *P. goldsteinii* was positively associated with change in L-threonine (r = 0.70, *p* = 0.011), vitamin A (r = 0.86, *p* < 0.001), myricetin (r = 0.72, *p* = 0.008), gallic acid (r = 0.76, *p* = 0.004), and luteolin (r = 0.72, *p* = 0.008); the change in *L. delbrueckii* was positively associated with the change in L-aspartate (r = 0.71, *p* = 0.010) and vitamin A (r = 0.62, *p* = 0.032); and the change in *B. vulgatus* was positively associated with change in vitamin A (r = 0.60, *p* = 0.039) in the FG group compared with the CK group.

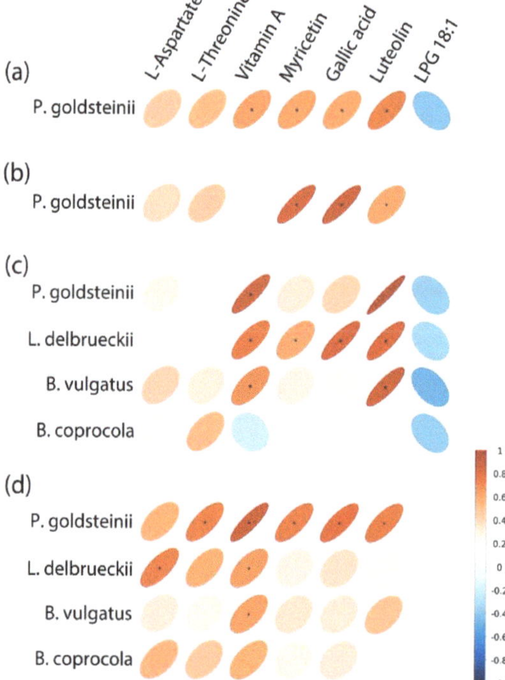

Figure 8. Significant associations between changes in metabolites and changes in abundance of species as measured by Spearman correlations: (**a**) N vs. CK; (**b**) NG vs. CK; (**c**) F vs. CK; (**d**) FG vs. CK.

According to the current situation with regard to basic metabolism, intestinal microbiota were able to enhance nutrient uptake in the four treatment groups, in contrast with the CK group. However, different treatments would cause different results. "Histidine metabolism" and "Glycine, serine, and threonine metabolism" belong to amino acid metabolism. As a result of increased amino acid metabolism, the capacity for protein digestion and absorption was enhanced, greatly contributing to basic growth metabolism [54]. Our results showed that, compared with the CK group, F increased gallic acid significantly. Gallic acid showed a positive correlation with *L. delbrueckii* and was involved in aminobenzoate degradation. Therefore, F might regulate aminobenzoate degradation by mediating *L. delbrueckii*. Compared with the CK group, L-threonine and L-aspartate were significantly increased in the FG group. L-threonine was positively associated with *P. goldsteinii* and involved in "Porphyrin and chlorophyll metabolism" and "Glycine, serine, and threonine metabolism". L-aspartate was positively associated with *L. delbrueckii* and involved in "Histidine metabolism" and "Glycine, serine, and threonine metabolism". Therefore, FG might regulate "Porphyrin and chlorophyll metabolism" and "Glycine, serine, and threonine metabolism" by mediating *P. goldsteinii* as well as regulating "Histidine metabolism" and "Glycine, serine, and threonine metabolism" by mediating *L. delbrueckii*.

4. Conclusions

In conclusion, the quality of *M. aquaticum* as a feed additive was improved by *L. plantarum* BW2013 solid-state fermentation. High-throughput sequencing and metabolomic results showed that *M. aquaticum* and *L. plantarum* BW2013, as new feed additives, could promote the intestinal health of mice by modulating microbiota composition and regulating fecal metabolic profiles. Intakes of fermented *M. aquaticum* and *L. plantarum* BW2013, especially, may have potential for intestinal protection as well as anti-inflammatory, anti-obesity, and anti-cancer benefits by increasing populations of beneficial microorganisms (*Parabacteroides, P. goldsteinii, L. delbrueckii, B.vulgatus,* and *B.coprocola*) and decreasing populations of harmful microorganisms (*Proteobacteria*). Meanwhile, FG might regulate "Porphyrin and chlorophyll metabolism" and "Glycine, serine, and threonine metabolism" by mediating *P. goldsteinii* as well as regulating "Histidine metabolism" and "Glycine, serine, and threonine metabolism" by mediating *L. delbrueckii*. Moreover, the correlation analysis of gut microbiota and metabolites showed that *P. goldsteinii* has a positive correlation with L-threonine, vitamin A, myricetin, gallic acid, and luteolin; *L. delbrueckii* has a positive correlation with L-aspartate and vitamin A; and *B. vulgatus* has a positive correlation with vitamin A. This study could be used as a reference for future developments of beneficial feed additives for animals. Future work may include safety assessments of *M. aquaticum* for humans.

5. Patents

There is a patent (202111462076.2, China) resulting from the work reported in this manuscript.

Author Contributions: Writing—original draft, writing—review and editing, and data curation, Y.L. (Yueyang Li); investigation and data curation, Y.L. (Yuxi Ling); supervision, J.L.; writing—review and editing, M.Z.; project administration, funding acquisition, and writing—review and editing, Z.L.; formal analysis and methodology, Z.B., Z.W. (Zhenlong Wu) and R.X.; conceptualization and visualization, Z.W. (Zhichao Wu), Y.W. and Q.Z. All authors have read and agreed to the published version of the manuscript.

Funding: This paper was supported by the Beijing Natural Science Foundation (grant number 6173033), the Beijing Union University Foundation (grant number 12213611605-001), Academic Research Projects of Beijing Union University (grant number ZK70202003), and Internal Trade Food Science and Technology (Beijing) Co., Ltd. Cooperation Projects (grant number 202116).

Institutional Review Board Statement: This study followed the recommendations of the Animal Welfare Committee of Beijing Union University (Beijing, China). The protocol was approved by the Animal Welfare Committee of Beijing Union University (protocol code: 20191101).

Informed Consent Statement: Not applicable.

Data Availability Statement: All 16S rRNA sequences were submitted to the National Center for Biotechnology Information (NCBI) Sequence Read Archive with the accession number PRJNA801077.

Conflicts of Interest: The authors declare no conflict of interest.

References

1. Colzi, I.; Lastrucci, L.; Rangoni, M.; Coppi, A.; Gonnelli, C. Using *Myriophyllum aquaticum* (Vell.) Verdc. to Remove Heavy Metals from Contaminated Water: Better Dead or Alive? *J. Environ. Manag.* **2018**, *213*, 320–328. [CrossRef] [PubMed]
2. Luo, P.; Liu, F.; Zhang, S.; Li, H.; Chen, X.; Wu, L.; Jiang, Q.; Xiao, R.; Wu, J. Evaluating Organics Removal Performance from Lagoon-Pretreated Swine Wastewater in Pilot-Scale Three-Stage Surface Flow Constructed Wetlands. *Chemosphere* **2018**, *211*, 286–293. [CrossRef] [PubMed]
3. Liu, F.; Zhang, S.; Luo, P.; Zhuang, X.; Chen, X.; Wu, J. Purification and Reuse of Non-Point Source Wastewater via *Myriophyllum*-Based Integrative Biotechnology: A Review. *Bioresour. Technol.* **2018**, *248*, 3–11. [CrossRef]
4. Lastrucci, L.; Lazzaro, L.; Dell'Olmo, L.; Foggi, B.; Cianferoni, F. Impacts of *Myriophyllum aquaticum* Invasion in a Mediterranean Wetland on Plant and Macro-Arthropod Communities. *Plant Biosyst.* **2017**, *152*, 427–435. [CrossRef]
5. Lastrucci, L.; Valentini, E.; Dell'Olmo, L.; Vietina, B.; Foggi, B. Hygrophilous Vegetation and Habitats of Conservation Interest in the Area of the Lake Porta (Tuscany, Central Italy). *Atti Soc. Toscana Sci. Nat. Mem. Ser. B* **2015**, *122*, 131–146.
6. Gareau, M.G.; Sherman, P.M.; Walker, W.A. Probiotics and the Gut Microbiota in Intestinal Health and Disease. *Nat. Rev. Gastroenterol. Hepatol.* **2010**, *7*, 503–514. [CrossRef]
7. Liu, Y.; Gao, Y.; Ma, F.; Sun, M.; Mu, G.; Tuo, Y. The Ameliorative Effect of *Lactobacillus plantarum* Y44 Oral Administration on Inflammation and Lipid Metabolism in Obese Mice Fed with a High Fat Diet. *Food Funct.* **2020**, *11*, 5024–5039. [CrossRef]
8. Kim, J.S.; Lee, Y.H.; Kim, Y.I.; Ahmadi, F.; Oh, Y.K.; Park, J.M.; Kwak, W.S. Effect of Microbial Inoculant or Molasses on Fermentative Quality and Aerobic Stability of Sawdust-Based Spent Mushroom Substrate. *Bioresour. Technol.* **2016**, *216*, 188–195. [CrossRef]
9. Ni, K.; Wang, F.; Zhu, B.; Yang, J.; Zhou, G.; Pan, Y.; Tao, Y.; Zhong, J. Effects of Lactic Acid Bacteria and Molasses Additives on the Microbial Community and Fermentation Quality of Soybean Silage. *Bioresour. Technol.* **2017**, *238*, 706–715. [CrossRef]
10. Pan, T.; Xiang, H.; Diao, T.; Ma, W.; Shi, C.; Xu, Y.; Xie, Q. Effects of Probiotics and Nutrients Addition on the Microbial Community and Fermentation Quality of Peanut Hull. *Bioresour. Technol.* **2019**, *273*, 144–152. [CrossRef]
11. Puntillo, M.; Gaggiotti, M.; Oteiza, J.M.; Binetti, A.; Massera, A.; Vinderola, G. Potential of Lactic Acid Bacteria Isolated From Different Forages as Silage Inoculants for Improving Fermentation Quality and Aerobic Stability. *Front. Microbiol.* **2020**, *11*, 3091. [CrossRef] [PubMed]
12. Wang, Y.; Du, W.; Lei, K.; Wang, B.; Wang, Y.; Zhou, Y.; Li, W. Effects of Dietary *Bacillus licheniformis* on Gut Physical Barrier, Immunity, and Reproductive Hormones of Laying Hens. *Probiotics Antimicrob. Proteins* **2017**, *9*, 292–299. [CrossRef] [PubMed]
13. Zeng, Y.; Zeng, D.; Zhang, Y.; Ni, X.Q.; Wang, J.; Jian, P.; Zhou, Y.; Li, Y.; Yin, Z.Q.; Pan, K.C.; et al. *Lactobacillus plantarum* BS22 Promotes Gut Microbial Homeostasis in Broiler Chickens Exposed to Aflatoxin B1. *J. Anim. Physiol. Anim. Nutr.* **2018**, *102*, e449–e459. [CrossRef] [PubMed]
14. Zhong, H.; Deng, L.; Zhao, M.; Tang, J.; Liu, T.; Zhang, H.; Feng, F. Probiotic-Fermented Blueberry Juice Prevents Obesity and Hyperglycemia in High Fat Diet-Fed Mice in Association with Modulating the Gut Microbiota. *Food Funct.* **2020**, *11*, 9192–9207. [CrossRef]
15. Kim, H.Y.; Park, K.Y. Clinical Trials of Kimchi Intakes on the Regulation of Metabolic Parameters and Colon Health in Healthy Korean Young Adults. *J. Funct. Foods* **2018**, *47*, 325–333. [CrossRef]
16. Nsenga Kumwimba, M.; Dzakpasu, M.; Li, X. Potential of Invasive Watermilfoil (*Myriophyllum* Spp.) to Remediate Eutrophic Waterbodies with Organic and Inorganic Pollutants. *J. Environ. Manag.* **2020**, *270*, 110919. [CrossRef]
17. Rajoka, M.I.; Ahmed, S.; Hashmi, A.S.; Athar, M. Production of Microbial Biomass Protein from Mixed Substrates by Sequential Culture Fermentation of *Candida utilis* and *Brevibacterium lactofermentum*. *Ann. Microbiol.* **2012**, *62*, 1173–1179. [CrossRef]
18. Wuyts, S.; van Beeck, W.; Allonsius, C.N.; van den Broek, M.F.; Lebeer, S. Applications of Plant-Based Fermented Foods and Their Microbes. *Curr. Opin. Biotechnol.* **2020**, *61*, 45–52. [CrossRef]
19. Zha, M.; Li, K.; Zhang, W.; Sun, Z.; Kwok, L.Y.; Menghe, B.; Chen, Y. Untargeted Mass Spectrometry-Based Metabolomics Approach Unveils Molecular Changes in Milk Fermented by *Lactobacillus plantarum* P9. *LWT Food Sci. Technol.* **2021**, *140*, 110759. [CrossRef]
20. Wen, J.J.; Li, M.Z.; Gao, H.; Hu, J.L.; Nie, Q.X.; Chen, H.H.; Zhang, Y.L.; Xie, M.Y.; Nie, S.P. Polysaccharides from Fermented *Momordica charantia* L. with *Lactobacillus plantarum* NCU116 Ameliorate Metabolic Disorders and Gut Microbiota Change in Obese Rats. *Food Funct.* **2021**, *12*, 2617–2630. [CrossRef]
21. Guedes, C.M.; Gonçalves, D.; Rodrigues, M.A.M.; Dias-da-Silva, A. Effects of a Saccharomyces Cerevisiae Yeast on Ruminal Fermentation and Fibre Degradation of Maize Silages in Cows. *Anim. Feed Sci. Technol.* **2008**, *145*, 27–40. [CrossRef]
22. Marcó, A.; Rubio, R.; Compañó, R.; Casals, I. Comparison of the Kjeldahl Method and a Combustion Method for Total Nitrogen Determination in Animal Feed. *Talanta* **2002**, *57*, 1019–1026. [CrossRef]
23. Manirakiza, P.; Covaci, A.; Schepens, P. Comparative Study on Total Lipid Determination Using Soxhlet, Roese-Gottlieb, Bligh & Dyer, and Modified Bligh & Dyer Extraction Methods. *J. Food Compos. Anal.* **2001**, *14*, 93–100.
24. Rodríguez, R.; Jiménez, A.; Fernández-Bolaños, J.; Guillén, R.; Heredia, A. Dietary Fibre from Vegetable Products as Source of Functional Ingredients. *Trends Food Sci. Technol.* **2006**, *17*, 3–15. [CrossRef]

25. Fuentes-Soriano, P.; Bellido-Milla, D.; García-Guzmán, J.J.; Hernández-Artiga, M.P.; Gallardo-Bernal, J.J.; Palacios-Santander, J.M.; Espada-Bellido, E. A Simple Phosphorus Determination in Walnuts and Assessment of the Assimilable Fraction. *Talanta* **2019**, *204*, 57–62. [CrossRef]

26. Chahdoura, H.; Morales, P.; Barreira, J.C.M.; Barros, L.; Fernández-Ruiz, V.; Ferreira, I.C.F.R.; Achour, L. Dietary Fiber, Mineral Elements Profile and Macronutrients Composition in Different Edible Parts of *Opuntia microdasys* (Lehm.) Pfeiff and *Opuntia macrorhiza* (Engelm.). *LWT Food Sci. Technol.* **2015**, *64*, 446–451. [CrossRef]

27. Edgar, R.C. UPARSE: Highly Accurate OTU Sequences from Microbial Amplicon Reads. *Nat. Methods* **2013**, *10*, 996–998. [CrossRef]

28. Quast, C.; Pruesse, E.; Yilmaz, P.; Gerken, J.; Schweer, T.; Yarza, P.; Peplies, J.; Glöckner, F.O. The SILVA Ribosomal RNA Gene Database Project: Improved Data Processing and Web-Based Tools. *Nucleic Acids Res.* **2013**, *41*, 590–596. [CrossRef]

29. Cao, Y.; Liu, Y.; Dong, Q.; Wang, T.; Niu, C. Alterations in the Gut Microbiome and Metabolic Profile in Rats Acclimated to High Environmental Temperature. *Microb. Biotechnol.* **2022**, *15*, 276–288. [CrossRef]

30. Wang, J.; Cao, F.; Su, E.; Zhao, L.; Qin, W. Improvement of Animal Feed Additives of Ginkgo Leaves through Solid-State Fermentation Using *Aspergillus niger*. *Int. J. Biol. Sci.* **2018**, *14*, 736. [CrossRef]

31. Shin, N.R.; Whon, T.W.; Bae, J.W. Proteobacteria: Microbial Signature of Dysbiosis in Gut Microbiota. *Trends Biotechnol.* **2015**, *33*, 496–503. [CrossRef] [PubMed]

32. Zhou, L.; Zhang, M.; Wang, Y.; Dorfman, R.G.; Liu, H.; Yu, T.; Chen, X.; Tang, D.; Xu, L.; Yin, Y.; et al. *Faecalibacterium prausnitzii* Produces Butyrate to Maintain Th17/Treg Balance and to Ameliorate Colorectal Colitis by Inhibiting Histone Deacetylase 1. *Inflamm. Bowel Dis.* **2018**, *24*, 1926–1940. [CrossRef] [PubMed]

33. Staley, C.; Kaiser, T.; Vaughn, B.P.; Graiziger, C.; Hamilton, M.J.; Kabage, A.J.; Khoruts, A.; Sadowsky, M.J. Durable Long-Term Bacterial Engraftment Following Encapsulated Fecal Microbiota Transplantation to Treat *Clostridium difficile* Infection. *mBio* **2019**, *10*, e01586-19. [CrossRef] [PubMed]

34. Wu, T.R.; Lin, C.S.; Chang, C.J.; Lin, T.L.; Martel, J.; Ko, Y.F.; Ojcius, D.M.; Lu, C.C.; Young, J.D.; Lai, H.C. Gut Commensal *Parabacteroides goldsteinii* Plays a Predominant Role in the Anti-Obesity Effects of Polysaccharides Isolated from *Hirsutella sinensis*. *Gut* **2019**, *68*, 248–262. [CrossRef] [PubMed]

35. Elean, M.; Albarracín, L.; Cataldo, P.G.; Londero, A.; Kitazawa, H.; Saavedra, L.; Villena, J.; Hebert, E.M. New Immunobiotics from Highly Proteolytic *Lactobacillus delbrueckii* Strains: Their Impact on Intestinal Antiviral Innate Immune Response. *Benef. Microbes* **2020**, *11*, 375–390. [CrossRef] [PubMed]

36. Lee, N.Y.; Joung, H.C.; Kim, B.K.; Kim, B.Y.; Park, T.S.; Suk, K.T. *Lactobacillus lactis* CKDB001 Ameliorate Progression of Nonalcoholic Fatty Liver Disease through of Gut Microbiome: Addendum. *Gut Microbes* **2020**, *12*, 1829449. [CrossRef]

37. Yoshida, N.; Emoto, T.; Yamashita, T.; Watanabe, H.; Hayashi, T.; Tabata, T.; Hoshi, N.; Hatano, N.; Ozawa, G.; Sasaki, N.; et al. *Bacteroides vulgatus* and *Bacteroides dorei* Reduce Gut Microbial Lipopolysaccharide Production and Inhibit Atherosclerosis. *Circulation* **2018**, *138*, 2486–2498. [CrossRef]

38. Makki, K.; Deehan, E.C.; Walter, J.; Bäckhed, F. The Impact of Dietary Fiber on Gut Microbiota in Host Health and Disease. *Cell Host Microbe* **2018**, *23*, 705–715. [CrossRef]

39. Jiang, T.; Gao, X.; Wu, C.; Tian, F.; Lei, Q.; Bi, J.; Xie, B.; Wang, H.Y.; Chen, S.; Wang, X. Apple-Derived Pectin Modulates Gut Microbiota, Improves Gut Barrier Function, and Attenuates Metabolic Endotoxemia in Rats with Diet-Induced Obesity. *Nutrients* **2016**, *8*, 126. [CrossRef]

40. Wexler, H.M. Bacteroides: The Good, the Bad, and the Nitty-Gritty. *Clin. Microbiol. Rev.* **2007**, *20*, 593–621. [CrossRef]

41. Kitahara, M.; Sakamoto, M.; Ike, M.; Sakata, S.; Benno, Y. *Bacteroides plebeius* Sp. Nov. and *Bacteroides coprocola* Sp. Nov., Isolated from Human Faeces. *Int. J. Syst. Evol. Microbiol.* **2005**, *55*, 2143–2147. [CrossRef] [PubMed]

42. Wang, J.; Ji, H.; Wang, S.; Liu, H.; Zhang, W.; Zhang, D.; Wang, Y. Probiotic *Lactobacillus plantarum* Promotes Intestinal Barrier Function by Strengthening the Epithelium and Modulating Gut Microbiota. *Front. Microbiol.* **2018**, *9*, 1953. [CrossRef] [PubMed]

43. Macfarlane, G.T.; Cummings, J.H.; Allison, C. Protein Degradation by Human Intestinal Bacteria. *Microbiology* **1986**, *132*, 1647–1656. [CrossRef] [PubMed]

44. Kang, P.; Liu, Y.; Zhu, H.; Li, S.; Shi, H.; Chen, F.; Leng, W.; Pi, D.; Hou, Y.; Yi, D. The Effect of Aspartate on the Energy Metabolism in the Liver of Weanling Pigs Challenged with Lipopolysaccharide. *Eur. J. Nutr.* **2014**, *54*, 581–588. [CrossRef] [PubMed]

45. Li, Z.; Wang, X.; Zhang, T.; Si, H.; Nan, W.; Xu, C.; Guan, L.; Wright, A.D.G.; Li, G. The Development of Microbiota and Metabolome in Small Intestine of Sika Deer (*Cervus nippon*) from Birth to Weaning. *Front. Microbiol.* **2018**, *9*, 4. [CrossRef] [PubMed]

46. Xiao, S.; Li, Q.; Hu, K.; He, Y.; Ai, Q.; Hu, L.; Yu, J. Vitamin A and Retinoic Acid Exhibit Protective Effects on Necrotizing Enterocolitis by Regulating Intestinal Flora and Enhancing the Intestinal Epithelial Barrier. *Arch. Med. Res.* **2018**, *49*, 1–9. [CrossRef] [PubMed]

47. Song, X.; Tan, L.; Wang, M.; Ren, C.; Guo, C.; Yang, B.; Ren, Y.; Cao, Z.; Li, Y.; Pei, J. Myricetin: A Review of the Most Recent Research. *Biomed. Pharmacother.* **2021**, *134*, 111017. [CrossRef]

48. Li, D.; Lv, B.; Wang, D.; Xu, D.; Qin, S.; Zhang, Y.; Chen, J.; Zhang, W.; Zhang, Z.; Xu, F. Network Pharmacology and Bioactive Equivalence Assessment Integrated Strategy Driven Q-Markers Discovery for Da-Cheng-Qi Decoction to Attenuate Intestinal Obstruction. *Phytomedicine* **2020**, *72*, 153236. [CrossRef]

49. Wilsher, N.E.; Arroo, R.R.; Matsoukas, M.T.; Tsatsakis, A.M.; Spandidos, D.A.; Androutsopoulos, V.P. Cytochrome P450 CYP1 Metabolism of Hydroxylated Flavones and Flavonols: Selective Bioactivation of Luteolin in Breast Cancer Cells. *Food Chem. Toxicol.* **2017**, *110*, 383–394. [CrossRef]

50. Yan, J.; Zhang, G.; Pan, J.; Wang, Y. α-Glucosidase Inhibition by Luteolin: Kinetics, Interaction and Molecular Docking. *Int. J. Biol. Macromol.* **2014**, *64*, 213–223. [CrossRef]

51. Shen, J.; Li, P.; Liu, S.; Liu, Q.; Li, Y.; Zhang, Z.; Yang, C.; Hu, M.; Sun, Y.; He, C.; et al. The Chemopreventive Effects of Huangqin-Tea against AOM-Induced Preneoplastic Colonic Aberrant Crypt Foci in Rats and Omics Analysis. *Food Funct.* **2020**, *11*, 9634–9650. [CrossRef] [PubMed]

52. Ye, G.; Yang, B.C.; Gao, H.; Wu, Z.; Chen, J.; Ai, X.Y.; Huang, Q. Metabolomics Insights into Oleate-Induced Disorders of Phospholipid Metabolism in Macrophages. *J. Nutr.* **2021**, *151*, 503–512. [CrossRef] [PubMed]

53. Shama, S.; Liu, W. Omega-3 Fatty Acids and Gut Microbiota: A Reciprocal Interaction in Nonalcoholic Fatty Liver Disease. *Dig. Dis. Sci.* **2020**, *65*, 906–910. [CrossRef] [PubMed]

54. Wang, M.; Chen, Y.; Wang, Y.; Li, Y.; Zhang, X.; Zheng, H.; Ma, F.; Ma, C.W.; Lu, B.; Xie, Z.; et al. Beneficial Changes of Gut Microbiota and Metabolism in Weaned Rats with *Lactobacillus acidophilus* NCFM and *Bifidobacterium lactis* Bi-07 Supplementation. *J. Funct. Foods* **2018**, *48*, 252–265. [CrossRef]

Article

Assessing Hydrolyzed Gluten Content in Dietary Enzyme Supplements Following Fermentation

Ekaterina Khokhlova [1,†], Pyeongsug Kim [2,†], Joan Colom [1], Shaila Bhat [2], Aoife M. Curran [3], Najla Jouini [3], Kieran Rea [1,*], Christopher Phipps [2,*] and John Deaton [2,*]

[1] Deerland Ireland R&D Ltd., Food Science Buiding, University College Cork, T12 YT20 Cork, Ireland; ekhokhlova@deerland.com (E.K.); jcomas@deerland.com (J.C.)

[2] Deerland Probiotics and Enzymes, 3800 Cobb International Boulevard, Kennesaw, GA 30152, USA; pkim@deerland.com (P.K.); sbhat@deerland.com (S.B.)

[3] Shannon Applied Biotechnology Centre, Munster Technological University, T12 P928 Cork, Ireland; aoife.curran@mtu.ie (A.M.C.); najla.jouini@mtu.ie (N.J.)

* Correspondence: krea@deerland.com (K.R.); cphipps@deerland.com (C.P.); jdeaton@deerland.com (J.D.)

† These authors contributed equally to this work.

Abstract: Partially digested gluten fragments from grains including wheat, rye, spelt and barley are responsible for triggering an inflammatory response in the intestinal tract of Celiac Disease (CD) and Non-Celiac Gluten Sensitive (NCGS) individuals. Fermentation is an effective method to metabolize gluten, with enzymes from bacterial or fungal species being released to help in this process. However, the levels of gluten in commercially available enzymes, including those involved in gluten fermentation, are unknown. In this study we investigated gluten levels in commercially available dietary enzymes combined with assessing their effect on inflammatory response in human cell culture assays. Using antibodies that recognize different gluten epitopes (G12, R5, 2D4, MloBS and Skerritt), we employed ELISA and immunoblotting methodologies to determine gluten content in crude gluten, crude gliadin, pepsin-trypsin digested gluten and a selection of commercially available enzymes. We further investigated the effect of these compounds on inflammatory response in immortalized immune and intestinal human cell lines, as well as in peripheral blood mononuclear cells (PBMCs) from coeliac individuals. All tested supplemental enzyme products reported a gluten concentration that was equivalent to or below 20 parts per million (ppm) as compared with an intact wheat reference standard and a pepsin-trypsin digested standard. Similarly, the inflammatory response to IL-8 and TNF-α inflammatory cytokines in mammalian cell lines and PBMCs from coeliac individuals to the commercial enzymes was not significantly different to 20 ppm of crude gluten, crude gliadin or pepsin-trypsin digested gluten. This combined approach provides insight into the extent of gluten breakdown in the fermentation process and the safety of these products to gluten-sensitive individuals.

Keywords: gluten; fermentation; supplemental enzymes; IL-8; TNF-α; ELISA; immunoblotting

Citation: Khokhlova, E.; Kim, P.; Colom, J.; Bhat, S.; Curran, A.M.; Jouini, N.; Rea, K.; Phipps, C.; Deaton, J. Assessing Hydrolyzed Gluten Content in Dietary Enzyme Supplements Following Fermentation. *Fermentation* **2022**, *8*, 203. https://doi.org/10.3390/fermentation8050203

Academic Editors: Xian Zhang and Zhiming Rao

Received: 11 April 2022
Accepted: 27 April 2022
Published: 29 April 2022

Publisher's Note: MDPI stays neutral with regard to jurisdictional claims in published maps and institutional affiliations.

1. Introduction

Various food products including breads, cereals, beer, pasta, sauces and beverages as well as cosmetic and skincare products utilize gluten-containing grains in their manufacturing processes [1]. For some of these products, gluten can be reduced using wheat starch or wheat that has had the gluten washed out with water. For others, the gluten either cannot be removed, or, like those derived from fermentation, the wheat is used as substrate and broken down by microorganisms to form the final product [1]. During fermentation, enzymes from bacterial or fungal species are released to help in this process, and it has been suggested that fermentation and enzymatic hydrolysis hold the most potential to create novel hypo-/nonallergenic wheat products [1]. Bacterial species have a broad capacity to catabolize different carbohydrates, proteins and lipids as sources of energy. Their ability to break down these source substrates is mediated by enzymes whose expression is under tight regulation by promoter sequences in its genome that are activated by the presence of the source substate and its degradation products [2]. Ironically, removing the target substrate (gluten) from the fermentation process dramatically reduces the yield of enzymes required for the catabolism of the substrate (gluten), as the promoter regions of the genome that are responsible for the enzyme production are not activated in the bacteria [3,4]. Furthermore, many more supplemental enzymes including lipases, cellulases, proteases, peptidases and enzymes involved in carbohydrate degradation similarly have promoter regions that are responsive to wheat to stimulate their production. These enzymes have demonstrated health benefits in a number of important disciplines across the lifespan including digestion [5–8], cardiovascular health [9,10], exercise [11] and food intolerances [12–16]. The ability to use wheat as a starting substrate in the fermentation process but to have minimal gluten content is paramount for making efficacious products that can be used by a wide variety of consumers.

Celiac disease (CD) is a chronic autoimmune disorder characterized by fatigue, nausea and a range of gastrointestinal discomfort and complications initiated by exposure to dietary gluten. While CD affects 1 in 141 people in the United States [17,18], non-coeliac gluten sensitivity (NCGS) is thought to be much more prevalent, varying between 1–13% of the population [19]. Gluten is a mixture of complex proteins called gliadins and glutenins, rich in prolines and glutamines that are difficult to digest by intestinal enzymes [20]. This leads to a partial digestion of gluten proteins, generating immunogenic peptides that trigger an inflammatory response in the intestinal tract of CD and NCGS individuals. This recurring immune response to partially digested gluten fragments causes weakening of the integrity of the intestinal lining and shrinking of intestinal villi, causing difficulties in nutrient absorption and allowing gluten fragments to penetrate further into the body [21–24], leading to the associated symptoms. Currently, the only effective treatment for the alleviation of the gastointestinal symptoms in CD and NCGS is a strict, gluten-free diet [19] whereby the absence of immunogenic gluten-derived peptides allows for the healing of the intestinal barrier [25,26] and the resolution of symptoms.

The FDA defines the term 'gluten-free' as products that are not made with any gluten-containing grains or that have been refined to remove the gluten. To meet this criteria, refined products must be tested to show they contain less than 20 ppm of gluten. However, the recent ruling on gluten-free labeling of fermented foods by the FDA states that, since no appropriate test currently exists to quantify gluten in hydrolyzed matrices, they may not be labeled as gluten-free. This decision was based on the difficulty of quantifying levels of gluten-derived immunogenic peptides at the end point of the fermentation process, highlighting the importance of developing better or complementary methods to accurately quantitate gluten peptides and their immunogenic properties in these products [27]. A potential, unintended effect of this ruling from the FDA is the elimination of the gluten-free labeling of supplemental enzymes, including those taken by gluten-sensitive individuals in order to reduce gluten content from food and reduce symptoms [15,28,29].

Current competitive ELISA quantification of fermented gluten needs to be adapted to accurately identify the range of gluten peptides produced during the fermentation of gluten, as it varies across different grains, fermentation organisms and fermentation processes [30]. Moreover, it has yet to be shown that these specific peptides from these fermentation processes can elicit an immune response below 20 ppm as specified by the FDA guidance. These differences in gluten peptide compositions cause a lack of correlation with the calibration methods used in competitive ELISA, leading to inaccurate quantitation and subsequent immunogenicity measurements [31]. Additionally, some small peptides may not interact with the individual antibody used in the competitive ELISA, remaining undetected and potentially causing the same physiological effects for a gluten-sensitive individual. Therefore, multiple antibody testing techniques and a complementary detection method based on cell culture response to digested gluten should be used to determine the effects of such molecules. In this study, the gluten concentration in supplemental enzymes preferentially synthesized during gluten fermentation was assessed using multiplex ELISA and immunoblots using antibodies targeting multiple epitopes to assess gluten digestion. A reference pepsin-trypsin hydrolyzed gluten was assessed alongside the supplemental enzymes using these methodologies to determine gluten content and peptide profile. Finally, the potential immune response elicited by pepsin-trypsin hydrolyzed gluten and supplemental enzymes was investigated in immortalized immune and intestinal human cell lines, as well as in peripheral blood mononuclear cells (PBMCs) from coeliac individuals and compared with crude gliadin or crude gluten at a concentration of 20 ppm. This study adapts a proactive approach to addressing the outlined concerns of the limitations of quantifying gluten-derived products as they pertain to supplemental enzymes by utilizing a multiplex ELISA approach first published by Panda et al., 2017, combined with human cell culture models to assess immunopathogenic response [32].

2. Materials and Methods

2.1. Materials

All materials, unless otherwise stated, were purchased from Sigma-Aldrich (Atlanta, GA, USA and Arklow, Wicklow, Ireland). All buffers and solutions were prepared with Milli-Q water (resistivity 18.2 MΩ, Merck Millipore, Saint-Quentin-Fallavier, France).

2.2. Preparation of Samples for Multiplex and Immunoblotting

To prepare samples for ELISA and Immunoblotting, stock solutions of 100 mg/mL were prepared in 60% ethanol to solubilize gliadins and boiled for 20 min to remove residual enzyme activity. Samples were cooled to room temperature in an ice bath and centrifuged for 10 min at 5000 rpm before being further diluted for analysis.

2.3. Hydrolysis of Gluten Using Pepsin and Trypsin Complex

Gluten hydrolysis using the sequential catalyzation of pepsin and trypsin was conducted based on the method of Rio et al. 2021 [33]. The pepsin and trypsin complexes were provided by Deerland Enzymes and Probiotics and used in the enzyme-to-substrate ratio, 1:20 (w/w), during the hydrolysis procedure. To acidify the gluten, powdered gluten from wheat was added to 0.1 M sodium phosphate buffer (pH 2.0) to a final concentration of 20 mg/mL and was incubated at 37 °C on a rotating rocker at 150 rpm for 30 min. Pepsin (10,000 FCC Units/mg) was added into the acidified gluten suspension (10 mg/mL), and the mixture was incubated at 37 °C on a rotating rocker at 150 rpm for 3 h. An equal volume of 0.1 M sodium phosphate buffer (pH 8.0) was added to pepsin-gluten mixture, and the pH was increased to 8.0 using 50% NaOH. The trypsin complex (2500 USP Units/mg) was then added at 0.5 mg/mL. The mixture was incubated at 37 °C on a rotating rocker at 80 rpm overnight and boiled for 30 min to denature both enzymes. A sample was taken during each step, diluted to a 5% or 10% concentration (v/v) in 60% ethanol (to a final concentration of gluten at 0.5 or 1 mg/mL) and boiled for 10 min for the further analysis in ELISA and immunoblot.

In parallel, the PT-digested gluten was prepared for cell culture work based on the above preparation with minor variation. After the sample was diluted to 10% concentration (v/v) in 60% ethanol, the precipitation and concentration of PT-digested gluten was conducted using acetone based on the protocol of Thermo Scientific (See Appendix A): 4 parts cold acetone (−20 °C) were added to 1 part sample, vortexed and incubated at −20 °C overnight. The samples were then centrifuged at 3000 rpm for 30 min, the supernatant was discarded and the acetone was evaporated for one hour at room temperature to leave a PT-digested gluten pellet.

2.4. Multiplex Competitive ELISA

The antibodies and dilutions used for multiplex ELISA and immunoblotting are found below in Table 1.

Table 1. HRP-conjugated antibodies from commercial ELISA kits in this study.

Antibody		ELISA Kits	Manufacturer	Dilution	
				ELISA	Immunoblot
Gtox-G12	G12	Gluten Tox ELISA Competitive G12	Biomedal Diagnostics	to 30 ng/mL *	to 200 ng/mL
A-G12	G12	AgraQuant Gluten G12	Romer Labs	1 to 3	1 to 1
R5Sand	R5	RIDASCREEN Gliadin	R-BioPharm, AG	1 to 35	1 to 5
R5Comp	R5	RIDASCREEN Gliadin Competitive	R-BioPharm, AG	1 to 100	1 to 15
V10-R5	R5	Veratox for Gliadin R5 (Cat # 8510)	Neogen Corp.	1 to 15	1 to 10
V80-GL	USDA	Veratox for Gliadin (Cat # 8480)	Neogen Corp.	1 to 10	1 to 3
MI-GL	MIoBS	Wheat/Gluten (Gliadin) MIoBS	Morinaga Institution of Biological Sciences Inc	1 to 3	1 to 1
AllSK	Skerritt	AllerTek Gluten	ELISA Technologies Inc.	1 to 15	1 to 10
2D4	2D4	Microbiologique Gluten Sandwich	Pi BioScientific Inc.	1 to 10	1 to 1

* The Gtox-G12 antibody was provided at a known concentration in mg/mL and diluted to 30 ng/mL for analyses in the ELISA and 200 ng/mL for the immunoblots. For the other antibodies, the optimized dilution factor is provided, similar to work performed in Panda et al., 2017 [32].

The multiplex competitive ELISA for gluten detection were performed based on a competitive ELISA protocol [32].

Microtiter plate wells were coated with either 10 µg/mL or 20 µg/mL gluten antigen. The ELISAs coated with 10 µg/mL antigen were used for Gtox-G12, A-G12, R5Comp, V10-R5, V80-GL, MI-GL or ALLSK, while the ELISAs with 20 µg/mL gluten antigen were for R5Sand or 2D4. To prepare the antigen, a solution of 1 mg/mL wheat gluten was prepared in phosphate buffer (PBS) containing 0.1% Tween®20, rotated overnight at room temperature and then diluted in 1× coating buffer (pH 9.6, Appendix A) to a final concentration of 10 or 20 µg/mL. The coated plates were incubated overnight at room temperature in the dark. The plates were washed three times with wash buffer (PBS containing 0.05% Tween®20) and then wells were blocked with 150 µL/well with blocking buffer (PBS containing 1% bovine serum albumin) at 37 °C for one hour. The plates were washed an additional four times with the wash buffer after blocking.

A gluten standard (1 mg/mL) from wheat standard was prepared following the same preparation listed above and then diluted in UD buffer (105 mM sodium phosphate, 75 mM NaCl, 2% BSA, 0.05% Tween 20, pH 7.4) to generate serial 1:3 dilutions starting at 10 mg/L (10, 3.33, 1.11, 0.37, 0.12, 0.041 and 0 ppm) for ELISA standards [6]. Antibodies were diluted using PBS buffer, as indicated in Table 1. A total of 110 µL of diluted antibody-HRP conjugates and 110 µL of either gluten standards or samples were mixed at 37 °C for one hour at 50 rpm on a rotating rocker. Then, 100 µL of each mixture was transferred to the coated wells in duplicate and incubated at 37 °C for one hour with 50 rpm on a shaker. After washing the plate four times with the wash buffer, 100 µL of the 3,3′,5,5′-tetramethylbenzidine (TMB) substrate was added in each well and incubated at RT for 30 min in the dark. The reaction was stopped by the addition of 50 µL of 0.2 M sulfuric acid to each well followed by the measurement of absorbance at 450 nm using a microplate reader (Bio-Rad model 680). The ELISA performance and measurements were conducted in triplicate with duplicated samples for each trial.

2.5. Immunoblot Using Automated Capillary Electrophoretic-Based Immunoassays

Immunoblotting was performed using the Wes™ (WS-2450), capillary electrophoretic immunoassay with 12–230 kDa pre-filled plates, capillary cartridges and reagents from ProteinSimple (San Jose, CA, USA). The procedure and preparation of reagents including sample buffer for diluents, 5× fluorescent master mix, biotinylated ladder and luminol-S in peroxide were based on the protocol of Wes™ and Nelson 2017 [34].

Two sets of standards were prepared by serial dilution of the extracted 1 mg/mL wheat gluten with the provided 0.1× sample buffer, generating 50, 25, 20, 5 and 0 and 10, 3.33, 1.11, 0.37, 0.12 and 0 mg/L to measure high or low gluten concentration in a sample. All test samples were diluted to a final concentration of 1:200 with 0.1× sample buffer. Additionally, 5× fluorescent master mix was added to each standard or sample and incubated at 75 °C for 10 min. The combinations were briefly vortexed and centrifuged after cooling down on an ice bath [34]. The HRP-conjugated antibodies were diluted with antibody diluent 2, as detailed in Table 1. All prepared reagents, serial standards, diluted samples and antibodies were dispensed into the assay plate into the corresponding well with the volume, as stated in the protocol [34]. The capillary cartridges and assay plates were inserted, and after approximately 3-h running in Wes™, the peak values of samples and standards were analyzed by Compass for SW (ver. 6.0) (see Appendix A). For both ELISA and immune assays, the standard curves were constructed, and the gluten concentration in samples was calculated using the four-parameter logistic curve of the online tool available at ATTBioquest (Sunnyvale, CA, USA) (see Appendix A).

2.6. Preparation of Gluten and Gliadin and PT-Digested Gluten for Cell Culture

Water-soluble fractions of crude gluten and gliadin were prepared by the salt-induced disaggregation method [35], with some modifications. Briefly, 2 g of gluten and gliadin (Sigma, Ronkonkoma, NY, USA) were suspended in 50 mL 2% NaCl solution and stirred using a magnetic stirrer for 5 min at 250 rpm. The liquid was discarded, and washing rounds in 50 mL 2% NaCl solution were repeated four more times. Residual salt was removed by one round of washing with 50 mL 0.2% NaCl followed by one round of washing with 50 mL dH_2O (exposure time was reduced to 30 s). Ten milliliters of serum-free media (either MEM or RPMI) were added to the washed gluten and gliadin and stirred with a magnetic stirrer for four hours. The resulted suspensions were transferred to falcon tubes and centrifuged for 5 min at 6000× g. The supernatants were filter-sterilized using syringe-mounted filters with a 0.45 μm pore diameter (Thermo-Fisher, Dublin, Ireland). Total protein concentration was determined by a Bicinchoninic Acid Protein Assay Kit (Sigma), with serum-free MEM or RPMI media used as blank samples. Fetal bovine serum, glutamine, non-essential amino and antibiotics (as below for cell culture experiments) were added to gluten and gliadin dissolved in MEM or RPMI to obtain full culture media, containing 1 mg/mL of wheat storage proteins. The PT-Gluten was prepared as described earlier. Prior to use, the PT-digested gluten was prepared by suspending in serum-free culture media (either MEM for Caco-2 work or RPMI for THP-1 cells) to either 20 ppm (20 mg/L) or 500 ppm (500 mg/L).

2.7. Preparation of Supplemental Enzymes for Cell Culture

Freeze-dried samples of supplemental enzymes were weighed in 15 mL falcon tubes, prediluted in serum free RPMI medium to a concentration of 10,000 ppm (10 g/L) and boiled for 10 min at 100 °C to inactivate the enzymes. After that, the tubes were placed on ice for 20 min. Sediment formed during manipulations was removed by centrifugation (6000 rpm, 10 min). Enzymes preparations were immediately taken for THP-1 cells treatment and adjusted to 20 ppm (20 mg/L) prior to cell work, as described below.

2.8. Cell Culture

The human colorectal adenocarcinoma cell line Caco-2 and the human monocytic cell line THP-1 were obtained from the DSMZ-German Collection of Microorganisms and Cell Cultures. All Caco-2 cell line work was performed by Shannon ABC (Munster Technological University). Cells were propagated using the standard technique in a 5% CO_2 atmosphere at 37 °C. Caco-2 were cultured in Minimum Essential Medium Eagle supplemented with 10% Fetal Bovine Serum, 2 mM glutamine, 1% non-essential amino acids, 100 U/mL penicillin and 100 μg/mL streptomycin (MEM and FBS—Sigma, additives—Lonza, Manchester,

UK). THP-1 cells were grown in RPMI 1640 media supplemented with 2 mM L-glutamine, 10% Fetal Bovine Serum, 100 U/mL penicillin and 100 µg/mL streptomycin (full media).

2.9. Treatment Scheme

Prior to experiment, 5×10^5 THP-1 cells were transferred into 1.5 mL Eppendorf tubes and centrifuged at $150 \times g$ for 15 min. The culture medium was discarded, and the cells were resuspended in full media containing 20 ppm and 500 ppm of gluten, gliadin and PT-digested gluten and 20 ppm supplemental enzyme preparations. Four hours after stimulation, the cells were harvested by centrifugation ($8000 \times g$ for 5 min), lysed by adding 300 µL of Lysis buffer (Monarch Total RNA Miniprep Kit, New England Biolabs, NEB) and stored at $-80\ °C$ for further PCR analysis. In a parallel experiment, culture supernatants were collected 18 h after stimulation and stored at $-80\ °C$ until further analysis for protein concentrations.

Caco-2 cells were cultured on 48-well plates (Corning) at a density of 3×10^4 cells/well. Prior to experiment, the cells were washed once with DPBS and subjected to 20 ppm and 500 ppm of gluten, gliadin and PT-digested gluten treatment. Four hours after stimulation, cell supernatants were removed and 300 µL of Lysis buffer (Monarch Total RNA Miniprep Kit, New England Biolabs, NEB) was added to the cells, which were then stored at $-80\ °C$ for further PCR analysis. In a parallel experiment, 18 h post-treatment, supernatants were collected and stored at $-80\ °C$ for subsequent protein determination.

2.10. PBMCs Culture Conditions and Treatment

Peripheral blood mononuclear cells (PBMCs) from the peripheral blood of a healthy donor and a celiac disease patient were purchased from AccuCell®, Bethesda, MA, United States. Upon receipt, the cells were thawed and resuspended in freshly made RPMI 1640 media containing 2 mM L-glutamine, 10% Fetal Bovine Serum, 100 U/mL penicillin, 100 µg/mL streptomycin and 2 µg/mL amphotericin. The cells were cultured for 48 h at a concentration of 1×10^6 cell/mL.

The exposure experiment has been done in the same manner as the THP-1 treatment (see above). Briefly, the cells were centrifuged at $150 \times g$ for 15 min and culture media was replaced with media containing 20 ppm and 500 ppm of gluten, gliadin and PT-digested gluten. Forty-eight hours after treatment, cell suspensions were collected and cell culture supernatants were obtained by centrifugation and stored at $-80\ °C$ until analysis for IL-8 concentration.

2.11. Cell Viability Assays

The cell viability assays were done using the CyQUANT™ XTT Cell Viability kit (Invitrogen™, Thermo-Fisher, Dublin, Ireland). Twenty-four hours after applying crude gliadin, gluten or the PT-digested samples, THP-1 and Caco-2 cells were stained with 0.3 mg/mL solution of XTT (sodium 3′-[1-(phenylaminocarbonyl)-3,4-tetrazolium]-bis (4-methoxy6-nitro) benzene sulfonic acid hydrate) for 4 h according to the protocol. Absorbance was then measured at 450 and 595 nm.

2.12. RNA Isolation and Quantitative Real-Time PCR Analysis

Total RNA was extracted from cell lysates using the Monarch Total RNA Miniprep Kit (NEB). Reverse transcription reaction was performed using the Luna script RT Supermix kit (NEB). Real-time PCR reactions were set up using pre-made FastStart Universal SYBR Green Master mix and appropriate primer pairs (all primers Table 2 were tested for linearity and amplification efficiency) at a concentration of 200 nM using generated cDNA as a template. The reactions were performed using the following program: initial denaturation 95 °C 5 min, denaturation 94 °C 20 s, annealing 60 °C 20 s, extension 72 °C 20 s (40 cycles). The specificity of reaction products was confirmed by melting temperature analysis (from 70 °C to 95 °C in 0.5 °C/15 s increments). The quantification of target transcripts was done using RPL5 or GAPDH as a normalizing house-keeping gene.

Table 2. Primer design for PCR targets.

Target	Forward Primer	Reverse Primer	Reference
IL-8	CAGTTTTGCCAAGGAGTGCT	CAACCCTCTGCACCCAGTTT	[36]
TNF-A	GCCAGAGGGCTGATTAGAGA	TCTTCTGCCTGCTGCACTT	[37]
RPL5	GGTCTCTGTTCCGCAGGATG	CAGTTTTACCCTCTCGTCGTCT	[38]
GAPDH	CTTTGACGCTGGGGCTGGCATT	TTGTGCTCTTGCTGGGGCTGGT	[39]

2.13. ELISA Cytokine Determination

The concentrations of IL-8 in cell culture supernatants were determined by sandwich ELISA: Human IL-8/CXCL8 ELISA Kit (RAB0319, Sigma) according to the manufacturer's instruction.

2.14. Statistical Analysis

All cell culture experiments were performed in two biological replicates and three technical replicates per experiment. The results are expressed as means ± SD. The values were compared using two-way ANOVA with Sidak post-hoc test for inflammatory markers and with one-way ANOVA followed by Dunnett's post-hoc test for viability assay, and the differences were considered significant where $p < 0.05$.

3. Results

3.1. Multiplex ELISA and Immunoblotting of Enzyme Fermentation Products

A multiplex ELISA approach was utilized to measure the amount of fermented gluten peptides in the concentrated form of supplemental enzymes in a similar manner to previous publications [32]. All tested supplemental enzyme products reported a gluten concentration that was equivalent to or below 20 ppm (20 mg/L) as compared with an intact wheat reference standard—except for the bacterial protease (Table 3), which we previously showed can interfere with competitive ELISA assays [40]. The wheat input, fermentation broth and final concentrated enzyme products showed a pronounced reduction throughout the fermentation process when tested with the various antibodies from the commercially available gluten detection kits. The composite standard curves (from 0.04 ppm to 10 ppm) can be found in Supplementary Figure S2.

In order to determine if there were any false positives and to create additional data sets for the quantitation of gluten in supplemental enzyme products, a multiplex immunoblotting approach was used (Table 4). Intact gluten standards were again used for calibration, and the banding pattern was used to determine the maximum size in which peaks should be included for quantitation, as some off-target and non-specific banding did occur at higher molecular weights. Similar to the ELISA method, the immunoblotting clearly demonstrates that the concentration of gluten-derived peptides in these supplemental enzyme products is negligible or below 20 ppm, as determined by the pepsin-trypsin digested sample. The individual immunoblots are shown in Supplementary Figures S3 and S4.

3.2. Effects of Gluten, Gliadin and Digested Gluten Standard on TNF-α and IL-8 Production in THP-1 and Caco-2 Cells

Prior to determining the effects of samples on inflammatory response, the viability of cells after exposure to crude gliadin, crude gluten or PT-digested gluten at 20 ppm (20 mg/L) and 500 ppm (500 mg/L) was assessed using the XTT method (Supplementary Figure S1). It was determined that cells from all groups were viable compared to media controls in Caco-2 ($F_{(6, 47)} = 1.742$, $p = 0.1322$) and THP-1 ($F_{(6, 35)} = 1.574$, $p = 0.1839$) cell lines.

Table 3. Gluten levels for the supplemental enzyme samples were negligible or lower than the PT-digested gluten sample (N = 6). The multiplex ELISAs using antibodies from commercially available kits specific for gluten detection were used to detect gluten levels in supplemental enzyme preparations. ND = non-detectable. NA = not available, as only one replicate was above the limit of detection.

Fermentation Product		Commercial ELISA Kit								
		Gtox-G12	A-G12	R5Sand	R5Comp	V10-R5	V80-GL	MI-GL	AllSK	2D4
						Antibody				
		G12	G12	R5	R5	R5	USDA	MIoBS	Skerritt	2D4
PT-Glu	Pepsin-trypsin hydrolysis	77.5 ± NA	33.5 ± NA	1.5 ± NA	1.0 ± NA	1.0 ± NA	2.5 ± NA	22.0 ± NA	28.0 ± NA	32 ± NA
Alpha-G	α-Galactosidase	ND	0.0 ± NA	0.36 ± 0.1	3.2 ± 3.0	ND	ND	5.3 ± 0.8	ND	1.63 ± 1.43
ALA	Acid Lactase	ND	0.5 ± NA	ND	1.7 ± 1.7	2.0 ± 0.2	ND	2.2 ± 1.2	ND	0.1 ± NA
ASP	Acid Stable Protease	ND	0.4 ± NA	0.4 ± 0.1	2.9 ± 0.5	4.3 ± 0.2	2.4 ± NA	1.6 ± 0.8	ND	3.8 ± 3.5
FL	Fungal Lactase	ND	0.4 ± NA	0.0 ± 0.0	2.3 ± 0.1	4.9 ± 0.2	ND	3.4 ± 1.8	1.42 ± NA	3.4 ± 3.3
FLC	Fungal Lactase Conc.	ND	0.8 ± NA	0.4 ± 0.4	1.1 ± 0.6	8.0 ± 3.3	0.2 ± NA	6.7 ± 0.8	ND	2.2 ± 2.1
PI	Peptidase I	ND	1.8 ± 0.1	0.8 ± 0.6	3.1 ± 0.5	0.4 ± 0.3	0.6 ± NA	2.9 ± 2.4	ND	3.4 ± 2.9
PIC	Peptidase I Conc.	ND	1.4 ± 0.6	0.3 ± 0.1	1.4 ± 0.2	0.7 ± NA	2.4 ± 1.3	1.4 ± 2.5	ND	2.4 ± 1.5
PII	Peptidase II	ND	0.8 ± NA	0.2 ± 0.1	2.3 ± 0.1	6.5 ± 4.6	ND	2.8 ± 0.1	ND	4.5 ± 4.1
PIIC	Peptidase II Conc.	ND	0.8 ± 0.2	3.9 ± 3.1	4.0 ± 0.1	1.4 ± 0.3	0.2 ± NA	3.7 ± 2.8	ND	3.5 ± 2.56
Alpha-G-DE	α-Galactosidase II	ND	0.8 ± NA	0.5 ± NA	2.3 ± 0.1	1.2 ± NA	0.1 ± NA	3.0 ± 1.4	ND	4.3 ± 3.4
FPA	Fungal Protease A	ND	0.8 ± 0.2	1.0 ± 0.7	6.8 ± 8.0	4.6 ± 4.8	6.7 ± 2.0	3.2 ± 1.0	1.0 ± NA	4.3 ± 4.0
BP	Bacterial Protease	137 ± 4.0	85.5 ± 42.8	9.5 ± 4.1	3.1 ± 1.6	18.3 ± 1.0	14.9 ± 2.1	4.9 ± 2.1	1.8 ± 0.0	7.7 ± 2.9
LipAN	Lipase AN	ND	0.1 ± 0.1	4.1 ± 3.4	2.9 ± 0.2	3.2 ± 0.9	1.7 ± 1.7	3.0 ± 2.4	ND	4.8 ± 3.3
CellAN	Cellulase AN	ND	0.5 ± 0.4	0.4 ± 0.4	2.5 ± 0.3	1.2 ± 1.0	1.6 ± NA	2.4 ± 2.1	ND	3.7 ± 3.7

Table 4. Gluten levels for the supplemental enzyme samples were negligible or lower than the PT-digested gluten standard (PT-Glu) for the majority of antibodies, as determined by immunoblotting (N = 6). ND = non-detectable.

Fermentation Product		Commercial ELISA Kit								
		Gtox-G12	A-G12	R5Sand	R5Comp	V10-R5	V80-GL	MI-GL	AllSK	2D4
						Antibody				
		G12	G12	R5	R5	R5	USDA	MIoBS	Skerritt	2D4
PT-Glu	Pepsin-trypsin hydrolysis	ND	ND	ND	ND	36.5	ND	ND	ND	1.5
Alpha-G	α-Galactosidase	ND	ND	ND	ND	ND	ND	ND	ND	677.08
ALA	Acid Lactase	ND	ND	ND	ND	ND	ND	ND	ND	ND
ASP	Acid Stable Protease	ND	ND	ND	ND	26.65	ND	ND	ND	ND
FL	Fungal Lactase	ND	ND	ND	95.86	ND	ND	ND	ND	ND
FLC	Fungal Lactase Conc.	ND	ND	ND	ND	ND	ND	ND	ND	ND
PI	Peptidase I	ND	ND	ND	ND	ND	ND	ND	606.66	ND
PIC	Peptidase I Conc.	ND	ND	ND	ND	ND	ND	ND	ND	ND
PII	Peptidase II	ND	ND	ND	ND	ND	ND	ND	ND	519.23
PIIC	Peptidase II Conc.	ND	ND	ND	ND	ND	ND	ND	ND	ND
Alpha-G-DE	α-Galactosidase II	ND	ND	0.72	ND	ND	ND	ND	ND	ND
FPA	Fungal Protease A	ND	ND	ND	ND	ND	ND	ND	ND	ND
BP	Bacterial Protease	ND	ND	ND	ND	ND	ND	ND	ND	ND
LipAN	Lipase AN	ND	ND	ND	ND	29.53	ND	ND	ND	ND
CellAN	Cellulase AN	ND	ND	ND	ND	ND	ND	ND	ND	ND

To compare the pro-inflammatory potential of crude gliadin, crude gluten and pepsin-trypsin digested gluten standard, we assessed their effect on IL-8 and TNF-α production in THP-1 and Caco-2 cell lines (Figure 1). Stimulation with 20 ppm (20 mg/L) of the whole molecules and the digested gluten standard had no effect on the IL-8 mRNA level in either THP-1 or Caco-2 cells (Figure 1A,B). Challenging these mammalian cells with 500 ppm (500 mg/L) increased IL-8 gene expression in gluten and PT-digested gluten groups in Caco-2 and THP-1 cells and in the crude gliadin group in THP-1 cells. This increase in IL-8 mRNA expression was exacerbated in both Caco-2 and THP-1 cells in the PT-digested gluten group. [Dose $F_{(1, 66)}$ = 33.27, $p \leq 0.0001$; Treatment $F_{(2, 66)}$ = 3.322, p = 0.0422; Interaction $F_{(2, 66)}$ = 2.830, p = 0.0662 for Caco-2] [Dose $F_{(1, 66)}$ = 132.2, $p \leq 0.0001$; Treatment $F_{(2, 66)}$ = 42.62, $p \leq 0.0001$; Interaction $F_{(2, 66)}$ = 16.53, $p \leq 0.0001$ for THP-1].

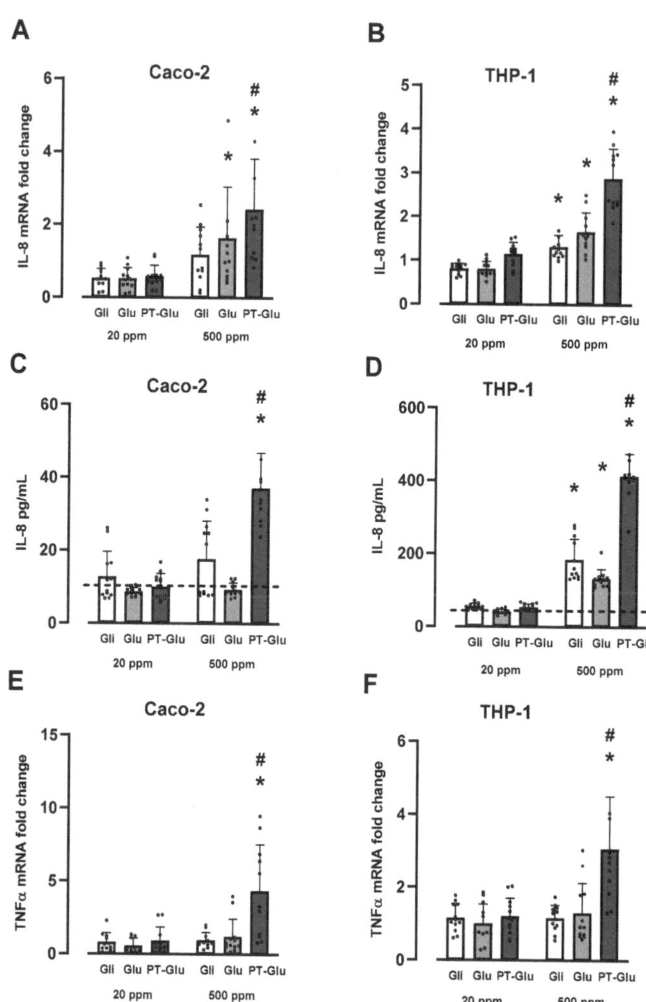

Figure 1. Differential effects of 20 ppm (20 mg/L) and 500 ppm (500 mg/L) of crude gliadin (Gli), crude gluten (Glu) or PT digested gluten (PT-Glu) on IL-8 gene expression (**A,B**), IL-8 protein secretion (**C,D**) and TNF-α gene expression (**E,F**) in Caco-2 and THP-1 cells, respectively. The dashed line represents effects of control media only for IL-8 protein expression. * represents significant difference as compared with the 20 ppm comparator group, $p < 0.05$. # represents significant difference as compared with other 500 ppm treatment groups, $p < 0.05$.

Similarly, 20 ppm of crude gliadin, crude gluten and PT-digested gluten standard did not significantly increase IL-8 protein levels in cell culture supernatants (Figure 1C,D) as compared with the control in Caco-2 ($F_{(3, 42)}$ = 1.753, p = 0.171) or THP-1 cells ($F_{(3, 44)}$ = 3.56, p = 0.0715). In Caco-2 cells, there was no significant increase in IL-8 secretion in response to 500 ppm of crude gliadin or crude gluten; however, the increase in protein production was still apparent in the PT-digested gluten group. In THP-1 cells, 500 ppm increased IL-8 levels in in all treatment groups. The IL-8 increase was most prominent in the PT-digested gluten group. [Dose $F_{(1, 66)}$ = 47.58, $p \leq 0.0001$; Treatment $F_{(2, 66)}$ = 28.77, $p \leq 0.0001$; Interaction $F_{(2, 66)}$ = 27.06, $p \leq 0.0001$ for Caco-2] and [Dose $F_{(1, 66)}$ = 511.3, $p \leq 0.0001$; Treatment $F_{(2, 66)}$ = 106.4, $p \leq 0.0001$; Interaction $F_{(2, 66)}$ = 95.27, $p \leq 0.0001$ for THP-1].

There was no significant difference between the PT-digested gluten and crude gliadin or gluten in TNF-α gene expression in either Caco-2 or THP-1 cells at 20 ppm concentration (Figure 1E,F). At 500 ppm, there was no significant increase in TNF-α gene expression in Caco-2 or THP-1 cells in response to either crude gliadin or crude gluten. However, PT-digested gluten significantly increased TNF-α gene expression in both Caco-2 and THP-1 cells. [Dose $F_{(1, 59)}$ = 13.59, p = 0.0005; Treatment $F_{(2, 59)}$ = 9.480, p = 0.0003; Interaction $F_{(2, 59)}$ = 7.340, p = 0.0014 for Caco-2] and [Dose $F_{(1, 66)}$ = 15.95, p = 0.0002; Treatment $F_{(2, 66)}$ = 13.00, $p \leq 0.0001$; Interaction $F_{(2, 66)}$ = 10.03, p = 0.0002 for THP-1].

3.3. Effects of Gluten, Gliadin, Digested Gluten Standard and Supplemental Enzymes on IL-8 Production in THP-1 Cells

Given that a stronger effect was observed in THP-1 cells, in IL-8 response, we further examined the effect of commercially available enzyme supplements on immune response with a focus on IL-8 protein expression in this cell line (Figure 2). At 500 ppm (500 mg/L), the crude gliadin, gluten and the pepsin-trypsin digested gluten evoked a significant increase in IL-8 response in these THP-1 cells ($F_{(16, 85)}$ = 4.541, p < 0.0001), as determined by one-way ANOVA followed by Dunnet's post-hoc test. There was no significant increase in IL-8 response with the administration of the commercially available enzyme supplements.

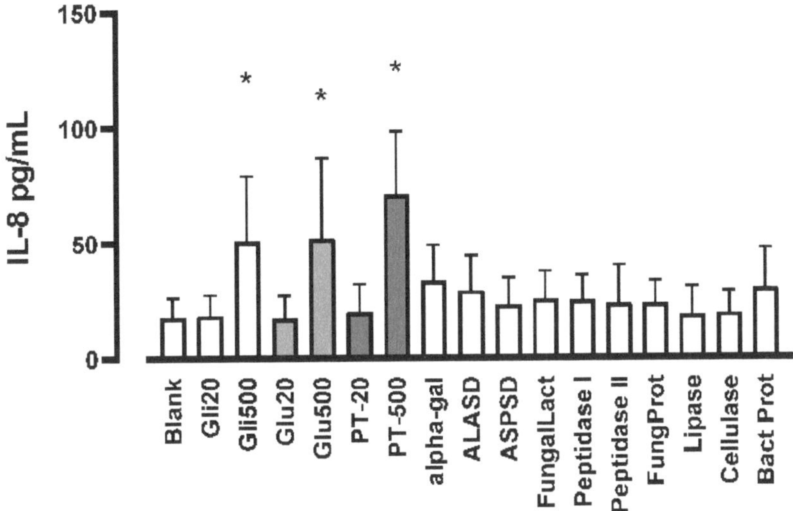

Figure 2. 500 ppm of crude gliadin, gluten and pepsin-trypsin digested gluten evoked an increase in IL-8 protein expression in THP-1 cells. There was no significant effect of commercially available enzyme supplements on IL-8 levels. * represents significant difference as compared with the 20 ppm comparator group, p < 0.05.

3.4. Effects of Gluten, Gliadin and Digested Gluten Standard on IL-8 Protein Production in Human Peripheral Blood Mononuclear Cells (PBMCs)

PBMCs from human healthy and coeliac individuals were assessed for their response to crude gluten, gliadin and pepsin-trypsin digested gluten at 20 ppm (20 mg/L) and 500 ppm (500 mg/L) (Figure 3). PBMCs were chosen, as they include lymphocytes (i.e., T cells, B cells and NK cells), monocytes and dendritic cells and so represent a collective immune response to sensitizing agents. In both healthy ($F(1, 30) = 515.5$, $p < 0.0001$) and coeliac ($F(1, 30) = 455.3$, $p < 0.0001$) PBMCs, 500 ppm of the respective treatments significantly increased IL-8 levels. In both the PBMCs from healthy individuals ($F(2, 30) = 11.75$, $p = 0.0002$) and from coeliac individuals ($F(2, 30) = 84.22$, $p < 0.0001$), the pepsin-trypsin digestion significantly decreased the inflammatory response as compared with the crude gluten or gliadin, and in healthy PBMCs, the response to crude gluten was lower than to crude gliadin.

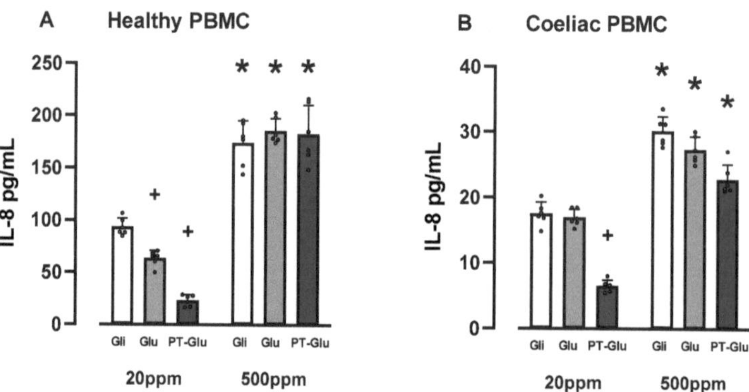

Figure 3. 500 ppm of crude gliadin, gluten and pepsin-trypsin digested gluten evoked an increase in IL-8 protein expression in PBMCs from both healthy and coeliac donors. Digestion by pepsin and trypsin significantly decreased IL-8 protein levels in both healthy and coeliac PBMCs. * represents significant difference as compared with the 20 ppm comparator group, $p < 0.05$; + represents significant difference as compared with all other groups, $p < 0.05$.

4. Discussion

Herein, we clearly demonstrate that the tested commercially available supplemental enzymes that are derived from wheat digestion do not have detectable levels of gliadin-, deamidated gliadin- and glutenin-derived epitopes from gluten proteins or peptides. Although intact gluten may be absent in the enzyme samples, the presence of small or large gluten-derived peptides equivalent to more than 20 ppm intact gluten that are not identifiable with these various epitopes may still be present in these samples. In these studies, we utilized wheat gluten, the same source as for the commercially available enzyme supplements, and while intact gluten may not be the perfect calibrant for competitive ELISAs for fermented goods, it demonstrates that the epitopes are in the proper range for analysis. We further demonstrate that neither the pepsin-trypsin digested gluten standard nor the supplemental enzymes elicit an inflammatory response in immortalized immune and intestinal human cell lines, as well as in peripheral blood mononuclear cells (PBMCs) from coeliac individuals at the threshold concentration of 20 ppm, as compared with both crude gluten and gliadin. This complementary approach provides strong rationale for the idea that the fermentation process effectively removes gluten and immunogenic peptides and lends credence to the acceptance of a gluten-free declaration specifically for hydrolyzed gluten products, including supplemental enzymes, derived from fermentation processes.

Historically, several qualitative and quantitative analytical methods are available for the detection of gluten in foods and fermented products including ELISAs, Western blotting, Mass spectrometry, DNA-based methodology and aptamer-based assays—all of which have their limitations [41]. Sandwich and competitive ELISAs are the most used methodologies with monoclonal antibodies such as Skerritt, USDA, 2D4, R5 and G12, as well as polyclonal antibodies such as MloBS. The polycloncal antibody (MloBS) yielded the highest results on average for the commercial products (excluding the bacterial protease false-positive) and thus may be the most conservative estimation for residual gluten concentration in fermented dietary enzymes for use in safety determination. By combining these methodologies with physiologically relevant bioassays, a more comprehensive assessment of gluten detection and potential immunopathogenesis was determined.

Cytokines are important mediators of inflammation and contribute to the pathogenesis of many inflammatory diseases associated with the gut. Numerous studies have demonstrated a role for IL-8 and TNF-α as markers for innate immune activation by gluten products in vitro [42–51] and in human studies [52–62]. The ELISA and immunoblotting methodologies tested in this study using gluten-specific epitopes suggest that the supplemental enzymes and the PT digested gluten had less than 20 ppm (20 mg/L) gluten (Tables 2 and 3). To compliment these findings, the digested gluten standard as a reference standard for the levels of gluten that would be in the supplemental enzyme groups was tested for the potential to elicit inflammatory response, using IL-8 and TNFα cytokines in Caco-2 intestinal cells and THP-1 human monocytes. We further assessed the response to commercially available supplemental enzymes to assess their effect on immune response. These samples were compared to crude gliadin or crude gluten preparations at the FDA denominated gluten-free concentration of 20 ppm and at a higher concentration of 500 ppm to validate the assays. The co-culturing of gliadin, gluten and PT-digested gluten did not affect cell viability in our cell lines (Supplementary Figure S1). Furthermore, the results from cell culture assays demonstrate that at 20 ppm concentration, the PT-digested gluten and the commercially available supplemental enzymes did not trigger any inflammatory response of IL-8 or TNF-α compared to the media itself or to crude gluten and crude gliadin (Figures 1 and 2). Therefore, the inflammatory response elicited by the detectable and non-detectable peptides generated in the PT-digested gluten standard is no different to crude gliadin or gluten molecules at a concentration of 20 ppm.

At the higher dose of 500 ppm, PT-digested gluten increased the mRNA levels for TNF-α and IL-8 as compared to the crude gliadin and gluten (Figures 1 and 2). This is in line with previous evidence in THP-1 cells where digested gliadin increased IL-8 and TNFα as compared to crude gliadin alone at 200 ppm [42]. In contrast, other studies in the Caco-2 cell line showed that PT-digested gluten did not exacerbate the effects of 1000 ppm crude gluten [63]. These observations suggest that there may be a threshold concentration of crude gliadin or gluten past which the effects of peptides from their hydrolyzation do not elicit further response. However, further dose response studies using crude and digested gliadin and gluten in these cell lines are required to confirm this observation. Regardless of this information, this PT-digested gluten and the supplemental enzymes elicit a similar negligible inflammatory response as 20 ppm crude gluten and crude gliadin.

PBMCs encompass a heterogeneous cell population with varying relative amounts of lymphocytes (T cells, B cells and NK cells), dendritic cells and monocytes. These cells are critical components of the innate and adaptive immune system that defends the body against infection and plays a role in sensitization to certain epitopes. Using human PBMCs we demonstrate that the hydrolyzation of gluten with pepsin-trypsin digest decreases the IL-8 protein response as compared with crude gliadin or gluten. During the fermentation process used in the synthesis of the commercial enzymes, the wheat undergoes a more extensive hydrolyzation process than that experienced by pepsin-trypsin digestion alone. Combining the data from the immunoblots and multiplex with the cell-based assays, and given that supplemental enzymes and fermentation by-products will be even further hydrolyzed, we did not further analyze the commercial enzymes in this assay. However, given that the fermentation process further hydrolyzes these commercially available enzymes and these 'neat' enzymes are further diluted before being commercially distributed, we can be confident that these compounds will have gluten content below 20 ppm or that any gluten-derived peptides not detectable by these antibody epitopes will not elicit an inflammatory response and thus can maintain their claim of being 'gluten-free'.

5. Conclusions

This study demonstrates that, by using these specific antibody epitopes, there is little or no gluten products in the supplemental enzymes following fermentation. Furthermore, these samples that may contain non-detected gluten-derived peptides do not elicit an inflammatory response in immortalized immune and intestinal human cell lines, as well as in peripheral blood mononuclear cells (PBMCs) from coeliac individuals at the threshold dose of 20 ppm. This complementary approach using multiplex ELISA, immunoblotting and cell-based assays provides strong rationale for the acceptance of supplemental enzymes in a gluten-free declaration. This is of clinical relevance, as many commercially available enzymes that are marketed for the alleviation of gastric discomfort and the alleviation of food sensitivities do so through the degradation of proteins and peptides (including gluten) to reduce sensitivity. Future work investigating the detection of gluten-derived peptides should consider appropriate reference calibrants for the more accurate detection and quantification of peptides in combination with cell-based assays to assess the potential to modulate the inflammatory tone of non-detectable products of starting gluten source hydrolysis.

Supplementary Materials: The following supporting information can be downloaded at: https://www.mdpi.com/article/10.3390/fermentation8050203/s1. Figure S1. Viability of Caco-2 and THP-1 cells. Figure S2. Comparison of Standard Curves of the ELISA Antibodies. Figure S3. Immunoblots of Supplemental Enzymes. Figure S4. Immunoblots of PT-digested products: Immunoblots from Wes using antibodies.

Author Contributions: Conceptualization, J.C., J.D. and C.P.; methodology, E.K., P.K., S.B., J.C., A.M.C., N.J. and C.P.; software, P.K. and C.P.; validation, E.K., P.K., J.C., A.M.C., N.J. and C.P.; formal analysis, E.K., P.K., J.C., K.R., A.M.C., N.J. and C.P.; investigation, E.K., P.K., J.C., A.M.C., N.J. and C.P.; data curation, E.K., P.K., J.C., A.M.C., N.J. and C.P.; writing—original draft preparation, K.R., E.K., P.K., J.C., C.P. and J.D.; writing—review and editing, E.K., P.K., J.C., A.M.C., N.J. and C.P.; supervision, K.R., C.P. and J.D.; project administration, J.D. All authors have read and agreed to the published version of the manuscript.

Funding: This research was funded by Deerland Probiotic and Enzymes, with the cell work preparation being conducted by Munster Technical University.

Conflicts of Interest: Deerland Probiotic and Enzymes were involved in the funding and design of the study; in the collection, analyses, and interpretation of the data; in the writing of the manuscript; and in the decision to publish the results.

Appendix A

Bio-Rad. Protocol: Competitive ELISA: Competitive ELISA Protocol | Bio-Rad (bio-rad-antibodies.com, accessed on 6 September 2021).

ProteinSimple. Simple Western Size Assay Development Guide, Revision 1. (2014); Wes_manual.book (proteinsimple.com, accessed on 17 August 2021).

ProteinSimple, Compass for Simple Western User Guide Revision D. (2021); Compass.book (proteinsimple.com, accessed on 1 September 2021).

AAT Bioquest, Inc. (18 November 2021). Quest Graph™ Four Parameter Logistic (4PL) Curve Calculator. Retrieved from https://www.aatbio.com/tools/four-parameter-logistic-4pl-curve-regression-online-calculator, accessed on 26 April 2022.

Thermo scientific. Acetone precipitation of proteins: Microsoft Word—TR0049.1.doc (thermofisher.com, accessed on 4 June 2021).

Kolde R. (2013). pheatmap: Pretty Heatmaps. R package version 0.7.7. http://CRAN.R-project.org/package=pheatmap, accessed on 1 September 2021.

R Core Team. (2015). R: A Language and Environment for Statistical Computing. 620 R Foundation for Statistical Computing, Vienna. http://www.R-project.org/, accessed on 11 August 2021.

References

1. Gao, H.; Jorgensen, R.; Raghunath, R.; Nagisetty, S.; Ng, P.K.W.; Gangur, V. Creating hypo-/nonallergenic wheat products using processing methods: Fact or fiction? *Compr. Rev. Food Sci. Food Saf.* **2021**, *20*, 6089–6115. [CrossRef] [PubMed]
2. Vu, V.; Farkas, C.; Riyad, O.; Bujna, E.; Kilin, A.; Sipiczki, G.; Sharma, M.; Usmani, Z.; Gupta, V.K.; Nguyen, Q.D. Enhancement of the enzymatic hydrolysis efficiency of wheat bran using the Bacillus strains and their consortium. *Bioresour. Technol.* **2021**, *343*, 126092. [CrossRef] [PubMed]
3. Dubos, R.J. The Adaptive Production of Enzymes by Bacteria. *Bacteriol. Rev.* **1940**, *4*, 1–16. [CrossRef]
4. Cezairliyan, B.; Ausubel, F.M. Investment in secreted enzymes during nutrient-limited growth is utility dependent. *Proc. Natl. Acad. Sci. USA* **2017**, *114*, E7796–E7802. [CrossRef] [PubMed]
5. Di Stefano, M.; Miceli, E.; Gotti, S.; Missanelli, A.; Mazzocchi, S.; Corazza, G.R. The effect of oral alpha-galactosidase on intestinal gas production and gas-related symptoms. *Dig. Dis. Sci.* **2007**, *52*, 78–83. [CrossRef] [PubMed]
6. Oben, J.; Kothari, S.C.; Anderson, M.L. An open label study to determine the effects of an oral proteolytic enzyme system on whey protein concentrate metabolism in healthy males. *J. Int. Soc. Sports Nutr.* **2008**, *5*, 10. [CrossRef]
7. Ianiro, G.; Pecere, S.; Giorgio, V.; Gasbarrini, A.; Cammarota, G. Digestive Enzyme Supplementation in Gastrointestinal Diseases. *Curr. Drug Metab.* **2016**, *17*, 187–193. [CrossRef]
8. Swami, O.C.; Shah, N.J. Functional dyspepsia and the role of digestive enzymes supplement in its therapy. *Int. J. Basic Clin. Pharmacol.* **2017**, *6*, 1035–1041. [CrossRef]
9. Wang, Y.H.; Li, S.A.; Huang, C.H.; Su, H.H.; Chen, Y.H.; Chang, J.T.; Huang, S.S. Sirt1 Activation by Post-ischemic Treatment with Lumbrokinase Protects Against Myocardial Ischemia-Reperfusion Injury. *Front. Pharmacol.* **2018**, *9*, 636. [CrossRef]
10. Altaf, F.; Wu, S.; Kasim, V. Role of Fibrinolytic Enzymes in Anti-Thrombosis Therapy. *Front. Mol. Biosci.* **2021**, *8*, 680397. [CrossRef]
11. Townsend, J.R.; Morimune, J.E.; Jones, M.D.; Beuning, C.N.; Haase, A.A.; Boot, C.M.; Heffington, S.H.; Littlefield, L.A.; Henry, R.N.; Marshall, A.C.; et al. The Effect of ProHydrolase((R)) on the Amino Acid and Intramuscular Anabolic Signaling Response to Resistance Exercise in Trained Males. *Sports* **2020**, *8*, 13. [CrossRef] [PubMed]
12. Biller, J.A.; King, S.; Rosenthal, A.; Grand, R.J. Efficacy of lactase-treated milk for lactose-intolerant pediatric patients. *J. Pediatr.* **1987**, *111*, 91–94. [CrossRef]
13. DiPalma, J.A.; Collins, M.S. Enzyme replacement for lactose malabsorption using a beta-D-galactosidase. *J. Clin. Gastroenterol.* **1989**, *11*, 290–293. [CrossRef] [PubMed]
14. Rosado, J.L.; Solomons, N.W.; Lisker, R.; Bourges, H. Enzyme replacement therapy for primary adult lactase deficiency. Effective reduction of lactose malabsorption and milk intolerance by direct addition of beta-galactosidase to milk at mealtime. *Gastroenterology* **1984**, *87*, 1072–1082. [CrossRef]
15. Wei, G.; Helmerhorst, E.J.; Darwish, G.; Blumenkranz, G.; Schuppan, D. Gluten Degrading Enzymes for Treatment of Celiac Disease. *Nutrients* **2020**, *12*, 2095. [CrossRef] [PubMed]
16. Deaton, J.; Cuentas, A.; Starnes, J. Tolerance and Efficacy of Glutalytic™: A Randomized, Double-Blind, Placebo Controlled Study. *J. Nutr. Food Sci.* **2018**, *8*, 1–7. [CrossRef]
17. Caio, G.; Volta, U.; Sapone, A.; Leffler, D.A.; De Giorgio, R.; Catassi, C.; Fasano, A. Celiac disease: A comprehensive current review. *BMC Med.* **2019**, *17*, 142. [CrossRef]
18. Rubio-Tapia, A.; Ludvigsson, J.F.; Brantner, T.L.; Murray, J.A.; Everhart, J.E. The prevalence of celiac disease in the United States. *Am. J. Gastroenterol.* **2012**, *107*, 1538–1544. [CrossRef]
19. Aziz, I.; Hadjivassiliou, M.; Sanders, D.S. The spectrum of noncoeliac gluten sensitivity. *Nat. Rev. Gastroenterol. Hepatol.* **2015**, *12*, 516–526. [CrossRef]
20. Wieser, H. Chemistry of gluten proteins. *Food Microbiol.* **2007**, *24*, 115–119. [CrossRef]
21. Clemente, M.G.; De Virgiliis, S.; Kang, J.S.; Macatagney, R.; Musu, M.P.; Di Pierro, M.R.; Drago, S.; Congia, M.; Fasano, A. Early effects of gliadin on enterocyte intracellular signalling involved in intestinal barrier function. *Gut* **2003**, *52*, 218–223. [CrossRef]
22. Abadie, V.; Jabri, B. IL-15: A central regulator of celiac disease immunopathology. *Immunol. Rev.* **2014**, *260*, 221–234. [CrossRef]
23. Leonard, M.M.; Sapone, A.; Catassi, C.; Fasano, A. Celiac Disease and Nonceliac Gluten Sensitivity: A Review. *JAMA* **2017**, *318*, 647–656. [CrossRef]
24. Catassi, C.; Fabiani, E.; Iacono, G.; D'Agate, C.; Francavilla, R.; Biagi, F.; Volta, U.; Accomando, S.; Picarelli, A.; De Vitis, I.; et al. A prospective, double-blind, placebo-controlled trial to establish a safe gluten threshold for patients with celiac disease. *Am. J. Clin. Nutr.* **2007**, *85*, 160–166. [CrossRef]
25. Halter, S.A.; Greene, H.L.; Helinek, G. Gluten-sensitive enteropathy: Sequence of villous regrowth as viewed by scanning electron microscopy. *Hum. Pathol.* **1982**, *13*, 811–818. [CrossRef]
26. Halter, S.A.; Greene, H.L.; Helinek, G. Scanning electron microscopy of small intestinal repair following treatment for gluten sensitive enteropathy. *Scan. Electron Microsc.* **1980**, *3*, 155–161.
27. Labeling, F.F. 21. CFR Part 101 [Docket No. FDA-2005-N-0404] RIN 0910-AG84. *Federal Regist.* **2013**, *78*, 47154–47179.

28. Montserrat, V.; Bruins, M.J.; Edens, L.; Koning, F. Influence of dietary components on Aspergillus niger prolyl endoprotease mediated gluten degradation. *Food Chem.* **2015**, *174*, 440–445. [CrossRef]

29. Scherf, K.A.; Wieser, H.; Koehler, P. Novel approaches for enzymatic gluten degradation to create high-quality gluten-free products. *Food Res. Int.* **2018**, *110*, 62–72. [CrossRef]

30. Panda, R.; Garber, E.A.E. Western blot analysis of fermented-hydrolyzed foods utilizing gluten-specific antibodies employed in a novel multiplex competitive ELISA. *Anal. Bioanal. Chem.* **2019**, *411*, 5159–5174. [CrossRef]

31. Rockendorf, N.; Meckelein, B.; Scherf, K.A.; Schalk, K.; Koehler, P.; Frey, A. Identification of novel antibody-reactive detection sites for comprehensive gluten monitoring. *PLoS ONE* **2017**, *12*, e0181566. [CrossRef]

32. Panda, R.; Boyer, M.; Garber, E.A.E. A multiplex competitive ELISA for the detection and characterization of gluten in fermented-hydrolyzed foods. *Anal. Bioanal. Chem.* **2017**, *409*, 6959–6973. [CrossRef]

33. Rivera Del Rio, A.; Keppler, J.K.; Boom, R.M.; Janssen, A.E.M. Protein acidification and hydrolysis by pepsin ensure efficient trypsin-catalyzed hydrolysis. *Food Funct.* **2021**, *12*, 4570–4581. [CrossRef]

34. Nelson, G.M.; Guynn, J.M.; Chorley, B.N. Procedure and Key Optimization Strategies for an Automated Capillary Electrophoretic-based Immunoassay Method. *J. Vis. Exp.* **2017**, *127*, e55911. [CrossRef]

35. Fu, B.X.; Sapirstein, H.D.; Bushuk, W. Salt-Induced Disaggregation/Solubilization of Gliadin and Glutenin Proteins in Water. *J. Cereal Sci.* **1996**, *24*, 241–246. [CrossRef]

36. Tsuge, M.; Hiraga, N.; Zhang, Y.; Yamashita, M.; Sato, O.; Oka, N.; Shiraishi, K.; Izaki, Y.; Makokha, G.N.; Uchida, T.; et al. Endoplasmic reticulum-mediated induction of interleukin-8 occurs by hepatitis B virus infection and contributes to suppression of interferon responsiveness in human hepatocytes. *Virology* **2018**, *525*, 48–61. [CrossRef]

37. Njock, M.S.; Cheng, H.S.; Dang, L.T.; Nazari-Jahantigh, M.; Lau, A.C.; Boudreau, E.; Roufaiel, M.; Cybulsky, M.I.; Schober, A.; Fish, J.E. Endothelial cells suppress monocyte activation through secretion of extracellular vesicles containing antiinflammatory microRNAs. *Blood* **2015**, *125*, 3202–3212. [CrossRef]

38. Steinberg-Shemer, O.; Keel, S.; Dgany, O.; Walsh, T.; Noy-Lotan, S.; Krasnov, T.; Yacobovich, J.; Quarello, P.; Ramenghi, U.; King, M.C.; et al. Diamond Blackfan Anemia: A Nonclassical Patient with Diagnosis Assisted by Genomic Analysis. *J. Pediatr. Hematol. Oncol.* **2016**, *38*, e260–e262. [CrossRef]

39. Chekhonin, V.P.; Lebedev, S.V.; Volkov, A.I.; Pavlov, K.A.; Ter-Arutyunyants, A.A.; Volgina, N.E.; Savchenko, E.A.; Grinenko, N.F.; Lazarenko, I.P. Activation of expression of brain-derived neurotrophic factor at the site of implantation of allogenic and xenogenic neural stem (progenitor) cells in rats with ischemic cortical stroke. *Bull. Exp. Biol. Med.* **2011**, *150*, 515–518. [CrossRef]

40. Lacorn, M.; Lindeke, S.; Siebeneicher, S.; Weiss, T. Commercial ELISA Measurement of Allergens and Gluten: What We Can Learn from Case Studies. *J. AOAC Int.* **2018**, *101*, 102–107. [CrossRef]

41. Panda, R.; Garber, E.A.E. Detection and Quantitation of Gluten in Fermented-Hydrolyzed Foods by Antibody-Based Methods: Challenges, Progress, and a Potential Path Forward. *Front. Nutr.* **2019**, *6*, 97. [CrossRef]

42. Jelinkova, L.; Tuckova, L.; Cinova, J.; Flegelova, Z.; Tlaskalova-Hogenova, H. Gliadin stimulates human monocytes to production of IL-8 and TNF-alpha through a mechanism involving NF-kappaB. *FEBS Lett.* **2004**, *571*, 81–85. [CrossRef]

43. Lammers, K.M.; Khandelwal, S.; Chaudhry, F.; Kryszak, D.; Puppa, E.L.; Casolaro, V.; Fasano, A. Identification of a novel immunomodulatory gliadin peptide that causes interleukin-8 release in a chemokine receptor CXCR3-dependent manner only in patients with coeliac disease. *Immunology* **2011**, *132*, 432–440. [CrossRef]

44. Kumar, S.; Prenner, S. Dyspepsia and Increased Levels of Liver Enzymes in a 24-Year-Old Man. *Gastroenterology* **2019**, *157*, 27–28. [CrossRef]

45. De Palma, G.; Nadal, I.; Collado, M.C.; Sanz, Y. Effects of a gluten-free diet on gut microbiota and immune function in healthy adult human subjects. *Br. J. Nutr.* **2009**, *102*, 1154–1160. [CrossRef]

46. Cinova, J.; Palova-Jelinkova, L.; Smythies, L.E.; Cerna, M.; Pecharova, B.; Dvorak, M.; Fruhauf, P.; Tlaskalova-Hogenova, H.; Smith, P.D.; Tuckova, L. Gliadin peptides activate blood monocytes from patients with celiac disease. *J. Clin. Immunol.* **2007**, *27*, 201–209. [CrossRef]

47. Capozzi, A.; Vincentini, O.; Gizzi, P.; Porzia, A.; Longo, A.; Felli, C.; Mattei, V.; Mainiero, F.; Silano, M.; Sorice, M.; et al. Modulatory Effect of Gliadin Peptide 10-mer on Epithelial Intestinal CACO-2 Cell Inflammatory Response. *PLoS ONE* **2013**, *8*, e66561. [CrossRef]

48. Palova-Jelinkova, L.; Rozkova, D.; Pecharova, B.; Bartova, J.; Sediva, A.; Tlaskalova-Hogenova, H.; Spisek, R.; Tuckova, L. Gliadin fragments induce phenotypic and functional maturation of human dendritic cells. *J. Immunol.* **2005**, *175*, 7038–7045. [CrossRef]

49. Grover, J.; Chhuneja, P.; Midha, V.; Ghia, J.E.; Deka, D.; Mukhopadhyay, C.S.; Sood, N.; Mahajan, R.; Singh, A.; Verma, R.; et al. Variable Immunogenic Potential of Wheat: Prospective for Selection of Innocuous Varieties for Celiac Disease Patients via in vitro Approach. *Front. Immunol.* **2019**, *10*, 84. [CrossRef]

50. Tuckova, L.; Novotna, J.; Novak, P.; Flegelova, Z.; Kveton, T.; Jelinkova, L.; Zidek, Z.; Man, P.; Tlaskalova-Hogenova, H. Activation of macrophages by gliadin fragments: Isolation and characterization of active peptide. *J. Leukoc. Biol.* **2002**, *71*, 625–631.

51. Vincentini, O.; Maialetti, F.; Gonnelli, E.; Silano, M. Gliadin-dependent cytokine production in a bidimensional cellular model of celiac intestinal mucosa. *Clin. Exp. Med.* **2015**, *15*, 447–454. [CrossRef]

52. Goel, G.; Tye-Din, J.A.; Qiao, S.W.; Russell, A.K.; Mayassi, T.; Ciszewski, C.; Sarna, V.K.; Wang, S.; Goldstein, K.E.; Dzuris, J.L.; et al. Cytokine release and gastrointestinal symptoms after gluten challenge in celiac disease. *Sci. Adv.* **2019**, *5*, eaaw7756. [CrossRef]

53. Goel, G.; Daveson, A.J.M.; Hooi, C.E.; Tye-Din, J.A.; Wang, S.; Szymczak, E.; Williams, L.J.; Dzuris, J.L.; Neff, K.M.; Truitt, K.E.; et al. Serum cytokines elevated during gluten-mediated cytokine release in coeliac disease. *Clin. Exp. Immunol.* **2020**, *199*, 68–78. [CrossRef]

54. Tye-Din, J.A.; Skodje, G.I.; Sarna, V.K.; Dzuris, J.L.; Russell, A.K.; Goel, G.; Wang, S.; Goldstein, K.E.; Williams, L.J.; Sollid, L.M.; et al. Cytokine release after gluten ingestion differentiates coeliac disease from self-reported gluten sensitivity. *United Eur. Gastroenterol. J.* **2020**, *8*, 108–118. [CrossRef]

55. Di Sabatino, A.; Giuffrida, P.; Fornasa, G.; Salvatore, C.; Vanoli, A.; Naviglio, S.; De Leo, L.; Pasini, A.; De Amici, M.; Alvisi, C.; et al. Innate and adaptive immunity in self-reported nonceliac gluten sensitivity versus celiac disease. *Dig. Liver Dis.* **2016**, *48*, 745–752. [CrossRef]

56. Di Sabatino, A.; Corazza, G.R. True Nonceliac Gluten Sensitivity in Real Patients. *Clin. Gastroenterol. Hepatol.* **2016**, *14*, 168–169. [CrossRef]

57. Brottveit, M.; Beitnes, A.C.; Tollefsen, S.; Bratlie, J.E.; Jahnsen, F.L.; Johansen, F.E.; Sollid, L.M.; Lundin, K.E. Mucosal cytokine response after short-term gluten challenge in celiac disease and non-celiac gluten sensitivity. *Am. J. Gastroenterol.* **2013**, *108*, 842–850. [CrossRef]

58. Heydari, F.; Rostami-Nejad, M.; Moheb-Alian, A.; Mollahoseini, M.H.; Rostami, K.; Pourhoseingholi, M.A.; Aghamohammadi, E.; Zali, M.R. Serum cytokines profile in treated celiac disease compared with non-celiac gluten sensitivity and control: A marker for differentiation. *J. Gastrointest. Liver Dis.* **2018**, *27*, 241–247. [CrossRef]

59. Manavalan, J.S.; Hernandez, L.; Shah, J.G.; Konikkara, J.; Naiyer, A.J.; Lee, A.R.; Ciaccio, E.; Minaya, M.T.; Green, P.H.; Bhagat, G. Serum cytokine elevations in celiac disease: Association with disease presentation. *Hum. Immunol.* **2010**, *71*, 50–57. [CrossRef]

60. Chowers, Y.; Marsh, M.N.; De Grandpre, L.; Nyberg, A.; Theofilopoulos, A.N.; Kagnoff, M.F. Increased proinflammatory cytokine gene expression in the colonic mucosa of coeliac disease patients in the early period after gluten challenge. *Clin. Exp. Immunol.* **1997**, *107*, 141–147. [CrossRef]

61. Street, M.E.; Volta, C.; Ziveri, M.A.; Zanacca, C.; Banchini, G.; Viani, I.; Rossi, M.; Virdis, R.; Bernasconi, S. Changes and relationships of IGFS and IGFBPS and cytokines in coeliac disease at diagnosis and on gluten-free diet. *Clin. Endocrinol.* **2008**, *68*, 22–28. [CrossRef]

62. Lahat, N.; Shapiro, S.; Karban, A.; Gerstein, R.; Kinarty, A.; Lerner, A. Cytokine profile in coeliac disease. *Scand J. Immunol.* **1999**, *49*, 441–446. [CrossRef]

63. Mohan Kumar, B.V.; Vijaykrishnaraj, M.; Kurrey, N.K.; Shinde, V.S.; Prabhasankar, P. Prolyl endopeptidase-degraded low immunoreactive wheat flour attenuates immune responses in Caco-2 intestinal cells and gluten-sensitized BALB/c mice. *Food Chem. Toxicol.* **2019**, *129*, 466–475. [CrossRef]

Article

Correlation Analysis of Microbiota and Volatile Flavor Compounds of Caishiji Soybean Paste

Jing Cai [1,†], Yueting Han [1,†], Wei Wu [1], Xuefeng Wu [1], Dongdong Mu [1], Shaotong Jiang [1] and Xingjiang Li [1,2,*]

[1] Key Laboratory for Agricultural Products Processing, School of Food and Biological Engineering, Hefei University of Technology, No.193 Tunxi Road, Hefei 230009, China; hfut-caijing@hotmail.com (J.C.); 2021111362@mail.hfut.edu.cn (Y.H.); wuwei123@mail.hfut.edu.cn (W.W.); wuxuefeng@hfut.edu.cn (X.W.); d.mu@hfut.edu.cn (D.M.); jiangshaotong1954@163.com (S.J.)
[2] Anhui Caishiji Foods Co., Ltd., Ma'anshan 243000, China
*   Correspondence: lixingjiang@hfut.edu.cn; Tel./Fax: +86-551-62901507
†   These authors contributed equally to this work.

Abstract: Microbial diversity plays a crucial part in the fermentation of Caishiji soybean paste (CSP). In the current study, the microbiota and volatile flavor compounds (VFCs) in CSP were identified through Illumina MiSeq sequencing and headspace gas chromatography–mass spectrometry. Five bacterial (*Bacillus, Tetragenococcus, Salinivibrio, Halomonas,* and *Staphylococcus*) and four fungal genera (*Aspergillus, Debaryomyces, Nigrospora,* and *Curvularia*) were revealed as dominant among the entire microbiome of CSP. More than 70 VFCs, including 8 acids, 15 esters, 8 alcohols, 14 aldehydes, 4 ketones, 5 phenols, and 20 miscellaneous VFCs were detected during the fermentation process. A total of 12 kinds of VFCs were identified in the odor activity value (OAV) analysis. The results of the correlation analysis between microbiota and VFCs indicated that *Bacillus, Tetragenococcus, Staphylococcus,* and *Aspergillus* were the main microbiota affecting the flavor of CSP. These results may serve as a reference for enhancing the quality of CSP.

Keywords: Caishiji soybean paste; volatile flavor compounds; Illumina MiSeq sequencing; correlations

Citation: Cai, J.; Han, Y.; Wu, W.; Wu, X.; Mu, D.; Jiang, S.; Li, X. Correlation Analysis of Microbiota and Volatile Flavor Compounds of Caishiji Soybean Paste. *Fermentation* **2022**, *8*, 196. https://doi.org/10.3390/ fermentation8050196

Academic Editors: Xian Zhang and Zhiming Rao

Received: 8 April 2022
Accepted: 23 April 2022
Published: 27 April 2022

Publisher's Note: MDPI stays neutral with regard to jurisdictional claims in published maps and institutional affiliations.

1. Introduction

Soybean paste [1,2], *sufu* [3–5], soy sauce [6–8], and *douchi* [9–11] are traditional bean products consumed in China and other East Asian countries. Among them, in China, soybean paste products are indispensable condiments with unique flavors and abundant nutriments [12–14]. Soybean paste can be produced using three fermentation methods: traditional spontaneous fermentation, low-salt, solid-state fermentation, and high-salt, diluted-state fermentation. Traditional spontaneous fermentation is the common fermentation method currently adopted by most enterprises. Caishiji soybean paste (CSP) from Ma'anshan, Anhui Province, has been famous for hundred years. High-quality soybeans, cinnamon, anise, licorice, sugar, salt, and other additives are used to make traditional fermented CSP. Volatile flavor compounds (VFCs) determine the unique flavors of soybean pastes. The nutrition and flavor in the soybean paste determine its quality [15]. Many studies have performed the flavor analysis in soybean pastes. Jo et al. [16] analyzed the VFCs of Korean *doenjang* and found that the VFCs in traditional *doenjang* were mostly acids, aldehydes, phenols, pyrazines, and furans, whereas the VFCs in commercial *doenjang* were mostly ethyl esters, maltol, and ethanol. Lin et al. [17] investigated the nonvolatile organic acids and amino acids in Pixian bean paste and concluded that citric acid, glutamic acid, methionine, and proline were the main flavor compounds in the paste. Zhao et al. [18] compared the VFCs of Chinese soybean pastes and identified that there were seven kinds of VFCs in the naturally fermented sample, and these were higher than that in the inoculated samples. Using a complicated biological chain and non-biological reactions, the macromolecular substances, such as protein, starch, and lipids, in soybean paste raw materials were decomposed to produce multifarious secondary metabolites and tiny molecules, including organic acids, amino acids, esters, alcohols, ketones, aldehydes, and acids.

The diversity of microorganisms plays a key role in soybean paste fermentation, especially in flavor formation. An analysis result of microbial proteins in *dajiang* revealed that the composition and variety of microbial proteins involved in the processes of protein synthesis, glycometabolism, and nucleic acid synthesis are different [19]. Hao et al. [20] studied the content of biogenic amines and the bacterial diversity in the naturally fermented farmhouse sauce as well as the interactions between biogenic amines and microorganisms during fermentation. The core microbiota, VFCs, and correlations between them were evaluated, and the analysis results may be used to improve the quality of *doubanjiang* [21]. Controlling the flavor is difficult due to the complexity of the microbial community throughout the fermenting procedure; thus, more attention should be paid to the correlation between the microbial community and VFCs [22].

There are few studies on the correlation between microorganisms and the flavor of soybean paste. Soybean paste types have distinct flavors, depending on the area in which they are produced and the technologies with which they are produced. In this work, in order to reveal the unique nutriment, flavors, and microbiota in CSP, the VFCs and microbiota in CSP at different fermentation stages were identified using headspace gas chromatography–mass spectrometry (HS-GC-MS) [23] and MiSeq sequencing [24], respectively. Additionally, the correlations between them were determined through correlation analysis.

2. Materials and Methods

2.1. Sample Preparation

The CSP samples were separated into six independent batches according to their fermentation stage: the beginning or the first, second, third, sixth, or eighth month of fermentation (CG0, CG1, CG2, CG3, CG6, and CG8, respectively). The samples were obtained from Anhui Ma'anshan Caishiji Foods, China. A multipoint sampling method was utilized (Figure S1), and the samples were kept at 4 °C for further analysis.

2.2. Identification of VFCs through HS-GC-MS

The fiber (50/30 μm DVB/CAR/PDMS) (Supelco, Inc., Bellefonte, PA, USA) was inserted into the headspace of a 20 mL glass vial containing a 2.0 g sample of CSP and 20 μL of 2,4,6-trimethylpyridine; the vial was sealed with a Teflon cover, and the sample was incubated at 70 °C for 60 min [25]. The GC-MS system (HP6890/5975C; Agilent, CA, USA) was used to analyze the VFCs according to previously reported methods with some modifications [26]. The initial oven temperature was 40 °C, which was retained for 5 min, raised by 5 °C/min to 60 °C, retained for 5 min, raised by 2 °C/min to 120 °C, retained for 5 min, raised by 10 °C/min to 250 °C, and retained for 3 min. The split ratio was 5:1. The temperature of the quadrupoles and transfer line was 150 °C and 250 °C, respectively. The energy of the electron hit was 70 eV, and the scanning range was 45–450 m/z.

2.3. Qualitative and Quantitative Analysis of VFCs

The retention index (RI) of C_7-C_{40} normal paraffin under the same GC-MS conditions was used to ascertain the RIs of all the VFCs, and they were compared with relevant reference values. Each RI was calculated as follows:

$$RI = 100n + 100 \times \frac{t_R - t_{Rn}}{t_{Rn+1} - t_{Rn}} \tag{1}$$

where n and $n + 1$ are the numbers of normal paraffin carbon atoms before and after the addition of the VFCs, respectively. t_{Rn} and t_{Rn+1} are n-alkane retention times, and t_R is the retention time of the unknown in the chromatography (min; $t_{Rn} < t_R < t_{Rn+1}$).

According to the internal standard content, the VFC content of each sample was computed as follows:

$$C = \frac{A_x \times C_0 \times V}{A_0 \times m \times 1000} \tag{2}$$

where C is the VFC content (μg/kg), A_x is the peak area of the VFCs (AU·min), C_0 is the mass concentration of the internal standard (0.996 μg/μL), A_0 is the peak area of the internal standard (AU·min), V is the injection amount of internal standard (μL), and m is the quality of the sample (g).

To evaluate the contribution of the VFCs to the flavor in CSP, the odor activity value (OAV) was computed as follows:

$$OAV = \frac{C}{T} \tag{3}$$

where C is the VFC content (μg/kg), and T is the VFC detection threshold (μg/kg).

2.4. DNA Extraction, PCR Amplification, and Illumina MiSeq Sequencing Analysis

DNA extraction, PCR amplification, and Illumina MiSeq sequencing experiments were conducted on the basis of the previous work [27] with minor modifications (Omega Bio-tek, Norcross, GA, USA). The V_4–V_5 regions of bacterial 16S rRNA genes were amplified using the universal primers, 338 F and 806 R (GeneAmp 9700, Applied Biosystems, Foster City, CA, USA).

For each reaction, the following were used: a 20 μL mixture containing 4 μL of 5 × FastPfu Buffer, 2 μL of 2.5-mM deoxynucleotide triphosphate, 0.8 μL of forward primer (5 μM), 0.8 μL of reverse primer (5 μM), 0.4 μL of FastPfu Polymerase, 0.2 μL of bovine serum albumin, 20 μL of ddH$_2$O, and 10 ng of template DNA. The response conditions were as follows: initial denaturation 95 °C for 3 min; for bacteria, 35 amplification cycles of 95 °C for 15 s; for fungus, 35 amplification cycles of 95 °C for 15 s; the last 55 °C for 30 s, 72 °C for 45 s, 72 °C for 10 min (Majorbio, Shanghai, China). The PCR products were extracted from 2% agarose gel, and amplicons were quantified using a QuantiFluor-ST fluorometer (Promega, Beijing, China). After amplification, the PCR products were sequenced on the Illumina MiSeq platform (Illumina, San Diego, CA, USA).

2.5. Data Analysis

The aligned sequences were clustered into operational taxonomic units (OTUs) at a 97% similarity threshold using the USERCH sequence analysis tool. The taxonomic classifications of the OTUs were studied by comparing the sequences with those in the SILVA and UNITE databases. The Sobs index and Circos data were calculated using Mothur software. R values were used to determine the Pearson correlation coefficients, and correlations of microbiota and VFCs were visualized by heat map. All the data are presented as the averages (±SD) of three repeated tests. The SPSS software package (version 22.0; SPSS, IBM, Armonk, NY, USA) and the Origin software package (version 7.0; OriginLab, Northampton, MA, USA) were used for all statistical analyses.

3. Results

3.1. VFC Changes during CSP Fermentation

More than 70 VFCs (including 8 acids, 15 esters, 8 alcohols, 14 aldehydes, 4 ketones, 5 phenols, and 20 miscellaneous VFCs) were detected during the CSP fermentation process (Table 1). As indicated in the Venn diagram in Figure 1a, 17 of the VFCs were common in CSP at all stages, and there were also their own unique VFCs in each stage samples (the quantity and composition of the VFCs are presented in Figures S1 and S2). Among the VFCs identified, esters, aldehydes, and alcohols were the main VFCs in CSP. All eight acids were detected in CG1, and there were only octanoic acid and nonanoic acid in CG6 and CG8. Esters, including amyl acetate, α-pentyl-γ-butyrolactone, ethyl palmitate, and diethyl phthalate, were the most abundant VFCs throughout the CSP fermentation process, whereas in the late stages of CSP fermentation, the relative abundance of the alcohols and aldehydes was the highest. Furthermore, the analyzed VFC results revealed that aldehydes of which isovaleraldehyde, 2-methylbutyraldehyde, benzaldehyde, and phenylacetaldehyde were common in all fermentation stages, and they obviously contributed to the flavor of the CSP. In summary, all the VFCs identified in the CSP fermentation process exhibited distinct qualitative and quantitative characteristics, which may lead to diverse flavors of CSP.

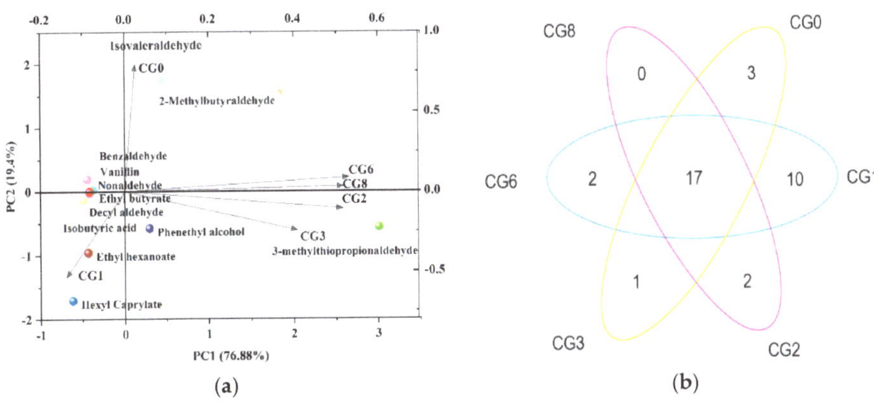

Figure 1. Venn (**a**) and PCA (**b**) analysis of characteristic VFCs of Caishiji soybean paste in different fermentation stages. The samples were coded according to Section 2.

To assess the contribution of characteristic VFCs to the overall flavor of CSP, the OAV analysis was conducted according to the odor threshold values of the VFCs. The results revealed that CSP contains 12 kinds of VFCs: one acid, three esters, one alcohol, and seven aldehydes (OAV ≥ 1, Table 2). Based on the OAV analysis results, the principal component analysis (PCA) was used to investigate the contribution rates of the characteristic VFCs to CSP flavor (Figure 1b). The two-dimension PCA model (F1 and F2) described 77.2% of the total variance, while the first and second principal components (F1 and F2) explained 57.8% and 19.4%, respectively. Isovaleraldehyde, 2-methylbutyraldehyde, and 3-methylthiopropionaldehyde greatly contributed to the flavor of CSP and were, therefore, included in the later correlation analysis.

3.2. Dynamics of Microbiota during CSP Fermentation

Illumina MiSeq sequencing was conducted to investigate the microbial diversity during the CSP fermentation. The aligned sequences were clustered into OTUs at a 97% similarity threshold. In total, 16 and 3 species of bacteria and fungi, respectively, were common to all stages during the CSP fermentation, and the species were most abundant in CG3 and CG6 (Figure 2 and Figure S3). Regarding the α-diversity indices employed in this study, the Chao index and the Shannon index were used to quantify microbial richness and diversity, respectively. The red mark (Figure 2) indicates significant similarities and differences among the microbiota present at different stages of CSP fermentation. The samples exhibited higher microbial diversity at later stages of fermentation: bacterial diversity increased sharply in the CG3, CG6, and CG8 phases, and fungi diversity increased sharply in the CG6 and CG8 phases; the increasing trends in the other stages were relatively gently (Figure 2). Consistent with the dynamic changes in microbiota at different stages of CSP fermentation, the diversity of the bacteria was higher than that of the fungi (Tables S1 and S2).

Table 1. Volatile flavor compounds of Caishiji soybean paste in different fermentation stages by GC-MS.

T_R (s)	Compounds	RI	RI*	Content (µg/kg)					
				CG-0	CG-1	CG-2	CG-3	CG-6	CG-8
	Acids (8)								
3.223	Acetic acid		-	ND	14.50 ± 1.83	ND	ND	ND	ND
10.340	Isobutyric acid		-	ND	2.78 ± 1.67	ND	ND	ND	ND
18.196	Isovaleric acid	839.57	841	ND	3.89 ± 0.82	ND	ND	ND	ND
28.196	Hexanoic acid	972.62	967	ND	4.81 ± 0.55	ND	ND	ND	ND
40.684	Benzoic acid	1153.64	-	ND	0.70 ± 0.06	ND	ND	ND	ND
40.901	Octanoic acid	1176.25	1167	ND	2.40 ± 0.30	ND	ND	1.68 ± 0.30	7.51 ± 0.78
43.442	Phenylacetic acid	1247.63	-	ND	0.31 ± 0.08	ND	ND	ND	ND
43.859	Nonanoic acid	1275.3	1280	ND	0.57 ± 0.16	ND	ND	0.82 ± 0.27	1.19 ± 0.15
	Esters (15)								
4.6465	Ethyl acetate	-	-	1.65 ± 2.00	ND	ND	7.24 ± 8.23	ND	ND
15.778	Ethyl butyrate	782.65	803	ND	ND	4.92 ± 1.61	ND	ND	ND
24.001	Amyl acetate	884.36	-	9.41 ± 1.58	0.86 ± 0.19	19.15 ± 8.75	103.82 ± 2.18	69.30 ± 33.72	22.37 ± 8.07
30.496	Ethyl hexanoate	984.83	1001	ND	6.97 ± 1.56	ND	7.45 ± 4.59	ND	ND
41.529	Ethyl benzoate	1161.25	1170	0.73 ± 0.16	ND	ND	ND	ND	ND
41.929	Butyl hexanoate	1375.85	-	ND	26.72 ± 13.98	ND	2.00 ± 0.13	0.99 ± 0.34	10.70 ± 3.64
42.096	Ethyl octoate	1203.45	-	ND	26.72 ± 13.98	ND	3.74 ± 0.50	0.99 ± 0.34	10.70 ± 3.64
46.328	α-pentyl-γ-butyrolactone	1284.99	-	1.65 ± 0.31	4.31 ± 0.58	0.67 ± 0.02	0.76 ± 0.13	2.76 ± 0.57	3.53 ± 0.43
46.535	Hexyl hexanoate	1373.11	1390	0.41 ± 0.11	7.12 ± 2.13	ND	5.77 ± 1.27	0.56 ± 0.19	4.66 ± 1.45
46.679	Ethyl caprate	-	-	ND	ND	ND	0.54 ± 0.13	ND	ND
47.174	Ethyl linoleate	1772.96	-	14.29 ± 0.67	14.15 ± 4.36	12.09 ± 0.99	ND	ND	ND
47.417	Ethyl tridecanoate	1930.57	-	7.60 ± 0.89	ND	ND	ND	ND	ND
48.672	Ethyl palmitate	1367.58	1924	45.12 ± 1.58	9.23 ± 6.27	33.39 ± 1.07	36.32 ± 5.21	41.30 ± 2.96	36.22 ± 3.62
49.614	Hexyl Caprylate	-	-	ND	2.78 ± 0.91	ND	0.92 ± 0.82	ND	ND
49.892	Diethyl phthalate	1647.27	-	10.42 ± 3.06	2.11 ± 0.78	1.07 ± 0.21	14.30 ± 0.71	24.23 ± 8.84	0.80 ± 0.26
	Alcohols (8)								
11.519	Isoamyl alcohol	693.31	734	ND	ND	ND	4.09 ± 1.37	1.32 ± 0.13	0.88 ± 0.31

Table 1. *Cont.*

T_R (s)	Compounds	RI	RI*	Content (µg/kg)					
				CG-0	CG-1	CG-2	CG-3	CG-6	CG-8
11.782	2-Methyl-1-butanol	695.63	-	ND	ND	ND	0.73 ± 0.24	0.49 ± 0.07	2.77 ± 0.25
13.552	1-Pentanol	765.98	820	ND	ND	ND	16.83 ± 5.69	0.97 ± 0.10	13.19 ± 2.75
14.703	2,3-Butanediol	746.31	-	6.91 ± 1.95	ND	ND	ND	32.20 ± 6.14	1.03 ± 0.36
29.550	1-nonen-3-ol	1054.76	-	ND	ND	ND	ND	8.01 ± 3.36	0.64 ± 0.12
38.819	Maltitol	1060.48	-	ND	ND	ND	ND	5.85 ± 0.44	ND
39.113	Phenethyl alcohol	1130.98	1121	ND	ND	ND	5.86 ± 2.09	0.95 ± 0.20	ND
46.676	1-Tetradecanol	1550.62	-	ND	10.5 ± 0.33	7.56 ± 2.34	4.94 ± 0.65	5.81 ± 0.49	6.50 ± 4.42
Aldehydes (14)									
7.102	Isovaleraldehyde	643.51	656	2.32 ± 0.11	4.13 ± 0.23	5.28 ± 0.23	15.48 ± 0.67	25.29 ± 1.77	1.22 ± 0.15
7.566	2-Methylbutyraldehyde	645.27	665	1.24 ± 0.05	2.31 ± 0.12	3.22 ± 1.17	12.46 ± 0.36	23.33 ± 1.58	16.29 ± 1.83
15.894	Hexanal	804.15	800	ND	ND	0.88 ± 0.10	0.70 ± 0.23	1.17 ± 0.40	3.26 ± 1.58
18.216	Furfural	-	-	ND	ND	4.30 ± 0.82	ND	ND	ND
23.841	3-methylthiopropionaldehyd-e	857.39	-	ND	ND	2.05 ± 0.57	2.24 ± 0.80	2.67 ± 0.22	1.76 ± 0.59
28.292	Benzaldehyde	975.41	961	7.84 ± 0.22	2.16 ± 1.04	38.45 ± 4.60	106.11 ± 40.00	15.04 ± 5.00	24.29 ± 3.67
34.439	Phenylacetaldehyde	1032.39	1045	10.79 ± 3.70	13.27 ± 0.38	12.18 ± 1.16	26.54 ± 9.60	18.02 ± 4.99	23.03 ± 1.08
36.693	Phenylglyoxal monohydrate	1217.09	-	ND	ND	ND	ND	4.87 ± 1.06	6.35 ± 1.88
38.419	Nonaldehyde	1101.97	1107	ND	ND	2.85 ± 0.31	4.64 ± 1.53	2.12 ± 0.23	9.87 ± 2.26
42.528	Decyl aldehyde	1179.84	1206	ND	ND	1.90 ± 0.24	9.30 ± 6.35	0.89 ± 0.19	3.19 ± 0.79
43.136	2,4-Dimethylbenzaldehyde	1208.59	-	ND	ND	1.18 ± 0.20	60.17 ± 15.07	1.09 ± 0.18	ND
47.087	Vanillin	1392.63	1418	1.34 ± 0.64	1.62 ± 0.54	0.81 ± 0.10	19.15 ± 0.96	ND	ND
48.035	5-methyl-2-phenyl-2-hexena-l	1505.88	-	ND	ND	ND	4.84 ± 1.57	12.00 ± 1.27	12.72 ± 4.26
50.017	FEMA 2763	1601.29	1611	ND	ND	ND	ND	0.74 ± 0.26	0.69 ± 0.23
Ketones (4)									
3.775	2-Butanone	-	-	ND	0.50 ± 1.34	ND	ND	ND	ND
23.154	(E)-hex-3-en-2-one	760.36	-	2.09 ± 0.22	ND	1.29 ± 0.42	1.00 ± 0.35	ND	ND

Table 1. *Cont.*

T_R (s)	Compounds	RI	RI*	Content (µg/kg)					
				CG-0	CG-1	CG-2	CG-3	CG-6	CG-8
33.996	1-(1-Ethyl-2,3-dimethyl-cyclopent-2-enyl)-ethanone	-	-	ND	ND	36.85 ± 12.85	51.16 ± 9.24	50.10 ± 16.57	29.37 ± 3.87
45.340	4'-Hydroxy-3'-methylacetophenone	1363.58	-	128.97 ± 2.77	44.44 ± 8.65	43.07 ± 5.97	26.40 ± 12.09	15.64 ± 5.77	6.88 ± 1.95
	Phenols (5)								
43.295	2-Naphthol	1454.96	-	ND	ND	ND	ND	0.90 ± 0.32	2.66 ± 0.83
44.479	4-Hydroxy-3-methoxystyren-e	1302.59	-	ND	413.18 ± 1.54	ND	ND	ND	36.27 ± 1.80
46.021	2,6-Dimethoxyphenol	1276.32	-	ND	1.57 ± 0.13	ND	ND	ND	15.94 ± 3.88
46.175	Eugenol	1380.17	1374	ND	0.39 ± 0.10	ND	ND	ND	ND
48.665	2,4-Di-tert-butylphenol	1555.65	1513	ND	1.28 ± 0.20	ND	ND	ND	ND
	Miscellaneous (20)								
4.042	2-Methylfuran	604.13	604	ND	ND	ND	ND	1.84 ± 0.62	ND
25.389	3,4-Lutidine	901.47	882	1.39 ± 0.48	1.93 ± 0.15	1.55 ± 0.51	0.76 ± 0.08	6.68 ± 0.12	3.77 ± 1.47
28.651	Mesitylene	1020.21	1013	195.72 ± 54.49	3.78 ± 0.98	146.68 ± 23.14	83.90 ± 7.92	231.64 ± 30.57	5.72 ± 0.77
29.892	cis-Anethol	-	-	6.53 ± 0.85	9.71 ± 1.18	8.28 ± 0.96	9.69 ± 1.03	4.87 ± 0.34	3.71 ± 0.29
30.161	1,2-Dimethyl-5-vinylpyrrole	945.18	-	422.06 ± 6.32	382.18 ± 14.58	465.86 ± 17.94	472.03 ± 52.02	367.79 ± 15.75	385.85 ± 15.45
30.535	2-propylpyridine	984.11	-	25.12 ± 2.41	ND	ND	ND	19.33 ± 0.31	ND
30.671	1,2,4-Trimethylenecyclohex-ane	921.39	-	ND	ND	575.54 ± 202.11	304.60 ± 64.54	908.03 ± 72.36	607.52 ± 55.99
31.002	4-Isopropylaniline	-	-	ND	ND	9.72 ± 3.23	ND	ND	ND
32.265	1-Isopentyl-1-cyclohexene	-	-	6.24 ± 2.58	ND	33.27 ± 5.30	ND	ND	ND
32.459	2-Isopropyltoluene	1040.66	1021	12.85 ± 4.52	ND	7.66 ± 2.51	164.24 ± 30.31	12.17 ± 3.44	59.24 ± 17.58
33.987	1-decylcyclohexene	1680.32	-	39.30 ± 12.45	7.86 ± 0.06	72.66 ± 24.57	42.54 ± 12.45	ND	5.34 ± 3.01
35.496	2-Acetyl pyrrole	1032.69	-	ND	ND	ND	ND	4.72 ± 0.52	99.31 ± 32.57
39.475	1,2,3,4-Tetramethylbenzene	1129.47	-	1.57 ± 0.10	ND	ND	8.02 ± 2.65	1.22 ± 0.31	2.73 ± 0.61
42.000	Cyclododecane	1422.4	-	ND	0.54 ± 0.12	ND	ND	0.97 ± 0.22	0.47 ± 0.16
42.361	Naphthalene	1178.73	1191	ND	0.72 ± 0.22	ND	ND	ND	ND

Table 1. *Cont.*

T_R (s)	Compounds	RI	RI*	Content (µg/kg)					
				CG-0	CG-1	CG-2	CG-3	CG-6	CG-8
42.846	2,3-Dihydrobenzofuran	1043.32	-	45.55 ± 0.98	96.94 ± 9.18	21.19 ± 0.51	18.09 ± 4.24	7.40 ± 2.38	4.80 ± 1.90
45.081	Indole	1294.51	1294	1.40 ± 0.19	2.06 ± 0.12	ND	ND	ND	ND
46.800	Tetradecane	-	-	6.62 ± 1.27	0.69 ± 0.09	ND	16.59 ± 4.93	ND	ND
47.667	Hexadecane	-	-	3.33 ± 1.06	ND	ND	ND	ND	ND
48.419	Pentadecane	1515.22	1500	5.42 ± 0.23	5.25 ± 0.11	2.93 ± 1.39	7.14 ± 4.88	5.86 ± 1.20	6.46 ± 0.22

Values represent means ± SD. RI is the retention index obtained in this study, RI* is the literature value; ND means the compound is not detected.

Table 2. Odor activity values (OAVs) of Caishiji soybean paste in different fermentation stages.

T_R (s)	Compounds	Threshold (µg/kg)	Content (µg/kg)					
			CG-0	CG-1	CG-2	CG-3	CG-6	CG-8
10.340	Isobutyric acid	1.00	-	2.78 ± 1.68	-	-	-	-
15.778	Ethyl butyrate	1.00	-	-	-	4.92 ± 1.61	-	-
30.496	Ethyl hexanoate	0.50	-	13.93 ± 3.12	-	14.90 ± 9.18	-	-
49.614	Hexyl Caprylate	0.10	-	27.77 ± 9.07	-	9.24 ± 0.82	-	-
39.113	Phenethyl alcohol	0.10	-	-	-	58.63 ± 20.88	9.49 ± 2.01	-
7.102	Isovaleraldehyde	2.00	1.30 ± 0.05	2.07 ± 0.12	2.64 ± 0.12	7.74 ± 0.34	12.65 ± 0.89	0.62 ± 0.07
7.566	2-Methylbutyraldehyde	1.00	1.24 ± 0.05	2.31 ± 0.12	3.22 ± 1.17	12.46 ± 0.37	23.33 ± 1.58	16.29 ± 1.83
23.841	3-methylthiopropionaldehyde	0.04	-	-	51.25 ± 14.25	56.00 ± 19.81	66.75 ± 5.50	43.67 ± 14.25
28.292	Benzaldehyde	50.00	0.16 ± 0.01	0.04 ± 0.02	0.77 ± 0.09	2.12 ± 0.80	0.40 ± 0.10	0.49 ± 0.08
38.419	Nonaldehyde	3.50	-	-	0.81 ± 0.09	1.02 ± 0.18	0.61 ± 0.07	2.82 ± 0.65
42.528	Decyl aldehyde	5.00	-	-	0.38 ± 0.05	2.53 ± 0.64	0.18 ± 0.04	0.64 ± 0.16
47.087	Vanillin	10.00	0.13 ± 0.07	0.16 ± 0.06	0.08 ± 0.01	2.25 ± 0.61	-	-

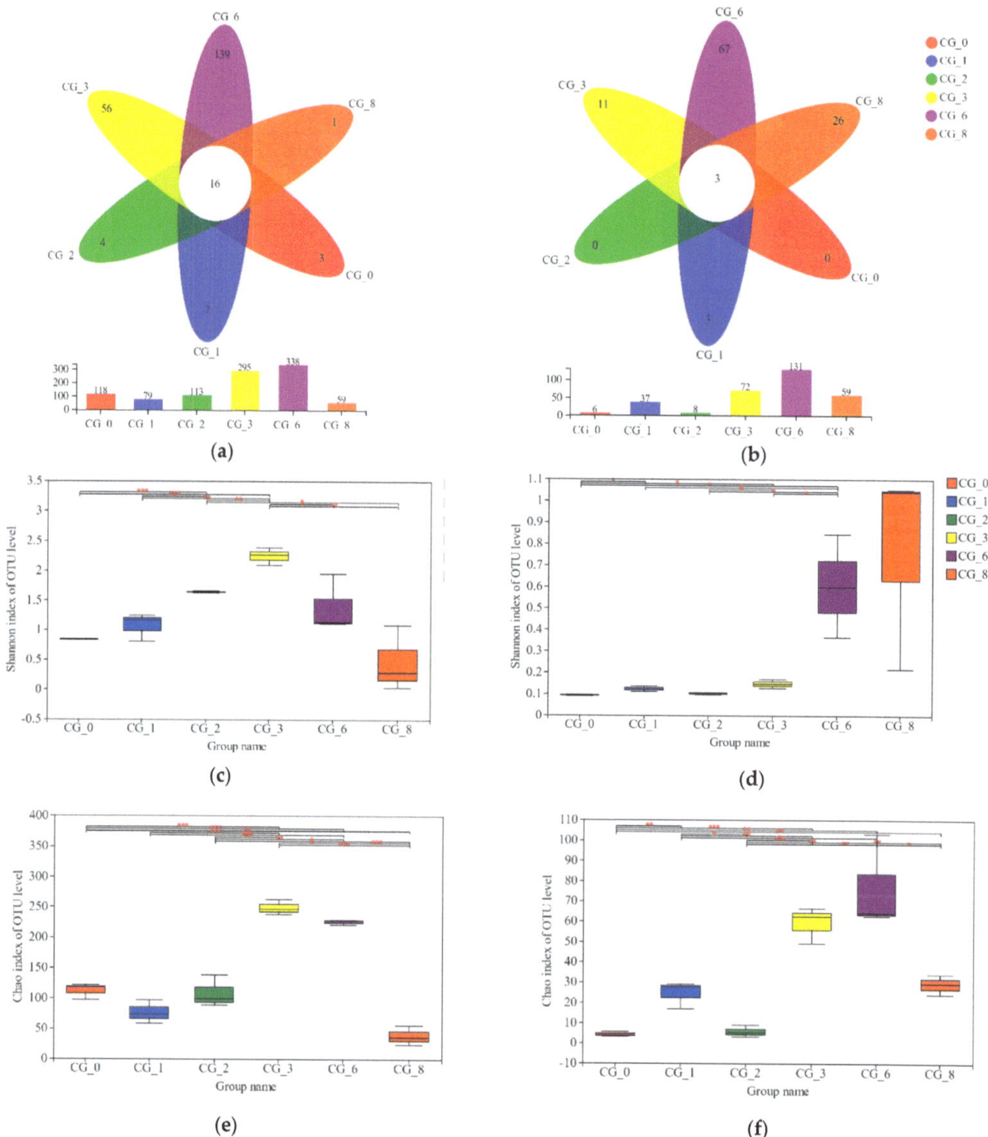

Figure 2. α- diversity analysis of bacterial (**a**,**c**,**e**) and fungi (**b**,**d**,**f**) of Caishiji soybean paste at different fermentation stages. ($0.01 < p \leq 0.05$ marked as *, $0.001 < p \leq 0.01$ marked as **, $p \leq 0.001$ marked as ***).

Circos was used to construct a circle graph to visualize the distribution of dominant species in each sample and among the different samples. The bacterial and fungal genera with the 20 highest abundances were obtained. Genera with a relative abundance higher than 4% were classified as dominant. There were five dominant bacterial genera (*Bacillus*, *Tetragenococcus*, *Salinivibrio*, *Halomonas*, and *Staphylococcus*) and four dominant fungal genera (*Aspergillus*, *Debaryomyces*, *Nigrospora*, and *Curvularia*), which were detected during the CSP fermentation process (Figure 3b). In CG0, *Tetragenococcus* (83.72%) and *Staphylococcus* (5.48%) were the dominant bacterial genera, and *Aspergillus* ($\geq$99.9%) was the dominant fungal genus. The relative abundances of *Tetragenococcus*, *Staphylococcus*, and *Aspergillus* in CG1 and CG2 were 29.9–67.34%, 5.32–7.38%, and 99.9%, respectively. Some emerging bacterial genera, such as *Klebsiella* (5.47%) and unclassified *Bacillus* (5.06%), were also detected in these samples. The dominant bacterial genera in CG3 and CG6 were *Bacillus* (24.97–66.79%), *Salinivibrio* (9.53–39.55%), *Halomonas* (6.36–10.50%), and *Tetragenococcus* (7.24%), while the dominant fungal genera were *Aspergillus* (≥90.0%) and *Debaryomyces* (5.37%). The relative abundance of *Bacillus*, *Salinivibrio*, and *Aspergillus* in CG8 were 88.51%, 9.99%, and 88.58%, respectively. *Orenia* (9.55%), an emerging bacterial genus, was also identified in CG8. All the results suggested that the composition and species of microbiome in CSP were changing during the fermentation process, which may affect the quality and flavor of CSP.

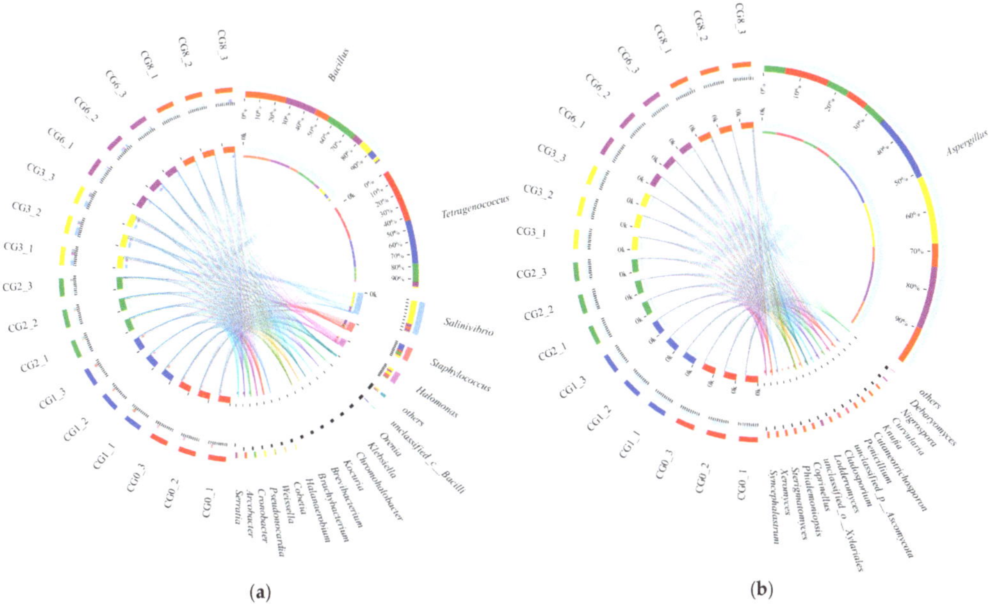

(a) (b)

Figure 3. Circos of bacteria (**a**) and fungi (**b**) of Caishiji soybean paste in different fermentation stages. The right half circle represents the distribution of the species in different samples at the taxonomic level, the outer ribbon represents the species, the inner ribbon color represents the different groups, and the length represents the distribution of the sample in a species.

3.3. Correlation Analysis of Microbiota and Characteristic VFCs

In this study, *R* values and Pearson correlation coefficients were calculated to analyze the connections between microbiota and characteristic VFCs in CSP. The bacterial and fungal genera with the 10 highest abundances were included in the correlation analysis. Redundancy analysis (RDA) was conducted to evaluate the complex relationships among the microbiota, VFCs, and CSP fermentation stages. On the basis of the PCA results, isovaleraldehyde, 2-methylbutyraldehyde, and 3-methylthiopropionaldehyde were selected for the RDA. The results revealed that compared with isovaleraldehyde and 3-methylthiopropionaldehyde, the microorganisms exerted a greater effect on 2-methylbutyraldehyde in CG3, CG6, and CG8. These three characteristic VFCs were positively correlated (Figure 4c,d).

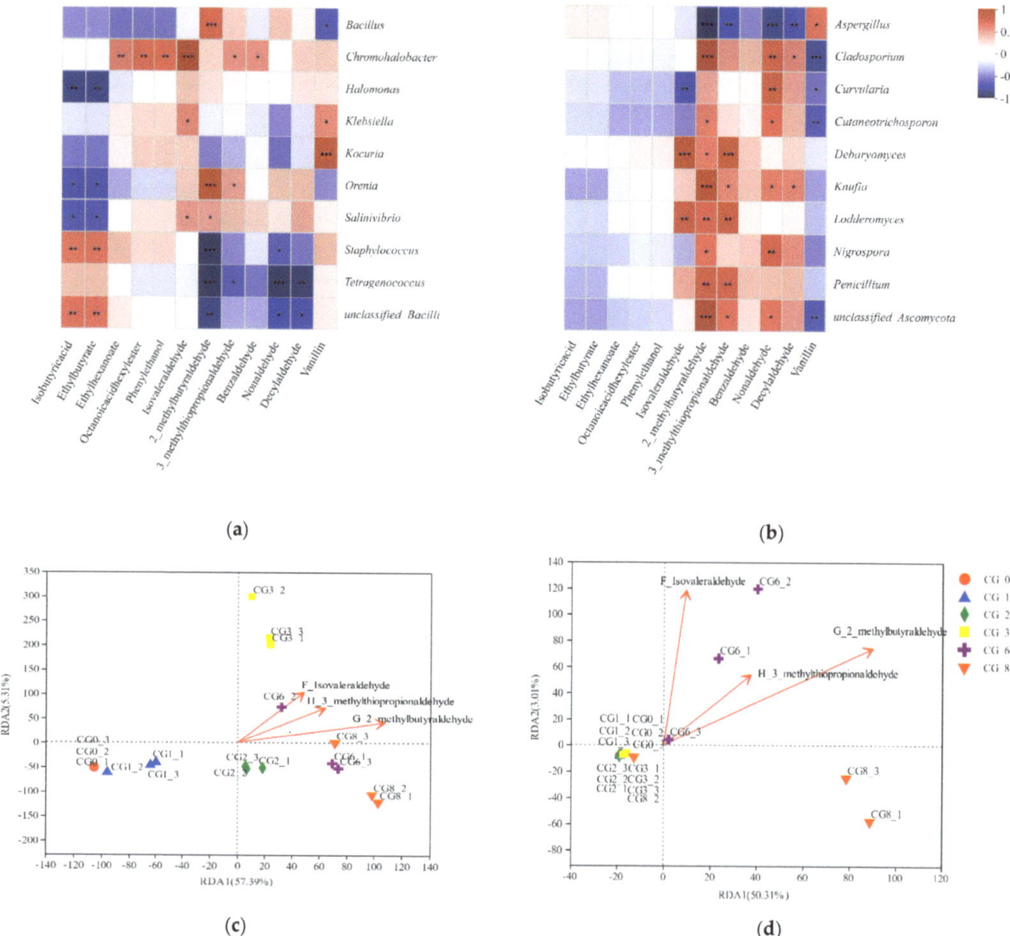

(a)

(b)

(c)

(d)

Figure 4. Heat-map (**a**,**b**) and RDA (**c**,**d**) correlation analysis of microbiota and characteristic VFCs of Caishiji bean paste in different fermentation stages. (*R* value was displayed in different colors in the heat-map. If the *p* value was less than 0.05, it is marked with *, the legend on the right side was the color range of different R values. * $0.01 < p \leq 0.05$, ** $0.001 < p \leq 0.01$, *** $p \leq 0.001$. In RDA, the red arrow indicates the flavor substance, and the length of the arrow indicates the impact degree of the flavor substance on the species; the angle between the environmental factors arrows represents positive and negative correlation (acute angle: positive correlation; obtuse angle: negative correlation; right angle: no correlation); the distance from the projection point to the origin represents the relative impact of the flavor substance on the distribution of the sample community).

4. Discussion

VFCs and microbiota strongly affect the flavor of soybean paste. In this study, VFCs and microbiota in CSP at different fermentation stages (CG0, CG1, CG2, CG3, CG6, CG8) were detected to identify potential correlations between them. More than 70 types of VFCs, including 12 types of characteristic VFCs (based on OAV), were detected. Among them, esters, alcohols, and aldehydes were the most prevalent in the CSP fermentation process.

The contributions of VFCs to CSP flavor depend not only on their concentrations but also on their odor threshold values [22]. Esters (especially short-chain esters), which are formed by a series of esterification reactions, are vital aromatic compounds in fermented foods, and it can endow fermented foods with sweetness and fruitiness. In the present study, 15 esters were detected during the CSP fermentation process. Among them, amyl acetate, α-pentyl-γ-butyrolactone, and ethyl palmitate, which have also been identified in other fermented soybean products, were common to CSP at differing fermentation stages. Based on the results of OAV analysis, ethyl butyrate, ethyl caproate, and hexyl caproate were the esters that most strongly affected the flavor of CSP. Ethyl butyrate, which produces a beany flavor [28], was detected only in CG3, and ethyl hexanoate and hexyl caprylate, which produce fruity flavors [28], were detected in CG1 and CG3. Most of the alcohols, which can endow CSP with pleasantly sweet and floral flavors, were formed in the later fermentation period. Isoamyl alcohol, 2-methyl-1-butanol, 1-pentanol, 2,3-butanediol, and 1-nonyl-3-ol were detected in CG6 and CG8. Phenethyl alcohol, which produces aromatic flavors [29], most strongly affected the flavor of CG3 according to the results of the OAV analysis. Aldehydes, including monoolefinic aldehyde, dialdehyde, and saturated aldehyde, which are mainly produced by the lipoxygenase-catalyzed decomposition of linoleic and linolenic acids, can endow fermented foods with sweet, fruity, nutty, and caramel flavors [30]. Fourteen aldehydes were detected in CSP, most of which were formed during the late stages of the fermentation process. The results of the OAV analysis indicated that seven characteristic aldehydes affected the flavor of CSP. Isovaleraldehyde, 2-methylbutyraldehyde, and benzaldehyde, which produce fruity flavors, stimulating odors, and cherry like flavors, respectively [29], were detected in each fermentation stage; 3-methylthiopropionaldehyde, which had the highest OAV value, can endow CSP with a strong roasted-potato flavor [29]. Although the other VFCs had lower thresholds, they may also endow CSP with strong flavors.

Microbiotas are the main contributors to soybean paste fermentation [31,32]. The fermentation of soybean paste involves a variety of bacteria and fungi. In addition, soybean pastes with distinct microbiota exhibit distinct flavors and qualities. In some previous studies, *Bacillus*, *Aspergillus*, *Tetragenococcus*, *Salinivibrio*, *Halomonas*, and *Staphylococcus* were the most abundant genera in CSP (Figure 3). During the fermentation process, the relative abundance of *Bacillus* increased, while that of *Staphylococcus* decreased or even disappeared completely. *Tetragenococcus* is closely related to the production of biogenic amines, which directly affect the flavor and safety of soybean paste; therefore, regulation of *Tetragenococcus* content is essential [1]. *Aspergillus* was the predominant fungal genus in CSP samples; it is found in soy sauce, douchi, and vinegar [33]. Compared with the other genera identified in this study, *Aspergillus* exhibits a higher tolerance to dryness. The analysis of microbial dynamics revealed that bacteria were involved in the decomposition of materials in the CSP fermentation, whereas most of the fungi were presented in the late stages of fermentation. The fermentation of soybean paste takes place mostly in anaerobic and high-humidity environments. Compared with bacteria, fungi may disrupt metabolic activity under dry and aerobic conditions; thus, fungal communities play a more important role in fermented bean paste.

The relationships between microbiota and VFCs in traditional fermented foods have been researched; however, such studies have produced limited relevant data. In the present study, we analyzed the complex connections between the microbiota and 12 characteristic VFCs during the CSP fermentation. As indicated in Figure 4, the Pearson's correlation coefficients (r) and significance (p) were calculated, and the bacteria and fungi with the 10 highest abundances were obtained. The 2-Methylbutyraldehyde was significantly positively correlated with *Bacillus* and *Orenia* (r > 0.7) but significantly negatively correlated with *Tetragenococcus* and *Staphylococcus*. *Kocuria* had a positive correlation with vanillin, and *Halomonas* was negatively correlated with isobutyric acid and ethyl butyrate (Figure 3a). The 2-Methylbutyraldehyde, 3-methylthiopropionaldehyde, nonanaldehyde, and decyl aldehyde were significantly negatively correlated with *Aspergillus*, but they were positively correlated with other fungi, such as *Debaryomyces*, *Nigrospora*, *Curvularia*, and *Penicillium*. Vanillin was positively correlated with *Aspergillus* but negatively correlated with *Cladosporium*, *Curvularia*, and *Cutaneotrichosporon* (Figure 3b). Through correlation analysis, microbiotas were determined to be the main factors affecting the generation of flavor compounds during the CSP fermentation. For example, *Bacillus* and *Orenia* may produce undesirable flavor compounds during CSP fermentation, whereas *Aspergillus*, *Etragenococcus*, and *Staphylococcus* may produce flavor-enhancing compounds. Through detection and analysis, the connections between microbiota and characteristic VFCs were determined. In follow-up studies, these correlations should be verified in application and potentially used to further advancements in "omics" technologies.

5. Conclusions

MiSeq sequencing technology is widely used, and correlation analysis of microbiota and VFCs has attracted considerable attention. In the present study, the complex influences of microbiota on the formation of VFCs during CSP fermentation were analyzed. The results indicated that *Bacillus*, *Tetragenococcus*, *Staphylococcus*, and *Aspergillus* were the main microbiota affecting the flavor of CSP. Although correlation studies involving many fermented foods have been conducted, the results of such studies still lack practical applicability. Therefore, follow-up research should focus on such practical applications to improve the sensory features of fermented soybean paste, and the pastes with consistent flavor profiles may be produced.

Supplementary Materials: The following supporting information can be downloaded at: https://www.mdpi.com/article/10.3390/fermentation8050196/s1, Table S1: Alpha diversity index of bacteria of Caishiji soybean paste in different fermentation stages, Table S2: Alpha diversity index of fungi of Caishiji soybean paste in different fermentation stages, Figure S1: Multi-point sampling (a) and numbers of different VFCs (b) of Caishiji soybean paste in different fermentation stages, Figure S2: Contents of main volatile components of Caishiji soybean paste in different fermentation stages (a: alcohols, b: esters, c: aldehydes, d: miscellaneous), Figure S3: Simpson curves of the bacteria (a) and fungi (b) of Caishiji soybean paste in different fermentation stages.

Author Contributions: Conceptualization: X.L.; methodology: Y.H. and W.W.; software: J.C.; formal analysis: J.C. and Y.H.; investigation: J.C. and Y.H.; resources: X.L.; data curation: D.M.; writing (original draft preparation): J.C. and Y.H.; writing (review and editing): X.L. and X.W.; visualization: X.W. and D.M.; supervision: X.L. and S.J.; project administration: X.L.; funding acquisition: X.L. All authors have read and agreed to the published version of the manuscript.

Funding: This research was funded by [the Project of Anhui Province] grant number [201903a06020034, 201903a06020007, 2108085MC123], and [the Fundamental Research Funds for the Central Universities of China] grant number [PA2021KCPY0048].

Institutional Review Board Statement: Not applicable.

Informed Consent Statement: Not applicable.

Data Availability Statement: The sequences from this study were deposited into the NCBI Sequence Read Archive (SRA) database (accession number:SRP255336; NCBI BioProject PRJNA623258).

Acknowledgments: This work was supported by the Project of Anhui Province [grant numbers 201903a06020034, 201903a06020007, 2108085MC123], and the Fundamental Research Funds for the Central Universities of China [grant number PA2021KCPY0048].

Conflicts of Interest: This manuscript has not been published or presented elsewhere in part or in entirety and is not under consideration by another journal. The authors declare no conflict of interest.

References

1. Byun, B.Y.; Mah, J.H. Occurrence of Biogenic Amines in *Miso*, Japanese Traditional Fermented Soybean Paste. *J. Food Sci.* **2012**, *77*, T216–T223. [CrossRef] [PubMed]
2. Kim, T.W.; Lee, J.H.; Kim, S.E.; Park, M.H.; Chang, H.C.; Kim, H.Y. Analysis of Microbial Communities in *Doenjang*, a Korean Fermented Soybean Paste, Using Nested PCR-Denaturing Gradient Gel Electrophoresis. *Int. J. Food Microbiol.* **2009**, *131*, 265–271. [CrossRef] [PubMed]
3. Chung, H.Y.; Fung, P.K.; Kim, J.S. Aroma Impact Components in Commercial Plain Sufu. *J. Agric. Food Chem.* **2005**, *53*, 1684–1691. [CrossRef]
4. Feng, Z.; Gao, W.; Ren, D.; Chen, X.; Li, J.J. Evaluation of Bacterial Flora during the Ripening of Kedong Sufu, a Typical Chinese Traditional Bacteria-Fermented Soybean Product. *J. Sci. Food Agric.* **2013**, *93*, 1471–1478. [CrossRef] [PubMed]
5. Han, B.Z.; Rombouts, F.M.; Robert Nout, M.J. A Chinese Fermented Soybean Food. *Int. J. Food Microbiol.* **2001**, *65*, 1–10. [CrossRef]
6. Lioe, H.N.; Selamat, J.; Yasuda, M. Soy Sauce and Its Umami Taste: A Link from the Past to Current Situation. *J. Food Sci.* **2010**, *75*, R71–R76. [CrossRef]
7. Wei, Q.Z.; Wang, H.B.; Chen, Z.X.; Lv, Z.J.; Xie, Y.F.; Lu, F.P. Profiling of Dynamic Changes in the Microbial Community during the Soy Sauce Fermentation Process. *Appl. Microbiol. Biotechnol.* **2013**, *97*, 9111–9119. [CrossRef]
8. Yang, Y.; Deng, Y.; Jin, Y.L.; Liu, Y.X.; Xia, B.X.; Sun, Q. Dynamics of Microbial Community during the Extremely Long-Term Fermentation Process of A Traditional Soy Sauce. *J. Sci. Food Agric.* **2017**, *97*, 3220–3227. [CrossRef]
9. Chen, T.T.; Wang, M.J.; Jiang, S.Y.; Xiong, S.Q.; Zhu, D.C.; Wei, H. Investigation of the Microbial Changes during Koji-Making Process of Douchi by Culture-Dependent Techniques and PCR-DGGE. *Int. J. Food Sci. Technol.* **2011**, *46*, 1878–1883. [CrossRef]
10. Chen, T.T.; Wang, M.J.; Li, S.J.; Wu, Q.L.; Wei, H. Molecular Identification of Microbial Community in Surface and Undersurface Douchi during Postfermentation. *J. Food Sci.* **2014**, *79*, M653–M658. [CrossRef]
11. Wang, H.; Yin, L.J.; Cheng, Y.Q.; Li, L.T. Effect of Sodium Chloride on the Color, Texture, and Sensory Attributes of Douchi during Post-Fermentation. *Int. J. Food Eng.* **2012**, *8*, 22. [CrossRef]
12. Jung, K.O.; Park, S.Y.; Park, K.Y. Longer Aging Time Increases the Anticancer and Antimetastatic Properties of *Doenjang*. *Nutrition* **2006**, *22*, 539–545. [CrossRef] [PubMed]
13. Park, K.Y.; Jung, K.O.; Rhee, S.H.; Choi, Y.H. Antimutagenic Effects of *Doenjang* (Korean Fermented Soypaste) and Its Active Compounds. *Mutat. Res. Lett.* **2003**, *523*, 43–53. [CrossRef]
14. Shin, Z.I.; Yu, R.; Park, S.A.; Chung, D.K.; Ahn, C.W.; Nam, H.S.; Kim, K.S.; Lee, H.J. His-His-Leu, An Angiotensin I Converting Enzyme Inhibitory Peptide Derived from Korean Soybean Paste, Exerts Antihypertensive Activity in Vivo. *J. Agric. Food Chem.* **2001**, *49*, 3004–3009. [CrossRef]
15. Liu, J.J.; Han, B.Z.; Deng, S.H.; Sun, S.P.; Chen, J.Y. Changes in Proteases and Chemical Compounds in the Exterior and Interior of Sufu, a Chinese Fermented Soybean Food, during Manufacture. *LWT Food Sci. Technol.* **2018**, *87*, 210–216. [CrossRef]
16. Jo, Y.J.; Cho, I.H.; Song, C.K.; Shin, H.W.; Kim, Y.S. Comparison of Fermented Soybean Paste (*Doenjang*) Prepared by Different Methods Based on Profiling of Volatile Compounds. *J. Food Sci.* **2011**, *76*, C368–C379. [CrossRef]
17. Lin, H.B.; Yu, X.Y.; Fang, J.X.; Lu, Y.H.; Liu, P.; Xing, Y.G.; Wang, Q.; Che, Z.M.; He, Q. Flavor Compounds in Pixian Broad-Bean Paste: Non-Volatile Organic Acids and Amino Acids. *Molecules* **2018**, *23*, 1299. [CrossRef]
18. Zhao, J.X.; Dai, X.J.; Liu, X.M.; Zhang, H.; Tang, J.; Chen, W. Comparison of Aroma Compounds in Naturally Fermented and Inoculated Chinese Soybean Pastes by GC-MS and GC-Olfactometry Analysis. *Food Control* **2011**, *22*, 1008–1013. [CrossRef]
19. Zhang, P.; Zhang, P.F.; Xie, M.X.; An, F.Y.; Qiu, B.S.; Wu, R. Metaproteomics of Microbiota in Naturally Fermented Soybean Paste, *Da-Jiang*. *J. Food Sci.* **2018**, *83*, 1342–1349. [CrossRef] [PubMed]
20. Hao, Y.; Sun, B. Analysis of Bacterial Diversity and Biogenic Amines Content during Fermentation of Farmhouse Sauce from Northeast China. *Food Control.* **2020**, *108*, 106861. [CrossRef]
21. Li, Z.; Dong, L.; Huang, Q.; Wang, X. Bacterial Communities and Volatile Compounds in Doubanjiang, a Chinese Traditional Red Pepper Paste. *J. Appl. Microbiol.* **2016**, *120*, 1585–1594. [CrossRef] [PubMed]
22. Xie, C.Z.; Zeng, H.Y.; Wang, C.X.; Xu, Z.M.; Qin, L.K. Volatile Flavour Components, Microbiota and Their Correlations in Different Sufu, a Chinese Fermented Soybean Food. *J. Appl. Microbiol.* **2018**, *125*, 1761–1773. [CrossRef] [PubMed]
23. Cocolin, L.; Alessandria, V.; Dolci, P.; Gorra, R.; Rantsiou, K. Culture Independent Methods to Assess the Diversity and Dynamics of Microbiota during Food Fermentation. *Int. J. Food Microbiol.* **2013**, *167*, 29–43. [CrossRef] [PubMed]
24. Rodriguez-Burruezo, A.; Kollmannsberger, H.; Carmen Gonzalez-Mas, M.; Nitz, S.; Nuez, F. HS-SPME Comparative Analysis of Genotypic Diversity in the Volatile Fraction and Aroma-Contributing Compounds of Capsicum Fruits from the *Annuum-Chinense-Frutescens* Complex. *J. Agric. Food Chem.* **2010**, *58*, 4388–4400. [CrossRef]
25. Li, X.Y.; Zhao, C.S.; Zheng, C.; Liu, J.; Vu, V.H.; Wang, X.D.; Sun, Q. Characteristics of Microbial Community and Aroma Compounds in Traditional Fermentation of Pixian Broad Bean Paste as Compared to Industrial Fermentation. *Int. J. Food Prop.* **2018**, *20*, S2520–S2531. [CrossRef]
26. Kang, K.M.; Baek, H.H. Aroma Quality Assessment of Korean Fermented Red Pepper Paste (*Gochujang*) by Aroma Extract Dilution Analysis and Headspace Solid-Phase Microextraction-Gas Chromatography-Olfactometry. *Food Chem.* **2014**, *145*, 488–495. [CrossRef]

27. Ye, J.B.; Yan, J.; Zhang, Z.; Yang, Z.C.; Liu, X.Z.; Zhou, H.; Wang, G.F.; Hao, H.; Ma, K.; Ma, Y.P.; et al. The Effects of Threshing and Redrying on Bacterial Communities that Inhabit the Surface of Tobacco Leaves. *Appl. Microbiol. Biotechnol.* **2017**, *101*, 4279–4287. [CrossRef]

28. Moy, Y.S.; Lu, T.J.; Chou, C.C. Volatile Components of the Enzyme-Ripened Sufu, a Chinese Traditional Fermented Product of Soy Bean. *J. Biosci. Bioeng.* **2012**, *113*, 196–201. [CrossRef]

29. Beaulieu, J.C.; Stein-Chisholm, R.E. HS-GC-MS Volatile Compounds Recovered in Freshly Pressed 'Wonderful' Cultivar and Commercial Pomegranate Juices. *Food Chem.* **2016**, *190*, 643–656. [CrossRef]

30. Zhao, J.; Wang, M.; Xie, J.C.; Zhao, M.Y.; Hou, L.; Liang, J.J.; Wang, S.; Cheng, J. Volatile Flavor Constituents in the Pork Broth of Black-Pig. *Food Chem.* **2017**, *226*, 51–60. [CrossRef]

31. Jung, J.Y.; Lee, S.H.; Jeon, C.O. Microbial Community Dynamics during Fermentation of Doenjang-Meju, Traditional Korean Fermented Soybean. *Int. J. Food Microbiol.* **2014**, *185*, 112–120. [CrossRef] [PubMed]

32. Lee, J.H.; Kim, T.W.; Lee, H.; Chang, H.C.; Kim, H.Y. Determination of Microbial Diversity in *Meju*, Fermented Cooked Soya Beans, Using Nested PCR-Denaturing Gradient Gel Electrophoresis. *Lett. Appl. Microbiol.* **2010**, *51*, 388–394. [CrossRef] [PubMed]

33. Wang, Z.M.; Lu, Z.M.; Shi, J.S.; Xu, Z.H. Exploring Flavour-Producing Core Microbiota in Multispecies Solid-State Fermentation of Traditional Chinese Vinegar. *Sci. Rep.* **2016**, *6*, 26818. [CrossRef] [PubMed]

Article

Controlling the Formation of Foams in Broth to Promote the Co-Production of Microbial Oil and Exopolysaccharide in Fed-Batch Fermentation

Yan-Feng Guo [1,†], Meng-Qi Wang [2,†], Yi-Lei Wang [1,*], Hong-Tao Wang [2] and Jian-Zhong Xu [2,*]

[1] College of Agriculture and Bioengineering, Heze University, 2269 University Road, Heze 274015, China; 123gyf002@163.com

[2] The Key Laboratory of Industrial Biotechnology, Ministry of Education, School of Biotechnology, Jiangnan University, 1800# Lihu Road, Wuxi 214122, China; 6200208003@stu.jiangnan.edu.cn (M.-Q.W.); 200208107@stu.jiangnan.edu.cn (H.-T.W.)

* Correspondence: wangyilei@hezeu.edu.cn (Y.-L.W.); xujianzhong@jiangnan.edu.cn (J.-Z.X.); Tel./Fax: +86-530-5662087 (Y.-L.W.); +86-510-85329312 (J.-Z.X.)

† These authors contributed equally to this work.

Abstract: A large amount of foam is generated in the production of microbial oil and exopolysaccharide (EPS) by *Sporidiobolus pararoseus* JD-2, which causes low efficiency in fermentation. In this study, we aimed to reduce the negative effects of foams on the co-production of oil and EPS by controlling the formation of foams in broth. As we have found, the formation of foams is positively associated with cell growth state, air entrapment, and properties of broth. The efficient foam-control method of adding 0.03% (*v*/*v*) of the emulsified polyoxyethylene polyoxypropylene pentaerythritol ether (PPE) and feeding corn steep liquor (CSL) at 8–24 h with speed of 0.02 L/h considerably improved the fermentation performance of *S. pararoseus* JD-2, and significantly increased the oil and EPS concentrations by 8.7% and 12.9%, respectively. The biomass, oil, and EPS concentrations were further increased using a foam backflow device combined with adding 0.03% (*v*/*v*) of the emulsified PPE and feeding CSL at 8–24 h, which reached to 62.3 ± 1.8 g/L, 31.2 ± 0.8 g/L, and 10.9 ± 0.4 g/L, respectively. The effective strategy for controlling the formation of foams in fermentation broth reported here could be used as a technical reference for producing frothing products in fed-batch fermentation.

Keywords: *Sporidiobolus pararoseus*; foam control; corn steep liquor feeding; microbial oil; exopolysaccharide

Citation: Guo, Y.-F.; Wang, M.-Q.; Wang, Y.-L.; Wang, H.-T.; Xu, J.-Z. Controlling the Formation of Foams in Broth to Promote the Co-Production of Microbial Oil and Exopolysaccharide in Fed-Batch Fermentation. *Fermentation* **2022**, *8*, 68. https://doi.org/10.3390/fermentation8020068

Academic Editor: Thaddeus Ezeji

Received: 5 January 2022
Accepted: 30 January 2022
Published: 7 February 2022

Publisher's Note: MDPI stays neutral with regard to jurisdictional claims in published maps and institutional affiliations.

1. Introduction

Microbial oil, a type of biodiesel, can be obtained from renewable raw materials, by oleaginous microorganisms, with many advantages, e.g., short production time and low pollution of the environment [1–3]. However, high production cost limits the broader application of microbial oils [4]. In order to decrease the production cost of microbial oil, many strategies have been applied: (1) genetic modification of the biosynthetic pathway of microbial oil in oleaginous microorganism [5,6]; (2) use of the cheaper raw materials as feedstock [7,8]; (3) optimization of the medium components and culture methods [9–11]; (4) mixed culture of oleaginous yeast and oleaginous microalgae [12,13]; (5) co-production of microbial oil and high value-added products, e.g., exopolysaccharide (EPS) [14]. In 2010, we obtained an oleaginous yeast (i.e., *Sporidiobolus pararoseus* JD-02, CCTCC M2010326), which could be used to co-produce microbial oil, carotenoid, and EPS [4,15,16]. Carotenoid and EPS, as high value-added products, have been widely used in the food industry, pharmaceutical industry, and chemical industry [17,18]. In previous studies, we tried to increase the production of microbial oil, carotenoid, and EPS by optimization of media components and culture conditions [4,15,16], and by limitation of ammonia-nitrogen supply [11]. Although these methods were used to increase the production of these value-added products with a satisfactory result, large amounts of foams led to an increase in the escape of fermentation broth and a decrease in the utilization rate of equipment. Therefore, how to control the formation of foams in fermentation broth has become the key problem, which when solved will be beneficial to increase the utilization rate of feedstock and equipment, as well as to cut production cost.

Foams are comprised of thousands of bubbles in liquid caused by mechanical or chemical factors [19]. Foaming is considered a "general nuisance" in industrial fermentation because the fermentation process provides the essential conditions for foam formation [20]. There are two essential conditions for foam formation and stability: the external force and the property of solution [21]. Many factors affect the stability and formation of foams: the air entrapment in solution (e.g., gas flow rate and stirring frequency), the compositions and viscosity of media (e.g., pH, concentration of proteins and sugars, as well as presence of surfactant), the growth state of microbial cells (e.g., logarithmic phase, stable phase, and death phase), and the concentration of metabolites and surface-active substances (e.g., cresotic acid, rhamnolipid, and saponin) [19,22–24]. The final foam volume depends on the complex interplay of four processes: bubble formation, bubble–atmosphere coalescence, bubble breakup into tiny bubbles, and bubble–bubble coalescence to increase bubble size [21]. A small amount of bubbles helps to increase media oxygen transfer, but excessive foam leads to a decrease in the utilization rate of equipment and to an increase in the escape of fermentation broth [19]. In order to relieve the negative effects of foams, foam control systems are widely used in industrial fermentation. For example, reasonably adjusting the media components and the culture conditions can be used to prevent the formation of foams [11,19,25]. The most common strategy is to use antifoaming equipment or antifoaming agents to crush the pre-existing foams, and thus avoid the abundant accumulation of foams [26,27]. In addition, Zaky et al. has reported that seawater can also be used to control the formation of foams in the production of biofuel [28,29]. Although these methods on controlling foams have achieved positive results, foaming results from complex interactions among the aforementioned factors. Therefore, the best foam control method is still needed to optimize and achieve the best efficiency of an industrial fermentation process (i.e., high carbon yield, final titer, and productivity).

The aim of the work presented here was to control the formation of foams in a medium, and thus promote the co-production of microbial oil and EPS by *S. pararoseus* JD-2 in fed-batch fermentation. To do this, the relationships between foams and the key factors involved in foaming were first discussed. Subsequently, different strategies were used to control the formation of foams in fed-batch fermentation, including screening of defoamers, optimization of adding ways of CSL, as well as use of foam backflow devices. After controlling the formation of foams according to the strategies reported in this study, *S. pararoseus* JD-2 produced 31.2 ± 0.8 g/L of oil and 10.9 ± 0.4 g/L of EPS in fed-batch fermentation. The effective strategy for controlling the formation of foams in fermentation broth reported here could be used as a technical reference for producing frothing products in fed-batch fermentation.

2. Materials and Methods

2.1. Strain and Culture Conditions

Strain *S. pararoseus* JD-2 (CCTCC M2010326) was used as a fermentation strain for co-producing microbial oil and EPS, which was isolated from bean-based sauce [15]. The YPD medium (Yeast extract 10 g/L, Peptone 20 g/L, Dextrose 20 g/L) was used for activating *S. pararoseus* JD-2. Unless stated otherwise, *S. pararoseus* JD-2 was cultivated at 28 °C and pH 6.0 for 72 h.

The fed-batch fermentation was performed in a 7-L fermenter (KF-7 l, Korea Fermenter Co., Inchon, Korea) containing 3 L of the medium with an inoculum size of 10% (*v/v*), and the inoculum was obtained from a seed culture grown to logarithmic phase (about 10 h). The seed medium was prepared according to the description reported by Wang et al. [11]. The initial culture medium used for fermentation consisted of (per liter): 120 g glucose, 20 g corn steep liquor (CSL; purchased from Shandong Shouguang Juneng Golden Corn Co., Ltd., Shouguang, China), 1.2 g $(NH_4)_2SO_4$, 1 g K_2HPO_4, and 0.1 g Na_2SO_4. The dissolved oxygen level and temperature were set at 20% and 28 °C, respectively. The 800 g/L sterile glucose solution was used to maintain the glucose concentration at ~15 g/L by adjusting the feeding rate. Additionally, the medium was adjusted to pH 6.0 with 20% (m/v) NaOH.

2.2. Extraction of Microbial Oil and EPS

A sample was taken from the fermenter and then centrifuged at 10,000 rpm for 20 min (Sorvall LYNX4000, ThermoFisher Scientific, Waltham, MA, USA). The cell pellets were used to extract microbial oil and the culture supernatants were used to extract EPS. The entire processes of extracting oil and EPS were referred to in previous reports [11,16].

2.3. Analyzing the Performance of Defoamer

Analyzing the performance of defoamer was based on the principle of the previous methods reported by Tamura et al. [30]. Then, 300 mL of fermentation broth was added into a graduated cylinder (range 1000 mL) and then blew air with a speed of 1 L/min. The schematic diagram of the foam forming device is presented in Figure 1a. The time it took foams to reach 700 mL was used to reflect the foaming ability of broth. Subsequently, 300 μL of defoamer with 10 times more dilution was added into the foam forming device, and the time of foam fading away was used to reflect the defoaming ability of the defoamer.

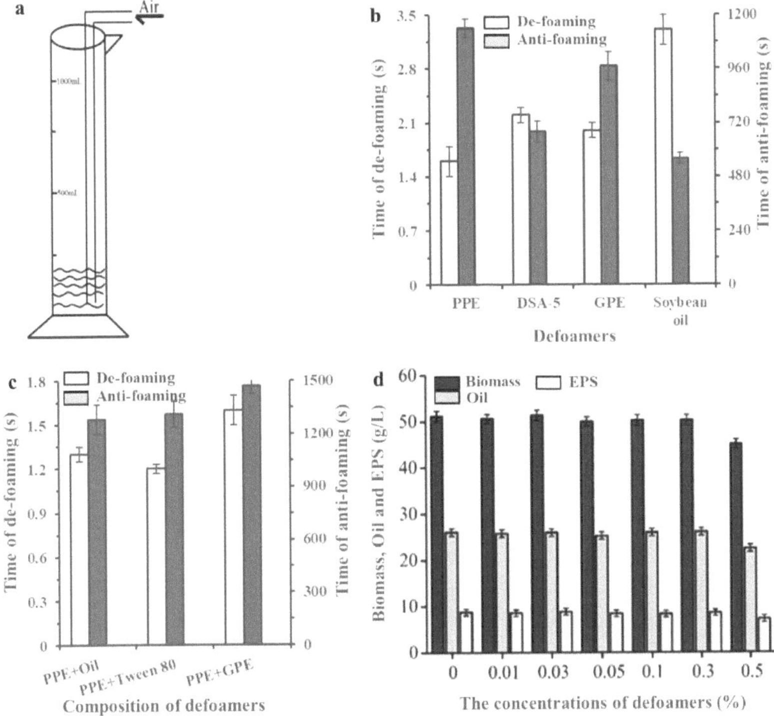

Figure 1. The effects of defoamers on defoaming, antifoaming and fermentation performance of *S. pararoseus* JD-2. (**a**) Simple foam forming device. (**b**) The antifoaming abilities of different defoaming:the DSA-5 represents the defoamer for bean products. (**c**) Effects of different synergistic methods on PPE defoaming ability (**d**) Effect of defoamer on *S. pararoseus* JD-2 fermentation. The data represent mean values and standard deviations obtained from three independent cultivations.

In order to analyze the antifoaming ability of defoamer, 300 mL of fermentation broth and 300 µL of defoamer with 10 times more dilution were added into a graduated cylinder (range 1000 mL) and then blew air with speed of 1 L/min. The time that that foams reached 400 mL was used to reflect the antifoaming ability of defoamer. In addition, different strategies were used to enhance defoaming ability of defoamer, e.g., defoamer plus soybean oil with 1:1.5, defoamer plus Tween 80 with 100:1, and mixed defoamer with 1:1.

2.4. The Mode of Corn Steep Liquor (CSL) Feeding

Four modes of CSL feeding were performed in fed-batch fermentation by *S. pararoseus* JD-2 (Table 1). The total CSL in the medium at different modes of CSL feeding was identical, and the total CSL was added into the fermentation medium at different incubation times and with different concentrations and feeding rates. It should be noted that the initial concentration of CLS in the media is inconsistent at different modes of CLS feeding, from 5 g/L to 20 g/L.

Table 1. Methods of feeding organic nitrogen source in *S. pararoseus* fed-batch fermentation.

Mode	Loading Volume (L)	Initial Defoamer Concentration (%)	Initial CSL Concentration (g/L)	Time of CSL Feeding (h)	Speed of CSL Feeding [1] (L/h)
I	3.0	0	20	—	—
II	2.7	0.1	10	24–36	0.03
III	2.7	0.1	5	8–24	0.02
IV	2.7	0.1	10	8–24	0.02

[1] The concentration of CSL used feeding in Mode II, Mode III, and Mode IV is 110 g/L, 155 g/L, and 110 g/L, respectively.

2.5. Analytical Methods

Then, 200 µL of samples were taken from the shake flasks or fermenter every 4 hours. These samples were used to measure the biomass using a spectrophotometer at 600 nm after 25 times more dilution, and to analyze the concentration of microbial oil and EPS. According to our previous description [11,16], the concentration of microbial oil and EPS was detected by weight after extraction. The analyses of biomass and concentration of microbial oil and EPS were performed in triplicate.

2.6. Statistical Analysis

The experiments in this study were independently carried out at least three times, and data are expressed as mean and standard deviation ($\pm$SD). Student's *t* test was used to compare statistical difference among the groups of experiment data.

3. Results and Discussion

3.1. The Relationships between Foams and the Key Factors Involved in Foaming in Fed-Batch Ferementation by S. pararoseus JD-2

As mentioned earlier, many factors affect the stability and formation of foam, e.g., the air entrapment in solution, the compositions and viscosity of the media, and the growth state of microbial cells [19,22–24]. In order to discuss the relationships between foams and the key factors involved in foaming, the cell growth state, the air entrapment, and the properties of broth were investigated. The foam formation occurred before 36 h of the whole fermentation period in fed-batch fermentation, especially before 24 h (Figure 2). As shown in Figure 2a, *S. pararoseus* JD-2 was at the early stable growth phase before 36 h, indicating that cell growth state is positively associated with foam formation. Similar results were also found in previous studies, in which the foam volume increased with the increase in cell growth rate [31,32]. Since *S. pararoseus* JD-2 is an aerobe [15], more oxygen was needed to maintain the cell growth with high rate. Therefore, the fast agitation speed and the high ventilatory capacity were needed to meet the oxygen supply at the early stable growth phase (Figure 2b). Based on the previous results reported by Conceicao et al. [33], static submerged cultivation was beneficial to surfactant production because of the decrease in foam formation. Therefore, air entrapment is also positively associated with foam formation (Figure 2b). Vardar-Sukan pointed out that the concentration of salts, proteins, and sugars in media affects foam formation [19], and this may be why foams were observed to have rapidly formed at a high concentration of protein (Figure 2c). In addition, the foam volume increased with the increase in EPS concentration and apparent viscosity of broth (Figure 2c). The results reported by Dai et al. indicated that the high viscosity of pre-hydrolysate causes the serious foam formation during air-aerated and agitated processes [34]. It should be noted that foams gradually faded away despite the high apparent viscosity after 40 h (Figure 2c). This is possibly because of the reduced agitation speed and ventilatory capacity. This theory is supported by previous results reported by Gong et al. [35], in which reducing aeration eases foaming at the later stages of fermentation.

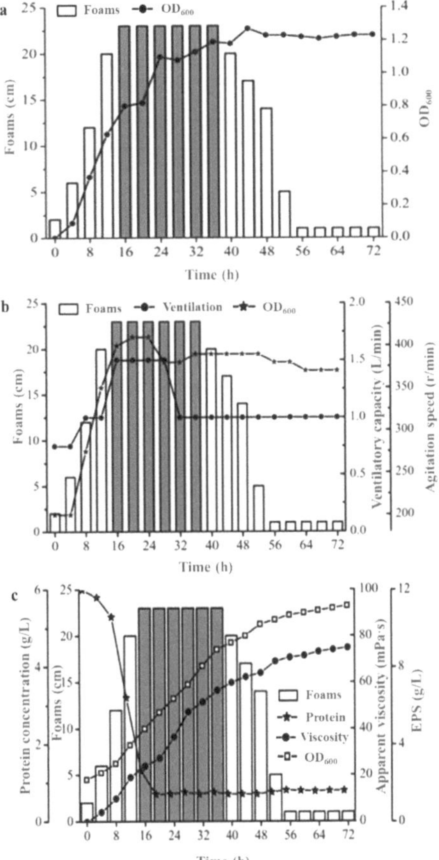

Figure 2. Causes of severe foaming in fed-batch fermentation by *S. pararoseus* JD-2. (**a**) The relationship between defoamer and frothing. (**b**) The relationship between ventilation and frothing. (**c**) The relationship between the properties of broth and frothing. The dark histograms represent the high of foams excess of fermenters. The data represent mean values and standard deviations obtained from three independent cultivations.

3.2. Optimization of Defoamer to Enhance Defoaming Ability and to Improve Efficiency of Fermentation Process

As mentioned above, foam formation is positively associated with the cell growth state, the air entrapment, and the properties of broth (Figure 2). Foam formation seems inevitable in agitated submerge fermentation, especially for producing biosurfactant, e.g., rhamnolipids [20,36]. However, excessive foams can do great harm to the normal fermentation process and to the best fermentation efficiency [19]. Therefore, natural or synthetic defoamers were usually used to prevent the formation of foam and/or to crush the pre-existing foams [19,20]. In order to obtain the best defoamer used to avoid the abundant accumulation of foams in fermentation by *S. pararoseus* JD-2, we investigated the defoaming and antifoaming abilities of one natural defoamer and three synthetic defoamers. As is expected, different defoamers showed the different defoaming abilities and antifoaming abilities (Figure 1b). Among these test defoamers, polyoxyethylene polyoxypropylene pentaerythritol ether (PPE) showed the greatest defoaming capacity and the longest foam-inhibiting time, and the natural defoamer (i.e., soybean oil) showed the worst defoaming abilities and antifoaming abilities (Figure 1b). PPE, as a safe food additive, has been widely used in industrial fermentation because of the high thermal stability, the chemical stability, and the best defoaming capacity [37]. It should be noted that the defoaming abilities and antifoaming abilities could be enhanced using three synergistic methods, i.e., carrier addition method (i.e., PPE plus soybean oil with 1:1.5), emulsifying method (i.e., PPE plus Tween 80 with 100:1), and combination method (PPE plus Polyoxypropylene oxyethylene glycol ether (GPE) with 1:1) (Figure 1c). As shown in Figure 1c, the emulsified PPE using Tween 80 showed the shortest time of defoaming and the longest time of foam-inhibiting. Emulsification promotes the substance into the other incompatible substance in liquid [38], and this may be why the emulsified PPE showed the best defoaming abilities and antifoaming abilities. Therefore, the emulsified PPE was used as the preferred defoamer for defoaming and antifoaming during fermentation by *S. pararoseus* JD-2 in the next study.

Although addition of defoamer can avoid the abundant accumulation of foams, defoamer negatively affects dissolved oxygen level in fermentation broth, thus restricting the fermentation performance of production strains [39]. Furthermore, the addition of the defoamer will be detrimental to the extraction and purification of target products [40]. These findings are confirmed once again in our results. As can be seen from Figure 1d, the biomass, microbial oil, and EPS concentrations obviously decreased during the addition of more than 0.3% (v/v) of the emulsified PPE. The addition of 0.03% (v/v) of the emulsified PPE in fermentation broth resulted in 51.5 ± 1.7 g/L of biomass, 25.6 ± 1.2 g/L of oil, and 9.5 ± 0.6 g/L of EPS, which is the best condition for cultivation of *S. pararoseus* JD-2 (Figure 1d). Thus, 0.03% (v/v) of the emulsified PPE was used to control foam formation in fed-batch fermentation by *S. pararoseus* JD-2 in the next study.

3.3. The Effects of the Mode of Corn Steep Liquor Feeding on Fermentation Performance of S. pararoseus JD-2

The concentration of proteins in media is one of the key factors in foam formation [19]. Corn steep liquors (CSL), one of the most commonly used complex organic nitrogen sources, are rich in proteins, sugars, inorganic salts, and vitamins [41]. Therefore, we investigated the effects of CSL on fermentation performance of *S. pararoseus* JD-2. The effect of the pH of CSL on foam formation was first investigated. The foaming time increased with the increase in pH, whereas the time of the disappearance of foams decreased with the increase in pH (Table 2). As far as we know, protein solubility is closely associated with the pH in solution [42]. We speculated that the proteins in CSL decreased with the increase in pH because of the protein deposition, thus limiting the formation of foams. Given the importance of protein and the optimized pH for cell growth, the best pH in CSL was set at 6.

Table 2. The foaming ability and bubble-holding ability of corn steep liquor with different pH.

pH of CSL [1]	Time of Foaming (s) [2]	Time of Defoaming (s)
4	23	120
5	30	100
6	55	50
7	*	5
8	*	3

[1] The concentration of CSL is 20 g/L. [2] "*" represents no foaming.

Next, four modes of CSL feeding were investigated to further improve the fermentation performance of *S. pararoseus* JD-2 (Table 1). The modes of CSL feeding significantly affected the fermentation performance of *S. pararoseus* JD-2, including escape volume, biomass, microbial oil, and EPS concentrations (Figure 3). The highest escape volumes (i.e., 1.3 L) were found during one-time addition of the overall CSL (i.e., Mode I). By contrast, only 0.3 L broth escaped from the fermenter during feeding CSL at 8–24 h with speed of 0.02 L/h (i.e., Mode IV), which was down by 76.9% as compared with one-time addition (Figure 3a). In addition, the biomass (i.e., 58.0 ± 1.1 g/L vs. 52.4 ± 1.2 g/L), microbial oil concentration (i.e., 27.5 ± 1.2 g/L vs. 25.3 ± 1.0 g/L), and EPS concentration (i.e., 10.5 ± 0.8 g/L vs. 9.3 ± 0.7 g/L) in Mode IV were increased by 10.7%, 8.7%, and 12.9% as compared with in Mode I, respectively (Figure 3b–d). Similar results were also found in Liu's reports, in which feeding trypsin resulted in lower formation of foams and higher L-glutamic acid production [43]. Interestingly, although there were no escape volumes during feeding CSL at 24–36 h with a speed of 0.03 L/h (i.e., Mode II), the biomass, microbial oil, and EPS concentrations were obviously inferior to the other three modes (Figure 3b–d). This is probably due to the nutrient deficiencies for cell growth in the early fermentation stage.

3.4. Using the Foams Backflow Device to Increase the Utility Ratio of Feedstock in Fed-Batch Fermentation by S. pararoseus JD-2

As mentioned above (Figure 3a), there is still fermented liquid leakage in Mode IV. In order to increase the utility ratio of feedstock in fed-batch fermentation by *S. pararoseus* JD-2, a foam backflow device was used to recycle the foams during fed-batch fermentation at the condition of feeding CSL at 8–24 h with speed of 0.02 L/h. The schematic diagram of the foam backflow device is shown in Figure 4a. The excess foams were entered into the collection bottle of the device and crushed, and then the liquor in the collection bottle was pumped into the fermenter by a peristaltic pump. As a control, the fermentation progress at the condition of the one-time addition of the overall CSL combined with the foam backflow device was also investigated. As can be seen from Table 3, using the foam backflow device was beneficial to increase biomass, microbial oil, and EPS concentrations. As compared with only feeding CSL at 8–24 h with speed of 0.02 L/h, the biomass, microbial oil, and EPS concentrations were increased by 7.4%, 13.5%, and 3.8% at the condition of feeding CSL at 8–24 h with speed of 0.02 L/h combined with foams backflow device, respectively. Our results are consistent with the previous results [36,44,45]. In addition, Anic et al. pointed out that application of foam adsorption increased the rhamnolipid yield from glucose feed [36]. It is worth noting that using the foam backflow device has no significant effect on the increase in biomass and products yielded during excess foams formation. As can be seen from Figure 4b, the cell growth was obviously inhibited at the condition of one-time addition of the overall CSL combined with the foam backflow device. One of the main reasons may be nutrient deficiencies because of the large amounts of fermented liquid leakage. Therefore, how to improve the efficiency of foam breakers is also important for controlling the formation of foams in fed-batch fermentation [45].

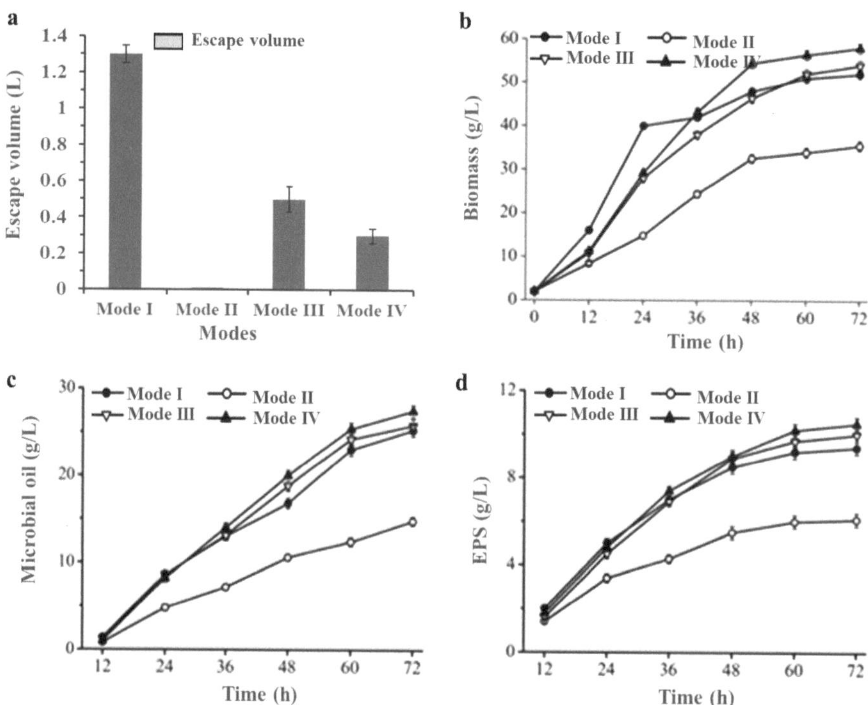

Figure 3. Effects of CSL feeding method on the fermentation of *S. pararoseus* JD-2. (**a**) The volumes of broth leakage in different modes of CSL feeding. (**b**) The biomass in different modes of CSL feeding. (**c**) The microbial oil concentration in different modes of CSL feeding. (**d**) The EPS concentration in different modes of CSL feeding. The data represent mean values and standard deviations obtained from three independent cultivations.

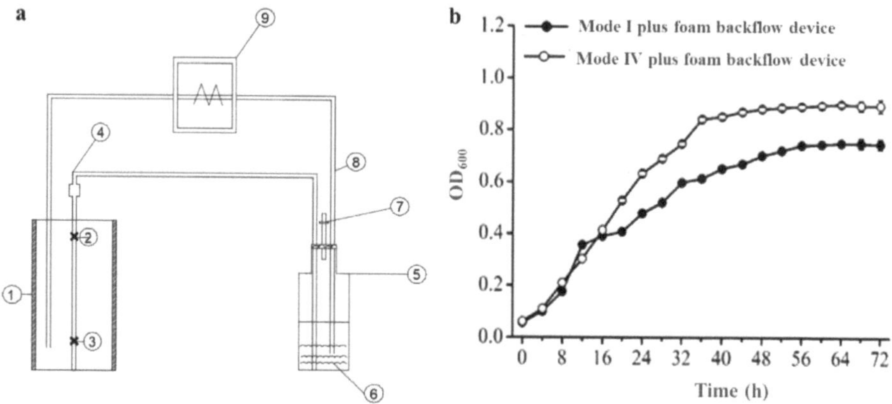

Figure 4. Effect of foam backflow device on the fermentation of *S. pararoseus* JD-2. (**a**) Foam backflow device. ① Fermenter, ② Antifoaming paddle, ③ Stirring paddle, ④ Exhaust pipe, ⑤ Collection bottle, ⑥ Solution without foams, ⑦ Off-gases, ⑧ Feeding pipe, ⑨ Peristaltic pump. (**b**) Effect of foam backflow device on cell growth of *S. pararoseus* JD-2. The data represent mean values and standard deviations obtained from three independent cultivations.

Table 3. Effect of foam return device on the fermentation of *S. pararoseus* JD-2.

Feeding Mode	Biomass (g/L)	Microbial Oil (g/L)	EPS (g/L)
Mode I plus foam backflow device	52.2 ± 1.6	25.6 ± 0.7	9.1 ± 0.3
Mode IV plus foam backflow device	62.3 ± 1.8	31.2 ± 0.8	10.9 ± 0.4

4. Conclusions

Foaming is considered a "general nuisance" in industrial fermentation because excessive foaming can lead to a decrease in the utilization rate of equipment and to an increase in the escape of fermentation broth. In this study, we pointed out that foam formation in fed-batch fermentation by *S. pararoseus* JD-2 is positively associated with the cell growth state, the air entrapment, and the properties of broth. We found that different defoamers and modes of CSL feeding show the different effects on the formation of foams. The addition of 0.03% (*v/v*) of the emulsified PPE using Tween 80 in fermentation broth showed the best defoaming abilities and antifoaming abilities. In addition, feeding CSL at 8–24 h with speed of 0.02 L/h resulted in only 0.3 L of fermented liquid leakage, and increased biomass, microbial oil and EPS concentrations. The foam backflow device once again proved beneficial for fed-batch fermentation. Under such efficient foam-control, *S. pararoseus* JD-2 produced 31.2 ± 0.8 g/L of microbial oil and 10.9 ± 0.4 g/L of EPS, which increased by 23.3% and 17.2%, respectively, in comparison to uncontrolled foaming.

Author Contributions: Methodology, Y.-F.G., H.-T.W. and M.-Q.W.; investigation, Y.-F.G., H.-T.W. and M.-Q.W.; data curation, Y.-F.G. and M.-Q.W.; writing—original draft preparation, Y.-F.G.; writing—review and editing, J.-Z.X.; supervision, J.-Z.X. and Y.-L.W.; project administration, J.-Z.X. and Y.-L.W.; funding acquisition, J.-Z.X. and Y.-L.W. All authors have read and agreed to the published version of the manuscript.

Funding: This research was funded by the Key Laboratory of Industrial Biotechnology, Ministry of Education, Jiangnan University, grant number KLIB-KF 201706 and KLIB-KF 202004, the Project of Shandong Province Higher Educational Science and Technology Program, grant number J17KA125. The APC was funded by Qingchuang Science and Technology Support Program of Shandong Provincial College.

Institutional Review Board Statement: Not applicable.

Informed Consent Statement: Not applicable.

Data Availability Statement: Not applicable.

Acknowledgments: The authors would like to acknowledge Wei-Guo Zhang from School of Biotechnology, Jiangnan University for data analysis.

Conflicts of Interest: The authors declare no conflict of interest.

References

1. Caporusso, A.; Capece, A.; De Bari, I. Oleaginous Yeasts as Cell Factories for the Sustainable Production of Microbial Lipids by the Valorization of Agri-Food Wastes. *Fermentation* **2021**, *7*, 50. [CrossRef]
2. Shaigani, P.; Awad, D.; Redai, V.; Fuchs, M.; Haack, M.; Mehlmer, N.; Brueck, T. Oleaginous yeasts- substrate preference and lipid productivity: A view on the performance of microbial lipid producers. *Microb. Cell Factories* **2021**, *20*, 220–237. [CrossRef]
3. Zhang, L.; Song, Y.; Wang, Q.; Zhang, X. Culturing rhodotorula glutinis in fermentation-friendly deep eutectic solvent extraction liquor of lignin for producing microbial lipid. *Bioresour. Technol.* **2021**, *337*, 125475. [CrossRef]
4. Han, M.; Xu, J.Z.; Liu, Z.M.; Qian, H.; Zhang, W.G. Co-production of microbial oil and exopolysaccharide by the oleaginous yeast *Sporidiobolus pararoseus* grown in fed-batch culture. *RSC Adv.* **2018**, *8*, 3348–3356. [CrossRef]
5. Ledesma-Amaro, R. Microbial oils: A customizable feedstock through metabolic engineering. *Eur. J. Lipid Sci. Technol.* **2014**, *117*, 141–144. [CrossRef]
6. Qiao, K.; Wasylenko, T.M.; Zhou, K.; Xu, P.; Stephanopoulos, G. Lipid production in *Yarrowia lipolytica* is maximized by engineering cytosolic redox metabolism. *Nat. Biotechnol.* **2017**, *35*, 173–177. [CrossRef]

7. Manowattana, A.; Techapun, C.; Watanabe, M.; Chaiyaso, T. Bioconversion of biodiesel-derived crude glycerol into lipids and carotenoids by an oleaginous red yeast Sporidiobolus pararoseus KM281507 in an airlift bioreactor. *J. Biosci. Bioeng.* **2018**, *125*, 59–66. [CrossRef]

8. Sathiyamoorthi, E.; Kumar, P.; Kim, B.S. Lipid production by Cryptococcus albidus using biowastes hydrolysed by indigenous microbes. *Bioprocess Biosyst. Eng.* **2019**, *42*, 687–696. [CrossRef]

9. Pawar, P.; Odaneth, A.A.; Vadgama, R.; Lali, A.M. Simultaneous lipid biosynthesis and recovery for oleaginous yeast *Yarrowia lipolytica*. *Biotechnol. Biofuels* **2019**, *12*, 237–254. [CrossRef]

10. Sakarika, M.; Kornaros, M. Kinetics of growth and lipids accumulation in Chlorella vulgaris during batch heterotrophic cultivation: Effect of different nutrient limitation strategies. *Bioresour. Technol.* **2017**, *243*, 356–365. [CrossRef]

11. Wang, H.; Hu, B.; Liu, J.; Qian, H.; Xu, J.; Zhang, W. Co-production of lipid, exopolysaccharide and single-cell protein by *Sporidiobolus pararoseus* under ammonia nitrogen-limited conditions. *Bioprocess Biosyst. Eng.* **2020**, *43*, 1403–1414. [CrossRef]

12. Zeng, Y.; Xie, T.; Li, P.; Jian, B.; Li, X.; Xie, Y.; Zhang, Y. Enhanced lipid production and nutrient utilization of food waste hydrolysate by mixed culture of oleaginous yeast *Rhodosporidium toruloides* and oleaginous microalgae Chlorella vulgaris. *Renew. Energy* **2018**, *126*, 915–923. [CrossRef]

13. Arora, N.; Patel, A.; Mehtani, J.; Pruthi, P.A.; Pruthi, V.; Poluri, K.M. Co-culturing of oleaginous microalgae and yeast: Paradigm shift towards enhanced lipid productivity. *Environ. Sci. Pollut. Res.* **2019**, *26*, 16952–16973. [CrossRef]

14. Liu, G.; Miao, X. Switching cultivation for enhancing biomass and lipid production with extracellular polymeric substance as co-products in Heynigia riparia SX01. *Bioresour. Technol.* **2017**, *227*, 214–220. [CrossRef] [PubMed]

15. Han, M.; He, Q.; Zhang, W.G. Carotenoids production in different culture conditions by *Sporidiobolus pararoseus*. *Prep. Biochem. Biotech.* **2012**, *42*, 293–303. [CrossRef] [PubMed]

16. Han, M.; Xu, Z.-Y.; Du, C.; Qian, H.; Zhang, W.-G. Effects of nitrogen on the lipid and carotenoid accumulation of oleaginous yeast Sporidiobolus pararoseus. *Bioprocess Biosyst. Eng.* **2016**, *39*, 1425–1433. [CrossRef] [PubMed]

17. Gupta, A.K.; Seth, K.; Maheshwari, K.; Baroliya, P.K.; Meena, M.; Kumar, A.; Vinayak, V. Harish Biosynthesis and extraction of high-value carotenoid from algae. *Front. Biosci.* **2021**, *26*, 171–190. [CrossRef]

18. Rehm, B.H.A. Bacterial polymers: Biosynthesis, modifications and applications. *Nat. Rev. Microbiol.* **2010**, *8*, 578–592. [CrossRef]

19. Vardar-Sukan, F. Foaming: Consequences, prevention and destruction. *Biotechnol. Adv.* **1998**, *16*, 913–948. [CrossRef]

20. Junker, B. Foam and its mitigation in fermentation systems. *Biotechnol. Progr.* **2007**, *23*, 767–784. [CrossRef]

21. Denkov, N.; Tcholakova, S.; Politova-Brinkova, N. Physicochemical control of foam properties. *Curr. Opin. Colloid Interface Sci.* **2020**, *50*, 101376. [CrossRef]

22. Lesov, I.; Tcholakova, S.; Kovadjieva, M.; Saison, T.; Lamblet, M.; Denkov, N. Role of Pickering stabilization and bulk gelation for the preparation and properties of solid silica foams. *J. Colloid Interface Sci.* **2017**, *504*, 48–57. [CrossRef]

23. Liao, S.; Ghosh, A.; Becker, M.D.; Abriola, L.M.; Cápiro, N.L.; Fortner, J.D.; Pennell, K.D. Effect of rhamnolipid biosurfactant on transport and retention of iron oxide nanoparticles in water-saturated quartz sand. *Environ. Sci. Nano* **2020**, *8*, 311–327. [CrossRef]

24. Politova, N.; Tcholakova, S.; Valkova, Z.; Golemanov, K.; Denkov, N.D. Self-regulation of foam volume and bubble size during foaming via shear mixing. *Colloids Surf. A Physicochem. Eng. Asp.* **2018**, *539*, 18–28. [CrossRef]

25. Cheng, X.-H.; Wang, K.; Cheng, N.-Q.; Mi, S.-Y.; Sun, L.-S.; Yeh, J.-T. The control of expansion ratios and cellular structure of supercritical CO_2-aid thermoplastic starch foams using crosslinking agents and nano-silica particles. *J. Polym. Res.* **2021**, *28*, 35–43. [CrossRef]

26. Akter, M.M.; Theary, K.; Kalkornsurapranee, E.; Prabhakar, C.S.; Thaochan, N. The effects of methyl eugenol, cue lure and plant essential oils in rubber foam dispenser for controlling Bactrocera dorsalis and Zeugodacus cucurbitae. *Asian J. Agric. Biol.* **2021**, *9*, 356–367. [CrossRef]

27. Jin, Y.; Li, J.; Wu, S.; Zhou, F. Comparison of polyurethane foam dressing and hydrocolloid dressing in patients with pressure ulcers A randomized controlled trial protocol. *Medicine* **2021**, *100*, e24165. [CrossRef]

28. Zaky, A.S. Introducing a Marine Biorefinery System for the Integrated Production of Biofuels, High-Value-Chemicals, and Co-Products: A Path Forward to a Sustainable Future. *Processes* **2021**, *9*, 1841. [CrossRef]

29. Zaky, A.S.; Carter, C.E.; Meng, F.; French, C.E. A preliminary life cycle analysis of bioethanol production using seawater in a coastal biorefinery setting. *Processes* **2021**, *9*, 1399. [CrossRef]

30. Tamura, T.; Kageyama, M.; Kaneko, Y.; Kishino, T.; Nikaido, M. Direct Observation of Foam Film Rupture by Several Types of Antifoams Using a Scanning Laser Microscope. *J. Colloid Interface Sci.* **1999**, *213*, 179–186. [CrossRef]

31. Valdés-Velasco, L.M.; Favela-Torres, E.; Théatre, A.; Arguelles-Arias, A.; Saucedo-Castañeda, J.G.; Jacques, P. Relationship between lipopeptide biosurfactant and primary metabolite production by Bacillus strains in solid-state and submerged fermentation. *Bioresour. Technol.* **2021**, *345*, 126556. [CrossRef]

32. Xu, N.; Liu, S.; Xu, L.; Zhou, J.; Xin, F.; Zhang, W.; Qian, X.; Li, M.; Dong, W.; Jiang, M. Enhanced rhamnolipids production using a novel bioreactor system based on integrated foam-control and repeated fed-batch fermentation strategy. *Biotechnol. Biofuels* **2020**, *13*, 80–89. [CrossRef] [PubMed]

33. Conceição, K.S.; Almeida, M.D.A.; Sawoniuk, I.C.; Marques, G.D.; Faria-Tischer, P.C.D.S.; Tischer, C.A.; Vignoli, J.A.; Camilios-Neto, D. Rhamnolipid production by Pseudomonas aeruginosa grown on membranes of bacterial cellulose supplemented with corn bran water extract. *Environ. Sci. Pollut. Res.* **2020**, *27*, 30222–30231. [CrossRef] [PubMed]

34. Dai, L.; Jiang, W.; Zhou, X.; Xu, Y. Enhancement in xylonate production from hemicellulose pre-hydrolysate by powdered activated carbon treatment. *Bioresour. Technol.* **2020**, *316*, 123944. [CrossRef] [PubMed]
35. Gong, Z.; Peng, Y.; Wang, Q. Rhamnolipid production, characterization and fermentation scale-up by Pseudomonas aeruginosa with plant oils. *Biotechnol. Lett.* **2015**, *37*, 2033–2038. [CrossRef]
36. Anic, I.; Apolonia, I.; Franco, P.; Wichmann, R. Production of rhamnolipids by integrated foam adsorption in a bioreactor system. *AMB Express* **2018**, *8*, 122. [CrossRef] [PubMed]
37. Sharma, S.C.; Tsuchiya, K.; Sakai, K.; Sakai, H.; Abe, M.; Komura, S.; Sakamoto, K.; Miyahara, R. Formation and Characterization of Microemulsions Containing Polymeric Silicone. *Langmuir* **2008**, *24*, 7658–7662. [CrossRef] [PubMed]
38. Liu, L.; Xiang, N.; Ni, Z.; Huang, X.; Zheng, J.; Wang, Y.; Zhang, X. Step emulsification: High-throughput production of monodisperse droplets. *Biotechniques* **2020**, *68*, 114–116. [CrossRef]
39. He, Y. High cell density production of *Deinococcus radiodurans* under optimized conditions. *J. Ind. Microbiol. Biotechnol.* **2009**, *36*, 539–546. [CrossRef]
40. Routledge, S.J. Beyond de-foaming: The effects of antifoams on bioprocess productivity. *Comput. Struct. Biotechnol. J.* **2012**, *3*, e201210001. [CrossRef]
41. Maleki-Kakelar, M.; Azarhoosh, M.J.; Senji, S.G.; Aghaeinejad-Meybodi, A. Urease production using corn steep liquor as a low-cost nutrient source by *Sporosarcina pasteurii*: Biocementation and process optimization via artificial intelligence approaches. *Environ. Sci. Pollut. Res.* **2021**, 1–15. [CrossRef] [PubMed]
42. Paker, I.; Jaczynski, J.; Matak, K.E. Calcium hydroxide as a processing base in alkali-aided pH-shift protein recovery process. *J. Sci. Food Agric.* **2017**, *97*, 811–817. [CrossRef] [PubMed]
43. Liu, X. Effects of proteases on L-glutamic acid fermentation. *Bioengineered* **2019**, *10*, 646–658. [CrossRef]
44. Anic, I.; Nath, A.; Franco, P.; Wichmann, R. Foam adsorption as an ex situ capture step for surfactants produced by fermentation. *J. Biotechnol.* **2017**, *258*, 181–189. [CrossRef] [PubMed]
45. Long, X.; Shen, C.; He, N.; Zhang, G.; Meng, Q. Enhanced rhamnolipids production via efficient foam-control using stop valve as a foam breaker. *Bioresour. Technol.* **2017**, *224*, 536–543. [CrossRef]

MDPI AG
Grosspeteranlage 5
4052 Basel
Switzerland
Tel.: +41 61 683 77 34

Fermentation Editorial Office
E-mail: fermentation@mdpi.com
www.mdpi.com/journal/fermentation